4G LTE -Advanced Pro 和通向 5G 之路 第 3 版
（影印版）

4G, LTE - Advanced Pro and The Road to 5G, 3E

U0397317

4G LTE -Advanced Pro 和通向 5G 之路 第 3 版

（影印版）

4G, LTE - Advanced Pro and The Road to 5G, 3E

Erik Dahlman,Stefan Parkvall,Johan Sköld 著

南京　东南大学出版社

图书在版编目(CIP)数据

4G LTE‐Advanced Pro 和通向 5G 之路：第 3 版：英文/(瑞典)埃里克·达尔曼(Erik Dahlman),(瑞典)斯蒂芬·帕克威尔(Stefan Parkvall),(瑞典)约翰·斯科德(Johan Sköld)著. —影印本. —南京：东南大学出版社,2017.9

书名原文：4G, LTE‐Advanced Pro and The Road to 5G,3E

ISBN 978‐7‐5641‐7177‐3

Ⅰ.①4… Ⅱ.①埃… ②斯… ③约… Ⅲ.①无线电通信‐移动网‐研究‐英文 Ⅳ.①TN929.5

中国版本图书馆 CIP 数据核字(2017)第 121088 号

图字：10‐2017‐086 号

4G LTE‐Advanced Pro 和通向 5G 之路 第 3 版(影印版)

出版发行：东南大学出版社
地 址：南京四牌楼 2 号 邮编：210096
出 版 人：江建中
网 址：http://www.seupress.com
电子邮件：press@seupress.com
印 刷：常州市武进第三印刷有限公司
开 本：787 毫米×980 毫米 16 开本
印 张：38.5
字 数：859 千字
版 次：2017 年 9 月第 1 版
印 次：2017 年 9 月第 1 次印刷
书 号：ISBN 978‐7‐5641‐7177‐3
定 价：128.00 元

本社图书若有印装质量问题，请直接与营销部联系。电话(传真)：025‐83791830

Contents

Preface

LTE has become the most successful mobile wireless broadband technology, serving over one billion users as of the beginning of 2016 and handling a wide range of applications. Compared to the analog voice-only systems 25 years ago, the difference is dramatic. Although LTE is still at a relatively early stage of deployment, the industry is already well on the road toward the next generation of mobile communication, commonly referred to as the fifth generation or 5G. Mobile broadband is, and will continue to be, an important part of future cellular communication, but future wireless networks are to a large extent also about a significantly wider range of use cases and a correspondingly wider range of requirements.

This book describes LTE, developed in 3GPP (*Third-Generation Partnership Project*) and providing true fourth-generation (4G) broadband mobile access, as well as the new radio-access technology 3GPP is currently working on. Together, these two technologies will provide 5G wireless access.

Chapter 1 provides a brief introduction, followed by a description of the standardization process and relevant organizations such as the aforementioned 3GPP and ITU in Chapter 2. The frequency bands available for mobile communication are also be covered, together with a discussion on the process for finding new frequency bands.

An overview of LTE and its evolution is found in Chapter 3. This chapter can be read on its own to get a high-level understanding of LTE and how the LTE specifications evolved over time. To underline the significant increase in capabilities brought by the LTE evolution, 3GPP introduced the names LTE-Advanced and LTE-Advanced Pro for some of the releases.

Chapters 4—11 cover the basic LTE structure, starting with the overall protocol structure in Chapter 4 and followed by a detailed description of the physical layer in Chapters 5—7. The remaining Chapters 8—11, cover connection setup and various transmission procedures, including multi-antenna support.

Some of the major enhancements to LTE introduced over time is covered in Chapters 12—21, including carrier aggregation, unlicensed spectrum, machine-type communication, and device-to-device communication. Relaying, heterogeneous deployments, broadcast/multicast services, and dual connectivity multi-site coordination are other examples of enhancements covered in these chapters.

Radio frequency (RF) requirements, taking into account spectrum flexibility and multi-standard radio equipment, is the topic of Chapter 22.

Chapters 23 and 24 cover the new radio access about to be standardized as part of 5G. A closer look on the requirements and how they are defined is the topic of Chapter 23, while Chapter 24 digs into the technical realization.

Finally, Chapter 25 concludes the book and the discussion on 5G radio access.

Acknowledgments

We thank all our colleagues at Ericsson for assisting in this project by helping with contributions to the book, giving suggestions and comments on the content, and taking part in the huge team effort of developing LTE and the next generation of radio access for 5G.

The standardization process involves people from all parts of the world, and we acknowledge the efforts of our colleagues in the wireless industry in general and in 3GPP RAN in particular. Without their work and contributions to the standardization, this book would not have been possible.

Finally, we are immensely grateful to our families for bearing with us and supporting us during the long process of writing this book.

Abbreviations and Acronyms

3GPP	Third-generation partnership project
AAS	Active antenna systems
ACIR	Adjacent channel interference ratio
ACK	Acknowledgment (in ARQ protocols)
ACLR	Adjacent channel leakage ratio
ACS	Adjacent channel selectivity
AGC	Automatic gain control
AIFS	Arbitration interframe space
AM	Acknowledged mode (RLC configuration)
A-MPR	Additional maximum power reduction
APT	Asia-Pacific telecommunity
ARI	Acknowledgment resource indicator
ARIB	Association of radio industries and businesses
ARQ	Automatic repeat-request
AS	Access stratum
ATC	Ancillary terrestrial component
ATIS	Alliance for telecommunications industry solutions
AWGN	Additive white Gaussian noise
BC	Band category
BCCH	Broadcast control channel
BCH	Broadcast channel
BL	Bandwidth-reduced low complexity
BM-SC	Broadcast multicast service center
BPSK	Binary phase-shift keying
BS	Base station
BW	Bandwidth
CA	Carrier aggregation
CACLR	Cumulative adjacent channel leakage ratio
CC	Component carrier
CCA	Clear channel assessment
CCCH	Common control channel
CCE	Control channel element
CCSA	China Communications Standards Association
CDMA	Code-division multiple access
CE	Coverage enhancement
CEPT	European Conference of Postal and Telecommunications Administrations
CGC	Complementary ground component
CITEL	Inter-American Telecommunication Commission
C-MTC	Critical MTC
CN	Core network
CoMP	Coordinated multi-point transmission/reception
CP	Cyclic prefix
CQI	Channel-quality indicator

CRC	Cyclic redundancy check
C-RNTI	Cell radio-network temporary identifier
CRS	Cell-specific reference signal
CS	Capability set (for MSR base stations)
CSA	Common subframe allocation
CSG	Closed Subscriber Group
CSI	Channel-state information
CSI-IM	CSI interference measurement
CSI-RS	CSI reference signals
CW	Continuous wave
D2D	Device-to-device
DAI	Downlink assignment index
DCCH	Dedicated control channel
DCH	Dedicated channel
DCI	Downlink control information
DCF	Distributed coordination function
DFS	Dynamic frequency selection
DFT	Discrete Fourier transform
DFTS-OFDM	DFT-spread OFDM (DFT-precoded OFDM)
DIFS	Distributed interframe space
DL	Downlink
DL-SCH	Downlink shared channel
DM-RS	Demodulation reference signal
DMTC	DRS measurements timing configuration
DRS	Discovery reference signal
DRX	Discontinuous reception
DTCH	Dedicated traffic channel
DTX	Discontinuous transmission
DwPTS	Downlink part of the special subframe (for TDD operation)
ECCE	Enhanced control channel element
EDCA	Enhanced distributed channel access
EDGE	Enhanced data rates for GSM evolution; enhanced data rates for global evolution
eIMTA	Enhanced Interference mitigation and traffic adaptation
EIRP	Effective isotropic radiated power
EIS	Equivalent isotropic sensitivity
EMBB	Enhanced MBB
eMTC	Enhanced machine-type communication
eNB	eNodeB
eNodeB	E-UTRAN NodeB
EPC	Evolved packet core
EPDCCH	Enhanced physical downlink control channel
EPS	Evolved packet system
EREG	Enhanced resource-element group
ETSI	European Telecommunications Standards Institute
E-UTRA	Evolved UTRA

E-UTRAN	Evolved UTRAN
EVM	Error vector magnitude
FCC	Federal Communications Commission
FDD	Frequency division duplex
FD-MIMO	Full-dimension multiple input—multiple output
FDMA	Frequency-division multiple access
FEC	Forward error correction
FeICIC	Further enhanced intercell interference coordination
FFT	Fast Fourier transform
FPLMTS	Future public land mobile telecommunications systems
FSTD	Frequency-switched transmit diversity
GB	Guard band
GERAN	GSM/EDGE radio access network
GP	Guard period (for TDD operation)
GPRS	General packet radio services
GPS	Global positioning system
GSM	Global system for mobile communications
GSMA	GSM Association
HARQ	Hybrid ARQ
HII	High-interference indicator
HSFN	Hypersystem frame number
HSPA	High-speed packet access
HSS	Home subscriber server
ICIC	Intercell interference coordination
ICNIRP	International Commission on Non-Ionizing Radiation Protection
ICS	In-channel selectivity
IEEE	Institute of Electrical and Electronics Engineers
IFFT	Inverse fast Fourier transform
IMT-2000	International Mobile Telecommunications 2000 (ITU's name for the family of 3G standards)
IMT-2020	International Mobile Telecommunications 2020 (ITU's name for the family of 5G standards)
IMT-Advanced	International Mobile Telecommunications Advanced (ITU's name for the family of 4G standards).
IOT	Internet of things
IP	Internet protocol
IR	Incremental redundancy
IRC	Interference rejection combining
ITU	International Telecommunications Union
ITU-R	International Telecommunications Union—Radio communications sector
KPI	Key performance indicator
LAA	License-assisted access
LAN	Local area network
LBT	Listen before talk
LCID	Logical channel identifier
LDPC	Low-density parity check code
LTE	Long-term evolution

MAC	Medium access control
MAN	Metropolitan area network
MBB	Mobile broadband
MBMS	Multimedia broadcast—multicast service
MBMS-GW	MBMS gateway
MB-MSR	Multi-band multi-standard radio (base station)
MBSFN	Multicast—broadcast single-frequency network
MC	Multi-carrier
MCCH	MBMS control channel
MCE	MBMS coordination entity
MCG	Master cell group
MCH	Multicast channel
MCS	Modulation and coding scheme
METIS	Mobile and wireless communications Enablers for Twenty—twenty (2020) Information Society
MIB	Master information block
MIMO	Multiple input—multiple output
MLSE	Maximum-likelihood sequence estimation
MME	Mobility management entity
M-MTC	Massive MTC
MPDCCH	MTC physical downlink control channel
MPR	Maximum power reduction
MSA	MCH subframe allocation
MSI	MCH scheduling information
MSP	MCH scheduling period
MSR	Multi-standard radio
MSS	Mobile satellite service
MTC	Machine-type communication
MTCH	MBMS traffic channel
MU-MIMO	Multi-user MIMO
NAK	Negative acknowledgment (in ARQ protocols)
NAICS	Network-assisted interference cancelation and suppression
NAS	Non-access stratum (a functional layer between the core network and the terminal that supports signaling)
NB-IoT	Narrow-band internet of things
NDI	New data indicator
NGMN	Next-generation mobile networks
NMT	Nordisk MobilTelefon (Nordic Mobile Telephony)
NodeB	A logical node handling transmission/reception in multiple cells; commonly, but not necessarily, corresponding to a base station
NPDCCH	Narrowband PDCCH
NPDSCH	Narrowband PDSCH
NS	Network signaling
OCC	Orthogonal cover code
OFDM	Orthogonal frequency-division multiplexing

OI	Overload indicator
OOB	Out-of-band (emissions)
OSDD	OTA sensitivity direction declarations
OTA	Over the air
PA	Power amplifier
PAPR	Peak-to-average power ratio
PAR	Peak-to-average ratio (same as PAPR)
PBCH	Physical broadcast channel
PCCH	Paging control channel
PCFICH	Physical control format indicator channel
PCG	Project Coordination Group (in 3GPP)
PCH	Paging channel
PCID	Physical cell identity
PCRF	Policy and charging rules function
PDC	Personal digital cellular
PDCCH	Physical downlink control channel
PDCP	Packet data convergence protocol
PDSCH	Physical downlink shared channel
PDN	Packet data network
PDU	Protocol data unit
P-GW	Packet-data network gateway (also PDN-GW)
PHICH	Physical hybrid-ARQ indicator channel
PHS	Personal handy-phone system
PHY	Physical layer
PMCH	Physical multicast channel
PMI	Precoding-matrix indicator
PRACH	Physical random access channel
PRB	Physical resource block
P-RNTI	Paging RNTI
ProSe	Proximity services
PSBCH	Physical sidelink broadcast channel
PSCCH	Physical sidelink control channel
PSD	Power spectral density
PSDCH	Physical sidelink discovery channel
P-SLSS	Primary sidelink synchronization signal
PSM	Power-saving mode
PSS	Primary synchronization signal
PSSCH	Physical sidelink shared channel
PSTN	Public switched telephone networks
PUCCH	Physical uplink control channel
PUSCH	Physical uplink shared channel
QAM	Quadrature amplitude modulation
QCL	Quasi-colocation
QoS	Quality-of-service
QPP	Quadrature permutation polynomial

QPSK	Quadrature phase-shift keying
RAB	Radio-access bearer
RACH	Random-access channel
RAN	Radio-access network
RA-RNTI	Random-access RNTI
RAT	Radio-access technology
RB	Resource block
RE	Resource element
REG	Resource-element group
RF	Radio frequency
RI	Rank indicator
RLAN	Radio local area networks
RLC	Radio link control
RNTI	Radio-network temporary identifier
RNTP	Relative narrowband transmit power
RoAoA	Range of angle of arrival
ROHC	Robust header compression
R-PDCCH	Relay physical downlink control channel
RRC	Radio-resource control
RRM	Radio resource management
RS	Reference symbol
RSPC	Radio interface specifications
RSRP	Reference signal received power
RSRQ	Reference signal received quality
RV	Redundancy version
RX	Receiver
S1	Interface between eNodeB and the evolved packet core
S1-c	Control-plane part of S1
S1-u	User-plane part of S1
SAE	System architecture evolution
SBCCH	Sidelink broadcast control channel
SCG	Secondary cell group
SCI	Sidelink control information
SC-PTM	Single-cell point to multipoint
SDMA	Spatial division multiple access
SDO	Standards developing organization
SDU	Service data unit
SEM	Spectrum emissions mask
SF	Subframe
SFBC	Space—frequency block coding
SFN	Single-frequency network (in general, see also MBSFN); system frame number (in 3GPP).
S-GW	Serving gateway
SI	System information message
SIB	System information block
SIB1-BR	SIB1 bandwidth reduced

SIC	Successive interference combining
SIFS	Short interframe space
SIM	Subscriber identity module
SINR	Signal-to-interference-and-noise ratio
SIR	Signal-to-interference ratio
SI-RNTI	System information RNTI
SL-BCH	Sidelink broadcast channel
SL-DCH	Sidelink discovery channel
SLI	Sidelink identity
SL-SCH	Sidelink shared channel
SLSS	Sidelink synchronization signal
SNR	Signal-to-noise ratio
SORTD	Spatial orthogonal-resource transmit diversity
SR	Scheduling request
SRS	Sounding reference signal
S-SLSS	Secondary sidelink synchronization signal
SSS	Secondary synchronization signal
STCH	Sidelink traffic channel
STBC	Space—time block coding
STC	Space—time coding
STTD	Space—time transmit diversity
SU-MIMO	Single-user MIMO
TAB	Transceiver array boundary
TCP	Transmission control protocol
TC-RNTI	Temporary C-RNTI
TDD	Time-division duplex
TDMA	Time-division multiple access
TD-SCDMA	Time-division-synchronous code-division multiple access
TF	Transport format
TPC	Transmit power control
TR	Technical report
TRP	Time repetition pattern; transmission reception point
TRPI	Time repetition pattern index
TS	Technical specification
TSDSI	Telecommunications Standards Development Society, India
TSG	Technical Specification Group
TTA	Telecommunications Technology Association
TTC	Telecommunications Technology Committee
TTI	Transmission time interval
TX	Transmitter
TXOP	Transmission opportunity
UCI	Uplink control information
UE	User equipment (the 3GPP name for the mobile terminal)
UEM	Unwanted emissions mask
UL	Uplink

UL-SCH	Uplink shared channel
UM	Unacknowledged mode (RLC configuration)
UMTS	Universal mobile telecommunications system
UpPTS	Uplink part of the special subframe, for TDD operation
URLLC	Ultra-reliable low-latency communication
UTRA	Universal terrestrial radio access
UTRAN	Universal terrestrial radio-access network
VoIP	Voice-over-IP
VRB	Virtual resource block
WARC	World Administrative Radio Congress
WAS	Wireless access systems
WCDMA	Wideband code-division multiple access
WCS	Wireless communications service
WG	Working group
WiMAX	Worldwide interoperability for microwave access
WLAN	Wireless local area network
WMAN	Wireless metropolitan area network
WP5D	Working Party 5D
WRC	World Radio communication Conference
X2	Interface between eNodeBs.
ZC	Zadoff-Chu

INTRODUCTION

Mobile communication has become an everyday commodity. In the last decades, it has evolved from being an expensive technology for a few selected individuals to today's ubiquitous systems used by a majority of the world's population.

The world has witnessed four generations of mobile-communication systems, each associated with a specific set of technologies and a specific set of supported use cases, see Figure 1.1. The generations and the steps taken between them are used here as background to introduce the content of this book. The rest of the book focuses on the latest generations that are deployed and under consideration, which are fourth generation (4G) and fifth generation (5G).

1.1 1G AND 2G—VOICE-CENTRIC TECHNOLOGIES

The first-generation (1G) systems were the analog voice-only mobile-telephony systems of the 1980s, often available on a national basis with limited or no international roaming. 1G systems include NMT, AMPS, and TACS. Mobile communication was available before the 1G systems, but typically on a small scale and targeting a very selected group of people.

The second-generation (2G) systems appeared in the early 1990s. Examples of 2G technologies include the European-originated GSM technology, the American IS-95/CDMA and IS-136/TDMA technologies, and the Japanese PDC technology. The 2G systems were

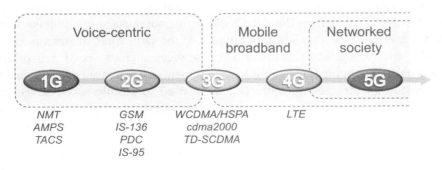

FIGURE 1.1

Cellular generations.

4G, LTE-Advanced Pro and The Road to 5G. http://dx.doi.org/10.1016/B978-0-12-804575-6.00001-7

still voice centric, but thanks to being all-digital provided a significantly higher capacity than the previous 1G systems. Over the years, some of these early technologies have been extended to also support (primitive) packet data services. These extensions are sometimes referred to as 2.5G to indicate that they have their roots in the 2G technologies but have a significantly wider range of capabilities than the original technologies. EDGE is a well-known example of a 2.5G technology. GSM/EDGE is still in widespread use in smart-phones but is also frequently used for some types of machine-type communication such as alarms, payment systems, and real-estate monitoring.

1.2 3G AND 4G—MOBILE BROADBAND

During the 1990s, the need to support not only voice but also data services had started to emerge, driving the need for a new generation of cellular technologies going beyond voice-only services. At this time in the late 1990s, 2G GSM, despite being developed within Europe, had already become a de facto global standard. To ensure global reach also for 3G tech-nologies it was realized that the 3G development had to be carried out on a global basis. To facilitate this, the *Third-Generation Partnership Project* (3GPP) was formed to develop the 3G WCDMA and TD-SCDMA technologies, see Chapter 2 for further details. Shortly af-terward, the parallel organization 3GPP2 was formed to develop the competing 3G cdma2000 technology, an evolution of the 2G IS-95 technology.

The first release of WCDMA (release 99[1]) was finalized in 1999. It included circuit-switched voice and video services, and data services over both packet-switched and circuit-switched bearers.

The first major enhancements to WCDMA came with the introduction of *High Speed Downlink Packet Access* (HSDPA) in release 5 followed by *Enhanced Uplink in release 6,* collectively known as *High Speed Packet Access (HSPA)* [61]. HSPA, sometimes referred to as 3.5G, allowed for a "true" mobile-broadband experience with data rates of several Mbit/s while maintaining the compatibility with the original 3G specifications. With the support for mobile broadband, the foundation for the rapid uptake of smart phones such as the iPhone and the wide range of Android devices were in place. Without the wide availability of mobile broadband for the mass market, the uptake of smart phone usage would have been signifi-cantly slower and their usability severely limited. At the same time, the massive use of smart phones and a wide range of packet-data-based services such as social networking, video, gaming, and online shopping translates into requirements on increased capacity and improved spectral efficiency. Users getting more and more used to mobile services also raise their expectations in terms of experiencing increased data rates and reduced latency. These needs

---

[1]For historical reasons, the first 3GPP release is named after the year it was frozen (1999), while the following releases are numbered 4, 5, 6, and so on.

were partly handled by a continuous, and still ongoing, evolution of HSPA, but it also triggered the discussions on 4G technology in the mid-2000s.

The 4G LTE technology was from the beginning developed for packet-data support and has no support for circuit-switched voice, unlike the 3G where HSPA was an "add-on" to provide high-performance packet data on top of an existing technology. Mobile broadband services were the focus, with tough requirements on high data rates, low latency, and high capacity. Spectrum flexibility and maximum commonality between FDD and TDD solutions were other important requirements. A new core network architecture was also developed, known as *Enhanced Packet Core* (EPC), to replace the architecture used by GSM and WCDMA/HSPA. The first version of LTE was part of release 8 of the 3GPP specifications and the first commercial deployment took place in late 2009, followed by a rapid and worldwide deployment of LTE networks.

One significant aspect of LTE is the worldwide acceptance of a single technology, unlike previous generations for which there has been several competing technologies, see Figure 1.2. Having a single, universally accepted technology accelerates development of new services and reduces the cost for both users and network operators.

Since its commercial introduction in 2009, LTE has evolved considerably in terms of data rates, capacity, spectrum and deployment flexibility, and application range. From macrocentric deployments with peak data rates of 300 Mbit/s in 20 MHz of contiguous, licensed spectrum, the evolution of LTE can in release 13 support multi-Gbit/s peak data rates through improvements in terms of antenna technologies, multisite coordination, exploitation of fragmented as well as unlicensed spectrum and densified deployments just to mention a few areas. The evolution of LTE has also considerably widened the use cases beyond mobile broadband by, for example, improving support for massive machine-type communication and introducing direct device-to-device communication.

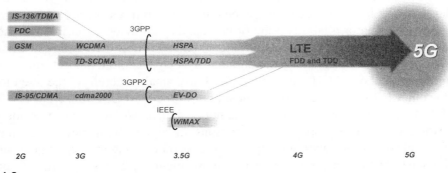

FIGURE 1.2

Convergence of wireless technologies.

1.3 5G—BEYOND MOBILE BROADBAND—NETWORKED SOCIETY

Although LTE is still at a relatively early stage of deployment, the industry is already well on the road towards the next generation of mobile communication, commonly referred to as fifth generation or 5G.

Mobile broadband is, and will continue to be, an important part of future cellular communication, but future wireless networks are to a large extent also about a significantly wider range of use cases. In essence, 5G should be seen as a platform enabling wireless connectivity to all kinds of services, existing as well as future not-yet-known services and thereby taking wireless networks beyond mobile broadband. Connectivity will be provided essentially anywhere, anytime to anyone and anything. The term *networked society* is sometimes used when referring to such a scenario where connectivity goes beyond mobile smartphones, having a profound impact on the society.

Massive machine-type communication, exemplified by sensor networks in agriculture, traffic monitoring, and remote management of utility equipment in buildings, is one type of non-mobile-broadband applications. These applications primarily put requirements on very low device power consumption while the data rates and amounts of data per device are modest. Many of these applications can already be supported by the LTE evolution.

Another example of non-mobile-broadband applications are *ultra-reliable and low-latency communications* (URLLC), also known as critical machine-type communication. Examples hereof are industrial automation, where latency and reliability requirements are very strict. Vehicle-to-vehicle communication for traffic safety is another example.

Nevertheless, mobile broadband will remain an important use case and the amount of traffic in wireless networks is increasing rapidly, as is the user expectation on data rates, availability, and latency. These enhanced requirements also need to be addressed by 5G wireless networks.

Increasing the capacity can be done in three ways: improved spectral efficiency, densified deployments, and an increased amount of spectrum. The spectral efficiency of LTE is already high and although improvements can be made, it is not sufficient to meet the traffic increase. Network densification is also expected to happen, not only from a capacity perspective, but also from a high-data-rate-availability point of view, and can provide a considerable increase in capacity although at the cost of finding additional antenna sites. Increasing the amount of spectrum will help, but unfortunately, the amount of not-yet-exploited spectrum in typical cellular bands, up to about 3 GHz, is limited and fairly small. Therefore, the attention has increased to somewhat higher frequency bands, both in the 3—6 GHz range but also in the range 6—30 GHz and beyond for which LTE is not designed, as a way to access additional spectrum. However, as the propagation conditions in higher frequency bands are less favorable for wide-area coverage and require more advanced antenna techniques such as beamforming, these bands can mainly serve as a complement to the existing, lower-frequency bands.

As seen from the discussion earlier, the range of requirements for 5G wireless networks are very wide, calling for a high degree of network flexibility. Furthermore, as many future

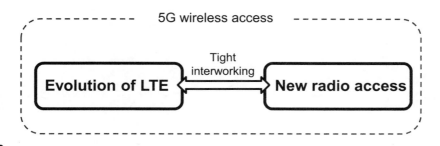

FIGURE 1.3

5G consisting of LTE evolution and a new radio-access technology.

applications cannot be foreseen at the moment, future-proofness is a key requirement. Some of these requirements can be handled by the LTE evolution, but not all, calling for a new radio-access technology to complement LTE evolution as illustrated in Figure 1.3.

1.4 OUTLINE

The remainder of this book describes the technologies for the 4G and 5G wireless networks.

Chapter 2 describes the standardization process and relevant organizations such as the aforementioned 3GPP and ITU. The frequency bands available for mobile communication is also be covered, together with a discussion on the process for finding new frequency bands.

An overview of LTE and its evolution is found in Chapter 3. This chapter can be read on its own to get a high-level understanding of LTE and how the LTE specifications evolved over time. To underline the significant increase in capabilities brought by the LTE evolution, 3GPP introduced the names LTE-Advanced and LTE-Advanced Pro for some of the releases.

Chapters 4—11 cover the basic LTE structure, starting with the overall protocol structure in Chapter 4 and followed by a detailed description of the physical layer in Chapters 5—7. The remaining Chapters 8—11, cover connection setup and various transmission procedures, including multi-antenna support.

Some of the major enhancements to LTE introduced over time is covered in Chapters 12—21, including carrier aggregation, unlicensed spectrum, machine-type communication, and device-to-device communication. Relaying, heterogeneous deployments, broadcast/ multicast services, dual connectivity multisite coordination are other examples of enhancements covered in these chapters.

RF requirements, taking into account spectrum flexibility and multi-standard radio equipment, is the topic of Chapter 22.

Chapters 23 and 24 cover the new radio access about to be standardized as part of 5G. A closer look on the requirements and how they are defined is the topic of Chapter 23, while Chapter 24 digs into the technical realization.

Finally, Chapter 25 concludes the book and the discussion on 5G radio access.

SPECTRUM REGULATION AND STANDARDIZATION FROM 3G TO 5G

2

The research, development, implementation, and deployment of mobile-communication systems are performed by the wireless industry in a coordinated international effort by which common industry specifications that define the complete mobile-communication system are agreed. The work depends also heavily on global and regional regulation, in particular for the spectrum use that is an essential component for all radio technologies. This chapter describes the regulatory and standardization environment that has been, and continues to be, essential for defining the mobile-communication systems.

2.1 OVERVIEW OF STANDARDIZATION AND REGULATION

There are a number of organizations involved in creating technical specifications and standards as well as regulation in the mobile-communications area. These can loosely be divided into three groups: standards developing organizations, regulatory bodies and administrations, and industry forums.

Standards developing organizations (SDOs) develop and agree on technical standards for mobile-communications systems, in order to make it possible for the industry to produce and deploy standardized products and provide interoperability between those products. Most components of mobile-communication systems, including base stations and mobile devices, are standardized to some extent. There is also a certain degree of freedom to provide proprietary solutions in products, but the communications protocols rely on detailed standards for obvious reasons. SDOs are usually nonprofit industry organizations and not government controlled. They often write standards within a certain area under mandate from governments(s), however, giving the standards a higher status.

There are nationals SDOs, but due to the global spread of communications products, most SDOs are regional and also cooperate on a global level. As an example, the technical specifications of GSM, WCDMA/HSPA, and LTE are all created by 3GPP (Third Generation Partnership Project) which is a global organization from seven regional and national SDOs in Europe (ETSI), Japan (ARIB and TTC), United States (ATIS), China (CCSA), Korea (TTA), and India (TSDSI). SDOs tend to have a varying degree of transparency, but 3GPP is fully transparent with all technical specifications, meeting documents, reports, and email reflectors publically available without charge even for nonmembers.

4G, LTE-Advanced Pro and The Road to 5G. http://dx.doi.org/10.1016/B978-0-12-804575-6.00002-9

Regulatory bodies and administrations are government-led organizations that set regulatory and legal requirements for selling, deploying, and operating mobile systems and other telecommunication products. One of their most important tasks is to control spectrum use and to set licensing conditions for the mobile operators that are awarded licenses to use parts of the radio frequency (RF) spectrum for mobile operations. Another task is to regulate "placing on the market" of products through regulatory certification, by ensuring that devices, base stations, and other equipment is type approved and shown to meet the relevant regulation.

Spectrum regulation is handled both on a national level by national administrations, but also through regional bodies in Europe (CEPT/ECC), Americas (CITEL), and Asia (APT). On a global level, the spectrum regulation is handled by the *International Telecommunications Union* (ITU). The regulatory bodies regulate what services the spectrum is to be used for and also set more detailed requirements such as limits on unwanted emissions from transmitters. They are also indirectly involved in setting requirements on the product standards through regulation. The involvement of ITU in setting requirements on the technologies for mobile communication is explained further in Section 2.2.

Industry forums are industry lead groups promoting and lobbying for specific technologies or other interests. In the mobile industry, these are often led by operators, but there are also vendors creating industry forums. An example of such a group is GSMA (GSM association) which is promoting mobile-communication technologies based on GSM, WCDMA, and LTE. Other examples of industry forums are *Next-Generation Mobile Networks* (NGMN) which is an operator group defining requirements on the evolution of mobile systems and *5G Americas*, which is a regional industry forum that has evolved from its predecessor 4G Americas.

Figure 2.1 illustrates the relation between different organizations involved in setting regulatory and technical conditions for mobile systems. The figure also shows the mobile industry view, where vendors develop products, place them on the market and negotiate with operators who procure and deploy mobile systems. This process relies heavily on the technical standards published by the SDOs, while placing products on the market also relies on certification of products on a regional or national level. Note that in Europe, the regional SDO (ETSI) is producing the so-called *Harmonized standards* used for product certification (through the "CE" mark), based on a mandate from the regulators. These standards are used for certification in many countries also outside of Europe.

2.2 ITU-R ACTIVITIES FROM 3G TO 5G
2.2.1 THE ROLE OF ITU-R

ITU-R is the radio communications sector of the ITU. ITU-R is responsible for ensuring efficient and economical use of the RF spectrum by all radio communication services. The different subgroups and working parties produce reports and recommendations that analyze and define the conditions for using the RF spectrum. The goal of ITU-R is to "ensure interference-free operations of radio communication systems," by implementing the *Radio*

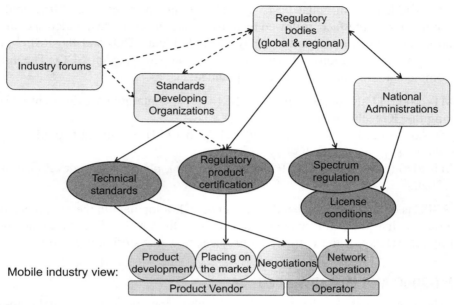

FIGURE 2.1

Simplified view of relation between regulatory bodies, standards developing organizations, industry forums, and the mobile industry.

Regulations and regional agreements. The Radio Regulations is an international binding treaty for how RF spectrum is used. A *World Radiocommunication Conference* (WRC) is held every 3—4 years. At WRC the Radio Regulations are revised and updated and in that way provide revised and updated use of RF spectrum across the world.

While the technical specification of mobile-communication technologies, such as LTE and WCDMA/HSPA is done within 3GPP, there is a responsibility for ITU-R in the process of turning the technologies into global standards, in particular for countries that are not covered by the SDOs are partners in 3GPP. ITU-R defines spectrum for different services in the RF spectrum, including mobile services and some of that spectrum is particularly identified for the so-called International Mobile Telecommunications (IMT) systems. Within ITU-R, it is *Working Party 5D* (WP5D) that has the responsibility for the overall radio system aspects of IMT systems, which, in practice, corresponds to the different generations of mobile-communication systems from 3G and onward. WP5D has the prime responsibility within ITU-R for issues related to the terrestrial component of IMT, including technical, operational, and spectrum-related issues.

WP5D does not create the actual technical specifications for IMT, but has kept the roles of defining IMT in cooperation with the regional standardization bodies and maintaining a set of recommendations and reports for IMT, including a set of *Radio Interface Specifications* (RSPC). These recommendations contain "families" of *radio interface technologies*

(RITs)—all included on an equal basis. For each radio interface, the RSPC contains an overview of that radio interface, followed by a list of references to the detailed specifications. The actual specifications are maintained by the individual SDO, and the RSPC provides references to the specifications transposed and maintained by each SDO. The following RSPC recommendations are in existence or planned:

- For IMT-2000: ITU-R Recommendation M.1457 [1] containing six different RITs including the 3G technologies.
- For IMT-Advanced: ITU-R Recommendation M.2012 [4] containing two different RITs where the most important is 4G/LTE.
- For IMT-2020 (5G): A new ITU-R Recommendation, planned to be developed in 2019—2020.

Each RSPC is continuously updated to reflect new development in the referenced detailed specifications, such as the 3GPP specifications for WCDMA and LTE. Input to the updates is provided by the SDOs and the Partnership Projects, nowadays primarily 3GPP.

2.2.2 IMT-2000 AND IMT-ADVANCED

Work on what corresponds to third generation of mobile communication started in the ITU-R already in the 1980s. First referred to as *Future Public Land Mobile Systems* (FPLMTS) it was later renamed IMT-2000. In the late 1990s, the work in ITU-R coincided with the work in different SDOs across the world to develop a new generation of mobile systems. An RSPC for IMT-2000 was first published in 2000 and included WCDMA from 3GPP as one of the RITs.

The next step for ITU-R was to initiate work on IMT-Advanced, the term used for systems that include new radio interfaces supporting new capabilities of systems beyond IMT-2000. The new capabilities were defined in a framework recommendation published by the ITU-R [2] and were demonstrated with the "van diagram" shown in Figure 2.2. The step into IMT-Advanced capabilities by ITU-R coincided with the step into 4G—the next generation of mobile technologies after 3G.

An evolution of LTE as developed by 3GPP was submitted as one candidate technology for IMT-Advanced. While actually being a new release (release 10) of the LTE specifications and thus an integral part of the continuous evolution of LTE, the candidate was named LTE-Advanced for the purpose of ITU-R submission. 3GPP also set up its own set of technical requirements for LTE-Advanced, with the ITU-R requirements as a basis.

The target of the ITU-R process is always harmonization of the candidates through consensus building. ITU-R determined that two technologies would be included in the first release of IMT-Advanced, those two being LTE and WirelessMAN-Advanced [3] based on the IEEE 802.16m specification. The two can be viewed as the "family" of IMT-Advanced technologies as shown in Figure 2.3. Note that, among these two technologies, LTE has emerged as the dominating 4G technology.

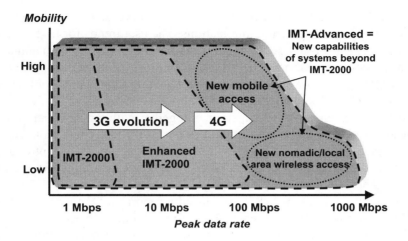

FIGURE 2.2

Illustration of capabilities of IMT-2000 and IMT-Advanced, based on the framework described in ITU-R Recommendation M.1645 [2].

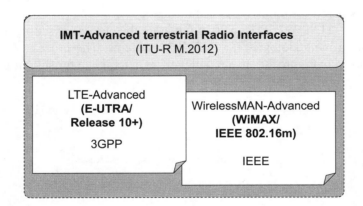

FIGURE 2.3

Radio interface technologies in IMT-Advanced.

2.2.3 IMT-2020

During 2012 to 2015, ITU-R WP5D set the stage for the next generation of IMT systems, named IMT-2020. It is to be a further development of the terrestrial component of IMT beyond the year 2020 and, in practice, corresponds to what is more commonly referred to as "5G," the fifth generation of mobile systems. The framework and objective for IMT-2020 is outlined in ITU-R Recommendation M.2083 [63], often referred to as the "Vision" recommendation. The recommendation provides the first step for defining the new developments of

IMT, looking at the future roles of IMT and how it can serve society, looking at market, user and technology trends, and spectrum implications. The user trends for IMT together with the future role and market leads to a set of usage scenarios envisioned for both human-centric and machine-centric communication. The usage scenarios identified are *Enhanced Mobile Broadband* (eMBB), *Ultra-Reliable and Low Latency Communications* (URLLC), and *Massive Machine-Type Communications* (MTC).

The need for an enhanced MBB experience, together with the new and broadened usage scenarios, leads to an extended set of capabilities for IMT-2020. The Vision recommendation [63] gives a first high-level guidance for IMT-2020 requirements by introducing a set of key capabilities, with indicative target numbers. The key capabilities and the related usage scenarios are further discussed in Chapter 23.

As a parallel activity, ITU-R WP5D produced a report on "Future technology trends of terrestrial IMT systems" [64], with focus on the time period 2015–2020. It covers trends of future IMT technology aspects by looking at the technical and operational characteristics of IMT systems and how they are improved with the evolution of IMT technologies. In this way, the report on technology trends relate to LTE release 13 and beyond, while the vision recommendation looks further ahead and beyond 2020. A report studying operation in frequencies above 6 GHz was also produced. Chapter 24 discusses some of the technology components considered for the new 5G radio access.

After WRC-15, ITU-R WP5D is in 2016 initiating the process of setting requirements and defining evaluation methodologies for IMT-2020 systems. The process will continue until mid-2017, as shown in Figure 2.4. In a parallel effort, a template for submitting an evaluation

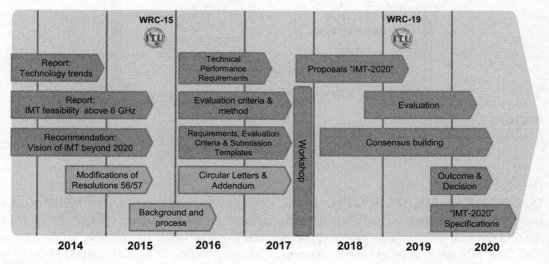

FIGURE 2.4

Work plan for IMT-2020 in ITU-R WP5D [4].

of candidate RITs will be created. External organizations are being informed of the process through a circular letter. After a workshop on IMT-2020 is held in 2017, the plan is to start the evaluation of proposals, aiming at an outcome with the RSPC for IMT-2020 being published early in 2020.

The coming evaluation of candidate RITs for IMT-2020 in ITU-R is expected to be conducted in a way similar to the evaluation done for IMT-Advanced, where the requirements were documented in Recommendation ITU-R M.2134 [28] and the detailed evaluation methodology in Recommendation ITU-R M.2135 [52]. The evaluation will be focused on the key capabilities identified in the VISION recommendation [63], but will also include other technical performance requirements. There are three fundamental ways that requirements are evaluated for a candidate technology:

- **Simulation**: This is the most elaborate way to evaluate a requirement and it involves system- or link-level simulations, or both, of the RIT. For system-level simulations, deployment scenarios are defined that correspond to a set of test environments, such as Indoor and Dense Urban. Requirements that are candidates for evaluation through simulation are for example spectrum efficiency and user-experienced data rate (for details on the key capabilities, see Chapter 23).
- **Analysis**: Some requirements can be evaluated through a calculation based on radio interface parameters. This applies for example in case of requirements on peak data rate and latency.
- **Inspection**: Some requirements can be evaluated by reviewing and assessing the functionality of the RIT. Examples of parameters that may be subject to inspection are bandwidth, handover functionality, and support of services.

Once the technical performance requirements and evaluation methodology are set up, the evaluation phase starts. Evaluation can be done by the proponent ("self-evaluation") or by an external evaluation group, doing partial or complete evaluation of one or more candidate proposals.

2.3 SPECTRUM FOR MOBILE SYSTEMS

There are a number of frequency bands identified for mobile use and specifically for IMT today. Many of these bands were first defined for operation with WCDMA/HSPA, but are now shared also with LTE deployments. Note that in the 3GPP specifications WCDMA/HSPA is referred to as *Universal Terrestrial Radio Access* (UTRA), while LTE is referred to as *Enhanced UTRA* (E-UTRA).

New bands are today often defined only for LTE. Both paired bands, where separated frequency ranges are assigned for uplink and downlink, and unpaired bands with a single shared frequency range for uplink and downlink, are included in the LTE specifications. Paired bands are used for Frequency Division Duplex (FDD) operation, while unpaired bands are used for Time Division Duplex (TDD) operation. The duplex modes of LTE are described further in Section 3.1.5. Note that some unpaired bands do not have any uplink specified. These "downlink only" bands are paired with the uplink of other bands through *carrier aggregation*, as described in Chapter 12.

An additional challenge with LTE operation in some bands is the possibility of using channel bandwidths up to 20 MHz with a single carrier and even beyond that with aggregated carriers.

Historically, the bands for the first and second generation of mobile services were assigned at frequencies around 800−900 MHz, but also in a few lower and higher bands. When 3G (IMT-2000) was rolled out, focus was on the 2 GHz band and with the continued expansion of IMT services with 3G and 4G, new bands were used at both lower and higher frequencies. All bands considered are up to this point below 6 GHz.

Bands at different frequencies have different characteristics. Due to the propagation properties, bands at lower frequencies are good for wide-area coverage deployments, both in urban, suburban, and rural environments. Propagation properties of higher frequencies make them more difficult to use for wide-area coverage, and higher-frequency bands have therefore to a larger extent been used for boosting capacity in dense deployments.

With new services requiring even higher data rates and high capacity in dense deployments, frequency bands above 6 GHz are being looked at as a complement to the frequency bands below 6 GHz. With the 5G requirements for extreme data rates and localized areas with very high area traffic capacity demands, deployment using much higher frequencies, even above 60 GHz, is considered. Referring to the wavelength, these bands are often called mm-wave bands.

2.3.1 SPECTRUM DEFINED FOR IMT SYSTEMS BY THE ITU-R

The global designations of spectrum for different services and applications are done within the ITU-R and are documented in the *ITU Radio Regulations* [65]. The *World Administrative Radio Congress* WARC-92 identified the bands 1885−2025 and 2110−2200 MHz as intended for implementation of IMT-2000. Of these 230 MHz of 3G spectrum, 2×30 MHz were intended for the satellite component of IMT-2000 and the rest for the terrestrial component. Parts of the bands were used during the 1990s for deployment of 2G cellular systems, especially in the Americas. The first deployments of 3G in 2001−2002 by Japan and Europe were done in this band allocation, and for that reason it is often referred to as the IMT-2000 "core band."

Additional spectrum for IMT-2000 was identified at the World Radiocommunication Conference[1] WRC-2000, where it was considered that an additional need for 160 MHz of spectrum for IMT-2000 was forecasted by the ITU-R. The identification includes the bands

---

[1]The World Administrative Radio Conference (WARC) was reorganized in 1992 and became the World Radiocommunication Conference (WRC).

used for 2G mobile systems at 806—960 and 1710—1885 MHz, and "new" 3G spectrum in the bands at 2500—2690 MHz. The identification of bands previously assigned for 2G was also recognition of the evolution of existing 2G mobile systems into 3G. Additional spectrum was identified at WRC'07 for IMT, encompassing both IMT-2000 and IMT-Advanced. The bands added were 450—470, 698—806, 2300—2400, and 3400—3600 MHz, but the applicability of the bands varies on a regional and national basis. At WRC'12 there were no additional spectrum allocations identified for IMT, but the issue was put on the agenda for WRC'15. It was also determined to study the use of the band 694—790 MHz for mobile services in Region 1 (Europe, Middle East, and Africa).

The somewhat diverging arrangement between regions of the frequency bands assigned to IMT means that there is not one single band that can be used for 3G and 4G roaming worldwide. Large efforts have, however, been put into defining a minimum set of bands that can be used to provide truly global roaming. In this way, multiband devices can provide efficient worldwide roaming for 3G and 4G devices.

2.3.2 FREQUENCY BANDS FOR LTE

LTE can be deployed both in existing IMT bands and in future bands that may be identified. The possibility of operating a radio access technology in different frequency bands is, in itself, nothing new. For example, 2G and 3G devices are multiband capable, covering bands used in the different regions of the world to provide global roaming. From a radio access functionality perspective, this has no or limited impact and the physical layer specifications such as the ones for LTE [24—27] do not assume any specific frequency band. What may differ, in terms of specification, between different bands are mainly the more specific RF requirements, such as the allowed maximum transmit power, requirements/limits on out-of-band (OOB) emission, and so on. One reason for this is that external constraints, imposed by regulatory bodies, may differ between different frequency bands.

The frequency bands where LTE will operate are in both paired and unpaired spectrum, requiring flexibility in the duplex arrangement. For this reason, LTE supports both FDD and TDD operation, will be discussed later.

Release 13 of the 3GPP specifications for LTE includes 32 frequency bands for FDD and 12 for TDD. The number of bands is very large and for this reason, the numbering scheme recently had to be revised to become future proof and accommodate more bands. The paired bands for FDD operation are numbered from 1 to 32 and 65 to 66 [38], as shown in Table 2.1, while the unpaired bands for TDD operation are numbered from 33 to 46, as shown in Table 2.2. Note that the frequency bands defined for UTRA FDD use the same numbers as the paired LTE bands, but are labeled with Roman numerals. Bands 15 and 16 are reserved for definition in Europe, but are presently not used. All bands for LTE are summarized in Figures 2.5 and 2.6, which also show the corresponding frequency allocation defined by the ITU-R.

Some of the frequency bands are partly or fully overlapping. In most cases this is explained by regional differences in how the bands defined by the ITU-R are implemented. At

Table 2.1 Paired Frequency Bands Defined by 3GPP for LTE

Band	Uplink Range (MHz)	Downlink Range (MHz)	Main Region(s)
1	1920—1980	2110—2170	Europe, Asia
2	1850—1910	1930—1990	Americas, Asia
3	1710—1785	1805—1880	Europe, Asia, Americas
4	1710—1755	2110—2155	Americas
5	824—849	869—894	Americas, Asia
6	830—840	875—885	Japan (only for UTRA)
7	2500—2570	2620—2690	Europe, Asia
8	880—915	925—960	Europe, Asia
9	1749.9—1784.9	1844.9—1879.9	Japan
10	1710—1770	2110—2170	Americas
11	1427.9—1447.9	1475.9—1495.9	Japan
12	698—716	728—746	United States
13	777—787	746—756	United States
14	788—798	758—768	United States
17	704—716	734—746	United States
18	815—830	860—875	Japan
19	830—845	875—890	Japan
20	832—862	791—821	Europe
21	1447.9—1462.9	1495.9—1510.9	Japan
22	3410—3490	3510—3590	Europe
23	2000—2020	2180—2200	Americas
24	1626.5—1660.5	1525—1559	Americas
25	1850—1915	1930—1995	Americas
26	814—849	859—894	Americas
27	807—824	852—869	Americas
28	703—748	758—803	Asia/Pacific
29	N/A	717—728	Americas
30	2305—2315	2350—2360	Americas
31	452.5—457.5	462.5—467.5	Americas
32	N/A	1452—1496	Europe
65	1920—2010	2110—2200	Europe
66	1710—1780	2110—2200	Americas
67	N/A	738—758	Europe

the same time, a high degree of commonality between the bands is desired to enable global roaming. A set of bands was first specified as bands for UTRA, with each band originating in global, regional, and local spectrum developments. The complete set of UTRA bands was then transferred to the LTE specifications in release 8 and additional ones have been added since then in later releases.

Table 2.2 Unpaired Frequency Bands Defined by 3GPP for LTE		
Band	**Frequency Range (MHz)**	**Main Region(s)**
33	1900–1920	Europe, Asia (not Japan)
34	2010–2025	Europe, Asia
35	1850–1910	(Americas)
36	1930–1990	(Americas)
37	1910–1930	–
38	2570–2620	Europe
39	1880–1920	China
40	2300–2400	Europe, Asia
41	2496–2690	United States
42	3400–3600	Europe
43	3600–3800	Europe
44	703–803	Asia/Pacific
45	1447–1467	Asia (China)
46	5150–5925	Global

Bands 1, 33, and 34 are the same paired and unpaired bands that were defined first for UTRA in release 99 of the 3GPPP specifications, also called the 2 GHz "core band." *Band 2* was added later for operation in the US PCS1900 band and *Band 3* for 3G operation in the GSM1800 band. The unpaired *Bands 35, 36, and 37* are also defined for the PCS1900 frequency ranges, but are not deployed anywhere today. *Band 39* is an extension of the unpaired Band 33 from 20 to 40 MHz for use in China. Band 45 is another unpaired band for LTE use in China.

Band 65 is an extension of Band 1 to 2 × 90 MHz for Europe. This means that in the upper part, which previously has been harmonized in Europe for *Mobile Satellite Services* (MSS), it will be for satellite operators to deploy a *Complementary Ground Component* (CGC) as a terrestrial LTE system integrated with a satellite network.

Band 4 was introduced as a new band for the Americas following the addition of the 3G bands at WRC-2000. Its downlink overlaps completely with the downlink of Band 1, which facilitates roaming and eases the design of dual Band 1 + 4 devices. *Band 10* is an extension of Band 4 from 2 × 45 to 2 × 60 MHz. *Band 66* is a further extension of the paired band to 2 × 70 MHz, with an additional 20 MHz at the top of the downlink band (2180–2200 MHz) intended as a supplemental downlink for LTE carrier aggregation with the downlink of another band.

Band 9 overlaps with Band 3, but is intended only for Japan. The specifications are drafted in such a way that implementation of roaming dual Band 3 + 9 devices is possible. The 1500-MHz frequency band is also identified in 3GPP for Japan as *Bands 11 and 21*. It is allocated globally to mobile service on a co-primary basis and was previously used for 2G in Japan.

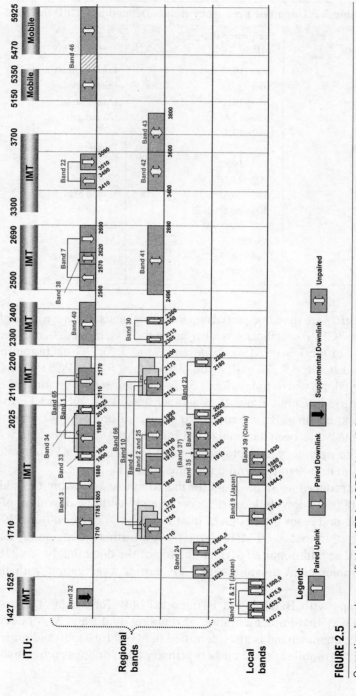

FIGURE 2.5

Operating bands specified for LTE in 3GPP above 1 GHz and the corresponding ITU-R allocation (regional or global).

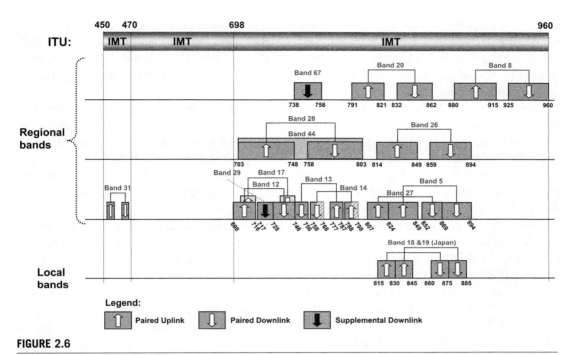

FIGURE 2.6

Operating bands specified for LTE in 3GPP below 1 GHz and the corresponding ITU-R allocation (regional or global).

With WRC-2000, the band 2500—2690 MHz was identified for IMT-2000, and it is identified as *Band 7* in 3GPP for FDD and *Band 38* for TDD operation in the "center gap" of the FDD allocation. The band has a slightly different arrangement in North America, where a US-specific *Band 41* is defined. *Band 40* is an unpaired band specified for the new frequency range 2300—2400 MHz identified for IMT and has a widespread allocation globally.

WRC-2000 also identified the frequency range 806—960 MHz for IMT-2000, complemented by the frequency range 698—806 MHz in WRC'07. As shown in Figure 2.6, several bands are defined for FDD operation in this range. Band 8 uses the same band plan as GSM900. *Bands 5, 18, 19, 26, and 27* overlap, but are intended for different regions. Band 5 is based on the US cellular band, while Bands 18 and 19 are restricted to Japan in the specifications. 2G systems in Japan had a very specific band plan and Bands 18 and 19 are a way of partly aligning the Japanese spectrum plan in the 810—960-MHz range to that in other parts of the world. Note that Band 6 was originally defined in this frequency range for Japan, but it is not used for LTE.

An extensive study was performed in 3GPP to create an extension of Band 5 (850 MHz), which is one of the bands with the most widespread deployment globally. The extension adds additional frequency ranges below the present Band 5 and is done with two new operating bands. Band 26 is the "Upper Extending 850-MHz" band, which encompasses the band 5 range, adding 2 × 10 MHz to create an extended 2 × 35-MHz band. Band 27 is the "Lower Extending 850-MHz" band which consists of the 2 × 17-MHz frequency range right below and adjacent to Band 5.

Bands 12, 13, 14, and 17 make up the first set of bands defined for what is called the *digital dividend*—that is, for spectrum previously used for broadcasting. This spectrum is partly migrated to be used by other wireless technologies, since TV broadcasting is migrating from analog to more spectrum-efficient digital technologies. Other regional band for the digital dividend is *Band 20* that is defined in Europe and *Band 28* for the Asia/Pacific region. An alternative unpaired arrangement in the Asia/Pacific region is the unpaired Band 44.

Band 29, 32, and 67 are "paired" bands that consist of a downlink without an identified uplink. The bands are intended for carrier aggregation with downlink carriers in other bands. Primarily, Band 29 can be paired with Band 2, 4, and 5 in the Americas and Band 32 and 67 can be paired with for example Band 20 in Europe.

The paired *Band 22* and unpaired *Band 42 and 43* are specified for the frequency range 3.4−3.8 GHz [39]. In Europe, a majority of countries already license the band at 3.4−3.6 GHz for both Fixed Wireless Access and mobile use, and there is a European spectrum decision for 3.4−3.8 GHz with "flexible usage modes" for deployment of fixed, nomadic, and mobile networks. In Japan, not only 3.4−3.6 GHz but also 3.6−4.2 GHz will be available to terrestrial mobile services in the future. The band 3.4−3.6 GHz has also been licensed for wireless access in Latin America.

The paired *Band 31* is the first 3GPP band defined in the 450-MHz range. Band 31 is specified for use of LTE in Brazil. Band 32 is an LTE band for the United States, also called the *Wireless Communication Service* (WCS) band.

Several Mobile Satellite Service operators in the United States are planning to deploy an *Ancillary Terrestrial Component* (ATC) using LTE. For this purpose two new frequency bands are defined, *Band 23* with 2 × 20-MHz band for the S-band MSS operators at 2 GHz and *Band 24* with 2 × 34-MHz band for the L-band MSS operators at 1.5 GHz.

Band 46 is a band in a frequency range at 5 GHz that is globally assigned for *Wireless Access Systems* (WAS) including Radio Local Area Networks (RLAN). The band is not fully assigned in any region, but parts are under study, see Section 17.1 for more details on the spectrum for LAA. Operation in the band is unlicensed. For LTE, the band is in Release 13 defined for what is called License-Assisted Access, where downlink operation in Band 46 is combined with licensed operation in other bands through downlink carrier aggregation.

2.3.3 NEW FREQUENCY BANDS

Additional frequency bands are continuously specified for UTRA and LTE. WRC'07 identified additional frequency bands for IMT, which encompasses both IMT-2000 and

IMT-Advanced. Several of the bands defined by WRC'07 are already available for LTE as described earlier, or will become available partly or fully for deployment on a global basis:

- *450–470 MHz* was identified for IMT globally. It is already allocated to mobile service globally, but it is only 20-MHz wide and has a number of different arrangements. LTE Band 31 is defined in this range.
- *698–806 MHz* was allocated to mobile service and identified IMT to some extent in all regions. Together with the band at 806–960 MHz identified at WRC-2000, it forms a wide frequency range from 698 to 960 MHz that is partly identified to IMT in all regions, with some variations. A number of LTE bands are defined in this frequency range.
- *2300–2400 MHz* was identified for IMT on a worldwide basis in all three regions. It is defined as LTE Bands 30 and 40.
- *3400–3600 MHz* was allocated to the mobile service on a primary basis in Europe and Asia and partly in some countries in the Americas. There is also satellite use in the bands today. It is defined as LTE Bands 22, 42, and 43.

For the frequency ranges below 1 GHz identified at WRC-07, 3GPP has already specified several operating bands, as shown in Figure 2.6. The bands with the widest use are Bands 5 and 8, while most of the other bands have regional or more limited use. With the identification of bands down to 698 MHz for IMT use and the switchover from analog to digital TV broadcasting, Bands 12, 13, 14, and 17 are defined in the United States, Band 20 in Europe, and Bands 28 and 44 in Asia/Pacific for the digital dividend.

Additional bands for IMT were identified at WRC'15, some of which are already bands defined for LTE:

- *470–698 MHz* was identified for IMT in some countries in the Americas, including the United States and Canada. Also some countries in the Asia–Pacific identified the bands fully or partly for IMT. In Europe and Africa, the use of this frequency range will be reviewed until WRC'23.
- *1427–1518 MHz*, also called the L-band, was identified for IMT globally. The band has been used for a long time in Japan and the LTE Bands 11, 21, and 32 are already defined for 3GPP in this frequency range.
- *3300–3700 MHz* is now identified for IMT at least in some regions or countries. The frequency range 3400–3600 MHz, which was identified already at WRC-07, is now identified globally for IMT. LTE Bands 22, 42, and 43 are in this range.
- *4800– 4990 MHz* was identified for IMT for a few countries in the Americas and Asia-Pacific.

2.4 SPECTRUM FOR 5G
2.4.1 NEW FREQUENCY BANDS TO BE STUDIED BY WRC

The frequency listings in the ITU Radio Regulations [65] do not directly list a band for IMT, but rather allocates a band for the mobile service with a footnote stating that the band is identified for use by administrations wishing to implement IMT. The identification is mostly by region, but is in some cases also specified on a per-country level. All footnotes mention "IMT" only, so there is no specific mentioning of the different generations of IMT. Once a band is assigned, it is therefore up to the regional and local administrations to define a band for IMT use in general or for specific generations. In many cases, regional and local assignments are "technology neutral" and allow for any kind of IMT technology.

This means that all existing IMT bands are potential bands for IMT-2020 (5G) deployment in the same way as they have been used for previous IMT generations. In addition, it is also expected that bands above 6 GHz will be used for IMT-2020. An agenda item has been set up for WRC'19 where additional spectrum will be considered, and studies will be conducted until WRC'19 to determine the spectrum needs for terrestrial IMT. Sharing and compatibility studies of IMT will also be performed for a set of specific bands in the range from 24.25 to 86 GHz as illustrated in Figure 2.7. A majority of the bands to be studied are already assigned to the mobile service on a primary basis, in most bands together with fixed and satellite services. These are:

- 24.25−27.5 GHz
- 37−40.5 GHz
- 42.5−43.5 GHz
- 45.5−47 GHz
- 47.2−50.2 GHz
- 50.4−52.6 GHz
- 66−76 GHz
- 81−86 GHz

There are also bands to be studied for IMT that are presently not allocated to the mobile service on a primary basis and where it will be investigated whether the allocation can be changed to include mobile:

- 31.8−33.4 GHz
- 40.5−42.5 GHz
- 47−47.2 GHz

Sharing studies between IMT as a mobile service and the other primary services in those bands will be a task for regional and national administrations together with the industry in

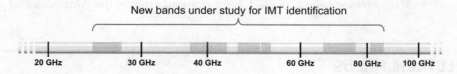

FIGURE 2.7

New IMT bands under study in ITU-R.

preparation for WRC'19. In some cases studies of adjacent services may be performed too. As an input to the studies, technical and operational characteristics of IMT are needed, which in this case implies the characteristics of IMT-2020.

2.4.2 **RF EXPOSURE ABOVE 6 GHZ**

With the expansion of the frequency ranges for 5G mobile communications to bands above 6 GHz, existing regulations on human exposure to RF *electromagnetic fields* (EMF) may restrict the maximum output power of user devices to levels significantly lower than what are allowed for lower frequencies.

International RF EMF exposure limits, for example, those recommended by the *International Commission on Non-Ionizing Radiation Protection* (ICNIRP) and those specified by the *Federal Communications Commission* (FCC) in the United States, have been set with wide safety margins to protect against excessive heating of tissue due to energy absorption. In the frequency range of 6 to 10 GHz, the basic limits change from being specified as specific absorption rate (W/kg) to incident power density (W/m^2). This is mainly because the energy absorption in tissue becomes increasingly superficial with increasing frequency, and thereby more difficult to measure.

It has been shown that for products intended to be used in close proximity of the body, there will be a discontinuity in maximum allowed output power as the transition is made from specific absorption rate to power-density-based limits [62]. To be compliant with ICNIRP exposure limits at the higher frequencies, the transmit power might have to be up to 10 dB below the power levels used for current cellular technologies. The exposure limits above 6 GHz appear to have been set with safety margins even larger than those used at lower frequencies, and without any obvious scientific justification.

For the lower-frequency bands, large efforts have been spent over the years to characterize the exposure and to set relevant limits. With a growing interest for utilizing frequency bands above 6 GHz for mobile communications, research efforts are likely to increase which eventually may lead to revised exposure limits. In the most recent RF exposure standards published by IEEE (C95.1-2005, C95.1-2010a), the inconsistency at the transition frequency is less evident. However, these limits have not yet been adopted in any national regulation, and it is important that also other standardization organizations and regulators work to address this issue. If not, this might have a large negative impact on coverage at higher frequencies, in particular for user equipment intended to be used near the body, such as wearables, tablets, and mobile phones, for which the maximum transmit power might be heavily limited by the current RF exposure regulations.

2.5 **3GPP STANDARDIZATION**

With a framework for IMT systems set up by the ITU-R, with spectrum made available by the WRC and with an ever-increasing demand for better performance, the task of specifying the actual mobile-communication technologies falls on organizations like 3GPP. More specifically, 3GPP writes the technical specifications for 2G GSM, 3G WCDMA/HSPA, and 4G LTE. 3GPP technologies are the most widely deployed in the world, with more than 90% of the world's 7.4 billion mobile subscriptions in Q4 2015 [54]. In order to understand how 3GPP works, it is important to also understand the process of writing specifications.

2.5.1 THE 3GPP PROCESS

Developing technical specifications for mobile communication is not a one-time job; it is an ongoing process. The specifications are constantly evolving trying to meet new demands for services and features. The process is different in the different fora, but typically includes the four phases illustrated in Figure 2.8:

1. *Requirements*, where it is decided what is to be achieved by the specification.
2. *Architecture*, where the main building blocks and interfaces are decided.
3. *Detailed specifications*, where every interface is specified in detail.
4. *Testing and verification*, where the interface specifications are proven to work with real-life equipment.

These phases are overlapping and iterative. As an example, requirements can be added, changed, or dropped during the later phases if the technical solutions call for it. Likewise, the technical solution in the detailed specifications can change due to problems found in the testing and verification phase.

The specification starts with the *requirements* phase, where it is decided what should be achieved with the specification. This phase is usually relatively short.

In the *architecture* phase, the architecture is decided—that is, the principles of how to meet the requirements. The architecture phase includes decisions about reference points and interfaces to be standardized. This phase is usually quite long and may change the requirements.

After the architecture phase, the *detailed specification* phase starts. It is in this phase the details for each of the identified interfaces are specified. During the detailed specification of the interfaces, the standards body may find that previous decisions in the architecture or even in the requirements phases need to be revisited.

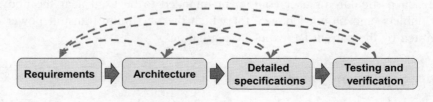

FIGURE 2.8

The standardization phases and iterative process.

Finally, the *testing and verification* phase starts. It is usually not a part of the actual specification , but takes place in parallel through testing by vendors and interoperability testing between vendors. This phase is the final proof of the specification. During the testing and verification phase, errors in the specification may still be found and those errors may change decisions in the detailed specification. Albeit not common, changes may also need to be made to the architecture or the requirements. To verify the specification, products are needed. Hence, the implementation of the products starts after (or during) the detailed specification phase. The testing and verification phase ends when there are stable test specifications that can be used to verify that the equipment is fulfilling the technical specification.

Normally, it takes about one year from the time when the specification is completed until commercial products are out on the market.

3GPP consists of three *Technical Specifications Groups* (TSGs)—see Figure 2.9—where TSG RAN (*Radio Access Network*) is responsible for the definition of functions, requirements, and interfaces of the Radio Access. It consists of six working groups (WGs):

1. RAN WG1, dealing with the physical layer specifications.
2. RAN WG2, dealing with the layer 2 and layer 3 radio interface specifications.
3. RAN WG3, dealing with the fixed RAN interfaces—for example, interfaces between nodes in the RAN—but also the interface between the RAN and the core network.
4. RAN WG4, dealing with the RF and *radio resource management* (RRM) performance requirements.
5. RAN WG5, dealing with the device conformance testing.
6. RAN WG6, dealing with standardization of GSM/EDGE (previously in a separate TSG called GERAN).

The work in 3GPP is carried out with relevant ITU-R recommendations in mind and the result of the work is also submitted to ITU-R as being part of IMT-2000 and IMT-Advanced. The organizational partners are obliged to identify regional requirements that may lead to options in the standard. Examples are regional frequency bands and special protection requirements local to a region. The specifications are developed with global roaming and circulation of devices in mind. This implies that many regional requirements in essence will be global requirements for all devices, since a roaming device has to meet the strictest of all regional requirements. Regional options in the specifications are thus more common for base stations than for devices.

The specifications of all releases can be updated after each set of TSG meetings, which occur four times a year. The 3GPP documents are divided into releases, where each release has a set of features added compared to the previous release. The features are defined in Work Items agreed and undertaken by the TSGs. The releases from release 8 and onward, with some main features listed for LTE, are shown in Figure 2.10. The date shown for each release is the day the content of the release was frozen. Release 10 of LTE is the first version approved by ITU-R as an IMT-Advanced technology and is therefore also the first release named *LTE-*

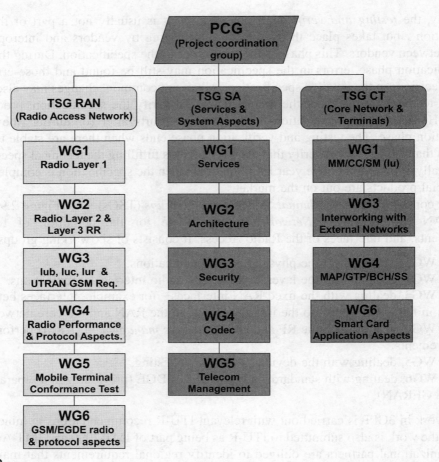

FIGURE 2.9

3GPP organization.

Advanced. From Release 13, the marketing name for LTE is changed to *LTE-Advanced Pro.* The content of the 3GPP releases for LTE is described with further details in Chapter 3.

The 3GPP Technical Specifications (TS) are organized in multiple series and are numbered TS XX.YYY, where XX denotes the number of the specification series and YYY is the number of the specification within the series. The following series of specifications define the radio access technologies in 3GPP:

- 25-series: Radio aspects for UTRA (WCDMA).
- 45-series: Radio aspects for GSM/EDGE.
- 36-series: Radio aspects for LTE, LTE-Advanced, and LTE-Advanced Pro.
- 37-series: Aspects relating to multiple radio access technologies.
- 38-series: Radio aspects for the next generation (5G).

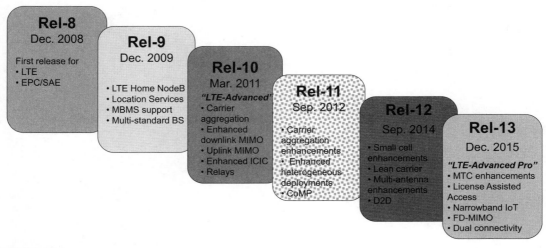

FIGURE 2.10

Releases of 3GPP specifications for LTE.

2.5.2 SPECIFYING 5G IN 3GPP

In parallel with the definition and evaluation of the next-generation access initiated in ITU-R, 3GPP has started to define the next-generation 3GPP radio access. A workshop on 5G radio access was held in 2014 and a process to define evaluation criteria for 5G was initiated with a second workshop in early 2015. The evaluation is planned to follow the same process that was used when LTE-Advanced was evaluated and submitted to ITU-R and approved as a 4G technology as part of IMT-Advanced. The evaluation and submission will follow the ITU-R time line described in Section 2.2.3. With reference to the four phases described in Figure 2.8, the 5G work in 3GPP is presently in the first phase of defining requirements. The 3GPP process is further described in Chapter 23.

3GPP TSG RAN is documenting scenarios, requirements, and evaluation criteria for the new 5G radio access in a new report TR 38.913 [66], which corresponds largely to the report TR 36.913 [29] that was developed for defining requirements on LTE-Advanced. As for the case of the IMT-Advanced evaluation, the corresponding 3GPP evaluation of the next-generation radio access could have a larger scope and may have stricter requirements than the ITU-R evaluation of candidate IMT-2020 RITs that is defined by ITU-R WP5D. It is essential that the ITU-R evaluation is kept at a reasonable complexity, in order to complete the work on time and for also external evaluation groups to be able to participate in the evaluation.

Further details on the 5G radio access and its possible components are given in Chapters 23 and 24.

LTE RADIO ACCESS: AN OVERVIEW

3

The work on LTE was initiated in late 2004 with the overall aim of providing a new radio-access technology focusing on packet-switched data only. The first phase of the 3GPP work on LTE was to define a set of performance and capability targets for LTE [6]. This included targets on peak data rates, user/system throughput, spectral efficiency, and control/user-plane latency. In addition, requirements were also set on spectrum flexibility, as well as on interaction/compatibility with other 3GPP radio-access technologies (GSM, WCDMA/HSPA, and TD-SCDMA).

Once the targets were set, 3GPP studies on the feasibility of different technical solutions considered for LTE were followed by development of detailed specifications. The first release of the LTE specifications, release 8, was completed in 2008 and commercial network operation began in late 2009. Release 8 has been followed by additional LTE releases, introducing additional functionality and capabilities in different areas, as illustrated in Figure 3.1. Of particular interest are release 10, being the first release of LTE-Advanced, and release 13, finalized late 2015 and the first release of LTE-Advanced Pro. Currently, as of this writing, 3GPP is working on LTE release 14.

In parallel to the development of LTE, there has also been an evolution of the overall 3GPP network architecture, termed *system architecture evolution* (SAE), including both the radio-access network and the core network. Requirements were also set on the architecture evolution, leading to a new flat radio-access-network architecture with a single type of node, the *eNodeB*,[1] as well as a new core-network architecture. An excellent description of the LTE-associated core-network architecture, the *evolved packet core* (EPC), can be found in [5].

The remaining part of this chapter provides an overview of LTE. The most important technology components of LTE release 8—including transmission schemes, scheduling, multi-antenna support, and spectrum flexibility—are presented, as well as the additional features and enhancements introduced in later releases up to and including release 13. The chapter can either be read on its own to get a high-level overview of LTE, or as an introduction to the subsequent chapters.

---

[1]eNodeB is a 3GPP term that can roughly be seen as being equivalent to a base station, see further Chapter 4.

4G, LTE-Advanced Pro and The Road to 5G. http://dx.doi.org/10.1016/B978-0-12-804575-6.00003-0

FIGURE 3.1

LTE and its evolution.

The following chapters, Chapters 4—22, provide a detailed description of the LTE radio-access technology. Chapter 4 provides an overview of the LTE protocol structure, including RLC, MAC, and the physical layer, explaining the logical and physical channels, and the related data flow. The time—frequency structure on which LTE is based is covered in Chapter 5, together with a brief overview of the LTE antenna-port concept. This is followed by a detailed description of the LTE physical-layer functionality for downlink and uplink transmission in Chapters 6 and 7, respectively. Chapter 8 contains a description of LTE retransmission mechanisms, followed by a discussion on the mechanisms available in LTE to support advanced scheduling and link adaptation in Chapter 9. Channel-state reporting to support scheduling, including handling of large antenna arrays, is covered in Chapter 10. Access procedures, necessary for a device (or a terminal, in 3GPP known as a *user equipment*, UE) to connect to the network, are the topic of Chapter 11.

The following chapters focus on some of the enhancements incorporated into LTE from release 10 onwards, starting with carrier aggregation (CA) in Chapter 12. Multi-point coordination/transmission is discussed in Chapter 13, followed by a discussion of heterogeneous deployments based on LTE in Chapter 14. Chapter 15 covers small-cell enhancements and dynamic TDD, Chapter 16 dual connectivity, and Chapter 17 operation in unlicensed spectrum. Relaying and broadcast/multicast are the topics of Chapters 18 and 19, respectively. Chapter 20 describes enhancements for improved support of machine-type communication while Chapter 21 focuses on direct device-to-device communication. The LTE part of the book is concluded with a discussion on the definition of radio-frequency (RF) requirements, taking into account the spectrum flexibility.

Finally, Chapters 23 and 24 provide an overview of the new 5G radio-access technology currently under discussion in 3GPP and Chapter 25 concludes the discussion on 5G radio access.

3.1 LTE BASIC TECHNOLOGIES

Release 8 is the first LTE release and forms the basis for the following releases. Due to time limitations for the release 8 work, some smaller features originally planned for release 8 was postponed for release 9, which thus can be seen as part of the basic LTE framework. In the following sections, the basic LTE technologies in release 8/9 are described.

3.1.1 TRANSMISSION SCHEME

The LTE downlink transmission scheme is based on conventional *orthogonal frequency-division multiplexing* (OFDM) [53], an attractive transmission scheme for several reasons. Due to the relatively long OFDM symbol time in combination with a cyclic prefix, OFDM provides a high degree of robustness against channel-frequency selectivity. Although signal corruption due to a frequency-selective channel can, in principle, be handled by equalization at the receiver side, the complexity of such equalization starts to become unattractively high for implementation in a device at larger bandwidths and especially in combination with advanced multi-antenna transmission schemes such as spatial multiplexing. Therefore, OFDM is an attractive choice for LTE for which a wide bandwidth and support for advanced multi-antenna transmission were key requirements.

OFDM also provides some additional benefits relevant for LTE:

- OFDM provides access to the frequency domain, thereby enabling an additional degree of freedom to the channel-dependent scheduler compared to time-domain-only scheduling used in major 3G systems.
- Flexible transmission bandwidth to support operation in spectrum allocations of different sizes is straightforward with OFDM, at least from a baseband perspective, by varying the number of OFDM subcarriers used for transmission. Note, however, that support of a flexible transmission bandwidth also requires flexible RF filtering, and so on, for which the exact transmission scheme is to a large extent irrelevant. Nevertheless, maintaining the same baseband-processing structure, regardless of the bandwidth, eases device development and implementation.
- Broadcast/multicast transmission, where the same information is transmitted from multiple base stations, is straightforward with OFDM as described in Chapter 19.

The LTE uplink is also based on OFDM transmission. However, to enable high-power amplifier efficiency on the device side, different means are taken to reduce the *cubic metric* [10] of uplink transmissions. Cubic metric is a measure of the amount of additional back-off needed for a certain signal waveform, relative to the back-off needed for some reference waveform. It captures similar properties as the more commonly known peak-to-average ratio but better represents the actual back-off needed in an implementation. Low cubic metric is achieved by preceding the OFDM modulator by a DFT precoder, leading to *DFT-spread OFDM* (DFTS-OFDM), see Chapter 7. Often the term DFTS-OFDM is used to describe the LTE uplink transmission scheme in general. However, it should be understood that DFTS-OFDM is only used for uplink data transmission. As described in more detail in later chapters, other means are used to achieve a low cubic metric for other types of uplink transmissions. Thus, the LTE uplink transmission scheme should be described as OFDM with different techniques, including DFT precoding for data transmission, being used to reduce the cubic metric of the transmitted signal.

The use of OFDM-based transmission for the LTE uplink allows for orthogonal separation of uplink transmissions also in the frequency domain. Orthogonal separation is in many cases beneficial as it avoids interference between uplink transmissions from different devices within the cell (*intra-cell interference*). Allocating a very large instantaneous bandwidth for transmission from a single device is not an efficient strategy in situations where the data rate is mainly limited by the available device transmit power rather than the bandwidth. In such situations a device can instead be allocated only a part of the total available bandwidth and other devices within the cell can be scheduled to transmit in parallel on the remaining part of the spectrum. In other words, the LTE uplink transmission scheme allows for both *time-division multiple access* (TDMA) and *frequency-division multiple access* (FDMA) to separate users.

3.1.2 CHANNEL-DEPENDENT SCHEDULING AND RATE ADAPTATION

One key characteristic of mobile radio communication is the large and typically rapid variations in the instantaneous channel conditions stemming from frequency-selective fading, distance-dependent path loss, and random interference variations due to transmissions in other cells and by other terminals. Instead trying to combat these variations through, for example, power control, LTE tries to exploit these variations through *channel-dependent scheduling* where the time—frequency resources are dynamically shared between users. Dynamic sharing of resources across the users is well matched to the rapidly varying resource requirements posed by packet-data communication and also enables several of the other key technologies on which LTE is based.

The scheduler controls, for each time instant, to which users the different parts of the shared resource should be assigned and the data rate to be used for each transmission. Thus, *rate adaptation* (i.e., trying to dynamically adjust the data rate to match the instantaneous channel conditions) can be seen as a part of the scheduling functionality.

However, even if the rate adaptation successfully selects an appropriate data rate, there is a certain likelihood of transmission errors. To handle transmission errors, *fast hybrid-ARQ with soft combining* is used in LTE to allow the device to rapidly request retransmissions of erroneously received data blocks and to provide a tool for implicit rate adaptation. Retransmissions can be rapidly requested after each packet transmission, thereby minimizing the impact on end-user performance from erroneously received packets. Incremental redundancy is used as the soft combining strategy, and the receiver buffers the soft bits to be able to perform soft combining between transmission attempts.

The scheduler is a key element and to a large extent determines the overall system performance, especially in a highly loaded network. Both downlink and uplink transmissions are subject to tight scheduling in LTE. A substantial gain in system capacity can be achieved if the channel conditions are taken into account in the scheduling decision, so-called *channel-dependent scheduling*, where transmission are directed to user with momentarily favorable channel conditions. Due to the use of OFDM in both the downlink and uplink transmission directions, the scheduler has access to both the time and frequency domains. In other words, the scheduler can, for each time instant and frequency region, select the user with the best channel conditions, as illustrated in Figure 3.2.

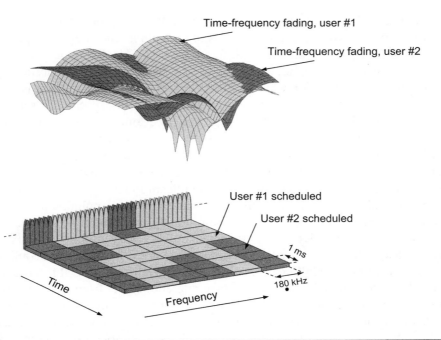

FIGURE 3.2

Downlink channel-dependent scheduling in time and frequency domains.

The possibility of channel-dependent scheduling in the frequency domain is particularly useful at low device speeds—in other words, when the channel is varying slowly in time. Channel-dependent scheduling relies on channel-quality variations between users to obtain a gain in system capacity. For delay-sensitive services, a time-domain-only scheduler may, due to the delay constraints, be forced to schedule a particular user, despite the channel quality not being at its peak. In such situations, exploiting channel-quality variations also in the frequency domain will help to improve the overall performance of the system. For LTE, scheduling decisions can be taken as often as once every 1 ms, and the granularity in the frequency domain is 180 kHz. This allows for relatively rapid channel variations in both the time and frequency domains to be tracked and utilized by the scheduler.

To support downlink scheduling, a device may provide the network with *channel-state* reports indicating the instantaneous downlink channel quality in both the time and frequency domains. The channel state is typically obtained by measuring on *reference signals* transmitted in the downlink. Based on the channel-state reports, also referred to as *channel-state information* (CSI), the downlink scheduler can assign resources for downlink transmission to different devices, taking the channel quality into account in the scheduling decision. In principle, a scheduled device can be assigned an arbitrary combination of 180 kHz wide *resource blocks* in each 1 ms scheduling interval.

As mentioned earlier in this chapter, the LTE uplink is based on orthogonal separation of different uplink transmissions, and it is the task of the uplink scheduler to assign resources in both the time and frequency domains to different devices. Scheduling decisions, taken once per 1 ms, control what set of devices are allowed to transmit within a cell during a given time interval and, for each device, on what frequency resources the transmission is to take place and what transmission parameters, including the data rate, to use. Similar scheduling strategies as in the downlink can be used, although there are some differences between the two. Fundamentally, the uplink power resource is *distributed* among the users, while in the downlink the power resource is *centralized* within the base station. Furthermore, the maximum uplink transmission power of a single terminal is typically significantly lower than the output power of a base station. This has a significant impact on the scheduling strategy. Unlike the downlink, where pure time division often can be used, uplink scheduling typically has to rely on sharing in the frequency domain in addition to the time domain, as a single terminal may not have sufficient power for efficiently utilizing the link capacity.

Channel conditions can also be taken into account in the uplink scheduling process, similar to the downlink scheduling. One possibility to acquire uplink CSI is through so-called *sounding*, where the terminal transmits a known reference signal from which the base station can assess the channel quality in the different parts of the spectrum. However, as is discussed in more detail in subsequent chapters, obtaining information about the uplink channel conditions may not be feasible or desirable in all situations. Therefore, different means to obtain *uplink diversity* are important as a complement in situations where uplink channel-dependent scheduling is not suitable.

3.1.3 INTER-CELL INTERFERENCE COORDINATION

LTE is designed to operate with a frequency reuse of one implying that the same carrier frequency can be used at neighboring transmission points. In particular, the basic control channels of LTE are designed to operate properly with the relatively low signal-to-interference ratio that may be experienced in a reuse-one deployment.

Fundamentally, having access to all available frequency resources at each transmission point is always beneficial. However, system efficiency and end-user quality are further improved if transmissions from neighboring transmission points can be coordinated in such a way that the most severe interference situations can be avoided.

Already the first release of LTE included explicit support for such coordination, in the release-8 context referred to as *inter-cell interference coordination* (ICIC), see Chapter 13. More specifically, the release-8 specifications defined a set of messages that can be exchanged between eNodeBs using the so-called *X2 interface*, see Chapter 4. These messages provide information about the interference situation experienced by the eNodeB issuing the message and can be used by a neighboring eNodeB receiving the message as input to its scheduling process, thereby providing a means for at least partly coordinating the transmissions and controlling the interference between cells of different eNodeBs. Especially severe interference situations may occur in so-called *heterogeneous network deployments* consisting of overlapping layers of base stations with large differences in the downlink transmission power. This is briefly discussed in Section 3.5, with a more in-depth discussion in Chapter 14.

3.1.4 MULTI-ANTENNA TRANSMISSION

Already from its first release, LTE has included support for different multi-antenna techniques as an integral part of the radio-interface specifications. In many respects, the use of multiple antennas is the key technology to reach many of the aggressive LTE performance targets. Multiple antennas can be used in different ways for different purposes:

- Multiple receive antennas can be used for receive diversity. For uplink transmissions, they have been used in many cellular systems for several years. However, as dual receive antennas are the baseline for all LTE devices,[2] the downlink performance is also improved. The simplest way of using multiple receive antennas is classical receive diversity to collect additional energy and suppress fading, but additional gains can be achieved in interference-limited scenarios if the antennas are used not only to provide diversity, but also to suppress interference.
- Multiple transmit antennas at the base station can be used for transmit diversity and different types of beam-forming. The main goal of beam-forming is to improve the received signal-to-interference-and-noise ratio (SINR) and, eventually, improve system capacity and coverage.
- *Spatial multiplexing*, sometimes referred to as multiple input, multiple output (MIMO) or more specifically single-user MIMO (SU-MIMO) using multiple antennas at both the transmitter and receiver, is supported by LTE. Spatial multiplexing results in an increased data rate, channel conditions permitting, in bandwidth-limited scenarios by creating several parallel "channels." Alternatively, by combining the spatial properties with the appropriate interference-suppressing receiver processing, multiple devices can transmit on the same time—frequency resource in order to improve the overall cell capacity. In 3GPP this is referred to as *multiuser MIMO*.

In general, the different multi-antenna techniques are beneficial in different scenarios. As an example, at relatively low SINR, such as at high load or at the cell edge, spatial multiplexing provides relatively limited benefits. Instead, in such scenarios multiple antennas at the transmitter side should be used to raise the SINR by means of beam-forming. On the other hand, in scenarios where there already is a relatively high SINR, for example, in small cells, raising the signal quality further provides relatively minor gains as the achievable data rates are then mainly bandwidth limited rather than SINR limited. In such scenarios, spatial multiplexing should instead be used to fully exploit the good channel conditions. The multi-antenna scheme used is under control of the base station, which therefore can select a suitable scheme for each transmission.

---

[2]For low-end MTC devices single-antenna operation is the baseline.

Release 9 enhanced the support for combining spatial multiplexing with beam-forming. Although the combination of beam-forming and spatial multiplexing was already possible in release 8, it was restricted to so-called *codebook-based precoding* (see Chapter 6). In release 9, the support for spatial multiplexing combined with so-called *non-codebook-based precoding* was introduced, allowing for improved flexibility in deploying various multi-antenna transmission schemes.

Up to four layers downlink spatial multiplexing was already supported from the first release of LTE. Later releases further extended the LTE multi-antenna capabilities as described in Section 3.4.

3.1.5 SPECTRUM FLEXIBILITY

A high degree of spectrum flexibility is one of the main characteristics of the LTE radio-access technology. The aim of this spectrum flexibility is to allow for the deployment of LTE radio access in difference frequency bands with various characteristics, including different duplex arrangements and different sizes of the available spectrum. Chapter 22 outlines further details of how spectrum flexibility is achieved in LTE.

3.1.5.1 Flexibility in duplex arrangements

One important part of the LTE requirements in terms of spectrum flexibility is the possibility to deploy LTE-based radio access in both paired *and* unpaired spectra. Therefore, LTE supports both frequency- and time-division-based duplex arrangements. *Frequency-division duplex* (FDD), as illustrated on the left in Figure 3.3, implies that downlink and uplink transmission take place in different, sufficiently separated, frequency bands. *Time-division duplex* (TDD), as illustrated on the right in Figure 3.3, implies that downlink and uplink

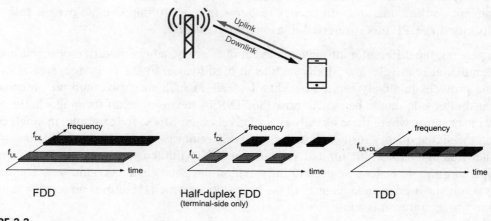

FIGURE 3.3

Frequency- and time-division duplex.

transmission take place in different, nonoverlapping time slots. Thus, TDD can operate in unpaired spectrum, whereas FDD requires paired spectrum. The required flexibility and resulting requirements to support LTE operation in different paired and unpaired frequency arrangements are further discussed in Chapter 22.

Operation in both paired and unpaired spectrum has been supported by 3GPP radio-access technologies even before the introduction of LTE by means of FDD-based WCDMA/HSPA in combination with TDD-based TD-SCDMA radio. However, this was then achieved by means of, at least in the details, relatively different radio-access technologies leading to additional effort and complexity when developing and implementing dual-mode devices capable of both FDD and TDD operation. LTE, on the other hand, supports both FDD and TDD *within a single radio-access technology*, leading to a minimum of deviation between FDD and TDD for LTE-based radio access. As a consequence of this, the overview of the LTE radio access provided in the following chapters is, to a large extent, valid for both FDD and TDD. In the case of differences between FDD and TDD, these differences are explicitly indicated. Furthermore, the TDD mode, also known as TD-LTE, is designed with coexistence between LTE (TDD) and TD-SCDMA in mind to simplify a gradual migration from TD-SCDMA to LTE.

LTE also supports *half-duplex* FDD at the device (illustrated in the middle of Figure 3.3). In half-duplex FDD, transmission and reception *at a specific device* are separated in both frequency and time. The base station still uses full-duplex FDD as it simultaneously may schedule *different* devices in uplink and downlink; this is similar to, for example, GSM operation. The main benefit with half-duplex FDD is the reduced device complexity as no duplex filter is needed in the device. This is especially beneficial in the case of multiband devices which otherwise would need multiple sets of duplex filters.

3.1.5.2 Bandwidth flexibility

An important characteristic of LTE is the support for a range of different transmission bandwidths on both downlink and uplink. The main reason for this is that the amount of spectrum available for LTE deployment may vary significantly between different frequency bands and also depending on the exact situation of the operator. Furthermore, the possibility of operating in different spectrum allocations gives the possibility for gradual migration of spectrum from other radio-access technologies to LTE.

LTE supports operation in a wide range of spectrum allocations, achieved by a flexible transmission bandwidth being part of the LTE specifications. To efficiently support very high data rates when spectrum is available, a wide transmission bandwidth is necessary. However, a sufficiently large amount of spectrum may not always be available, either due to the band of operation or due to a gradual migration from another radio-access technology, in which case LTE can be operated with a narrower transmission bandwidth. In such cases, the maximum achievable data rates will be reduced accordingly. As discussed in the following, the spectrum flexibility is further improved in later releases of LTE.

The LTE physical-layer specifications [24–27] are bandwidth agnostic and do not make any particular assumption on the supported transmission bandwidths beyond a minimum value. As is seen in the following, the basic radio-access specification, including the physical-layer and protocol specifications, allows for any transmission bandwidth ranging from roughly 1 MHz up to around 20 MHz. At the same time, at an initial stage, RF requirements are only specified for a limited subset of transmission bandwidths, corresponding to what is predicted to be relevant spectrum-allocation sizes and relevant migration scenarios. Thus, in practice the LTE radio-access technology supports a limited set of transmission bandwidths, but additional transmission bandwidths can easily be introduced by updating only the RF specifications.

3.1.6 MULTICAST AND BROADCAST SUPPORT

In situations where a large number of users want to receive the same information, for example, a TV news clip, information about the local weather conditions, or stock-market information, separate transmission of the information to each user may not be the most efficient approach. Instead, transmitting the same information once to all interested users may be a better choice. This is known as *broadcast*, or *multimedia broadcast–multicast services* (MBMS) in 3GPP, implying transmission of the same information to multiple receivers. In many cases the same information is of interest over a large area in which case identical signals can be transmitted from multiple cell sites with identical coding and modulation and with timing and frequency synchronized across the sites. From a device perspective, the signal will appear exactly as a signal transmitted from a single cell site and subject to multipath propagation, see Figure 3.4. Due to the OFDM robustness to multipath propagation, such multicell transmission, in 3GPP also referred to as *multicast/broadcast single-frequency network* (MBSFN) transmission, will then not only improve the received signal strength, but also eliminate the inter-cell interference. Thus, with OFDM, multicell broadcast/multicast throughput may eventually be limited by noise only and can then, in the case of small cells, reach extremely high values.

It should be noted that the use of MBSFN transmission for multicell broadcast/multicast assumes the use of tight synchronization and time alignment of the signals transmitted from different cell sites.

MBSFN support was part of the LTE work from the beginning, but due to time limitations, the support was not completed until the second release of LTE, release 9. There are also

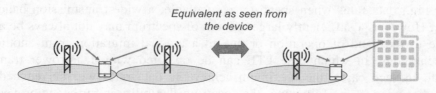

FIGURE 3.4

Equivalence between broadcast and multipath propagation from a device perspective.

enhancements for supporting multicast/broadcast services in a single cell in release 13, known as *single-cell point to multi-point* (SC-PTM).

3.1.7 POSITIONING

Positioning, as the name implies, refers to functionality in the radio-access network to determine the location of individual devices. Determining the position of a device can, in principle, be done by including a GPS receiver in the device. Although this is a quite common feature, not all devices include the necessary GPS receiver and there may also be cases when the GPS service is not available. LTE release 9 therefore introduced positioning support inherent in the radio-access network. By letting the device measure and report to the network the relative time of arrival of special reference signals transmitted regularly from different cell sites, the location of the device can be determined by the network.

3.2 LTE EVOLUTION

Releases 8 and 9 form the foundation of LTE, providing a highly capable mobile-broadband standard. However, to meet new requirements and expectations, the releases following the basic ones provides additional enhancements and features in different areas. Figure 3.5 illustrates some of the major areas in which LTE has been enhanced with details provided in the following. Table 3.1 captures the major enhancements per release.

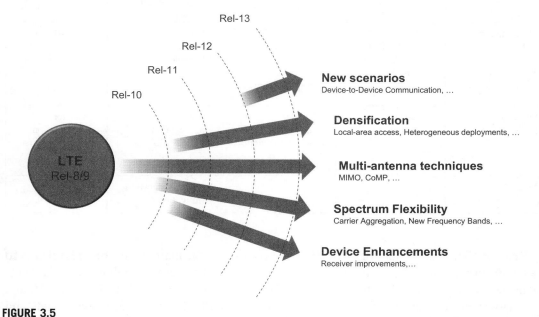

FIGURE 3.5

LTE evolution.

Table 3.1 Major LTE Features per Release

	Release 10	Release 11	Release 12	Release 13
New scenarios			• Device-to-device communication • Machine-type communication	• Enhancements for device-to-device communication • Enhancements for machine-type communication • Narrow-band IoT
Device enhancements		• Performance requirements for interference-rejection receivers	• Network-assisted interference cancellation • SIC, IRC for SU-MIMO	
Densification	• Heterogeneous deployments • Relaying	• Heterogeneous deployments	• Dual connectivity • Small-cell on/off • Dynamic TDD • Mobility enhancements in heterogeneous deployments	• Dual connectivity • License-assisted access
Multi-antenna techniques	• Downlink 8 × 8 MIMO • Uplink 4 × 4 MIMO	• CoMP with low-latency backhaul • Enhanced control-channel structures	• CoMP with nonideal backhaul	• FD-MIMO
Spectrum flexibility	• Carrier aggregation	• Carrier aggregation across different TDD configurations	• Carrier aggregation, FDD + TDD	• Carrier aggregation, up to 32 carriers • License-assisted access
Miscellaneous			• Smart congestion mitigation	• SC-PTM

Release 10 marks the start of the LTE evolution. One of the main targets of LTE release 10 was to ensure that the LTE radio-access technology would be fully compliant with the IMT-Advanced requirements, thus the name *LTE-Advanced* is often used for LTE release 10 and later. However, in addition to the ITU requirements, 3GPP also defined its own targets and requirements for LTE-Advanced [29]. These targets/requirements extended the ITU requirements both in terms of being more aggressive as well as including additional requirements. One important requirement was *backward compatibility*. Essentially this means that an earlier-release LTE device should be able to access a carrier supporting LTE release-10 functionality, although obviously not being able to utilize all the release-10 features of that carrier.

LTE release 10 was completed in late 2010 and implied enhanced LTE spectrum flexibility through carrier aggregation, further extended multi-antenna transmission, introduced support for relaying, and provided improvements in the area of interference coordination in heterogeneous network deployments.

Release 11 further extended the performance and capabilities of LTE. One of the most notable features of LTE release 11, finalized in late 2012, was radio-interface functionality for *coordinated multi-point* (CoMP) transmission and reception. Other examples of improvements in release-11 were carrier-aggregation enhancements, a new control-channel structure, and performance requirements for more advanced device receivers.

Release 12 was completed in 2014 and focused on small cells with features such as dual connectivity, small-cell on/off, and *enhanced Interference Mitigation and Traffic Adaptation* (eIMTA), as well as on new scenarios with introduction of direct device-to-device communication and provisioning of complexity-reduced machine-type communication.

Release 13, finalized at the end of 2015, marks the start of *LTE-Advanced Pro*. It is sometimes in marketing dubbed 4.5G and seen as an intermediate technology between 4G defined by the first releases of LTE and the upcoming new 5G air interface (see Chapters 23 and 24). License-assisted access (LAA) to support unlicensed spectrum as a complement to licensed spectrum, improved support for machine-type communication, and various enhancements in CA, multi-antenna transmission, and device-to-device communication are some of the highlights from release 13.

3.3 SPECTRUM FLEXIBILITY

Already the first release of LTE provides a certain degree of spectrum flexibility in terms of multi-bandwidth support and a joint FDD/TDD design. In later releases this flexibility was considerably enhanced to support higher bandwidths and fragmented spectrum using CA and access to unlicensed spectrum as a complement using license-assisted access (LAA).

3.3.1 CARRIER AGGREGATION

As mentioned earlier, the first release of LTE already provided extensive support for deployment in spectrum allocations of various characteristics, with bandwidths ranging from roughly 1 MHz up to 20 MHz in both paired and unpaired bands. With LTE release 10 the transmission bandwidth can be further extended by means of so-called carrier aggregation, where multiple *component carriers* are aggregated and jointly used for transmission to/from a single device. Up to five component carriers, possibly each of different bandwidth, can be aggregated in release 10, allowing for transmission bandwidths up to 100 MHz. All component carriers need to have the same duplex scheme and, in case of TDD, uplink–downlink configuration. In later releases, this requirement was relaxed, as well as the number of component carriers possible to aggregate was increased to 32, resulting in a total bandwidth of 640 MHz. Backward compatibility was ensured as each component carrier uses the release-8 structure. Hence, to a release-8/9 device each component carrier will appear as an LTE release-8 carrier, while a carrier-aggregation-capable device can exploit the total aggregated bandwidth, enabling higher data rates. In the general case, a different number of component carriers can be aggregated for the downlink and uplink. This is an important property from a device complexity point of view where aggregation can be supported in the downlink where very high data rates are needed without increasing the uplink complexity.

Component carriers do not have to be contiguous in frequency, which enables exploitation of *fragmented spectrum*; operators with a fragmented spectrum can provide high-data-rate services based on the availability of a wide overall bandwidth even though they do not possess a single wideband spectrum allocation.

From a baseband perspective, there is no difference between the cases in Figure 3.6, and they are all supported by LTE release 10. However, the RF-implementation complexity is vastly different with the first case being the least complex. Thus, although carrier aggregation is supported by the basic specifications, not all devices will support it. Furthermore, release 10 has some restrictions on carrier aggregation in the RF specifications, compared to what has been specified for physical layer and signaling, while in later releases there is support for CA within and between a much larger number of bands.

Release 11 provided additional flexibility for aggregation of TDD carriers. Prior to release 11, the same downlink–uplink allocation was required for all the aggregated carriers. This can be unnecessarily restrictive in case of aggregation of different bands as the configuration in each band may be given by coexistence with other radio-access technologies in that particular band. An interesting aspect of aggregating different downlink–uplink allocations is that the device may need to receive and transmit simultaneously in order to fully utilize both

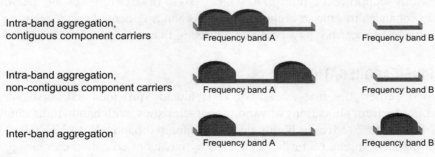

FIGURE 3.6

Carrier aggregation.

carriers. Thus, unlike previous releases, a TDD-capable device may, similarly to a FDD-capable device, need a duplex filter. Release 11 also saw the introduction of RF requirements for inter-band and noncontiguous intra-band aggregation, as well as support for an even larger set of inter-band aggregation scenarios.

Release 12 defined aggregations between FDD and TDD carriers, unlike earlier releases that only supported aggregation within one duplex type. FDD—TDD aggregation allows for efficient utilization of an operator's spectrum assets. It can also be used to improve the uplink coverage of TDD by relying on the possibility for continuous uplink transmission on the FDD carrier.

Release 13 increased the number of carriers possible to aggregate from 5 to 32, resulting in a maximum bandwidth of 640 MHz and a theoretical peak data rates around 25 Gbit/s in the downlink. The main motivation for increasing the number of subcarriers is to allow for very large bandwidths in unlicensed spectrum as is further discussed in conjunction with LAA.

The evolution of carrier aggregation is summarized in Figure 3.7 and further described in Chapter 12.

3.3.2 LICENSE-ASSISTED ACCESS

Originally, LTE was designed for licensed spectrum where an operator has an exclusive license for a certain frequency range. Licensed spectrum offers many benefits since the operator can plan the network and control the interference situation, but there is typically a cost associated with obtaining the spectrum license and the amount of licensed spectrum is limited. Therefore, using unlicensed spectrum as a *complement* to offer higher data rates and higher capacity in local areas is of interest. One possibility is to complement the LTE network with Wi-Fi, but higher performance can be achieved with a tighter coupling between licensed and unlicensed spectrum. LTE release 13 therefore introduced license-assisted access, where the carrier aggregation framework is used to aggregate downlink carriers in unlicensed frequency bands, primarily in the 5 GHz range, with carriers in licensed frequency bands as illustrated in Figure 3.8. Mobility, critical control signaling and services demanding high quality-of-service rely on carriers in the licensed spectrum while (parts of) less demanding traffic can be handled by the carriers using unlicensed spectrum. Operator-controlled

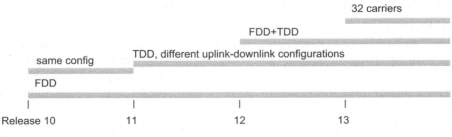

FIGURE 3.7

Evolution of carrier aggregation.

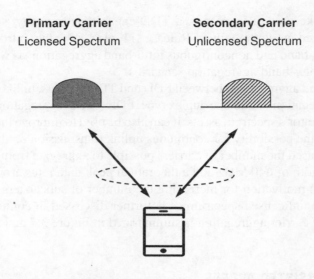

FIGURE 3.8

License-assisted access.

small-cell deployments is the target. Fair sharing of the spectrum resources with other systems, in particular Wi-Fi, is an important characteristic of LAA which therefore incudes a listen-before-talk mechanism.

Chapter 17 provides an in-depth discussion of license-assisted access.

3.4 MULTI-ANTENNA ENHANCEMENTS

Multi-antenna support has been enhanced over the different releases, increasing the number of transmission layers in the downlink to 8 and introducing uplink spatial multiplexing. Full-dimension MIMO (FD-MIMO) and two-dimensional beam-forming are other enhancements, as is the introduction of CoMP transmission.

3.4.1 EXTENDED MULTI-ANTENNA TRANSMISSION

In release 10, downlink spatial multiplexing was expanded to support up to eight transmission layers. This can be seen as an extension of the release-9 dual-layer beam-forming to support up to eight antenna ports and eight corresponding layers. Together with the support for carrier aggregation this enables downlink data rates up to 3 Gbit/s in 100 MHz of spectrum in release 10, increased to 25 Gbit/s in release 13 using 32 carriers, 8 layers spatial multiplexing and 256QAM.

Uplink spatial multiplexing of up to four layers was also introduced as part of LTE release 10. Together with the possibility for uplink carrier aggregation this allows for uplink data rates up to 1.5 Gbit/s in 100 MHz of spectrum. Uplink spatial multiplexing consists of a codebook-based scheme under control of the base station, which means that the structure can also be used for uplink transmitter-side beam-forming.

An important consequence of the multi-antenna extensions in LTE release 10 was the introduction of an enhanced downlink *reference-signal structure* that more extensively separated the function of channel estimation and the function of acquiring CSI. The aim of this was to better enable novel antenna arrangements and new features such as more elaborate multi-point coordination/transmission in a flexible way.

In release-13, improved support for a large number of antennas was introduced, in particular in terms of more extensive feedback of CSI. The larger degrees of freedom can be used for, for example, beam-forming in both elevation and azimuth and massive multiuser MIMO where several spatially separated devices are simultaneously served using the same time—frequency resource. These enhancements are sometimes termed full-dimension MIMO and form a step into massive MIMO with a very large number of steerable antenna elements.

The multi-antenna support is described as part of the general downlink processing in Chapter 6, uplink processing in Chapter 7, and the supporting CSI reporting mechanism in Chapter 10.

3.4.2 MULTI-POINT COORDINATION AND TRANSMISSION

As discussed earlier, the first release of LTE included specific support for coordination between transmission points, referred to as ICIC, as a means to at least partly control the interference between cells. However, the support for such coordination was significantly expanded as part of LTE release 11 including the possibility for much more dynamic coordination between transmission points.

In contrast to release-8 ICIC, which was limited to the definition of certain X2 messages to assist coordination between cells, the release 11 activities focused on radio-interface features and device functionality to assist different coordination means, including the support for channel-state feedback for multiple transmission points. Jointly these features and functionality go under the name CoMP transmission/reception. Refinement to the reference-signal structure was also an important part of the CoMP support, as was the enhanced control-channel structure introduced as part of release 11, see later.

Support for CoMP includes *multi-point coordination*, that is, when transmission to a device is carried out from one specific transmission point but where scheduling and link adaptation is coordinated between the transmission points, as well as *multi-point transmission* in which case transmission to a device can be carried out from multiple transmission points either in such a way that that transmission can switch dynamically between different transmission points (*Dynamic Point Selection*) or be carried out jointly from multiple transmission points (*Joint Transmission*), see Figure 3.9.

A similar distinction can be made for uplink where one can distinguish between (uplink) multi-point coordination and multi-point *reception*. In general, though uplink CoMP is

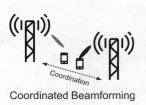
Coordinated Beamforming

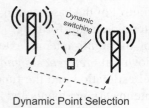
Dynamic Point Selection

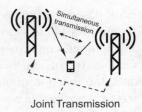
Joint Transmission

FIGURE 3.9

Different types of CoMP.

mainly a network implementation issue and has very little impact on the device and very little visibility in the radio-interface specifications.

The CoMP work in release 11 assumed "ideal" backhaul, in practice implying centralized baseband processing connected to the antenna sites using low-latency fiber connections. Extensions to relaxed backhaul scenarios with noncentralized baseband processing were introduced in release 12. These enhancements mainly consisted of defining new X2 messages for exchanging information about so-called CoMP hypotheses, essentially a potential resource allocation, and the associated gain/cost.

CoMP is described in more detail in Chapter 13.

3.4.3 ENHANCED CONTROL-CHANNEL STRUCTURE

In release 11, a new complementary control-channel structure was introduced in order to support ICIC and to exploit the additional flexibility of the new reference-signal structure not only for data transmission, which was the case in release 10, but also for control signaling. The new control-channel structure can thus be seen as a prerequisite for many CoMP schemes, although it is also beneficial for beam-forming and frequency-domain interference coordination. It is also used to support narrow-band operation for MTC enhancements in releases 12 and 13. A description of the enhanced control-channel structure is found in Chapter 6.

3.5 DENSIFICATION, SMALL CELLS, AND HETEROGENEOUS DEPLOYMENTS

Small cells and dense deployment has been in focus for several releases as means to provide very high capacity and data rates. Relaying, small-cell on/off, dynamic TDD, and heterogeneous deployments are some examples of enhancements over the releases. LAA, discussed in the area of spectrum flexibility, can also be seen as primarily an enhancement for small cells.

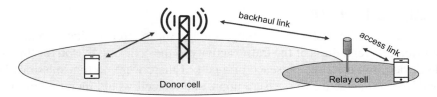

FIGURE 3.10

Example of relaying.

3.5.1 **RELAYING**

In the context of LTE, *relaying* implies that the device communicates with the network via a *relay node* that is *wirelessly connected* to a *donor cell* using the LTE radio-interface technology (see Figure 3.10). From a device point of view, the relay node will appear as an ordinary cell. This has the important advantage of simplifying the device implementation and making the relay node backward compatible—that is, also LTE release 8/9 devices can access the network via the relay node. In essence, the relay is a low-power base station wirelessly connected to the remaining part of the network, see Chapter 18 for more details.

3.5.2 **HETEROGENEOUS DEPLOYMENTS**

Heterogeneous deployments refer to deployments with a mixture of network nodes with different transmit power and overlapping geographic coverage (Figure 3.11). A typical example is a pico-node placed within the coverage area of a macrocell. Although such deployments were already supported in release 8, release 10 introduced new means to handle the interlayer interference that may occur between, for example, a pico-layer and the overlaid macro-layer as described in Chapter 14. The multi-point-coordination techniques introduced in release 11 further extends the set of tools for supporting heterogeneous deployments. Enhancements to improve mobility between the pico-layer and the macro-layer were introduced in release 12.

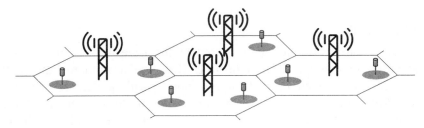

FIGURE 3.11

Example of heterogeneous deployment with low-power nodes inside macrocells.

3.5.3 SMALL-CELL ON-OFF

In LTE, cells are continuously transmitting cell-specific reference signals and broadcasting system information, regardless of the traffic activity in the cell. One reason for this is to enable idle-mode devices to detect the presence of a cell; if there were no transmissions from a cell there is nothing for the device to measure upon, and the cell would therefore not be detected. Furthermore, in a large macrocell deployment there is a relatively high likelihood of at least one device being active in a cell motivating continuous transmission of reference signals.

However, in a dense deployment with a large number of relatively small cells, the likelihood of not all cells serving device at the same time can be relatively high in some scenarios. The downlink interference scenario experienced by a device may also be more severe with devices experiencing very low signal-to-interference ratios due to interference from neighboring, potentially empty, cells, especially if there is a large amount of line-of-sight propagation. To address this, release 12 introduced mechanisms for turning on/off individual cells as a function of the traffic situation to reduce the average inter-cell interference and reduce power consumption. Chapter 15 describes these mechanisms in more detail.

3.5.4 DUAL CONNECTIVITY

Dual connectivity implies a device is simultaneously connected to two eNodeBs as opposed to the baseline case with a single eNodeB connected to the device. User-plane aggregation, where the device is receiving data transmission from multiple sites, separation of control and user planes, and uplink—downlink separation where downlink transmissions originates from a different node than the uplink reception node are some examples of the benefits with dual connectivity, see Chapter 16. To some extent it can be seen as carrier aggregation extended to the case of nonideal backhaul. The dual connectivity framework has also turned out to be very promising for integrating other radio-access schemes such as WLAN (wireless local-area network, for example Wi-Fi) into 3GPP networks. It will also play an important role in the 5G era for providing tight interworking between LTE and the new 5G radio-access technology as discussed in Chapter 23.

3.5.5 DYNAMIC TDD

In TDD, the same carrier frequency is shared in the time domain between uplink and downlink. The fundamental approach to this in LTE, as well as in many other TDD systems, is to statically split the resources in to uplink and downlink. Having a static split is a reasonable assumption in larger macrocells as there are multiple users and the aggregated per-cell load in uplink and downlink is relatively stable. However, with an increased interest in local-area deployments, TDD is expected to become more important compared to the situation for wide-area deployments to date. One reason is unpaired spectrum allocations being more common in higher-frequency bands not suitable for wide-area coverage.

Another reason is that many problematic interference scenarios in wide-area TDD networks are not present with below-rooftop deployments of small nodes. An existing wide-area FDD network could be complemented by a local-area layer using TDD, typically with low-output power per node.

To better handle the high traffic dynamics in a local-area scenario, where the number of devices transmitting to/receiving from a local-area access node can be very small, dynamic TDD is beneficial. In dynamic TDD, the network can dynamically use resources for either uplink or downlink transmissions to match the instantaneous traffic situation, which leads to an improvement of the end-user performance compared to the conventional static split of resources between uplink and downlink. To exploit these benefits, LTE release 12 includes support for dynamic TDD, or eIMTA as it the official name for this feature in 3GPP. More details on dynamic TDD can be found in Chapter 15.

3.5.6 WLAN INTERWORKING

The 3GPP architecture allows for integrating non-3GPP access, for example WLAN and also cdma2000 [74], into 3GPP networks. Essentially, these solutions connect the non-3GPP access to the EPC and are thus not visible in the LTE radio-access network. One drawback of this way of WLAN interworking is the lack of network control; the device may select Wi-Fi even if staying on LTE would provide a better user experience. One example of such a situation is when the Wi-Fi network is heavily loaded while the LTE network enjoys a light load. Release 12 therefore introduced means for the network to assist the device in the selection procedure. Basically, the network configures a signal-strength threshold controlling when the device should select LTE or Wi-Fi.

Release 13 provided further enhancements in the area of WLAN interworking with more explicit control from the LTE RAN on when a device should use Wi-Fi and when to use LTE. Furthermore, release 13 also includes LTE-WLAN aggregation where LTE and WLAN are aggregated at the PDCP level using a framework very similar to dual connectivity. Only downlink aggregation is currently supported.

3.6 DEVICE ENHANCEMENTS

Fundamentally, a device vendor is free to design the device receiver in any way as long as it supports the minimum requirements defined in the specifications. There is an incentive for the vendors to provide significantly better receivers as this could be directly translated into improved end-user data rates. However, the network may not be able to exploit such receiver improvements to their full extent as it might not know which devices have significantly better performance. Network deployments therefore need to be based on the minimum requirements. Defining performance requirements for more advanced receiver types to some extent alleviates this as the minimum performance of a device equipped with an advanced receiver is known. Both releases 11 and 12 saw a lot of focus on receiver improvements with cancellation of some overhead signals in release 11 and more generic schemes in release 12, including network-assisted interference cancellation and suppression (NAICS) where the network can provide the devices with information-assisting inter-cell interference cancellation.

3.7 NEW SCENARIOS

LTE was originally designed as a mobile broadband system, aiming at providing high data rates and high capacity over wide areas. The evolution of LTE has added features improving capacity and data rates, but also enhancements making LTE highly relevant also for new use cases, for example, massive machine-type communication. Operation in areas without network coverage, for example, in a disaster area, is another example, resulting in support for device-to-device communication being included in the LTE specifications in release 12.

3.7.1 DEVICE-TO-DEVICE COMMUNICATION

Cellular systems such as LTE are designed assuming that devices connect to a base station to communicate. In most cases this is an efficient approach as the server with the content of interest typically not being in the vicinity of the device. However, if the device is interested in communicating with a neighboring device, or just detect whether there is a neighboring device that is of interest, the network-centric communication may not be the best approach. Similarly, for public safety such as a first responder officer searching for people in need in a disaster situation, there is typically a requirement that communication should be possible also in absence of network coverage.

To address these situations, release 12 introduced network-assisted device-to-device communication using parts of the uplink spectrum. Two scenarios were considered when developing the device-to-device enhancements, in coverage as well as out-of-coverage communication for public safety, and in coverage discovery of neighboring devices for commercial use cases. More details on device-to-device communication in LTE can be found in Chapter 21.

In release 13, device-to-device communication was further enhanced with relaying solutions for extended coverage.

3.7.2 MACHINE-TYPE COMMUNICATION

Machine-type communication is a very wide term, basically covering all types of communication between machines. Although spanning a wide range of different applications, many of which are yet unknown, MTC applications can be divided into two main categories, massive MTC and critical MTC.

Examples of massive MTC scenarios are different types of sensors, actuators, and similar devices. These devices typically have to be of very low cost and have very low energy consumption enabling very long battery life. At the same time, the amount of data generated by each device is normally very small and very low latency is not a critical requirement. Critical MTC, on the other hand, corresponds to applications such as traffic safety/control or wireless connectivity for industrial processes, and in general where very high reliability and availability is required.

To better support massive MTC, several enhancements have been introduced starting with release 12 and the introduction of a new, low-end device category, category 0, supporting data rates up to 1 Mbit/s. A power-save mode for reduced device power consumption was also defined. Release 13 further improved the MTC support by defining category M1 with further extended coverage and support for 1.4 MHz device bandwidth, irrespective of the system bandwidth, to further reduce device cost. From a network perspective these devices are normal LTE devices, albeit with limited capabilities, and can be freely mixed with more capable LTE devices on a carrier.

Narrow-band Internet-of-Things, NB-IoT, is a parallel to the LTE track, to be completed in release 13. It targets even lower cost and data rates than category-M1, 250 kbit/s or less, in a bandwidth of 180 kHz, and even further enhanced coverage. Thanks to the use of OFDM with 15 kHz subcarrier spacing, it can be deployed inband on top of an LTE carrier, outband in a separate spectrum allocation, or in the guard bands of LTE, providing a high degree of flexibility for an operator. In the uplink, transmission on a single tone is supported to obtain very large coverage for the lowest data rates. NB-IoT uses the same family of higher-layer protocols (MAC, RLC, PDCP) as LTE, with extensions for faster connection setup applicable to both NB-IoT and category-M1, and can therefore easily be integrated into existing deployments.

Chapter 20 contains an in-depth description of massive MTC support in LTE.

3.8 DEVICE CAPABILITIES

To support different scenarios, which may call for different device capabilities in terms of data rates, as well as to allow for market differentiation in terms of low- and high-end devices with a corresponding difference in price, not all devices support all capabilities. Furthermore, devices from an earlier release of the standard will not support features introduced in later versions of LTE. For example, a release-8 device will not support carrier aggregation as this feature was introduced in release 10. Therefore, as part of the connection setup, the device indicates not only which release of LTE it supports, but also its capabilities within the release.

In principle, the different parameters could be specified separately, but to limit the number of combinations and avoid a parameter combination that does not make sense, a set of physical-layer capabilities are lumped together to form a UE category (UE, *User Equipment*, is the term used in 3GPP to denote a device). In total five different UE categories have been specified for LTE release 8/9, ranging from the low-end category 1 not supporting spatial multiplexing to the high-end category 5 supporting the full set of features in the release-8/9 physical-layer specifications. The categories are summarized in Table 3.2 (in simplified form with uplink and downlink categories merged in a single table; for the full set of details, see [30]). Note that, regardless of the category, a device is always capable of receiving single-stream transmissions from up to four antenna ports. This is necessary as the system information can be transmitted on up to four antenna ports.

In later releases, features such as carrier aggregation has been added calling for additional capability signaling compared to release 8/9 either in the form of additional UE categories or as separate capabilities. In order to be able to operate in networks following an earlier release a device has to be able to declare both release-8/9 category and categories for later releases.

Defining new categories for each foreseen combination of the maximum number of component carriers and maximum degree of spatial multiplexing could be done in

Table 3.2 UE Categories (Simplified Description)

		Downlink			Uplink	
Category	Release	Peak Rate (Mbit/s)	Maximun Number of MIMO Layers	Maximum Modulation	Peak Rate (Mbit/s)	Maximum Modulation
M1	13	0.2	1		0.14	
0	12	1	1	64QAM	1	16QAM
1	8	10	1	64QAM	5	16QAM
2	8	50	2	64QAM	25	16QAM
3	8	100	2	64QAM	50	16QAM
4	8	150	2	64QAM	50	16QAM
5	8	300	4	64QAM	75	64QAM
6	10	300	2 or 4	64QAM	50	16QAM
7	10	300	2 or 4	64QAM	100	16QAM
8	10	3000	8	64QAM	1500	64QAM
9	11	450	2 or 4	64QAM	50	16QAM
10	11	450	2 or 4	64QAM	100	16QAM
11	12	600	2 or 4	256QAM optional	50	16QAM
12	12	600	2 or 4	256QAM optional	100	16QAM
13	12	400	2 or 4	256QAM	150	64QAM
14	12	400	2 or 4	256QAM	100	16QAM
15	12	4000	8	256QAM		

principle, although the number of categories might become very large and which categories a device supports may be frequency-band dependent. Therefore, the maximum number of component carriers and degree of spatial multiplexing supported both in uplink and downlink, are signaled separately from the category number. Category-independent signaling is also used for several capabilities, especially for features added to LTE after the basic release 8/9. The duplexing schemes supported is one such example, and the support of UE-specific reference signals for FDD in release 8 is another. Whether the device supports other radio-access technologies, for example, GSM and WCDMA, is also declared separately.

RADIO-INTERFACE ARCHITECTURE

4

This chapter contains a brief overview of the overall architecture of an LTE radio-access network (RAN) and the associated core network (CN), followed by descriptions of the RAN user-plane and control-plane protocols.

4.1 OVERALL SYSTEM ARCHITECTURE

In parallel to the work on the LTE radio-access technology in 3GPP, the overall system architecture of both the radio-access network and the core network was revisited, including the split of functionality between the two networks. This work was known as the *system architecture evolution* (SAE) and resulted in a flat RAN architecture, as well as a new core-network architecture referred to as the *evolved packet core* (EPC). Together, the LTE RAN and the EPC are referred to as the *evolved packet system* (EPS).[1]

The RAN is responsible for all radio-related functionality of the overall network including, for example, scheduling, radio-resource handling, retransmission protocols, coding, and various multi-antenna schemes. These functions are discussed in detail in the subsequent chapters.

The EPC is responsible for functions not related to the radio access but needed for providing a complete mobile-broadband network. This includes, for example, authentication, charging functionality, and setup of end-to-end connections. Handling these functions separately, instead of integrating them into the RAN, is beneficial as it allows for several radio-access technologies to be served by the same CN.

Although this book focuses on the LTE RAN, a brief overview of the EPC, as well as how it connects to the RAN, is useful. For an excellent in-depth discussion of EPC, the reader is referred to [5].

4.1.1 CORE NETWORK

The EPC is a radical evolution from the GSM/GPRS core network used for GSM and WCDMA/HSPA. EPC supports access to the *packet-switched domain* only, with no access to the *circuit-switched domain*. It consists of several different types of nodes, some of which are briefly described in the following and illustrated in Figure 4.1.

---

[1]UTRAN, the WCDMA/HSPA RAN, is also part of the EPS.

4G, LTE-Advanced Pro and The Road to 5G. http://dx.doi.org/10.1016/B978-0-12-804575-6.00004-2

The *mobility management entity* (MME) is the control-plane node of the EPC. Its responsibilities include connection/release of bearers to a device, handling of IDLE to ACTIVE transitions, and handling of security keys. The functionality operating between the EPC and the device is sometimes referred to as the *non-access stratum* (NAS), to separate it from the *access stratum* (AS) which handles functionality operating between the device and the RAN.

The *serving gateway* (S-GW) is the user-plane node connecting the EPC to the LTE RAN. The S-GW acts as a mobility anchor when devices move between eNodeBs (see next section), as well as a mobility anchor for other 3GPP technologies (GSM/GPRS and HSPA). Collection of information and statistics necessary for charging is also handled by the S-GW.

The *packet data network gateway* (PDN gateway, P-GW) connects the EPC to the internet. Allocation of the IP address for a specific device is handled by the P-GW, as well as quality-of-service (QoS) enforcement according to the policy controlled by the PCRF (see later). The P-GW is also the mobility anchor for non-3GPP radio-access technologies, such as CDMA2000, connected to the EPC.

In addition, the EPC also contains other types of nodes such as *policy and charging rules function* (PCRF) responsible for QoS handling and charging, and the *home subscriber service*

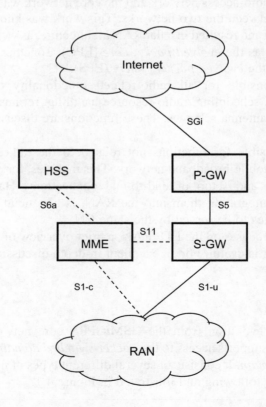

FIGURE 4.1

Core network architecture.

(HSS) node, a database containing subscriber information. There are also some additional nodes present with regard to network support of *multimedia broadcast multicast services* (MBMS) (see Chapter 19 for a more detailed description of MBMS, including the related architecture aspects).

It should be noted that the nodes discussed earlier are *logical* nodes. In an actual physical implementation, several of them may very well be combined. For example, the MME, P-GW, and S-GW could very well be combined into a single physical node.

4.1.2 RADIO-ACCESS NETWORK

The LTE RAN uses a flat architecture with a single type of node[2]—the *eNodeB*. The eNodeB is responsible for all radio-related functions in one or several cells. It is important to note that an eNodeB is a *logical* node and not a physical implementation. One common implementation of an eNodeB is a three-sector site, where a base station is handling transmissions in three cells, although other implementations can be found as well, such as one baseband processing unit to which a number of remote radio heads are connected. One example of the latter is a large number of indoor cells, or several cells along a highway, belonging to the same eNodeB. Thus, a base station is a *possible* implementation of, but not *the same* as, an eNodeB.

As can be seen in Figure 4.2, the eNodeB is connected to the EPC by means of the *S1 interface*, more specifically to the S-GW by means of the *S1 user-plane part*, S1-u, and to the

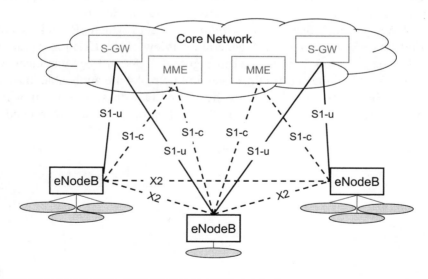

FIGURE 4.2

Radio-access network interfaces.

[2]The introduction of MBMS (see Chapter 19) in release 9 and relaying (see Chapter 18) in release 10 bring additional node types to the RAN.

MME by means of the *S1 control-plane part*, S1-c. One eNodeB can be connected to multiple MMEs/S-GWs for the purpose of load sharing and redundancy.

The *X2 interface*, connecting eNodeBs to each other, is mainly used to support active-mode mobility. This interface may also be used for multi-cell *radio-resource management* (RRM) functions such as *inter-cell interference coordination* (ICIC) discussed in Chapter 13. The X2 interface is also used to support lossless mobility between neighboring cells by means of packet forwarding.

The interface between the eNodeB to the device is known as the *Uu interface*. Unless dual connectivity is used, see Chapter 16, a device is connected to a single eNodeB as a time. There is also a *PC5 interface* defined for direct device-to-device communication, see Chapter 21.

4.2 RADIO PROTOCOL ARCHITECTURE

With the overall network architecture in mind, the RAN protocol architecture for the user and control planes can be discussed. Figure 4.3 illustrates the RAN protocol architecture (the MME is, as discussed in the previous section, not part of the RAN but is included in the figure for completeness). As seen in the figure, many of the protocol entities are common to the user and control planes. Therefore, although this section mainly describes the protocol architecture from a user-plane perspective, the description is in many respects also applicable to the control plane. Control-plane-specific aspects are discussed in Section 4.3.

The LTE RAN provides one or more *radio bearers* to which IP packets are mapped according to their QoS requirements. A general overview of the LTE (user-plane) protocol architecture for the downlink is illustrated in Figure 4.4. As will become clear in the subsequent discussion, not all the entities illustrated in Figure 4.4 are applicable in all situations. For example, neither MAC scheduling nor hybrid ARQ with soft combining is used for broadcast of the basic system information. The LTE protocol structure related to uplink

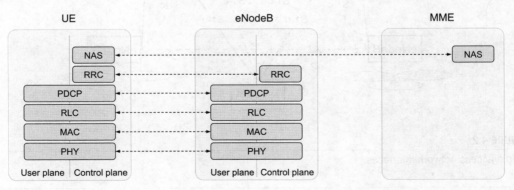

FIGURE 4.3

Overall RAN protocol architecture.

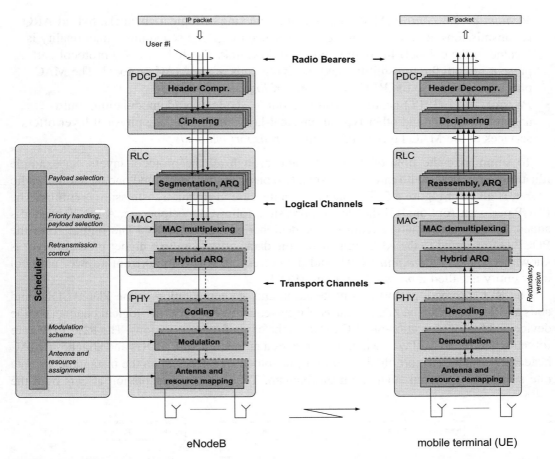

FIGURE 4.4

LTE protocol architecture (downlink).

transmissions is similar to the downlink structure in Figure 4.4, although there are some differences with respect to, for example, transport-format selection.

The different protocol entities of the RAN are summarized and described in more detail in the following sections:

- *Packet data convergence protocol* (PDCP) performs IP header compression, ciphering, and integrity protection. It also handles in-sequence delivery and duplicate removal in case of handover. There is one PDCP entity per radio bearer configured for a device.
- *Radio-link control* (RLC) is responsible for segmentation/concatenation, retransmission handling, duplicate detection, and in-sequence delivery to higher layers. The RLC provides services to the PDCP. There is one RLC entity per radio bearer configured for a device.

- *Medium-access control* (MAC) handles multiplexing of logical channels, hybrid-ARQ retransmissions, and uplink and downlink scheduling. The scheduling functionality is located in the eNodeB for both uplink and downlink. The hybrid-ARQ protocol part is present in both the transmitting and receiving ends of the MAC protocol. The MAC provides services to the RLC in the form of *logical channels*.
- *Physical layer* (PHY) handles coding/decoding, modulation/demodulation, multi-antenna mapping, and other typical physical-layer functions. The physical layer offers services to the MAC layer in the form of *transport channels*.

To summarize the flow of downlink data through all the protocol layers, an example illustration for a case with three IP packets, two on one radio bearer and one on another radio bearer, is given in Figure 4.5. The data flow in the case of uplink transmission is similar. The PDCP performs (optional) IP-header compression, followed by ciphering. A PDCP header is added, carrying information required for deciphering in the device. The output from the PDCP is forwarded to the RLC. In general, the data entity from/to a higher protocol layer is known as a *service data unit* (SDU) and the corresponding entity to/from a lower protocol layer entity is called a *protocol data unit* (PDU).

The RLC protocol performs concatenation and/or segmentation of the PDCP SDUs and adds an RLC header. The header is used for in-sequence delivery (per logical channel) in the device and for identification of RLC PDUs in the case of retransmissions. The RLC PDUs are forwarded to the MAC layer, which multiplexes a number of RLC PDUs and attaches a MAC header to form a transport block. The transport-block size depends on the instantaneous data rate selected by the link-adaptation mechanism. Thus, the link adaptation affects both the

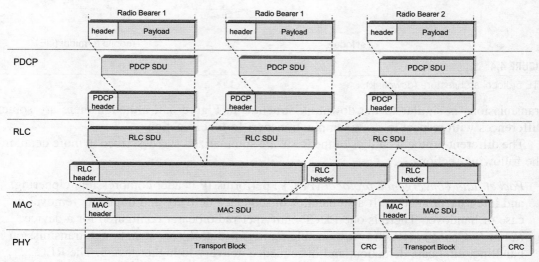

FIGURE 4.5

Example of LTE data flow.

MAC and RLC processing. Finally, the physical layer attaches a CRC to the transport block for error-detection purposes, performs coding and modulation, and transmits the resulting signal, possibly using multiple transmit antennas.

The remainder of the chapter contains an overview of the RLC, MAC, and physical layers. A more detailed description of the LTE physical-layer processing is given in Chapter 6 (downlink) and Chapter 7 (uplink), followed by descriptions of some specific uplink and downlink radio-interface functions and procedures in the subsequent chapters as well as some enhancements introduced after release 8/9.

4.2.1 PACKET-DATA CONVERGENCE PROTOCOL

The PDCP performs IP header compression to reduce the number of bits to transmit over the radio interface. The header-compression mechanism is based on robust header compression (ROHC) [31], a standardized header-compression algorithm also used for several other mobile-communication technologies. PDCP is also responsible for ciphering to protect against eavesdropping and, for the control plane, integrity protection to ensure that control messages originate from the correct source. At the receiver side, the PDCP performs the corresponding deciphering and decompression operations.

In addition, the PDCP also plays an important role for intra-eNodeB handover, handling in-sequence delivery and duplicate removal.[3] Upon handover, undelivered downlink data packets will be forwarded by the PDCP from the old eNodeB to the new eNodeB. The PDCP entity in the device will also handle retransmission of all uplink packets not yet delivered to the eNodeB as the hybrid-ARQ buffers are flushed upon handover.

4.2.2 RADIO-LINK CONTROL

The RLC protocol is responsible for segmentation/concatenation of (header-compressed) IP packets, also known as RLC SDUs, from the PDCP into suitably sized RLC PDUs. It also handles retransmission of erroneously received PDUs, as well as removal of duplicated PDUs. Finally, the RLC ensures in-sequence delivery of SDUs to upper layers. Depending on the type of service, the RLC can be configured in different modes to perform some or all of these functions.

Segmentation and concatenation, one of the main RLC functions, is illustrated in Figure 4.6. Depending on the scheduler decision, a certain amount of data is selected for transmission from the RLC SDU buffer, and the SDUs are segmented/concatenated to create the RLC PDU. Thus, for LTE the RLC PDU size varies *dynamically*. For high data rates, a large PDU size results in a smaller relative overhead, while for low data rates, a small PDU size is required as the payload would otherwise be too large. Hence, as the LTE data rates may

---

[3]Reordering is done in a similar way as the RLC reordering handling out-of-sequence PDUs from the hybrid-ARQ entity, see Chapter 8.

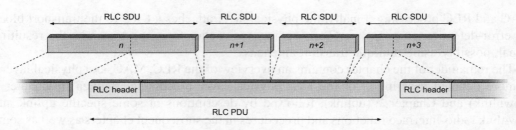

FIGURE 4.6

RLC segmentation and concatenation.

range from a few kbit/s up to several Gbit/s, dynamic PDU sizes are motivated for LTE in contrast to earlier mobile-communication technologies, which typically used a fixed PDU size. Since the RLC, scheduler, and rate-adaptation mechanisms are all located in the eNodeB, dynamic PDU sizes are easily supported for LTE. In each RLC PDU, a header is included, containing, among other things, a sequence number used for in-sequence delivery and retransmission handling.

The RLC retransmission mechanism is also responsible for providing error-free delivery of data to higher layers. To accomplish this, a retransmission protocol operates between the RLC entities in the receiver and transmitter. By monitoring the sequence numbers of the incoming PDUs, the receiving RLC can identify missing PDUs. Status reports are then fed back to the transmitting RLC entity, requesting retransmission of missing PDUs. Based on the received status report, the RLC entity at the transmitter can take appropriate action and retransmit the missing PDUs if needed.

Although the RLC is capable of handling transmission errors due to noise, unpredictable channel variations, and so on, error-free delivery is in most cases handled by the MAC-based hybrid-ARQ protocol. The use of a retransmission mechanism in the RLC may therefore seem superfluous at first. However, as is discussed in Section 4.2.3.3, this is not the case, and the use of both RLC- and MAC-based retransmission mechanisms is in fact well motivated by the differences in the feedback signaling.

The details of RLC are further described in Chapter 8.

4.2.3 MEDIUM-ACCESS CONTROL

The MAC layer handles logical-channel multiplexing, hybrid-ARQ retransmissions, and uplink and downlink scheduling. It is also responsible for multiplexing/demultiplexing data across multiple component carriers when carrier aggregation is used, see Chapter 12, and clear-channel assessment for license-assisted access, see Chapter 17.

4.2.3.1 Logical Channels and Transport Channels

The MAC provides services to the RLC in the form of *logical channels*. A logical channel is defined by the *type* of information it carries and is generally classified as a *control channel*, used for transmission of control and configuration information necessary for operating an LTE system, or as a *traffic channel*, used for the user data. The set of logical-channel types specified for LTE includes:

- The *broadcast control channel* (BCCH), used for transmission of *system information* from the network to all devices in a cell. Prior to accessing the system, a device needs to acquire the system information to find out how the system is configured and, in general, how to behave properly within a cell.
- The *paging control channel* (PCCH), used for paging of devices whose location on a cell level is not known to the network. The paging message therefore needs to be transmitted in multiple cells.
- The *common control channel* (CCCH), used for transmission of control information in conjunction with random access.
- The *dedicated control channel* (DCCH), used for transmission of control information to/ from a device. This channel is used for individual configuration of devices such as different handover messages.
- The *dedicated traffic channel* (DTCH), used for transmission of user data to/from a device. This is the logical channel type used for transmission of all uplink and non-MBSFN downlink user data.
- The *multicast control channel* (MCCH), used for transmission of control information required for reception of the MTCH (see later).
- The *single-cell multicast control channel* (SC-MCCH), used for transmission of control information for single-cell MTCH reception (see later).
- The *multicast traffic channel* (MTCH), used for downlink transmission of MBMS across multiple cells.
- The *single-cell multicast traffic channel* (SC-MTCH), used for downlink transmission of MBMS in a single cell.
- The *sidelink broadcast control channel* (SBCCH), used for sidelink synchronization.
- The *sidelink traffic channel* (STCH), used for sidelink communication (sidelink is the link for direct device-to-device communication, compare uplink and downlink for communication between a device and an eNodeB), see Chapter 21 for a description of device-to-device communication.

From the physical layer, the MAC layer uses services in the form of *transport channels*. A transport channel is defined by *how* and *with what characteristics* the information is transmitted over the radio interface. Data on a transport channel is organized into *transport blocks*. In each *transmission time interval* (TTI), at most one transport block of dynamic size is transmitted over the radio interface to/from a device in the absence of spatial multiplexing. In the case of spatial multiplexing (MIMO), there can be up to two transport blocks per TTI.

Associated with each transport block is a *transport format* (TF), specifying *how* the transport block is to be transmitted over the radio interface. The transport format includes information about the transport-block size, the modulation-and-coding scheme, and the antenna mapping. By varying the transport format, the MAC layer can thus realize different data rates. Rate control is also known as *transport-format selection*.

The following transport-channel types are defined for LTE:

- The *broadcast channel* (BCH) has a fixed transport format, provided by the specifications. It is used for transmission of parts of the BCCH system information, more specifically the so-called *master information block* (MIB), as described in Chapter 11.
- The *paging channel* (PCH) is used for transmission of paging information from the PCCH logical channel. The PCH supports *discontinuous reception* (DRX) to allow the device to save battery power by waking up to receive the PCH only at predefined time instants. The LTE paging mechanism is also described in Chapter 11.
- The *downlink shared channel* (DL-SCH) is the main transport channel used for transmission of downlink data in LTE. It supports key LTE features such as dynamic rate adaptation and channel-dependent scheduling in the time and frequency domains, hybrid ARQ with soft combining, and spatial multiplexing. It also supports DRX to reduce device power consumption while still providing an always-on experience. The DL-SCH is also used for transmission of the parts of the BCCH system information not mapped to the BCH. There can be multiple DL-SCHs in a cell, one per device[4] scheduled in this TTI, and, in some subframes, one DL-SCH carrying system information.
- The *multicast channel* (MCH) is used to support MBMS. It is characterized by a semi-static transport format and semi-static scheduling. In the case of multi-cell transmission using MBSFN, the scheduling and transport format configuration are coordinated among the transmission points involved in the MBSFN transmission. MBSFN transmission is described in Chapter 19.
- The *uplink shared channel* (UL-SCH) is the uplink counterpart to the DL-SCH—that is, the uplink transport channel used for transmission of uplink data.
- The *sidelink shared channel* (SL-SCH) is the transport channel used for sidelink communication as described in Chapter 21.
- The *sidelink broadcast channel* (SL-BCH) is used for sidelink synchronization, see Chapter 21.
- The *sidelink discovery channel* (SL-DCH) is used in the sidelink discovery process as described in Chapter 21.

In addition, the *random-access channel* (RACH) is also defined as a transport channel, although it does not carry transport blocks. Furthermore, the introduction of NB-IoT in release 13, see Chapter 20, resulted in a set of channels optimized for narrowband operation.

Part of the MAC functionality is multiplexing of different logical channels and mapping of the logical channels to the appropriate transport channels. The mapping between logical-

---

[4]For carrier aggregation, a device may receive multiple DL-SCHs, one per component carrier.

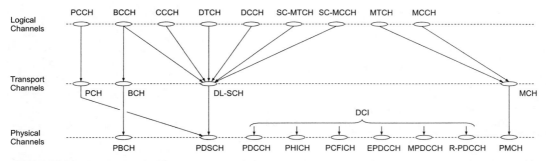

FIGURE 4.7

Downlink channel mapping.

channel types and transport channel types is given in Figure 4.7 for the downlink and Figure 4.8 for the uplink. The figures clearly indicate how DL-SCH and UL-SCH are the main downlink and uplink transport channels, respectively. In the figures, the corresponding physical channels, described later, are also included and the mapping between transport channels and physical channels is illustrated. For details on the mapping of sidelink channels see Chapter 21.

To support priority handling, multiple logical channels, where each logical channel has its own RLC entity, can be multiplexed into one transport channel by the MAC layer. At the receiver, the MAC layer handles the corresponding demultiplexing and forwards the RLC PDUs to their respective RLC entity for in-sequence delivery and the other functions handled by the RLC. To support the demultiplexing at the receiver, a MAC header, as shown in

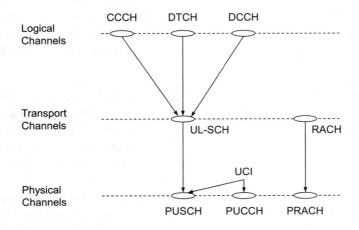

FIGURE 4.8

Uplink channel mapping.

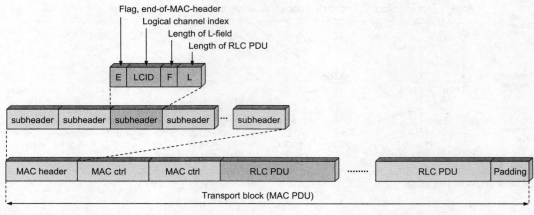

FIGURE 4.9

MAC header and SDU multiplexing.

Figure 4.9, is used. To each RLC PDU, there is an associated subheader in the MAC header. The subheader contains the identity of the logical channel (LCID) from which the RLC PDU originated and the length of the PDU in bytes. There is also a flag indicating whether this is the last subheader or not. One or several RLC PDUs, together with the MAC header and, if necessary, padding to meet the scheduled transport-block size, form one transport block which is forwarded to the physical layer.

In addition to multiplexing of different logical channels, the MAC layer can also insert the so-called *MAC control elements* into the transport blocks to be transmitted over the transport channels. A MAC control element is used for inband control signaling—for example, timing-advance commands and random-access response, as described in Sections 7.6 and 11.3, respectively. Control elements are identified with reserved values in the LCID field, where the LCID value indicates the type of control information. Furthermore, the length field in the subheader is removed for control elements with a fixed length.

4.2.3.2 Scheduling

One of the basic principles of LTE radio access is shared-channel transmission—that is, time–frequency resources are dynamically shared between users. The *scheduler* is part of the MAC layer (although often better viewed as a separate entity as illustrated in Figure 4.4) and controls the assignment of uplink and downlink resources in terms of so-called *resource-block pairs*. Resource-block pairs correspond to a time–frequency unit of 1 ms times 180 kHz, as described in more detail in Chapter 9.

The basic operation of the scheduler is so-called *dynamic* scheduling, where the eNodeB in each 1 ms interval takes a scheduling decision and sends scheduling information to the selected set of devices. However, there is also a possibility for semi-persistent scheduling

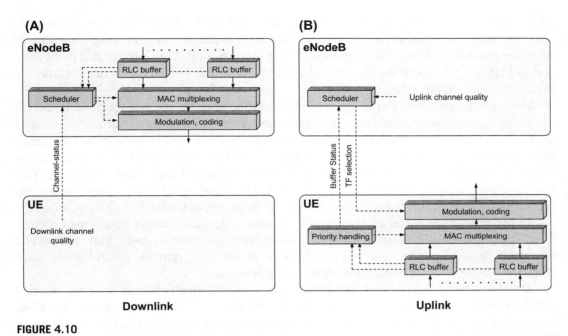

FIGURE 4.10

TF selection in (A) downlink and (B) uplink.

where a semi-static scheduling pattern is signaled in advance to reduce the control-signaling overhead.

Uplink and downlink scheduling are separated in LTE, and uplink and downlink scheduling decisions can be taken independently of each other (within the limits set by the uplink/downlink split in the case of half-duplex FDD operation).

The downlink scheduler is responsible for (dynamically) controlling which device(s) to transmit to and, for each of these devices, the set of resource blocks upon which the device's DL-SCH should be transmitted. Transport-format selection (selection of transport-block size, modulation scheme, and antenna mapping) and logical-channel multiplexing for downlink transmissions are controlled by the eNodeB, as illustrated in the left part of Figure 4.10. As a consequence of the scheduler controlling the data rate, the RLC segmentation and MAC multiplexing will also be affected by the scheduling decision. The outputs from the downlink scheduler can be seen in Figure 4.4.

The uplink scheduler serves a similar purpose, namely to (dynamically) control which devices are to transmit on their respective UL-SCH and on which uplink time—frequency resources (including component carrier). Despite the fact that the eNodeB scheduler determines the TF for the device, it is important to point out that the uplink scheduling decision is taken *per device* and not per radio bearer. Thus, although the eNodeB scheduler controls the payload of a scheduled device, the device is still responsible for selecting *from which radio bearer(s)* the data is taken. The device handles logical-channel multiplexing according to rules, the parameters of which can be configured by the eNodeB. This is illustrated in the right part of Figure 4.10, where the eNodeB scheduler controls the TF and the device controls the logical-channel multiplexing.

Although the scheduling strategy is implementation specific and not specified by 3GPP, the overall goal of most schedulers is to take advantage of the channel variations between devices and preferably schedule transmissions to a device on resources with advantageous channel conditions. A benefit of the use of OFDM in LTE is the possibility to exploit channel variations in both time and frequency domains through channel-dependent scheduling. This was mentioned earlier, in Chapter 9, and illustrated in Figure 3.2. For the larger bandwidths supported by LTE, where a significant amount of frequency-selective fading may occur, the possibility for the scheduler to also exploit frequency-domain channel variations becomes increasingly important compared to exploiting time-domain variations only. This is beneficial especially at low speeds, where the variations in the time domain are relatively slow compared to the delay requirements set by many services.

Downlink channel-dependent scheduling is supported through *channel-state information* (CSI), reported by the device to the eNodeB and reflecting the instantaneous downlink channel quality in the time and frequency domains, as well as information necessary to determine the appropriate antenna processing in the case of spatial multiplexing. In the up-link, the channel-state information necessary for uplink channel-dependent scheduling can be based on a *sounding reference signal* transmitted from each device for which the eNodeB wants to estimate the uplink channel quality. To aid the uplink scheduler in its decisions, the device can transmit buffer-status information to the eNodeB using a MAC message. This information can only be transmitted if the device has been given a valid scheduling grant. For situations when this is not the case, an indicator that the device needs uplink resources is provided as part of the uplink L1/L2 control-signaling structure (see Chapter 7).

4.2.3.3 Hybrid ARQ with Soft Combining

Hybrid ARQ with soft combining provides robustness against transmission errors. As hybrid-ARQ retransmissions are fast, many services allow for one or multiple retransmissions, thereby forming an implicit (closed-loop) rate-control mechanism. The hybrid-ARQ protocol is part of the MAC layer, while the actual soft combining is handled by the physical layer.[5]

Hybrid ARQ is not applicable for all types of traffic. For example, broadcast trans-missions, where the same information is intended for multiple devices, typically do not rely on hybrid ARQ.[6] Hence, hybrid ARQ is only supported for the DL-SCH and the UL-SCH, although its usage is optional.

---

[5]The soft combining is done before or as part of the channel decoding, which clearly is a physical-layer functionality.
[6]Autonomous retransmissions, where the transmitter repeats the same information several times without the receivers trans-mitting an acknowledgment, is possible and sometimes used to obtain a soft combining gain.

The LTE hybrid-ARQ protocol uses multiple parallel stop-and-wait processes. Upon reception of a transport block, the receiver makes an attempt to decode the transport block and informs the transmitter about the outcome of the decoding operation through a single acknowledgment bit indicating whether the decoding was successful or if a retransmission of the transport block is required. Clearly, the receiver must know to which hybrid-ARQ process a received acknowledgment is associated. This is solved by using the timing of the acknowledgment for association with a certain hybrid-ARQ process. Note that, in the case of TDD operation, the time relation between the reception of data in a certain hybrid-ARQ process and the transmission of the acknowledgment is also affected by the uplink/down-link allocation.

The use of multiple parallel hybrid-ARQ processes, illustrated in Figure 4.11, for each user can result in data being delivered from the hybrid-ARQ mechanism out of sequence. For example, transport block 5 in the figure was successfully decoded before transport block 1 which required retransmissions. In-sequence delivery of data is therefore ensured by the RLC layer.

Downlink retransmissions may occur at any time after the initial transmission—that is, the protocol is asynchronous—and an explicit hybrid-ARQ process number is used to indicate which process is being addressed. In an asynchronous hybrid-ARQ protocol, the retransmissions are in principle scheduled similarly to the initial transmissions. Uplink retransmissions, on the other hand, are based on a synchronous protocol, the retransmission occurs at a predefined time after the initial transmission and the process number can be implicitly derived. In a synchronous protocol the time instant for the retransmissions is fixed once the initial transmission has been scheduled, which must be accounted for in the scheduling operation. However, note that the scheduler knows from the hybrid-ARQ entity in the eNodeB whether a retransmission needs to be scheduled or not.

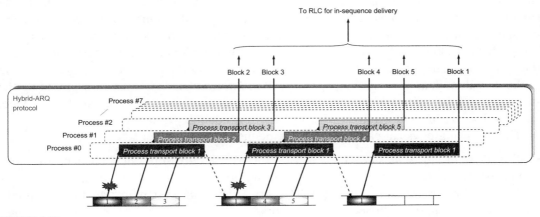

FIGURE 4.11

Multiple parallel hybrid-ARQ processes.

The hybrid-ARQ mechanism will rapidly correct transmission errors due to noise or unpredictable channel variations. As discussed earlier in this chapter, the RLC is also capable of requesting retransmissions, which at first sight may seem unnecessary. However, the reason for having two retransmission mechanisms on top of each other can be seen in the feedback signaling—hybrid ARQ provides fast retransmissions but due to errors in the feedback the residual error rate is typically too high, for example, for good TCP performance, while RLC ensures (almost) error-free data delivery but slower retransmissions than the hybrid-ARQ protocol. Hence, the combination of hybrid ARQ and RLC provides an attractive combination of small round-trip time and reliable data delivery. Furthermore, as the RLC and hybrid ARQ are located in the same node, tight interaction between the two is possible, as discussed in Chapter 8.

4.2.4 PHYSICAL LAYER

The physical layer is responsible for coding, physical-layer hybrid-ARQ processing, modulation, multi-antenna processing, and mapping of the signal to the appropriate physical time—frequency resources. It also handles mapping of transport channels to physical channels, as shown in Figures 4.7 and 4.8.

As mentioned in the introduction, the physical layer provides services to the MAC layer in the form of transport channels. Data transmission in downlink, uplink, and sidelink use the DL-SCH, UL-SCH, and SL-SCH transport-channel types, respectively. There is at most one or, in the case of spatial multiplexing,[7] two transport blocks per TTI on a DL-SCH, UL-SCH, or SL-SCH. In the case of carrier aggregation, there is one DL-SCH (or UL-SCH) per component carrier seen by the device.

A *physical channel* corresponds to the set of time—frequency resources used for transmission of a particular transport channel and each transport channel is mapped to a corresponding physical channel, as shown in Figures 4.7 and 4.8 for the downlink and uplink (the sidelink is covered in Chapter 21). In addition to the physical channels with a corresponding transport channel, there are also physical channels without a corresponding transport channel. These channels, known as L1/L2 control channels, are used for *downlink control information* (DCI), providing the device with the necessary information for proper reception and decoding of the downlink data transmission, *uplink control information* (UCI) used for providing the scheduler and the hybrid-ARQ protocol with information about the situation at the device, and *sidelink control information* (SCI) for handling sidelink transmissions.

The following physical-channel types are defined for LTE:

- The *physical downlink shared channel* (PDSCH) is the main physical channel used for unicast data transmission, but also for transmission of paging information.
- The *physical broadcast channel* (PBCH) carries part of the system information, required by the device in order to access the network.

---

[7]There is no spatial multiplexing defined for sidelink communication.

- The *physical multicast channel* (PMCH) is used for MBSFN transmission.
- The *physical downlink control channel* (PDCCH) is used for downlink control information, mainly scheduling decisions, required for reception of PDSCH, and for scheduling grants enabling transmission on the PUSCH.
- The *enhanced physical downlink control channel* (EPDCCH) was introduced in release 11. It essentially serves the same purpose as the PDCCH but allows for transmission of the control information in a more flexible way.
- The *MTC physical downlink control channel* (MPDCCH) was introduced in release 13 as part of the improved support for massive machine-type communication, see Chapter 20. In essence it is a variant of the EPDCCH.
- The *relay physical downlink control channel* (R-PDCCH) was introduced in release 10 and is used to carry L1/L2 control signaling on the donor-eNodeB-to-relay link.
- The *physical hybrid-ARQ indicator channel* (PHICH) carries the hybrid-ARQ acknowledgment to indicate to the device whether a transport block should be retransmitted or not.
- The *physical control format indicator channel* (PCFICH) is a channel providing the devices with information necessary to decode the set of PDCCHs. There is only one PCFICH per component carrier.
- The *physical uplink shared channel* (PUSCH) is the uplink counterpart to the PDSCH. There is at most one PUSCH per uplink component carrier per device.
- The *physical uplink control channel* (PUCCH) is used by the device to send hybrid-ARQ acknowledgments, indicating to the eNodeB whether the downlink transport block(s) was successfully received or not, to send channel-state reports aiding downlink channel-dependent scheduling, and for requesting resources to transmit uplink data upon. There is at most one PUCCH per device.
- The *physical random-access channel* (PRACH) is used for random access, as described in Chapter 11.
- The *physical sidelink shared channel* (PSSCH) is used for sidelink data transfer, see Chapter 21.
- The *physical sidelink control channel* (PSCCH), used for sidelink-related control information.
- The *physical sidelink discovery channel* (PSDCH), used for sidelink discovery.
- The *physical sidelink broadcast channel* (PSBCH), used to convey basic sidelink-related information between devices.

Note that some of the physical channels, more specifically the channels used for downlink control information (namely, PCFICH, PDCCH, PHICH, EPDCCH, and R-PDCCH), uplink control information (namely, PUCCH), and sidelink control information (namely, PSCCH) do not have a corresponding transport channel mapped to them.

The remaining downlink transport channels are based on the same general physical-layer processing as the DL-SCH, although with some restrictions in the set of features used. This is especially true for PCH and MCH transport channels. For the broadcast of system information on the BCH, a device must be able to receive this information channel as one of the first steps prior to accessing the system. Consequently, the transmission format must be known to the devices a priori, and there is no dynamic control of any of the transmission parameters from the MAC layer in this case. The BCH is also mapped to the physical resource (the OFDM time–frequency grid) in a different way, as described in more detail in Chapter 11.

For transmission of paging messages on the PCH, dynamic adaptation of the transmission parameters can, to some extent, be used. In general, the processing in this case is similar to the generic DL-SCH processing. The MAC can control modulation, the amount of resources, and the antenna mapping. However, as an uplink has not yet been established when a device is paged, hybrid ARQ cannot be used as there is no possibility for the device to transmit a hybrid-ARQ acknowledgment.

The MCH is used for MBMS transmissions, typically with single-frequency network operation by transmitting from multiple cells on the same resources with the same format at the same time. Hence, the scheduling of MCH transmissions must be coordinated between the cells involved and dynamic selection of transmission parameters by the MAC is not possible.

4.3 CONTROL-PLANE PROTOCOLS

The control-plane protocols are, among other things, responsible for connection setup, mobility, and security. Control messages transmitted from the network to the devices can originate either from the MME, located in the core network, or from the *radio resource control* (RRC), located in the eNodeB.

NAS control-plane functionality, handled by the MME, includes EPS bearer management, authentication, security, and different idle-mode procedures such as paging. It is also responsible for assigning an IP address to a device. For a detailed discussion about the NAS control-plane functionality, see [5].

The RRC is located in the eNodeB and is responsible for handling the RAN-related procedures, including:

- Broadcast of system information necessary for the device to be able to communicate with a cell. Acquisition of system information is described in Chapter 11.
- Transmission of paging messages originating from the MME to notify the device about incoming connection requests. Paging, discussed further in Chapter 11, is used in the RRC_IDLE state (described later) when the device is not connected to a particular cell. Indication of system-information updates is another use of the paging mechanism, as is public warning systems.
- Connection management, including setting up bearers and mobility within LTE. This includes establishing an RRC context—that is, configuring the parameters necessary for communication between the device and the RAN.
- Mobility functions such as cell (re)selection.

- Measurement configuration and reporting.
- Handling of device capabilities; when connection is established the device (UE) will announce its capabilities as all devices are not capable of supporting all the functionality described in the LTE specifications, as briefly discussed in Chapter 3.

RRC messages are transmitted to the device using *signaling radio bearers* (SRBs), using the same set of protocol layers (PDCP, RLC, MAC, and PHY) as described in Section 4.2. The SRB is mapped to the CCCH during establishment of connection and, once a connection is established, to the DCCH. Control-plane and user-plane data can be multiplexed in the MAC layer and transmitted to the device in the same TTI. The aforementioned MAC control elements can also be used for control of radio resources in some specific cases where low latency is more important than ciphering, integrity protection, and reliable transfer.

4.3.1 STATE MACHINE

In LTE, a device can be in two different states from an RRC perspective,[8] RRC_CON-NECTED and RRC_IDLE, as illustrated in Figure 4.12.

In RRC_CONNECTED, there is an RRC context established—that is, the parameters necessary for communication between the device and the RAN are known to both entities. The cell to which the device belongs is known, and an identity of the device, the *cell radio-network temporary identifier* (C-RNTI), used for signaling purposes between the device and the network, has been configured. RRC_CONNECTED is intended for data transfer to/from the device, but DRX can be configured in order to reduce device power consumption (DRX is described in further detail in Chapter 9). Since there is an RRC context established in the eNodeB in RRC_CONNECTED, leaving DRX and starting to receive/transmit data is relatively fast as no connection setup with its associated signaling is needed.

Although expressed differently in the specifications, RRC_CONNECTED can be thought of as having two substates, IN_SYNC and OUT_OF_SYNC, depending on whether the

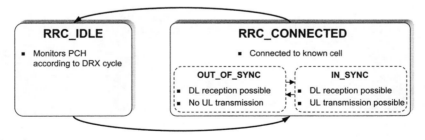

FIGURE 4.12

LTE states.

---

[8]There are also different core-network states for a device, but these are not described here.

uplink is synchronized to the network or not. Since LTE uses an orthogonal FDMA/TDMA-based uplink, it is necessary to synchronize the uplink transmission from different devices such that they arrive at the receiver at (about) the same time. The procedure for obtaining and maintaining uplink synchronization is described in Chapter 7, but in short the receiver measures the arrival time of the transmissions from each actively transmitting device and sends timing-correction commands in the downlink. As long as the uplink is synchronized, uplink transmission of user data and L1/L2 control signaling is possible. If no uplink transmission has taken place within a configurable time window and therefore timing alignment has not been possible, uplink synchronization cannot be guaranteed and the uplink is declared to be nonsynchronized. In this case, the device needs to perform a random-access procedure to restore uplink synchronization prior to transmission of uplink data or control information.

In RRC_IDLE, there is no RRC context in the RAN, and the device does not belong to a specific cell. No data transfer may take place as the device sleeps most of the time in order to reduce battery consumption. Uplink synchronization is not maintained and hence the only uplink transmission activity that may take place is random access, discussed in Chapter 11, to move to RRC_CONNECTED. When moving to RRC_CONNECTED the RRC context needs to be established in both the RAN and the device. Compared to leaving DRX this takes a somewhat longer time. In the downlink, devices in RRC_IDLE periodically wake up in order to receive paging messages, if any, from the network, as described in Chapter 11.

PHYSICAL TRANSMISSION RESOURCES

In Chapter 4, the overall LTE architecture was discussed, including an overview of the different protocol layers. Prior to discussing the detailed LTE downlink and uplink transmission schemes, a description of the basic time—frequency transmission resource of LTE is provided in this chapter. An overview of the concept of *antenna ports* is also provided.

5.1 OVERALL TIME—FREQUENCY STRUCTURE

OFDM is the basic transmission scheme for both the downlink and uplink transmission directions in LTE although, for the uplink, specific means are employed to reduce the cubic metric[1] of the transmitted signal, thereby allowing for improved efficiency for the device transmitter power amplifier. Thus, for uplink user data and higher-layer control signaling corresponding to transmission of the PUSCH physical channel, DFT precoding is applied before OFDM modulation, leading to *DFT-spread OFDM* or *DFTS-OFDM*, see further Chapter 7. As will be described in Chapter 7, for other uplink transmissions, such as the transmission of *L1/L2 control signaling* and different types of *reference-signal transmissions*, other measures are taken to limit the cubic metric of the transmitted signal.

The LTE OFDM subcarrier spacing equals 15 kHz for both downlink and uplink. The selection of the subcarrier spacing in an OFDM-based system needs to carefully balance overhead from the cyclic prefix against sensitivity to Doppler spread/shift and other types of frequency errors and inaccuracies. The choice of 15 kHz for the LTE subcarrier spacing was found to offer a good balance between these different constraints.

Assuming an FFT-based transmitter/receiver implementation, 15 kHz subcarrier spacing corresponds to a sampling rate $f_s = 15000 \, N_{FFT}$, where N_{FFT} is the FFT size. It is important to understand though that the LTE specifications do not in any way mandate the use of FFT-based transmitter/receiver implementations and even less so a particular FFT size or sampling rate. Nevertheless, FFT-based implementations of OFDM are common practice and an

---

[1]The cubic metric is a better measure than the peak-to-average ratio of the amount of additional back-off needed for a certain signal waveform, relative to the back-off needed for some reference waveform [10].

4G, LTE-Advanced Pro and The Road to 5G. http://dx.doi.org/10.1016/B978-0-12-804575-6.00005-4

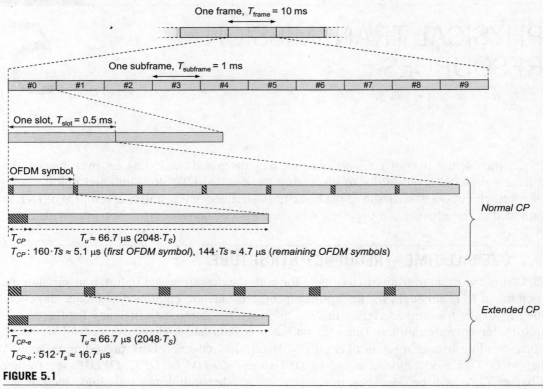

FIGURE 5.1

LTE time-domain structure.

FFT size of 2048, with a corresponding sampling rate of 30.72 MHz, is suitable for the wider LTE carrier bandwidths, such as bandwidths of the order of 15 MHz and above. However, for smaller carrier bandwidths, a smaller FFT size and a correspondingly lower sampling rate can very well be used.

In addition to the 15 kHz subcarrier spacing, a *reduced subcarrier spacing* of 7.5 kHz, with a corresponding OFDM symbol time that is twice as long, is also defined for LTE. The introduction of the reduced subcarrier spacing specifically targeted MBSFN-based multicast/broadcast transmissions (see Chapter 19). However, currently the 7.5 kHz subcarrier numerology is only partly implemented in the LTE specifications. The remaining discussions within this and the following chapters will assume the 15 kHz subcarrier spacing unless explicitly stated otherwise.

In the time domain, LTE transmissions are organized into *frames* of length 10 ms, each of which is divided into ten equally sized *subframes* of length 1 ms, as illustrated in Figure 5.1.

Each subframe consists of two equally sized *slots* of length $T_{slot} = 0.5$ ms, with each slot consisting of a number of OFDM symbols including cyclic prefix.[2] To provide consistent and exact timing definitions, different time intervals within the LTE specifications are defined as multiples of a basic time unit $T_s = 1/(15000 \cdot 2048)$ s. The basic time unit T_s can thus be seen as the sampling time of an FFT-based transmitter/receiver implementation with an FFT size equal to 2048. The time intervals outlined in Figure 5.1 can thus also be expressed as $T_{frame} = 307200\ T_s$, $T_{subframe} = 30720\ T_s$, and $T_{slot} = 15360\ T_s$ for the frame, subframe, and slot durations, respectively.

On a higher level, each frame is identified by a *System Frame Number* (SFN). The SFN is used to define different transmission cycles that have a period longer than one frame, for example, paging cycles, see Chapter 11. The SFN period equals 1024, thus the SFN repeats itself after 1024 frames or 10.24 s.

The 15 kHz LTE subcarrier spacing corresponds to a useful symbol time $T_u = 2048\ T_s$ or approximately 66.7 μs. The overall OFDM symbol time is then the sum of the useful symbol time and the cyclic prefix length T_{CP}. As illustrated in Figure 5.1, LTE defines two cyclic prefix lengths, the *normal* cyclic prefix and an *extended* cyclic prefix, corresponding to seven and six OFDM symbols per slot, respectively. The exact cyclic prefix lengths, expressed in the basic time unit T_s, are given in Figure 5.1. It can be noted that, in the case of the normal cyclic prefix, the cyclic prefix length for the first OFDM symbol of a slot is somewhat larger compared to the remaining OFDM symbols. The reason for this is simply to fill the entire 0.5 ms slot, as the number of basic time units T_s per slot (15360) is not divisible by seven.

The reasons for defining two cyclic prefix lengths for LTE are twofold:

- A longer cyclic prefix, although less efficient from a cyclic prefix-overhead point of view, may be beneficial in specific environments with extensive delay spread, for example, in very large cells. It is important to have in mind, though, that a longer cyclic prefix is not necessarily beneficial in the case of large cells, even if the delay spread is very extensive in such cases. If, in large cells, link performance is limited by noise rather than by signal corruption due to residual time dispersion not covered by the cyclic prefix, the additional robustness to radio-channel time dispersion, due to the use of a longer cyclic prefix, may not justify the corresponding additional energy overhead of a longer cyclic prefix.
- As is discussed in Chapter 19, the cyclic prefix in the case of MBSFN-based multicast/broadcast transmission should not only cover the main part of the actual channel time dispersion, but also the timing difference between the transmissions received from the cells involved in the MBSFN transmission. In the case of MBSFN operation, the extended cyclic prefix is therefore used.

It should be noted that different cyclic prefix lengths may be used for different subframes within a frame in case of MBSFN subframes. As discussed further in Chapter 19, MBSFN-

---

[2]This is valid for the "normal" downlink subframes. As described in Section 5.4, the *special subframe* present in case of TDD operation is not divided into two slots but rather into three fields.

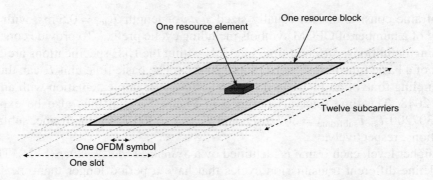

One resource element

One resource block

Twelve subcarriers

One OFDM symbol

One slot

FIGURE 5.2

The LTE physical time—frequency resource.

based multicast/broadcast transmission is always confined to a limited set of subframes, in which case the use of the extended cyclic prefix, with its associated additional cyclic prefix overhead, may only be applied to these subframes.[3]

A *resource element*, consisting of one subcarrier during one OFDM symbol, is the smallest physical resource in LTE. Furthermore, as illustrated in Figure 5.2, resource elements are grouped into *resource blocks*, where each resource block consists of 12 consecutive subcarriers in the frequency domain and one 0.5 ms slot in the time domain. Each resource block thus consists of $7 \cdot 12 = 84$ resource elements in the case of a normal cyclic prefix and $6 \cdot 12 = 72$ resource elements in the case of an extended cyclic prefix.

Although resource blocks are defined over one slot, the basic time-domain unit for dynamic scheduling in LTE is one subframe, consisting of two consecutive slots. The reason for defining the resource blocks over one slot is that *distributed downlink transmission* (described in Chapter 10) and *uplink frequency hopping* (described in Chapter 7) are defined on a slot basis. The minimum scheduling unit, consisting of two time-consecutive resource blocks within one subframe (one resource block per slot), can be referred to as a *resource-block pair*.

The LTE physical-layer specifications allow for a carrier to consist of any number of resource blocks in the frequency domain, ranging from a minimum of six resource blocks up to a maximum of 110 resource blocks. This corresponds to an overall transmission bandwidth ranging from roughly 1 MHz up to in the order of 20 MHz with very fine granularity and thus allows for a very high degree of LTE bandwidth flexibility, at least from a physical-layer-specification point of view. However, as mentioned in Chapter 3, LTE radio-frequency requirements are, at least initially, only specified for a limited set of transmission bandwidths, corresponding to a limited set of possible values for the number

---

[3]The extended cyclic prefix is then actually applied only to the so-called *MBSFN part* of the MBSFN subframes; see Section 5.2.

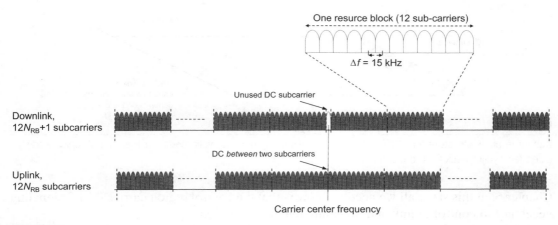

FIGURE 5.3

Frequency-domain structure for LTE.

of resource blocks within a carrier. Also note that, in LTE release 10 and later, the total bandwidth of the transmitted signal can be significantly larger than 20 MHz, up to 100 MHz in release 10 and 640 MHz in release 13, by aggregating multiple carriers, see Section 5.5.

The resource-block definition above applies to both the downlink and uplink transmission directions. However, there is a minor difference between the downlink and uplink in terms of where the carrier center frequency is located in relation to the subcarriers.

In the downlink (upper part of Figure 5.3), there is an unused *DC subcarrier* that coincides with the carrier center frequency. The reason why the DC subcarrier is not used for downlink transmission is that it may be subject to disproportionately high interference due to, for example, local-oscillator leakage.

On the other hand, in the uplink (lower part of Figure 5.3), no unused DC subcarrier is defined and the center frequency of an uplink carrier is located *between* two uplink sub-carriers. The presence of an unused DC-carrier in the center of the spectrum would have prevented the assignment of the entire cell bandwidth to a single device and while still retaining the assumption of mapping to consecutive inputs of the OFDM modulator, something that is needed to retain the low-cubic-metric property of the DFTS-OFDM modulation used for uplink data transmission.

5.2 NORMAL SUBFRAMES AND MBSFN SUBFRAMES

In LTE, each downlink subframe (and the DwPTS in the case of TDD; see Section 5.4.2 for a discussion of the TDD frame structure) is normally divided into a *control region*, consisting of the first few OFDM symbols, and a *data region*, consisting of the remaining part of the subframe. The usage of the resource elements in the two regions is discussed in detail in

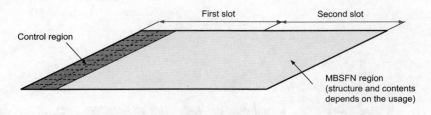

FIGURE 5.4

Resource-block structure for MBSFN subframes, assuming normal cyclic prefix for the control region and extended cyclic prefix for the MBSFN region.

Chapter 6; at this stage all we need to know is that the control region carries L1/L2 signaling necessary to control uplink and downlink data transmissions.

Additionally, already from the first release of LTE, so-called *MBSFN subframes* have been defined. The original intention with MBSFN subframes was, as indicated by the name, to support MBSFN transmission, as described in Chapter 19. However, MBSFN subframes have also been found to be useful in other contexts, for example, as part of relaying functionality as discussed in Chapter 18. Hence, MBSFN subframes are therefore better seen as a generic tool and not related to MBSFN transmission only.

An MBSFN subframe, illustrated in Figure 5.4, consists of a control region of length one or two OFDM symbols, which is in essence identical to its counterpart in a normal subframe, followed by an *MBSFN region* whose contents depend on the usage of the MBSFN subframe. The reason for keeping the control region also in MBSFN subframes is, for example, to be able to transmit control signaling necessary for uplink transmissions. All devices, from LTE release 8 and onward, are capable of receiving the control region of an MBSFN subframe. This is the reason why MBSFN subframes have been found useful as a generic tool to introduce, in a backward-compatible way, new types of signaling and transmission not part of an earlier release of the LTE radio-access specification. Such transmissions can be carried within the MBSFN region of the subframe and earlier-release devices, not recognizing these transmissions, will simply ignore them.

Information about the set of subframes that are configured as MBSFN subframes in a cell is provided as part of the system information (Chapter 11). In principle, an arbitrary pattern of MBSFN subframes can be configured with the pattern repeating after 40 ms.[4] However, as information necessary to operate the system (to be more specific, synchronization signals, system information, and paging, all of which are discussed in detail in later chapters) needs to be transmitted in order for devices to find and connect to a cell, subframes where such information is provided cannot be configured as MBSFN subframes. Therefore, subframes 0, 4, 5, and 9 for FDD and subframes 0, 1, 5, and 6 for TDD cannot be configured as MBSFN

---

[4]The reason for repetition time of 40 ms is that both the 10 ms frame length and the 8 ms hybrid-ARQ roundtrip time are factors in 40 ms which is important for many applications, for example, relaying; see further Chapter 18.

subframes, leaving the remaining six subframes of a frame as candidates for MBSFN subframes.

5.3 ANTENNA PORTS

Downlink multi-antenna transmission is a key technology of LTE. Signals transmitted from different antennas or signals subject to different, and for the receiver, unknown *multi-antenna precoders*, see Section 6.3, will experience different "radio channels" even if the set of antennas are located at the same site.[5]

In general, it is important for a device to understand what it can assume in terms of the relation between the radio channels experienced by different downlink transmissions. This is, for example, important in order for the device to be able to understand what reference signal(s) should be used for channel estimation for a certain downlink transmission. It is also important in order for the device to be able to determine relevant channel-state information, for example, for scheduling and link-adaptation purposes.

For this reason, the concept of *antenna port* has been introduced in the LTE specifications, defined such that *the channel over which a symbol on the antenna port is conveyed can be inferred from the channel over which another symbol on the same antenna port is conveyed.* Expressed differently, each individual downlink transmission is carried out from a specific antenna port, the identity of which is known to the device. Furthermore, the device can assume that two transmitted signals have experienced the same radio channel *if and only if* they are transmitted from the same antenna port.[6]

In practice, each antenna port can, at least for the downlink, be seen as corresponding to a specific reference signal. A device receiver can then assume that this reference signal can be used to estimate the channel corresponding to the specific antenna port. The reference signal can also be used by the device to derive detailed channel-state information related to the antenna port.

It should be understood that an antenna port is an abstract concept that does not necessarily correspond to a specific physical antenna:

- Two different signals may be transmitted in the same way from multiple physical antennas. A device receiver will then see the two signals as propagating over a single channel corresponding to the "sum" of the channels of the different antennas and the overall transmission would be seen as a transmission from a single antenna port being the same for the two signals.
- Two signals may be transmitted from the same set of antennas but with different, for the receiver unknown, antenna transmitter-side precoders. A receiver will have to see the

---

[5]An unknown transmitter-side precoder needs to be seen as part of the overall radio channel.
[6]For certain antenna ports, more specifically those that correspond to so-called demodulation reference signals, the assumption of same radio channel is valid only within a given subframe.

unknown antenna precoders as part of the overall channel implying that the two signals will appear as having been transmitted from two different antenna ports. It should be noted that if the antenna precoders of the two transmissions were known to be the same, the transmissions could be seen as originating from the same antenna port. The same would be true if the precoders were known to the receiver as, in that case, the precoders would not need to be seen as part of the radio channel.

5.3.1 QUASI-CO-LOCATED ANTENNA PORTS

Even if two signals have been transmitted from two different antennas, the channels experienced by the two signals may still have many *large-scale* properties in common. As an example, the channels experienced by two signals transmitted from two different antenna ports corresponding to different physical antennas at the same site will, even if being different in the details, typically have the same or at least similar large-scale properties, for example, in terms of Doppler spread/shift, average delay spread, and average gain. It can also be expected that the channels will introduce similar average delay. Knowing that the radio channels corresponding to two different antenna ports have similar large-scale properties can be used by the device receiver, for example, in the setting of parameters for channel estimation.

However, with the introduction of different types of multi-point transmission in LTE release 11, different downlink transmit antennas serving the same device may be much more geographically separated. In that case the channels of different antenna ports relevant for a device may differ even in terms of large-scale properties.

For this reason, the concept of *quasi-co-location* with respect to antenna ports was introduced as part of LTE release 11. A device receiver can assume that the radio channels corresponding to two different antenna ports have the same large-scale properties in terms of specific parameters such as average delay spread, Doppler spread/shift, average delay, and average gain *if and only if* the antenna ports are specified as being quasi-co-located. Whether or not two specific antenna ports can be assumed to be quasi-co-located with respect to a certain channel property is in some cases given by the LTE specification. In other cases, the device may be explicitly informed by the network by means of signaling if two specific antenna ports can be assumed to be quasi-co-located or not.

As the name suggests, in practice, two antenna ports will typically be "quasi-co-located" if they correspond to physical antennas at the same location while two antenna ports corresponding to antennas at different locations would typically not be "quasi-co-located." However, there is nothing explicitly stated in the specification about this and quasi-co-location is simply defined with regard to what can be assumed regarding the relation between the long-term channel properties of different antenna ports.

5.4 DUPLEX SCHEMES

Spectrum flexibility is one of the key features of LTE. In addition to the flexibility in transmission bandwidth, LTE also supports operation in both paired and unpaired spectrums

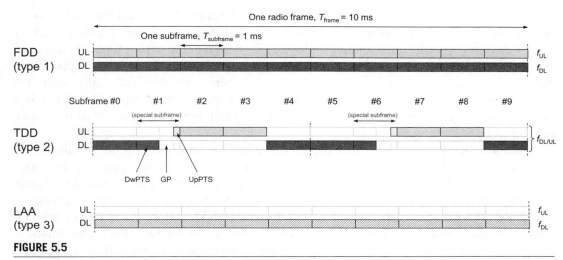

FIGURE 5.5

Uplink/downlink time—frequency structure in case of FDD and TDD.

by supporting both FDD- and TDD-based duplex operation with the time—frequency structures illustrated in Figure 5.5. This is supported through two slightly different frame structures, type 1 for FDD and type 2 for TDD. Operation in unlicensed spectrum was added in release 13, known as *license-assisted access*, and for this reason frame structure type 3 was introduced. Although the time-domain structure is, in most respects, the same for all three frame structures, there are some differences, most notably the presence of a *special subframe* in the case of frame structure type 2. The special subframe is used to provide the necessary guard time for downlink—uplink switching, as discussed in the following.

5.4.1 **FREQUENCY-DIVISION DUPLEX**

In the case of FDD operation (frame structure type 1) uplink and downlink are carried on different carrier frequencies, denoted f_{UL} and f_{DL} in the upper part of Figure 5.5. During each frame, there are thus ten uplink subframes and ten downlink subframes, and uplink and downlink transmission can occur simultaneously within a cell. Isolation between downlink and uplink transmissions is achieved by transmission/reception filters, known as duplex filters, and a sufficiently large *duplex separation* in the frequency domain.

Even if uplink and downlink transmission can occur simultaneously within a cell in the case of FDD operation, a device may be capable of *full-duplex* operation or only *half-duplex* operation for a certain frequency band, depending on whether or not it is capable of simultaneous transmission/reception. In the case of full-duplex capability, transmission and reception may also occur simultaneously at a device, whereas a device capable of only half-duplex operation cannot transmit and receive simultaneously. As mentioned in Chapter 3, half-duplex operation allows for simplified device implementation due to relaxed or no duplex filters. Simplified device implementation can be relevant in several cases, for example, devices for massive machine-type communication (see Chapter 20) where a low device cost is of uttermost importance. Another example is operation in certain frequency bands with a very narrow duplex gap with correspondingly challenging design of the duplex filters. In this case, full-duplex support can be *frequency-band dependent* such that a device may support only half-duplex operation in certain frequency bands while being capable of full-duplex operation in the remaining supported bands. It should be noted that full/half-duplex capability is a property of the *device*; the base station is operating in full duplex irrespective of the device capabilities.

From a network perspective, half-duplex operation has an impact on the sustained data rates that can be provided to/from a single mobile device as it cannot transmit in all uplink subframes. The cell capacity is hardly affected as typically it is possible to schedule different devices in uplink and downlink in a given subframe. No provisioning for guard periods is required as the network is still operating in full duplex and therefore is capable of simultaneous transmission and reception. The relevant transmission structures and timing relations are identical between full-duplex and half-duplex FDD and a single cell may therefore simultaneously support a mixture of full-duplex and half-duplex FDD devices. Since a half-duplex device is not capable of simultaneous transmission and reception, the scheduling decisions must take this into account and half-duplex operation can be seen as a scheduling restriction, is discussed in more detail in Chapter 9.

From a device perspective, half-duplex operation requires the provisioning of a guard period where the device can switch between transmission and reception, the length of which depends on implementation. LTE therefore supports two ways of providing the necessary guard period:

- *Half-duplex type A*, where a guard time is created by allowing the device to skip receiving the last OFDM symbol(s) in a downlink subframe immediately preceding an uplink subframe, as illustrated in Figure 5.6. Guard time for the uplink-to-downlink switch is handled by setting the appropriate amount of timing advance in the devices, implying that a base station supporting half-duplex devices need to apply a larger timing advance value for all devices compared to a full-duplex-only scenario. The type A mechanism is part of LTE since its creation.
- *Half-duplex type B*, where a whole subframe is used as guard between reception and transmission to allow for low-cost implementations with only a single oscillator that is retuned between uplink and downlink frequencies. Half-duplex type B was introduced in LTE release 12 to enable even lower cost devices for MTC applications; see Chapter 20 for more details.

5.4.2 TIME-DIVISION DUPLEX

In the case of TDD operation (frame structure type 2, middle part of Figure 5.5), there is a single carrier frequency and uplink and downlink transmissions are separated in the time domain on a cell basis. As seen in the figure, in each frame some subframes are allocated for

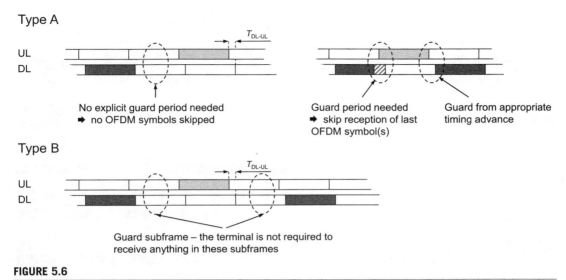

Type A

No explicit guard period needed
➡ no OFDM symbols skipped

Guard period needed
➡ skip reception of last OFDM symbol(s)

Guard from appropriate timing advance

Type B

Guard subframe – the terminal is not required to receive anything in these subframes

FIGURE 5.6

Guard time at the device for half-duplex FDD for type A (top) and type B (bottom).

uplink transmissions and some subframes are allocated for downlink transmission, with the switch between downlink and uplink occurring in a *special subframe* (subframe 1 and, for some uplink–downlink configurations, also subframe 6). Different asymmetries in terms of the amount of resources—that is, subframes—allocated for uplink and downlink transmission, respectively, are provided through the seven different uplink–downlink configurations illustrated in Figure 5.7. As seen in the figure, subframes 0 and 5 are always allocated for downlink transmission while subframe 2 is always allocated for uplink transmissions. The remaining subframes (except the special subframe; see the following paragraphs) can then be flexibly allocated for downlink or uplink transmission depending on the uplink–downlink configuration.

As a baseline, the same uplink–downlink configuration, provided as part of the system information and hence seldom changed, is used in each frame. Furthermore, to avoid severe interference between downlink and uplink transmissions in different cells, neighboring cells typically have the same uplink–downlink configuration. However, release 12 introduced the possibility to dynamically change the uplink–downlink configurations per frame, a feature that is further described in Chapter 15. This feature is primarily useful in small and relatively isolated cells where the traffic variations can be large and inter-cell interference is less of an issue.

As the same carrier frequency is used for uplink and downlink transmission, both the base station and the device need to switch from transmission to reception and vice versa. The switch between downlink and uplink occurs in the special subframe, which is split into three parts: a downlink part (DwPTS), a guard period (GP), and an uplink part (UpPTS).

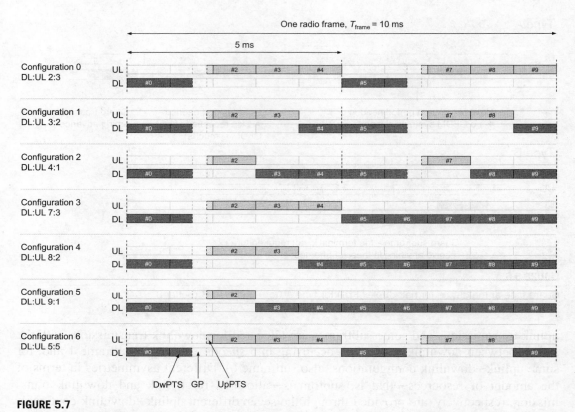

FIGURE 5.7

Different uplink—downlink configurations in case of TDD.

The DwPTS is in essence treated as a normal downlink subframe,[7] although the amount of data that can be transmitted is smaller due to the reduced length of the DwPTS compared to a normal subframe. The UpPTS, however, is not used for data transmission due to the very short duration. Instead, it can be used for channel sounding or random access. It can also be left empty, in which case it serves as an extra guard period.

An essential aspect of any TDD system is the possibility to provide a sufficiently large *guard period* (or guard time), where neither downlink nor uplink transmissions occur. This guard period is necessary for switching from downlink to uplink transmission and vice versa and, as already mentioned, is obtained from the special subframe. The required length of the guard period depends on several factors. First, it should be sufficiently large to provide the necessary time for the circuitry in base stations and the devices to switch from downlink to

---

[7]For the shortest DwPTS duration of three OFDM symbols, DwPTS cannot be used for PDSCH transmission.

uplink. Switching is typically relatively fast, of the order of 20 ms, and in most deployments does not significantly contribute to the required guard time.

Secondly, the guard time should also ensure that uplink and downlink transmissions do not interfere at the base station. This is handled by advancing the uplink timing at the devices such that, at the base station, the last uplink subframe before the uplink-to-downlink switch ends before the start of the first downlink subframe. The uplink timing of each device can be controlled by the base station by using the timing advance mechanism, as elaborated in Chapter 7. Obviously, the guard period must be large enough to allow the device to receive the downlink transmission and switch from reception to transmission before it starts the (timing-advanced) uplink transmission. In essence, some of the guard period of the special subframe is "moved" from the downlink-to-uplink switch to the uplink-to-downlink switch by the timing-advance mechanism. This is illustrated in Figure 5.8. As the timing advance is proportional to the distance to the base station, a larger guard period is required when operating in large cells compared to small cells.

Finally, the selection of the guard period also needs to take interference between base stations into account. In a multi-cell network, inter-cell interference from downlink transmissions in neighboring cells must decay to a sufficiently low level before the base station can start to receive uplink transmissions. Hence, a larger guard period than that motivated by the cell size itself may be required as the last part of the downlink transmissions from distant base stations may otherwise interfere with uplink reception. The amount of guard period depends on the propagation environments, but in some cases the inter-base-station interference is a non-negligible factor when determining the guard period.

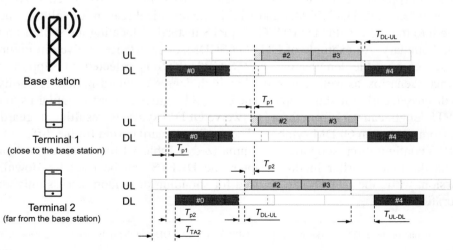

FIGURE 5.8

Timing relation for TDD operation.

Table 5.1 Resulting Guard Period in OFDM Symbols for Different DwPTS and UpPTS Lengths (Normal Cyclic Prefix)

DwPTS	12	11		10		9		6[a]	3	
GP	1	1	2	2	3	3	4	6	9	10
UpPTS	1	2	1	2	1	2	1	2	2	1

[a]*The 6:6:2 configuration was added in release 11 to improve efficiency when coexisting with some common TD-SCDMA configurations (devices prior to release 11 use 3:9:2).*

From the discussion in the preceding paragraphs, it is clear that a sufficient amount of configurability of the guard period is needed to meet different deployment scenarios. Therefore, a set of DwPTS/GP/UpPTS configurations is supported as shown in Table 5.1, where each configuration corresponds to a given length of the three fields in the special subframes. The DwPTS/GP/UpPTS configuration used in the cell is signaled as part of the system information.

5.4.3 LTE AND TD-SCDMA COEXISTENCE

In addition to supporting a wide range of different guard periods, an important aspect in the design of TDD in LTE was to simplify coexistence with, and migration from, systems based on the 3GPP TD-SCDMA standard.[8] Basically, to handle inter-system interference from two different but co-sited TDD systems operating close in frequency, it is necessary to align the switch-points between the two systems. Since LTE supports configurable lengths of the DwPTS field, the switch-points of LTE and TD-SCDMA (or any other TDD system) can be aligned, despite the different subframe lengths used in the two systems. Aligning the switch-points between TD-SCDMA and LTE is the technical reason for splitting the special subframe into the three fields DwPTS/GP/UpPTS instead of locating the switch-point at the subframe boundary. An example of LTE/TD-SCDMA coexistence is given in Figure 5.9.

The set of possible lengths of DwPTS/GP/UpPTS is selected to support common coexistence scenarios, as well as to provide a high degree of guard-period flexibility for the reasons discussed earlier in this chapter. The UpPTS length is one or two OFDM symbols and the DwPTS length can vary from three[9] to twelve OFDM symbols, resulting in guard periods ranging from one to ten OFDM symbols. The resulting guard period for the different DwPTS and UpPTS configurations supported is summarized in Table 5.1 for the case of normal cyclic prefix. As discussed earlier in this chapter, the DwPTS can be used for downlink data transmission, while the UpPTS can be used for sounding or random access only, due to its short duration.

---

[8]TD-SCDMA is one of three TDD modes defined by 3GPP for UTRA TDD and the only one having been deployed on a larger scale

[9]The smallest DwPTS length is motivated by the location of the primary synchronization signal in the DwPTS (see Chapter 11).

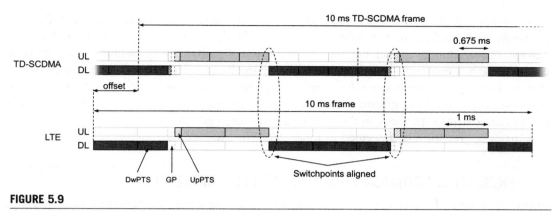

FIGURE 5.9

Coexistence between TD-SCDMA and LTE.

5.4.4 LICENSE-ASSISTED ACCESS

License-assisted access, that is exploiting unlicensed spectrum as a complement to, and assisted by, licensed spectrum, was introduced in release 13 targeting the 5 GHz band. This band is an unpaired band and hence TDD is the relevant duplex scheme. However, since listen before talk, that is, checking whether the spectrum resource is available prior to transmission, is required in some regions and highly beneficial from a Wi-Fi coexistence point of view, frame structure type 2 with its fixed split into uplink and downlink cannot be used. Furthermore, as unlicensed spectrum can be used for downlink only and not uplink in release 13, there is a need for a third frame structure suitable for starting downlink transmissions in any subframe subject to listen before talk. From most perspectives, frame structure type 3 has the same mapping of signals and channels as frame structure type 1.

License-assisted access is described in detail in Chapter 17.

5.5 CARRIER AGGREGATION

The possibility for *carrier aggregation* was introduced in LTE release 10 with enhancements in the following releases. In the case of carrier aggregation, multiple LTE carriers, each with a bandwidth up to 20 MHz, can be transmitted in parallel to/from the same device, thereby allowing for an overall wider bandwidth and correspondingly higher per-link data rates. In the context of carrier aggregation, each carrier is referred to as a *component carrier*[10] as, from an RF point of view, the entire set of aggregated carriers can be seen as a single (RF) carrier.

Up to five component carriers, possibly of different bandwidths up to 20 MHz, can be aggregated allowing for overall transmission bandwidths up to 100 MHz. In release 13 this

---

[10]In the specifications, the term "cell" is used instead of "component carrier," but as the term "cell" is something of a misnomer in the uplink case, the term "component carrier" is used.

was extended to 32 carriers allowing for an overall transmission bandwidth of 640 MHz, primarily motivated by the possibility for large bandwidths in unlicensed spectrum. A device capable of carrier aggregation may receive or transmit simultaneously on multiple component carriers. Each component carrier can also be accessed by an LTE device from earlier releases, that is, component carriers are *backward compatible*. Thus, in most respects and unless otherwise mentioned, the physical-layer description in the following chapters applies to each component carrier separately in the case of carrier aggregation.

Carrier aggregation is described in more detail in Chapter 12.

5.6 FREQUENCY-DOMAIN LOCATION OF LTE CARRIERS

In principle, an LTE carrier could be positioned anywhere within the spectrum and, actually, the basic LTE physical-layer specification does not say anything about the exact frequency location of an LTE carrier, including the frequency band. However, in practice, there is a need for restrictions on where an LTE carrier can be positioned in the frequency domain:

- In the end, an LTE device must be implemented and RF-wise such a device can only support certain frequency bands. The frequency bands for which LTE is specified to operate are discussed in Chapter 2.

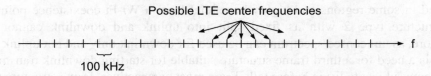

FIGURE 5.10

LTE carrier raster.

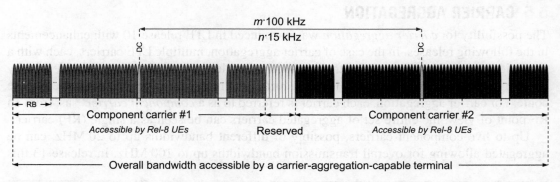

FIGURE 5.11

LTE carrier raster and carrier aggregation.

- After being activated, an LTE device has to search for a network-transmitted carrier within the frequency bands supported by the device. In order for that carrier search to not take an unreasonably long time, there is a need to limit the set of frequencies to be searched.

For this reason, it is assumed that, within each supported frequency band, LTE carriers may exist on a 100 kHz carrier raster or carrier grid, that is, the carrier center frequency can be expressed as $m \cdot 100$ kHz, where m is an integer (see in Figure 5.10).

In the case of carrier aggregation, multiple carriers can be transmitted to/from the same device. In order for the different component carriers to be accessible by earlier-release devices, each component carrier should fall on the 100 kHz carrier grid. However, in the case of carrier aggregation, there is an additional constraint that the carrier spacing between adjacent component carriers should be a multiple of the 15 kHz subcarrier spacing to allow transmission/reception with a single FFT.[11] Thus, in the case of carrier aggregation, the carrier spacing between the different component carriers should be a multiple of 300 kHz, the smallest carrier spacing being a multiple of both 100 kHz (the raster grid) and 15 kHz (the subcarrier spacing). A consequence of this is that there will always be a small gap between two component carriers, even when they are locateds as close as possible to each other, as illustrated in Figure 5.11.

---

[11]This is obviously only relevant for component carriers that are contiguous in the frequency domain. Furthermore, in case of independent frequency errors between component carriers, separate FFTs may be needed at the receiver.

DOWNLINK PHYSICAL-LAYER PROCESSING

In Chapter 4, the LTE radio-interface architecture was discussed with an overview of the functions and characteristics of the different protocol layers. Chapter 5 then gave an overview of the time-frequency structure of LTE transmissions including the structure of the basic OFDM time-frequency grid being the fundamental physical resource on both uplink and downlink. It also discussed the concept of antenna ports, especially relevant for the under-standing of multi-antenna and multi-point transmissions within LTE.

This chapter provides a more detailed description of the downlink physical-layer func-tionality including the transport-channel processing (Section 6.1), reference signals (Section 6.2), multi-antenna transmission (Section 6.3), and L1/L2 control signaling (Section 6.4). Chapter 7 provides a corresponding description for the *uplink* transmission direction. The later chapters go further into the details of some specific uplink and downlink functions and procedures.

6.1 TRANSPORT-CHANNEL PROCESSING

As described in Chapter 4, the physical layer provides services to the MAC layer in the form of transport channels. As also described, for the LTE downlink there are four different types of transport channels defined: the downlink shared channel (DL-SCH), the multicast channel (MCH), the paging channel (PCH), and the broadcast channel (BCH). This section provides a detailed description of the physical-layer processing applied to the DL-SCH, including the mapping to the physical resource—that is, to the resource elements of the OFDM time—frequency grid of the set of antenna ports to be used for the transmission. DL-SCH is the transport-channel type in LTE used for transmission of downlink user-specific higher-layer information, both user data and dedicated control information, as well as the main part of the downlink system information (see Chapter 11). The DL-SCH physical-layer processing is to a large extent applicable also to MCH and PCH transport channels, although with some additional constraints. On the other hand, as mentioned in Chapter 4, the physical-layer processing, and the structure in general, for BCH transmission is quite

4G, LTE-Advanced Pro and The Road to 5G. http://dx.doi.org/10.1016/B978-0-12-804575-6.00006-6

different. BCH transmission is described in Chapter 11 as part of the discussion on LTE system information.

6.1.1 PROCESSING STEPS

The different steps of the DL-SCH physical-layer processing are outlined in Figure 6.1. In the case of carrier aggregation—that is transmission on multiple component carriers in parallel to the same device—the transmissions on the different carriers correspond to separate transport channels with separate and essentially independent physical-layer processing. The transport-channel processing outlined in Figure 6.1 and the discussion as follows is thus valid also in the case of carrier aggregation.

Within each *transmission time interval* (TTI), corresponding to one subframe of length 1 ms, up to two transport blocks of dynamic size are delivered to the physical layer and transmitted over the radio interface for each component carrier. The number of transport

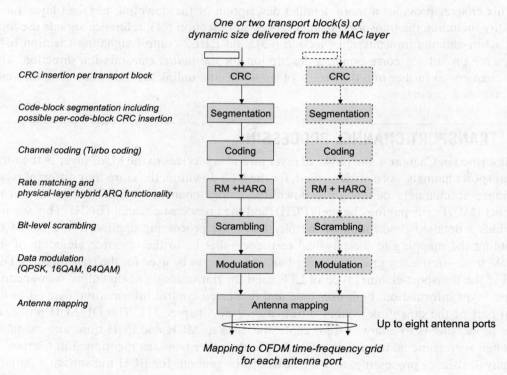

One or two transport block(s) of dynamic size delivered from the MAC layer

CRC insertion per transport block	CRC / CRC
Code-block segmentation including possible per-code-block CRC insertion	Segmentation / Segmentation
Channel coding (Turbo coding)	Coding / Coding
Rate matching and physical-layer hybrid ARQ functionality	RM +HARQ / RM + HARQ
Bit-level scrambling	Scrambling / Scrambling
Data modulation (QPSK, 16QAM, 64QAM)	Modulation / Modulation
Antenna mapping	Antenna mapping

Up to eight antenna ports

Mapping to OFDM time-frequency grid for each antenna port

FIGURE 6.1

Physical-layer processing for the downlink shared channel (DL-SCH).

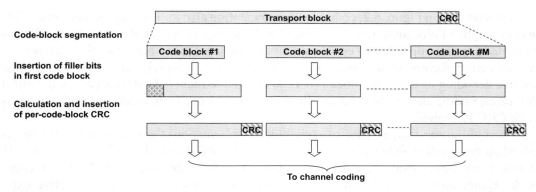

FIGURE 6.2

Code-block segmentation and per-code-block CRC insertion.

blocks transmitted within a TTI depends on the configuration of the multi-antenna transmission scheme (see Section 6.3):

- In the case of no *spatial multiplexing* there is at most a single transport block in a TTI.
- In the case of spatial multiplexing, with transmission on multiple *layers* in parallel to the same device, there are two transport blocks within a TTI.[1]

6.1.1.1 CRC Insertion per Transport Block

In the first step of the physical-layer processing, a 24-bit CRC is calculated for and appended to each transport block. The CRC allows for receiver-side detection of errors in the decoded transport block. The corresponding error indication can, for example, be used by the downlink hybrid-ARQ protocol as a trigger for requesting retransmissions.

6.1.1.2 Code-Block Segmentation and per-Code-Block CRC Insertion

The LTE Turbo-coder internal interleaver is only defined for a limited number of code-block sizes, with a maximum block size of 6144 bits. If the transport block, including the transport-block CRC, exceeds this maximum code-block size, *code-block segmentation* is applied before the Turbo coding as illustrated in Figure 6.2. Code-block segmentation implies that the transport block is segmented into smaller *code blocks*, the sizes of which should match the set of code-block sizes supported by the Turbo coder.

In order to ensure that a transport block of arbitrary size can be segmented into code blocks that match the set of available code-block sizes, the LTE specification includes the possibility to insert "dummy" *filler bits* at the head of the first code block. However, the set of transport-block sizes currently defined for LTE has been selected so that filler bits are not needed.

---

[1]This is true for initial transmissions. In the case of hybrid-ARQ retransmissions there may also be cases when a single transport block is transmitted over multiple layers as discussed, for example, in Section 6.3.

As can be seen in Figure 6.2, code-block segmentation also implies that an additional CRC (also of length 24 bits but different compared to the transport-block CRC described earlier) is calculated for and appended to each code block. Having a CRC per code block allows for early detection of correctly decoded code blocks and correspondingly early termination of the iterative decoding of that code block. This can be used to reduce the device processing effort and corresponding energy consumption. In the case of a single code block no additional code-block CRC is applied.

One could argue that, in case of code-block segmentation, the transport-block CRC is redundant and implies unnecessary overhead as the set of code-block CRCs should indirectly provide information about the correctness of the complete transport block. However, code-block segmentation is only applied to large transport blocks for which the relative extra overhead due to the additional transport-block CRC is small. The transport-block CRC also adds additional error-detection capabilities and thus further reduces the risk for undetected errors in the decoded transport block.

Information about the transport-block size is provided to the device as part of the scheduling assignment transmitted on the physical downlink control channel/enhanced physical downlink control channel (PDCCH/EPDCCH), as described in Section 6.4. Based on this information, the device can determine the code-block size and number of code blocks. The device receiver can thus, based on the information provided in the scheduling assignment, straightforwardly undo the code-block segmentation and recover the decoded transport blocks.

6.1.1.3 Channel Coding
Channel coding for DL-SCH (as well as for PCH and MCH) is based on Turbo coding [17], with encoding according to Figure 6.3. The encoding consists of two rate-1/2, eight-state

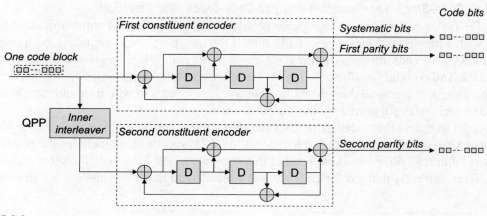

FIGURE 6.3

LTE Turbo encoder.

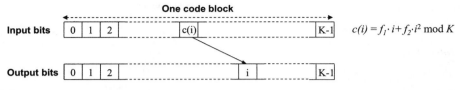

FIGURE 6.4

Principles of QPP-based interleaving.

constituent encoders, implying an overall code rate of 1/3, in combination with QPP-based[2] interleaving [32]. As illustrated in Figure 6.4, the QPP interleaver provides a mapping from the input (non-interleaved) bits to the output (interleaved) bits according to the function:

$$c(i) = f_1 \cdot i + f_2 \cdot i^2 \bmod K,$$

where i is the index of the bit at the output of the interleaver, $c(i)$ is the index of the same bit at the input of the interleaver, and K is the code-block/interleaver size. The values of the parameters f_1 and f_2 depend on the code-block size K. The LTE specification lists all supported code-block sizes, ranging from a minimum of 40 bits to a maximum of 6144 bits, together with the associated values for the parameters f_1 and f_2. Thus, once the code-block size is known, the Turbo-coder inner interleaving, as well as the corresponding de-interleaving at the receiver side, can straightforwardly be carried out.

A QPP-based interleaver is *maximum contention free* [33], implying that the decoding can be parallelized without the risk for contention when the different parallel processes are accessing the interleaver memory. For the very high data rates supported by LTE, the improved possibilities for parallel processing offered by QPP-based interleaving can substantially simplify the Turbo-encoder/decoder implementation.

6.1.1.4 Rate Matching and Physical-Layer Hybrid-ARQ Functionality

The task of the rate-matching and physical-layer hybrid-ARQ functionality is to extract, from the blocks of code bits delivered by the channel encoder, the exact set of code bits to be transmitted within a given TTI/subframe.

As illustrated in Figure 6.5, the outputs of the Turbo encoder (systematic bits, first parity bits, and second parity bits) are first separately interleaved. The interleaved bits are then inserted into what can be described as a circular buffer with the systematic bits inserted first, followed by alternating insertion of the first and second parity bits.

The bit selection then extracts consecutive bits from the circular buffer to an extent that matches the number of available resource elements in the resource blocks assigned for the transmission. The exact set of bits to extract depends on the *redundancy version* (RV) corresponding to different starting points for the extraction of coded bits from the circular buffer. As can be seen, there are four different alternatives for the RV. The transmitter/scheduler

---

[2]*QPP*, Quadrature permutation polynomial.

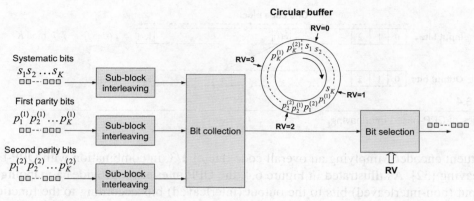

FIGURE 6.5

Rate-matching and hybrid-ARQ functionality.

selects the RV and provides information about the selection as part of the scheduling assignment (see Section 6.4.4).

Note that the rate-matching and hybrid-ARQ functionality operates on the full set of code bits corresponding to one transport block and not separately on the code bits corresponding to a single code block.

6.1.1.5 Bit-Level Scrambling

LTE downlink scrambling implies that the block of code bits delivered by the hybrid-ARQ functionality is multiplied (*exclusive-or* operation) by a bit-level *scrambling sequence*. Without downlink scrambling, the channel decoder at the device could, at least in principle, be equally matched to an interfering signal as to the target signal, thus being unable to properly suppress the interference. By applying different scrambling sequences for neighboring cells, the interfering signal(s) after descrambling is (are) randomized, ensuring full utilization of the processing gain provided by the channel code. Thus, the bit scrambling essentially serves the same purpose as the scrambling applied at chip level after the direct-sequence spreading in DS-CDMA-based systems such as WCDMA/HSPA. Fundamentally, channel coding can be seen as "advanced" spreading providing processing gain similar to direct-sequence spreading but also additional coding gain.

In LTE, downlink scrambling is applied to all transport channels as well as to the downlink L1/L2 control signaling. For all downlink transport-channel types except MCH the scrambling sequences differ between neighboring cells (*cell-specific scrambling*). The scrambling also depends on the identity of the device to which the transmission is intended, assuming that the data is intended for a specific device. In contrast, in the case of MBSFN-based transmission using MCH, the same scrambling should be applied to all cells taking part in the MBSFN transmission—that is, all cells within the so-called *MBSFN area* (see Chapter 19). Thus, in the case of MCH transmission the scrambling depends on the MBSFN area identity.

6.1.1.6 Data Modulation

The downlink data modulation transforms the block of scrambled bits to a corresponding block of complex modulation symbols. The set of modulation schemes supported for the LTE downlink includes QPSK, 16QAM, and 64QAM, corresponding to two, four, and six bits per modulation symbol respectively. Optional support for 256QAM, corresponding to eight bits per symbol, was added in release 12, primarily intended for small-cell environments where the achievable SNR can be relatively high.[3]

6.1.1.7 Antenna Mapping

The antenna mapping jointly processes the modulation symbols corresponding to the one or two transport blocks and maps the result to the set of antenna ports to be used for the transmission. The antenna mapping can be configured in different ways corresponding to different multi-antenna transmission schemes, including transmit diversity, beam-forming, and spatial multiplexing. As indicated in Figure 6.1, LTE supports simultaneous transmission using up to eight antenna ports depending on the exact multi-antenna transmission scheme. More details about LTE downlink multi-antenna transmission are provided in Section 6.3.

6.1.1.8 Resource-Block Mapping

The resource-block mapping takes the symbols to be transmitted on each antenna port and maps them to the set of available resource elements in the set of resource blocks assigned by the MAC scheduler for the transmission. As described in Chapter 5, each resource block consists of 84 resource elements (twelve subcarriers during seven OFDM symbols).[4] However, some of the resource elements within a resource block will not be available for the transport-channel transmission as they are occupied by

- different types of downlink reference signals as described in Section 6.2;
- downlink L1/L2 control signaling (one, two, or three OFDM symbols at the head of each subframe) as described in Section 6.4.[5]

Furthermore, as described in Chapter 11, within some resource blocks additional resource elements are reserved for the transmission of *synchronization signals* as well as for the PBCH physical channel which carries the BCH transport channel.

It should also be pointed out that for the so-called Transmission Mode 10, the possibility for more dynamic control of the PDSCH mapping has been introduced to support multi-point transmission. This is further discussed in Chapter 13 as part of the description of CoMP-related features introduced in LTE release 11. The introduction of license-assisted access in release 13 provides some additional flexibility in the PDSCH mapping as described in Chapter 17.

---

[3]For backward-compatibility reasons, 256QAM is not supported for PMCH when carrying MCCH (see Chapter 19 for a description of the MBMS channels).
[4]72 resource elements in case of extended cyclic prefix.
[5]In MBSFN subframes the control region is limited to a maximum of two OFDM symbols.

6.1.2 LOCALIZED AND DISTRIBUTED RESOURCE MAPPING

As already discussed in Chapter 3, when deciding what set of resource blocks to use for transmission to a specific device, the network may take the downlink channel conditions in both the time and frequency domains into account. Such time—frequency-domain channel-dependent scheduling, taking channel variations—for example, due to frequency-selective fading—into account, may significantly improve system performance in terms of achievable data rates and overall cell throughput.

However, in some cases downlink channel-dependent scheduling is not suitable to use or is not practically possible:

- For low-rate services such as voice, the feedback signaling associated with channel-dependent scheduling may lead to extensive relative overhead.
- At high mobility (high device speed), it may be difficult or even practically impossible to track the instantaneous channel conditions to the accuracy required for channel-dependent scheduling to be efficient.

In such situations, an alternative means to handle radio-channel frequency selectivity is to achieve frequency diversity by distributing a downlink transmission in the frequency domain.

One way to distribute a downlink transmission in the frequency domain, and thereby achieve frequency diversity, is to assign multiple nonfrequency-contiguous resource blocks for the transmission to a device. LTE allows for such *distributed resource-block allocation* by means of *resource allocation types 0 and 1* (see Section 6.4.6.1). However, although sufficient in many cases, distributed resource-block allocation by means of these resource-allocation types has certain drawbacks:

- For both types of resource allocations, the minimum size of the allocated resource can be as large as four resource-block pairs and may thus not be suitable when resource allocations of smaller sizes are needed.
- In general, both these resource-allocation methods are associated with a relatively large control-signaling overhead for the scheduling assignment, see Section 6.4.6

In contrast, *resource-allocation type 2* (Section 6.4.6.1) always allows for the allocation of a single resource-block pair and is also associated with a relatively small control-signaling overhead. However, resource allocation type 2 only allows for the allocation of resource blocks that are contiguous in the frequency domain. In addition, regardless of the type of resource allocation, frequency diversity by means of distributed resource-block allocation will only be achieved in the case of resource allocations larger than one resource-block pair.

In order to provide the possibility for distributed resource-block allocation in the case of resource-allocation type 2, as well as to allow for distributing the transmission of a single resource-block pair in the frequency domain, the notion of a *virtual resource block* (VRB) has been introduced for LTE.

What is being provided in the resource allocation is the resource allocation in terms of VRB pairs. The key to distributed transmission then lies in the mapping from VRB pairs to *physical resource block* (PRB) pairs—that is, to the actual physical resource used for transmission.

The LTE specification defines two types of VRBs: *localized* VRBs and *distributed* VRBs. In the case of localized VRBs, there is a direct mapping from VRB pairs to PRB pairs as illustrated in Figure 6.6.

However, in the case of distributed VRBs, the mapping from VRB pairs to PRB pairs is more elaborate in the sense that:

- consecutive VRBs are not mapped to PRBs that are consecutive in the frequency domain;
- even a single VRB pair is distributed in the frequency domain.

The basic principle of distributed transmission is outlined in Figure 6.7 and consists of two steps:

- A mapping from VRB pairs to PRB pairs such that consecutive VRB pairs are not mapped to frequency-consecutive PRB pairs (first step of Figure 6.7). This provides frequency diversity between consecutive VRB pairs. The spreading in the frequency domain is done by means of a block-based "interleaver" operating on resource-block pairs.
- A split of each resource-block pair such that the two resource blocks of the resource-block pair are transmitted with a certain frequency gap in between (second step of Figure 6.7). This also provides frequency diversity for a single VRB pair. This step can be seen as the introduction of frequency hopping on a slot basis.

Whether the VRBs are localized (and thus mapped according to Figure 6.6) or distributed (mapped according to Figure 6.7) is indicated as part of the scheduling assignment in the case

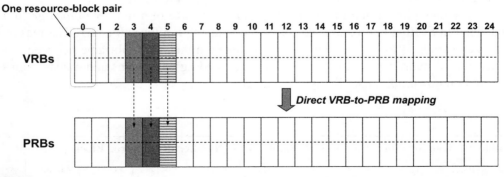

FIGURE 6.6

VRB-to-PRB mapping in case of localized VRBs. Figure assumes a cell bandwidth corresponding to 25 resource blocks.

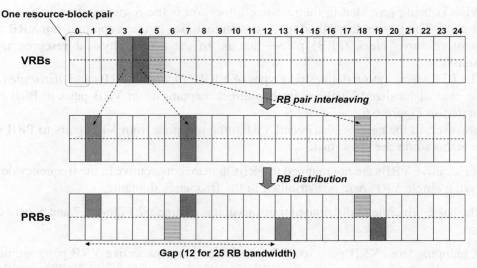

FIGURE 6.7

VRB-to-PRB mapping in case of distributed VRBs. Figure assumes a cell bandwidth corresponding to 25 resource blocks.

of type 2 resource allocation. Thus, it is possible to dynamically switch between distributed and localized transmission and also mix distributed and localized transmission for different devices within the same subframe.

The exact size of the frequency gap in Figure 6.7 depends on the overall downlink cell bandwidth according to Table 6.1. These gaps have been chosen based on two criteria:

1. The gap should be of the order of half the downlink cell bandwidth in order to provide good frequency diversity also in the case of a single VRB pair.

2. The gap should be a multiple of P^2, where P is the size of a *resource-block group* as defined in Section 6.4.6 and used for resource allocation types 0 and 1. The reason for this constraint is to ensure a smooth coexistence in the same subframe between distributed transmission as described in the preceding paragraphs and transmissions based on downlink allocation types 0 and 1.

Table 6.1 Gap Size for Different Cell Bandwidths (Number of Resource Blocks)									
Bandwidth	**6**	**7–8**	**9–10**	**11**	**12–19**	**20–26**	**27–44**	**45–63**	**64–79**	**80–110**
P	1	1	1	2	2	2	3	3	4	4
Gap size	3	4	5	4	8	12	18	27	32	48

Table 6.2 Second Gap Size for Different Cell Bandwidth (Only Applicable for Cell Bandwidths of 50 RBs and Beyond)

Bandwidth	50–63	64–110
Gap size	9	16

Due to the constraint that the gap size should be a multiple of P^2, the gap size will in most cases deviate from exactly half the cell bandwidth. In these cases, not all resource blocks within the cell bandwidth can be used for distributed transmission. As an example, for a cell bandwidth corresponding to 25 resource blocks (the example in Figure 6.7) and a corresponding gap size equal to 12 according to Table 6.1, the 25th resource-block pair cannot be used for distributed transmission. As another example, for a cell bandwidth corresponding to 50 resource blocks (gap size equal to 27 according to Table 6.1) only 46 resource blocks would be available for distributed transmission.

In addition to the gap size outlined in Table 6.1, for wider cell bandwidths (50 RBs and beyond), there is a possibility to use a second, smaller frequency gap with a size of the order of one-fourth of the cell bandwidth (see Table 6.2). The use of the smaller gap enables restriction of the distributed transmission to only a part of the overall cell bandwidth. Selection between the larger gap according to Table 6.1 and the smaller gap according to Table 6.2 is indicated by an additional bit in the resource allocation.

6.2 DOWNLINK REFERENCE SIGNALS

Downlink reference signals are predefined signals occupying specific resource elements within the downlink time–frequency grid. The LTE specification includes several types of downlink reference signals transmitted in different ways and intended to be used for different purposes by a receiving device:

- *Cell-specific reference signals* (CRS) are transmitted in every downlink subframe and in every resource block in the frequency domain. CRS are intended to be used by devices for channel estimation for coherent demodulation of all downlink physical channels except PMCH, PDSCH in case of *transmission modes* 7–10, and the EPDCCH control channel introduced in LTE release 11 (see Section 6.4).[6] CRS are also assumed to be used to acquire *channel-state information* (CSI) by devices configured in transmission modes 1–8. Finally, device measurements on CRS are assumed to be used as the basis for cell-selection and handover decisions.
- *Demodulation reference signals* (DM-RS), also sometimes referred to as *UE-specific reference signals*, are intended to be used by devices for channel estimation for coherent

---

[6]See Section 6.3.1 for more details on LTE *transmission modes*.

demodulation of PDSCH in case of transmission modes 7−10.[7] DM-RS are also to be used for demodulation of the EPDCCH physical channel. The alternative label "*UE-specific reference signals*" relates to the fact that a specific demodulation reference signal is typically intended to be used for channel estimation by a specific device (UE). The reference signal is then only transmitted within the resource blocks specifically assigned for PDSCH/EPDCCH transmission to that device.

- *CSI reference signals* (CSI-RS) are intended to be used by devices to acquire CSI. More specifically, CSI-RS are intended to be used to acquire CSI by devices configured in transmission modes 9 and 10. CSI-RS have a significantly lower time−frequency density, thus implying less overhead, and a higher degree of flexibility compared to the CRS.
- *MBSFN reference signals* are intended to be used by devices for channel estimation for coherent demodulation in case of MCH transmission using *MBSFN* (see Chapter 19 for more details on MCH transmission).
- *Positioning reference signals* were introduced in LTE release 9 to enhance *LTE positioning functionality*, more specifically to support the use of device measurements on multiple LTE cells to estimate the geographical position of the device. The positioning reference symbols of a certain cell can be configured to correspond to empty resource elements in neighboring cells, thus enabling high-SIR conditions when receiving neighbor-cell positioning reference signals.

6.2.1 CELL-SPECIFIC REFERENCE SIGNALS

CRS, introduced in the first release of LTE (release 8), are the most basic downlink reference signals in LTE. There can be one, two, or four CRS in a cell, defining one, two, or four corresponding antenna ports, referred to as antenna port 0 to antenna port 3 in the LTE specifications.

6.2.1.1 Structure of a Single Reference Signal

Figure 6.8 illustrates the structure of a single CRS. As can be seen, it consists of *reference symbols* of predefined values inserted within the first and third last[8] OFDM symbol of each slot and with a frequency-domain spacing of six subcarriers. Furthermore, there is a frequency-domain staggering of three subcarriers for the reference symbols within the third last OFDM symbol. Within each resource-block pair, consisting of 12 subcarriers during one 1 ms subframe, there are thus eight reference symbols.

In general, the values of the reference symbols vary between different reference-symbol positions and also between different cells. Thus, a CRS can be seen as a two-dimensional cell-specific sequence. The period of this sequence equals one 10 ms frame. Furthermore, regardless of the cell bandwidth, the reference-signal sequence is defined assuming the

---

[7]In the LTE specifications, these reference signals are actually referred to as *UE-specific reference signals*, although they are still "abbreviated" DM-RS.

[8]This corresponds to the fifth and fourth OFDM symbols of the slot for normal and extended cyclic prefixes, respectively.

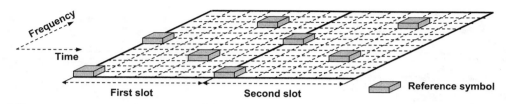

FIGURE 6.8

Structure of CRS within a pair of resource blocks.

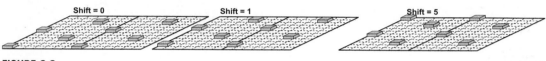

FIGURE 6.9

Different CRS frequency shifts.

maximum possible LTE carrier bandwidth corresponding to 110 resource blocks in the frequency domain. Thus, the basic reference-signal sequence has a length of 8800 symbols.[9] For cell bandwidths less than the maximum possible value, only the reference symbols within that bandwidth are actually transmitted. The reference symbols in the center part of the band will therefore be the same, regardless of the actual cell bandwidth. This allows for the device to estimate the channel corresponding to the center part of the carrier, where, for example, the basic system information of the cell is transmitted on the BCH transport channel, without knowing the cell bandwidth. Information about the actual cell bandwidth, measured as number of resource blocks, is then provided on the BCH.

There are 504 different reference-signal sequences defined for LTE, where each sequence corresponds to one of 504 different *physical-layer cell identities*. As described in more detail in Chapter 11, during the so-called *cell-search procedure* the device detects the physical-layer identity of the cell as well as the cell frame timing. Thus, from the cell-search procedure, the device knows the reference-signal sequence of the cell (given by the physical-layer cell identity) as well as the start of the reference-signal sequence (given by the frame timing).

The set of reference-symbol positions outlined in Figure 6.8 is only one of six possible *frequency shifts* for the CRS reference symbols, as illustrated in Figure 6.9. The frequency shift to use in a cell depends on the physical-layer identity of the cell such that each shift corresponds to 84 different cell identities. Thus, the six different frequency shifts jointly cover all 504 different cell identities. By properly assigning physical-layer cell identities to different cells, different reference-signal frequency shifts may be used in neighboring cells. This can be beneficial, for example, if the reference symbols are transmitted with higher energy compared

---

[9]Four reference symbols per resource block, 110 resource blocks per slot, and 20 slots per frame.

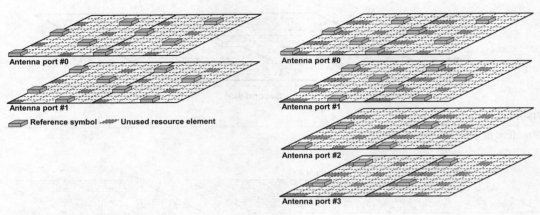

FIGURE 6.10

Structure of CRS in case of multiple reference signals: two reference signals corresponding to two antenna ports (left) and four reference signals corresponding to four antenna ports (right).

to other resource elements, also referred to as *reference-signal power boosting*, in order to improve the reference-signal SIR. If reference signals of neighboring cells were transmitted using the same time—frequency resource, the boosted reference symbols of one cell would be interfered by equally boosted reference symbols of all neighboring cells,[10] implying no gain in the reference-signal SIR. However, if different frequency shifts are used for the reference-signal transmissions of neighboring cells, the reference symbols of one cell will at least partly be interfered by nonreference symbols of neighboring cells, implying an improved reference-signal SIR in the case of reference-signal boosting.

6.2.1.2 Multiple Reference Signals

Figure 6.10 illustrates the reference-signal structure in the case of multiple, more specifically two and four, CRS, and corresponding multiple antenna ports, within a cell:[11]

- In the case of two reference signals within a cell (left part of Figure 6.10), the second reference signal is frequency multiplexed with the first reference signal, with a frequency-domain offset of three subcarriers.
- In the case of four reference signals (right part of Figure 6.10), the third and fourth reference signals are frequency multiplexed and transmitted within the *second* OFDM symbol of each slot, thus being time multiplexed with the first and second reference signals.

Obviously, the reference-symbol density for the third and fourth reference signals is lower, compared to the density of the first and second reference signals. The reason for this is to

---

[10]This assumes that the cell transmissions are frame-timing aligned.
[11]It is not possible to configure a cell with three CRS.

reduce the reference-signal overhead in the case of four reference signals. More specifically, while the first and second reference signals each correspond to a relative overhead of about 5% (4 reference symbols within a resource block consisting of a total of 84 resource elements), the relative overhead of the third and fourth reference signals is only half of that or about 2.5%. This obviously has an impact on the possibility for the device to track very fast channel variations. However, this can be justified based on an expectation that, for example, high-order spatial multiplexing will mainly be applied to scenarios with low mobility.

It can also be noted that in a resource element carrying reference signals for a certain transmission port, nothing is being transmitted on the antenna ports corresponding to the other reference signals. Thus, a CRS is not interfered by transmissions on other antenna ports. Multi-antenna transmission schemes, such as spatial multiplexing, to a large extent rely on good channel estimates to suppress interference between the different layers at the receiver side. However, in the channel estimation itself there is obviously no such suppression. Reducing the interference to the reference signals of an antenna port is therefore important in order to allow for good channel estimation, and corresponding good interference suppression, at the receiver side.

Note that, in MBSFN subframes, only the reference signals in the two first OFDM symbols of the subframe, corresponding to the control region of the MBSFN subframe, are actually transmitted. Thus, there is no transmission of CRS within the MBSFN part of the MBSFN subframe.

6.2.2 DEMODULATION REFERENCE SIGNALS

In contrast to CRS, a demodulation reference signal (DM-RS) is intended to be used for channel estimation by a specific device and is then only transmitted within the resource blocks assigned for transmission to that device.

DM-RS was supported already in the first release of LTE (release 8). However, the use of DM-RS was then limited to the demodulation of single-layer PDSCH transmission—that is, no spatial multiplexing—corresponding to transmission mode 7. In LTE release 9, transmission based on DM-RS was extended to support dual-layer PDSCH transmission corresponding to transmission mode 8, requiring up to two simultaneous reference signals (one for each layer). Transmission based on DM-RS was then further extended in LTE release 10 to support up to eight-layer PDSCH transmission (transmission mode 9 and, from release 11, also transmission mode 10), corresponding to up to eight reference signals.[12]

Actually, the dual-layer-supporting DM-RS structure introduced in LTE release 9 was not a straightforward extension of the release 8 single-layer-limited DM-RS structure but rather a new structure, supporting both single-layer and dual-layer transmission. Already at the time of finalizing LTE release 9 it was relatively clear that the LTE radio-access technology should

---

[12]Transmission mode 10 is a release 11 extension of transmission mode 9 introducing improved support for multi-point co-ordination/transmission (CoMP), see also Section 6.3.1.

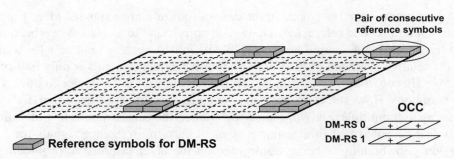

FIGURE 6.11

Structure of DM-RS for the case of one or two reference signals including size-two OCC to separate the two reference signals.

be further extended to support up to eight-layer spatial multiplexing in release 10. It was also quite clear that this extension would be difficult to achieve based on the release 8 DM-RS structure. Rather than extending the release 8 structure to support two reference signals and then introduce a completely new structure for release 10, it was instead decided to introduce a new, more future-proof structure already in release 9. Here we focus on the DM-RS structure introduced in LTE release 9 including the release 10 extension to support up to eight simultaneous reference signals.

The structure of the DM-RS for EPDCCH is very similar to that of DM-RS for PDSCH although with some limitations such as support for a maximum of four reference signals.

6.2.2.1 DM-RS for PDSCH

Figure 6.11 illustrates the DM-RS time–frequency structure for the case of one or two reference signals.[13] As can be seen, there are 12 reference symbols within a resource-block pair. In contrast to CRS, for which the reference symbols of one reference signal correspond to unused resource elements for other reference signals (see Figure 6.10), in the case of two DM-RS all 12 reference symbols in Figure 6.11 are transmitted for both reference signals. Interference between the reference signals is instead avoided by applying mutually orthogonal patterns, referred to as *orthogonal cover codes* (OCC), to pairs of consecutive reference symbols as illustrated in the lower right part of the figure.

Figure 6.12 illustrates the extended DM-RS structure introduced in LTE release 10 to support up to eight reference signals. In this case, there are up to 24 reference-symbol positions within a resource-block pair. The reference signals are frequency multiplexed in groups of up to four reference signals while, within each group, the up to four reference signals are separated by means of OCC spanning four reference symbols in the time domain (two pairs of consecutive reference symbols). It should be noted that orthogonality between

---

[13]In the case of TDD, the DM-RS structure is slightly modified in the DwPTS due to the shorter duration of the DwPTS compared with normal downlink subframes.

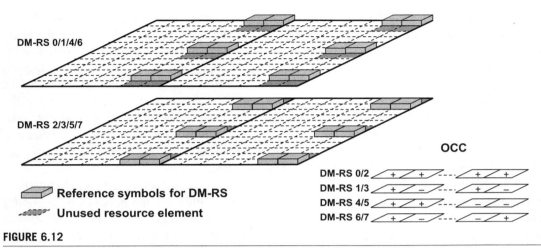

FIGURE 6.12

Demodulation reference signal structure for the case of more than two reference signals including size-four OCC to separate up to four reference signals.

the full set of eight reference signals requires that the channel does not vary over the four reference symbols spanned by the cover codes. As the four reference symbols that the cover codes span are not consecutive in time, this implies somewhat stronger constraints in terms of the amount of channel variations that can be tolerated without seriously impacting the reference-signal orthogonality. However, more than four DM-RS are only transmitted in case of spatial multiplexing with more than four layers, which is typically only applicable to low-mobility scenarios. Also note that the cover codes are defined such that, for four or less reference signals, orthogonality is achieved already over pairs of reference symbols. Thus, for three and four reference signals, the constraints on channel variations are the same as for two reference signals (Figure 6.11).

The up to eight different PDSCH DM-RS that can be configured for a device correspond to antenna port 7 up to antenna port 14 in the LTE specification, with antenna port 7 and antenna port 8 corresponding to the release 9 DM-RS supporting up to two-layers spatial multiplexing.[14]

The DM-RS *reference-signal sequence*—that is, the values to be taken by the DM-RS reference symbols—repeats itself every frame. Up to and including LTE release 10, the reference-signal sequence was independent of what device the DM-RS transmission was intended for but depended on the physical-layer cell identity. The reference-signal sequence thus differed between cells. Furthermore, there was the possibility to dynamically—that is, on a subframe basis—select between two different reference-signal sequences. Information about the selected sequence was then signaled to the device by means of a one-bit indicator in

---

[14]The single DM-RS supported already in release 8 corresponds to antenna port 5.

the scheduling assignment (see also Section 6.4.6). The reason for having the possibility to dynamically select between two reference-signal sequences was to be able to use the same resource block for PDSCH transmission to two different devices and rely on spatial separation, for example, by means of beam-forming, to separate the two transmissions. Such spatial separation, in 3GPP referred to as *multi-user multi-input—multi-output* (MU-MIMO), is typically not perfect in the sense that there will remain some interference between the transmissions. By applying different reference-signal sequences for the two spatially separated transmissions, interference randomization is achieved for the channel estimation. Downlink MU-MIMO is discussed in more detail in Section 6.3 together with the different means to separate the DM-RS of different transmissions.

However, in LTE release 11, the possibility for *device-specific* reference-signal sequences was introduced. This was done by introducing the possibility for the network to explicitly configure a device with a parameter that, if configured, should replace that cell identity when deriving the reference-signal sequence to be used by the device. If no device-specific parameter is configured, the device should assume cell-specific reference-signal sequences in line with releases 9/10 as previously discussed.[15]

The reason for introducing the possibility for device-specific reference-signal sequences was to be able to spatially separate significantly more devices within a cell. Especially in the case of so-called *shared-cell* heterogeneous deployments there may be situations with a large number of low-power transmission points, all being part of the same cell. In such a case one typically wants to be able to reuse the same physical resources—that is, the same resource blocks, for simultaneous PDSCH transmission to different devices from several of the transmission points. For robust channel estimation, the reference signals of each of these transmissions should preferably be based on unique reference-signal sequences, thus making device-specific reference signal sequences desirable. Heterogeneous deployments for LTE is extensively discussed in Chapter 14.

When DM-RS are transmitted within a resource block, PDSCH mapping to the time—frequency grid of the resource block will be modified to avoid the resource elements in which the reference signals are transmitted (the 12 and 24 resource elements in Figures 6.11 and 6.12, respectively). Although this modified mapping is not "understood" by earlier-release devices not supporting DM-RS, this is not a problem as DM-RS will only be transmitted in resource blocks that are scheduled for PDSCH transmission to devices of later releases supporting DM-RS and thus "understanding" the modified PDSCH mapping.

As the number of transmitted layers may vary dynamically, the number of transmitted DM-RS may also vary. Thus, the transmission may dynamically change between the DM-RS structures outlined in Figures 6.11 and 6.12, respectively. The device is informed about the number of transmitted layers (the "transmission rank") as part of the scheduling assignment and will thus know the DM-RS structure and associated PDSCH mapping for each subframe.

---

[15]*Cell-specific reference-signal sequences for DM-RS* should not be mixed up with CRS.

6.2.2.2 DM-RS for EPDCCH

As mentioned earlier, as part of LTE release 11, a new L1/L2 control channel structure was introduced based on the so-called *enhanced PDCCH* (EPDCCH). The EPDCCH is extensively described in Section 6.4.4. Here it can just be said that, in contrast to the legacy control-channel structure (PDCCH), the EPDCCH is transmitted within resource blocks in a similar way as PDSCH. Furthermore, in contrast to the PDCCH, EPDCCH demodulation is assumed to be based on DM-RS transmitted together with the EPDCCH, similar to the use of DM-RS for PDSCH.

The structure of DM-RS for EPDCCH is very similar to the PDSCH DM-RS structure described previously. Especially, the time—frequency structure of the EPDCCH DM-RS is the same as that for PDSCH. However, for EPDCCH, there can only be up to four DM-RS, compared to up to eight DM-RS in case of PDSCH transmission. Thus the four orthogonal covers corresponding to DM-RS 4—7 in Figure 6.11 are not supported for EPDCCH transmission. Furthermore, the EPDCCH reference-signal sequence is *always* device specific— that is, the device is explicitly configured with a parameter that is used to derive the reference-signal sequence. It should be noted that this configuration of the DM-RS reference-signal sequence for EPDCCH is done independently of the corresponding configuration for the PDSCH DM-RS.

The antenna ports corresponding to up to four DM-RS for EPDCCH are, in the LTE specifications, referred to as antenna port 107 to antenna port 110. It should be noted that, although up to four different DM-RS and corresponding antenna ports can be defined for EPDCCH, a specific EPDCCH is only transmitted from a single antenna port in case of localized transmission and two antenna ports in case of distributed transmission (see Section 6.4.4). Thus, in some sense it is somewhat misleading for the specification to talk about *up to four antenna ports* for EPDCCH as a device will only see one or two DM-RS-related antenna port(s).

6.2.3 CSI REFERENCE SIGNALS

CSI-RS were introduced in LTE release 10. CSI-RS are specifically intended to be used by devices to acquire CSI—for example, for channel-dependent scheduling, link adaptation, and transmission settings related to multi-antenna transmission. More specifically, CSI-RS were introduced to acquire CSI for devices configured with transmission mode 9 and 10,[16] but in later releases also serve other purposes.

As mentioned early in Section 6.2, the CRS, available since the first release of LTE, can also be used to acquire CSI. The direct reason to introduce CSI-RS was the introduction of support for up to eight-layers spatial multiplexing in LTE release 10 and the corresponding need for devices to be able to acquire CSI for, at least, up to eight antenna ports.

---

[16]The reason why CSI-RS are not used for transmission modes 7 and 8 despite the fact that these transmission modes assume DM-RS for channel estimation, was simply that these transmission modes were introduced in LTE releases 8 and 9, respectively, while CSI-RS was not introduced until LTE release 10.

However, there was also a more fundamental desire to separate two different functions of downlink reference signals, namely

- the function to acquire detailed channel estimates for coherent demodulation of different downlink transmissions;
- the function to acquire CSI for, for example, downlink link adaptation and scheduling.

For the early releases of LTE, both these functions relied on CRS. As a consequence, CRS has to be transmitted with high density in both time and frequency to support accurate channel estimation and coherent demodulation also for rapidly varying channels. At the same time, in order to allow for devices to acquire CSI at regular intervals, CRS has to be transmitted in every subframe regardless of whether or not there is any data transmission. For the same reason CRS is transmitted over the entire cell area and cannot be beam-formed in the direction of a specific device.

By introducing separate sets of reference signals for channel estimation and for the acquisition of CSI (DM-RS and CSI-RS, respectively) more opportunities for optimization and a higher degree of flexibility are achieved. The high-density DM-RS are only transmitted when there is data to transmit and can, for example, be subject to more or less arbitrary beam-forming. At the same time, CSI-RS provides a very efficient tool for deriving CSI for a more arbitrary number of network nodes and antenna ports. This is especially important for the support for multi-point coordination/transmission and heterogeneous deployments as is further discussed in Chapters 13 and 14, respectively.

6.2.3.1 CSI-RS Structure

The structure of the CSI-RS to be used by a device is given by a *CSI-RS configuration*. The possibility for up to eight CSI-RS in release 10 is directly related to the support for up to eight-layers spatial multiplexing and corresponding up to eight DM-RS. The number of CSI-RSs was increased to 16 in release 13 in order to better support two-dimensional beam-forming, see Chapter 10. It should be noted though that the antenna ports corresponding to CSI-RS *are not the same* as the antenna ports corresponding to DM-RS. Antenna ports corresponding to CSI-RS typically correspond to actual transmit antennas while antenna ports corresponding to DM-RS may include any antenna precoding applied at the transmitter side (see also Section 6.3.4). The antenna ports corresponding to CSI-RS are referred to as antenna port 15 up to antenna port 22 in the LTE specification, a number that was increased to 30 in release 13.

In the time domain, CSI-RS can be configured for transmission with different periodicity, ranging from a period of 5 ms (twice every frame) to 80 ms (every eighth frame). Furthermore, for a given CSI-RS periodicity, the exact subframe in which CSI-RS is transmitted can also be configured by means of a *subframe offset*.[17] In subframes in which CSI-RS is to be

---

[17]All up to 16 CSI-RS of a CSI-RS configuration are transmitted within the same set of subframes—that is, with the same period and subframe offset.

transmitted, it is transmitted in every resource block in the frequency domain. In other words, a CSI-RS transmission covers the entire cell bandwidth.

Within a resource-block pair, different resource elements can be used for CSI-RS transmission (illustrated by the 40 different resource elements colored by gray in Figure 6.13; for TDD there are even more possibilities). Exactly what set of resource elements is used for a certain CSI-RS then depends on the exact CSI-RS configuration. More specifically:

- In the case of a CSI-RS configuration consisting of one or two configured CSI-RS, a CSI-RS consists of two consecutive reference symbols, as illustrated in the upper part of Figure 6.13. In the case of two CSI-RS, the CS-RS are then separated by applying size-two OCC to the two reference symbols, similar to DM-RS. Thus, for the case of one or two CSI-RS, there is a possibility for 20 different CSI-RS configurations in a resource-block pair, two of which are illustrated in Figure 6.13.
- In the case of a CSI-RS configuration consisting of four/eight configured CSI-RS, the CSI-RS are pair-wise frequency multiplexed, as illustrated in the middle/lower part of Figure 6.13. For four/eight CSI-RS there is thus the possibility for ten/five different CSI-RS configurations.

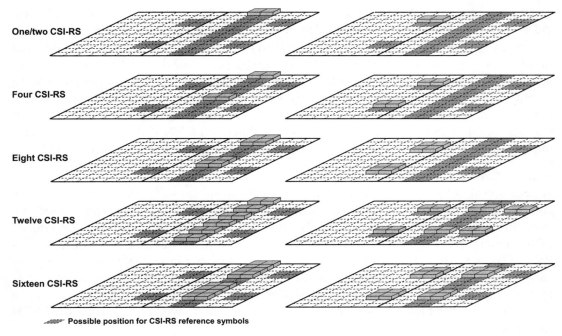

_...◀▦▬_ **Possible position for CSI-RS reference symbols**

FIGURE 6.13

Examples of reference-signal positions for different number of CSI-RS within a cell. In case of a single CSI-RS, the same structure as for two CSI-RS is used.

- In case of a CSI-RS configuration consisting of twelve or sixteen configured CSI-RS, a possibility introduced in release 13, aggregation of size-4 or size-8 CSI-RS are used. A configuration consisting of twelve CSI-RS is created by aggregating three size-4 CSI-RS configurations and a configuration consisting of 16 CSI-RS by aggregating two size-8 CSI-RS configurations. In other words, the resource mapping is similar to the four/eight antenna port case but more than one such configuration is used. In addition to size-two orthogonal cover code as done for eight and less configured CSI-RS, there is possibility to configure size-four OCC. The reason for the longer cover codes is to improve the possibilities to "borrow" power between CSI-RS, see Chapter 10.

To summarize, a CSI-RS configuration specifies

- the number of CSI-RS (1, 2, 4, 8, 12, or 16);
- the CSI-RS periodicity (5, 10, 20, 40, or 80 ms);
- the CSI-RS subframe offset within the CSI-RS period;
- the exact CSI-RS configuration within a resource block pair—that is, exactly what resource elements from the 40 possible resource elements (gray resource elements in Figure 6.13) are used for the CSI-RS in a resource block pair; and
- the size of the orthogonal cover code, two or four, in case of more than eight CSI-RS.

CSI-RS configurations are device-specific—meaning that each device is separately provided with a specific CSI-RS configuration that defines the number of CSI-RS to be used by the device and their detailed structure as described previously. Note though that this does not mean that a certain transmitted CSI-RS is only used by one single device. Even if each device is separately provided with its CSI-RS configuration, the configuration will, in practice, be identical for a group of, or even all, devices within a cell implying that the devices will, in practice, use the same set of CSI-RS to acquire CSI. However, the separate configuration of CSI-RS for different devices *allows* for devices within a cell to use different CSI-RS. This is important, for example, in case of shared-cell heterogeneous deployments, see Chapter 14.

6.2.3.2 CSI-RS and PDSCH Mapping

As mentioned in Section 6.2.2.1, when DM-RS are transmitted within a resource block, the corresponding resource elements on which the reference signals are transmitted are explicitly avoided when mapping PDSCH symbols to resource elements. This "modified" PDSCH mapping, which is not "understood" by earlier-release devices, is possible as DM-RS can be assumed to be transmitted only in resource blocks in which devices supporting such reference signals are scheduled—that is, devices based on LTE release 10 or later.[18] Expressed alternatively, an earlier-release device can be assumed never to be scheduled in a resource block in which DM-RS are transmitted and thus in which the modified PDSCH mapping is used.

---

[18]Partly also for devices of release 9, but then only for a maximum of two DM-RS.

The situation is different for CSI-RS. As a CSI-RS is transmitted within all resource blocks in the frequency domain, it would imply a strong scheduler constraint to assume that release 8/9 devices would never be scheduled in a resource block in which CSI-RS is transmitted. If the PDSCH mapping were modified to explicitly avoid the resource elements in which CSI-RS is transmitted, the mapping would not be recognized by a releases-8/9 device. Instead, in the case of resource blocks scheduled to release 8/9 devices, the PDSCH is mapped exactly according to release 8—that is, the mapping is not modified to avoid the resource elements on which CSI-RS is to be transmitted. The CSI-RS is then simply transmitted on top of the corresponding PDSCH symbols.[19] This will impact the PDSCH demodulation performance, as some PDSCH symbols will be highly corrupted. However, the remaining PDSCH symbols will not be impacted and the PDSCH will still be decodable, although with somewhat reduced performance.

On the other hand, if a release 10 device is scheduled in a resource block in which CSI-RS is transmitted, the PDSCH mapping is modified to explicitly avoid the resource elements on which the CSI-RS is transmitted, similar to DM-RS. Thus, if CSI-RS is transmitted in a resource block, the PDSCH mapping to that resource block will be somewhat different depending on the release of the device being scheduled in the resource block.

It should be noted that release 8 mapping also has to be used for transmission of, for example, system information and paging messages, as such transmissions must be possible to receive also by release 8/9 devices.

6.2.3.3 Zero-Power CSI-RS

As described, release 10 and beyond devices can assume that the PDSCH mapping avoids the resource elements corresponding to the set of CSI-RS configured for the device.

In addition to conventional CSI-RS, there is also the possibility to configure a device with a set of *zero-power CSI-RS* resources, where each zero-power CSI-RS has the same structure as a "conventional" (nonzero-power) CSI-RS:

- A certain periodicity (5, 10, 20, 40, or 80 ms).
- A certain subframe offset within the period.
- A certain configuration within a resource block pair.

The intention with the zero-power CSI-RS is simply to define additional CSI-RS resources to which the device should assume that PDSCH is not mapped. These resources may, for example, correspond to CSI-RS of other devices within the cell or within neighbor cells. They may also correspond to so-called CSI-IM resources as discussed in more detail in Chapter 10.

It should be noted that, despite the name, the zero-power CSI-RS resources may not necessarily be of zero power as they may, for example, correspond to "normal" (nonzero-

---

[19]In practice, the base station may instead not transmit PDSCH at all or, equivalently, transmit PDSCH with zero energy in these resource elements in order to avoid interference to the CSI-RS transmission. The key thing is that the mapping of the remaining PDSCH symbols is in line with release 8.

power) CSI-RS configured for other devices within the cell. The key point is that a device for which a certain zero-power CSI-RS resource has been configured should assume that PDSCH mapping avoids the corresponding resource elements.

6.2.4 QUASI-COLOCATION RELATIONS

Chapter 4 briefly discussed the concept of quasi-colocated antenna ports. As also mentioned, at least for the downlink an antenna port can be seen as corresponding to a specific reference signal. Thus it is important to understand what assumptions can be made regarding the relations, in terms of quasi-colocation, between downlink antenna ports corresponding to different reference signals.

Downlink antenna ports 0–3, corresponding to up to four CRS, can always be assumed to be jointly quasi-colocated. Similarly, antenna ports 7–14, corresponding to up to eight DM-RS, can also always be assumed to be jointly quasi-located. It should be pointed out though that the quasi-colocation assumption for DM-RS is only valid within a subframe. The reason for this restriction is to be able to switch a PDSCH transmission that relies on DM-RS between different transmission points on a subframe basis implying that quasi-colocation cannot be assumed *between* subframes even for a certain antenna port. Finally, antenna ports 15–30, corresponding to the up to sixteen CSI-RS of a specific CSI-RS configuration, can also always be assumed to be jointly quasi-colocated.

When it comes to quasi-colocation relations between antenna ports corresponding to different *types* of reference signals, for transmission modes 1–9 it can always be assumed that antenna ports 0–3 and 7–30—that is, CRS, DM-RS, and CSI-RS—are all *jointly* quasi-colocated. As a consequence, the only case when quasi-colocation can not necessarily be assumed for different types of reference signals is for the case of transmission mode 10. As also discussed in Section 6.3, transmission mode 10 was specifically introduced in LTE release 10 to support *multi-point* coordination/transmission. It is also in this case that the concept of quasi-colocation and lack thereof becomes relevant and, as indicated in Chapter 4, the concept of quasi-colocation was introduced in LTE release 11 for this specific reason. The specific aspects of quasi-colocation in case of transmission mode 10, and especially the quasi-colocation relation between the CSI-RS configured for a device and the set of DM-RS related to PDSCH transmission for that device, is discussed in Chapter 13 as part of a more detailed discussion on multi-point coordination and transmission.

6.3 MULTI-ANTENNA TRANSMISSION

As illustrated in Figure 6.14, multi-antenna transmission in LTE can, in general, be described as a mapping from the output of the data modulation to a set of antennas ports. The input to the antenna mapping thus consists of the modulation symbols (QPSK, 16QAM, 64QAM, and 256QAM) corresponding to the one or two transport blocks of a TTI.

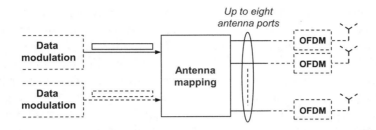

FIGURE 6.14

General structure for LTE downlink multi-antenna transmission. Modulation symbols corresponding to one or two transport blocks mapped to up to eight antenna ports.

The output of the antenna mapping is a set of symbols for each antenna port. These symbols are subsequently applied to the OFDM modulator—that is, mapped to the basic OFDM time—frequency grid corresponding to that antenna port.

6.3.1 TRANSMISSION MODES

The different multi-antenna transmission schemes correspond to different so-called *transmission modes*. There are currently ten different transmission modes defined for LTE. They differ in terms of the specific structure of the antenna mapping of Figure 6.14 but also in terms of what reference signals are assumed to be used for demodulation (CRS or DM-RS, respectively) and how CSI is acquired by the device and fed back to the network. Transmission mode 1 corresponds to single-antenna transmission while the remaining transmission modes correspond to different multi-antenna transmission schemes, including transmit diversity, beam-forming, and spatial multiplexing. Actually, LTE supports both beam-forming and spatial multiplexing as part of more general *antenna precoding*. Furthermore, there are two approaches to downlink antenna precoding—*codebook-based precoding* and *non-codebook-based precoding*. The reason for these specific names is further clarified below.

Transmission mode 10 is somewhat of a special case. As mentioned earlier, transmission mode 10 was introduced in LTE release 11 to support different means of dynamic multi-point coordination and transmission (see Chapter 13). From a device point-of-view, the downlink transmission in case of transmission mode 10 is identical to that of transmission mode 9—that is, the device will see an up-to-eight-layers PDSCH transmission and rely on DM-RS for channel estimation. One important difference between transmission mode 9 and transmission mode 10 lies in the acquisition and feedback of CSI where transmission mode 10 allows for more elaborate multi-point measurements and feedback based on CSI processes as is further discussed in Chapter 13. Another important difference lies in what a device can assume in terms of quasi-colocation relations between different types of antenna ports as mentioned in Section 6.2.4 and further discussed in Chapter 13.

It should be pointed out that transmission modes are only relevant for DL-SCH transmission. Thus, a certain transmission mode should not be seen as identical to a certain multi-antenna transmission configuration. Rather, a certain multi-antenna transmission scheme is applied to DL-SCH transmission when a device is configured in a certain transmission mode. The same multi-antenna transmission scheme may also be applied to other types of transmissions, such as transmission of BCH and L1/L2 control signaling.[20] However, this does not mean that the corresponding transmission mode is applied to such transmissions.

The following list summarizes the currently defined transmission modes and the associated multi-antenna transmission schemes. The different multi-antenna transmission schemes are described in more detail in the subsequent sections:

- *Transmission mode 1*: Single-antenna transmission
- *Transmission mode 2*: Transmit diversity
- *Transmission mode 3*: *Open-loop* codebook-based precoding in the case of more than one layer, transmit diversity in the case of rank-one transmission
- *Transmission mode 4*: *Closed-loop* codebook-based precoding
- *Transmission mode 5*: MU-MIMO version of transmission mode 4
- *Transmission mode 6*: Special case of closed-loop codebook-based precoding limited to single-layer transmission
- *Transmission mode 7*: Non-codebook-based precoding supporting single-layer PDSCH transmission
- *Transmission mode 8*: Non-codebook-based precoding supporting up to two layers (introduced in LTE release 9)
- *Transmission mode 9*: Non-codebook-based precoding supporting up to eight layers (extension of transmission mode 8, introduced in LTE release 10)
- *Transmission mode 10*: Extension of transmission mode 9 for enhanced support of different means of downlink multi-point coordination and transmission, also referred to as CoMP (introduced in LTE release 11)

In case of transmission modes 1–6, transmission is carried out from antenna ports 0–3. Thus the CRS are to be used for channel estimation. Transmission mode 7 corresponds to transmission on antenna port 5 while transmission modes 8–10 correspond to transmission on antenna ports 7–14 (in case of transmission mode 8 limited to antenna ports 7 and 8). Thus, for transmission modes 7–10, DM-RS are to be used for channel estimation.

In practice, devices configured for Transmission modes 1–8 can be assumed to rely on CRS to acquire CSI while, for transmission modes 9 and 10, CSI-RS should be used.

It should also be mentioned that, although a certain multi-antenna transmission scheme can be seen as being associated with a certain transmission mode, for transmission modes 3–10 there is a possibility for dynamic fallback to transmit diversity without implying that

---

[20]Actually, only single-antenna transmission and transmit diversity are specified for BCH and L1/L2 control signaling, although the EPDCCH can use non-codebook-based precoding.

the configured transmission mode is changed. One reason for this is to enable the use of smaller DCI formats when the full set of multi-antenna features associated with a certain transmission mode is not used. Another reason is to handle ambiguities about the transmission mode applied by the device during transmission mode reconfiguration as discussed in Section 6.4.5.

6.3.2 TRANSMIT DIVERSITY

Transmit diversity can be applied to any downlink physical channel. However, it is especially applicable to transmissions that cannot be adapted to time-varying channel conditions by means of link adaptation and/or channel-dependent scheduling, and thus for which diversity is more important. This includes transmission of the BCH and PCH transport channels, as well as L1/L2 control signaling. Actually, as already mentioned, transmit diversity is the *only* multi-antenna transmission scheme applicable to these channels. Transmit diversity is also used for transmission of DL-SCH when transmission mode 2 is configured. Furthermore, as also already mentioned, transmit diversity is a "fallback mode" for DL-SCH transmission when the device is configured in transmission mode 3 and higher. More specifically, a scheduling assignment using DCI format 1A (see Section 6.4.6) implies the use of transmit diversity regardless of the configured transmission mode.

Transmit diversity assumes the use of CRS for channel estimation. Thus, a transmit-diversity signal is always transmitted on the same antenna ports as the CRS (antenna ports 0−3). Actually, if a cell is configured with two CRS, transmit diversity for two antenna ports *must be used* for BCH and PCH, as well as for the L1/L2 control signaling on PDCCH.[21] Similarly, if four CRS are configured for the cell, transmit diversity for four antenna ports has to be used for the transmission of these channels. In this way, a device does not have to be explicitly informed about what multi-antenna transmission scheme is used for these channels. Rather, this is given implicitly from the number of CRS configured for a cell.[22]

6.3.2.1 Transmit Diversity for Two Antenna Ports

In the case of two antenna ports, LTE transmit diversity is based on *space-frequency block coding* (SFBC). As can be seen from Figure 6.15, SFBC implies that two consecutive modulation symbols S_i and S_{i+1} are mapped directly to frequency-adjacent resource elements on the first antenna port. On the second antenna port the frequency-swapped and transformed symbols $-S_{i+1}^*$ and S_i^* are mapped to the corresponding resource elements, where "*" denotes complex conjugate.

Figure 6.15 also indicates how the antenna ports on which a transmit-diversity signal is being transmitted correspond to the CRS, more specifically CRS 0 and CRS 1 in the case of two antenna ports. Note that one should not interpret this such that the CRS is specifically

---

[21]Note that this is not true for the EPDCCH control channel.
[22]Actually, the situation is partly the opposite—that is, the device blindly detects the number of antenna ports used for BCH transmission and, from that, decides on the number of CRS configured within the cell.

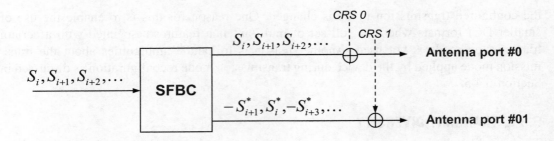

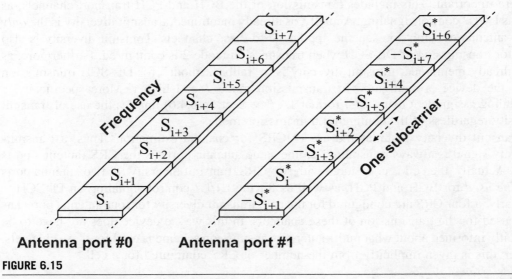

FIGURE 6.15

Transmit diversity for two antenna ports—SFBC.

transmitted for this transmit-diversity signal. There are, in practice, multiple transmissions on antenna ports 0 and 1, all of which rely on the corresponding CRS for channel estimation.

6.3.2.2 Transmit Diversity for Four Antenna Ports

In the case of four antenna ports, LTE transmit diversity is based on a combination of SFBC and *frequency-switched transmit diversity* (FSTD). As can be seen in Figure 6.16, combined SFBC/FSTD implies that pairs of modulation symbols are transmitted by means of SFBC with transmission alternating between pairs of antenna ports (antenna ports 0 and 2 and antenna ports 1 and 3, respectively). For the resource elements where transmission is on one pair of antenna ports, there is no transmission on the other pair of antenna ports. Thus, combined SFBC/FSTD in some sense operates on groups of four modulation symbols

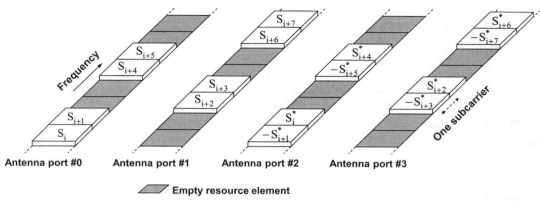

FIGURE 6.16

Transmit diversity for four antenna ports—combined SFBC/FSTD.

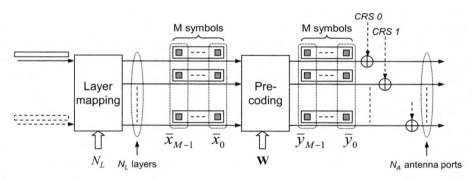

FIGURE 6.17

The basic structure of LTE codebook-based antenna precoding. The figure also indicates how CRS are applied after pre-coding.

and corresponding groups of four frequency-consecutive resource elements on each antenna port.

6.3.3 CODEBOOK-BASED PRECODING

The basic processing for codebook-based precoding is illustrated in Figure 6.17. The modulation symbols corresponding to one or two transport blocks are first mapped to N_L *layers*. The number of layers may range from a minimum of one layer up to a maximum number of layers equal to the number of antenna ports.[23] The layers are then mapped to the antenna ports

---

[23]In practice, the number of layers is also limited by, and should not exceed, the number of receive antennas available at the device.

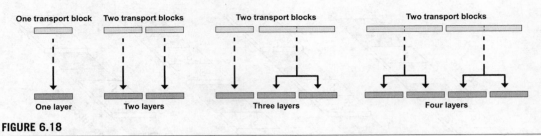

FIGURE 6.18

Transport-block-to-layer mapping for codebook-based antenna precoding (initial transmission).

by means of the *precoder*. As codebook-based precoding relies on the CRS for channel estimation, and there are at most four CRS in a cell, codebook-based precoding allows for a maximum of four antenna ports and, as a consequence, a maximum of four layers.

The mapping to layers is outlined in Figure 6.18 for the case of an initial transmission. There is one transport block in the case of a single layer ($N_L = 1$) and two transport blocks for two or more layers ($N_L > 1$). In the case of a hybrid-ARQ retransmission, if only one of two transport blocks needs to be retransmitted and that transport block was mapped to two layers for the initial transmission, the retransmission will also be carried out on two layers. Thus, in the case of a retransmission, a single transport block may also be transmitted using two layers.

The mapping to layers is such that the number of modulation symbols on each layer is the same and equal to the number of symbols to be transmitted on each antenna port. Thus, in the case of three layers, there should be twice as many modulation symbols corresponding to the second transport block (mapped to the second and third layers) compared to the first transport block (mapped to the first layer). This is ensured by the selection of an appropriate transport-block size in combination with the rate-matching functionality. In the case of four layers, the first transport block is mapped to the first and second layers while the second transport block is mapped to the third and fourth layers. In this case, the number of modulation symbols should thus be the same for the two transport blocks. For one transport block being mapped to two layers, the modulated symbols corresponding to the transport block are mapped to the layers in an alternating fashion—that is, every second modulation symbol is mapped to the first and second layer respectively.

In the case of multi-antenna precoding the number of layers is also often referred to as the *transmission rank*.[24] The transmission rank can vary dynamically, for example, based on the number of layers that can be supported by the channel. The latter is sometimes also referred to as the *channel rank*.

After layer mapping, a set of N_L symbols (one symbol from each layer) is linearly combined and mapped to the antenna ports. This combination/mapping can be described by a *precoder matrix* **W** of size $N_A \times N_L$, where N_A is the number of antenna ports which, for

---

[24]In the LTE specification, transmit diversity is actually also described as transmission using *multiple layers*. However, transmit diversity is still a *single-rank* transmission scheme.

codebook-based precoding equal two or four. More specifically, the vector $\bar{y}_i$ of size N_A, consisting of one symbol for each antenna port, is given by $\bar{y}_i = \mathbf{W} \cdot \bar{x}_i$, where the vector $\bar{x}_i$ of size N_L consists of one symbol from each layer. As the number of layers can vary dynamically, also the number of columns of the precoder matrix will vary dynamically. Specifically, in the case of a single layer, the precoder matrix $\mathbf{W}$ is a vector of size $N_A \times 1$ that provides beam-forming for a single modulation symbol.

Figure 6.17 also indicates how the CRS are applied after antenna precoding. Channel estimation based on the CRS will thus reflect the channel for each antenna port *not including the precoding*. As a consequence, the device receiver must have explicit knowledge about what precoding has been applied at the transmitter side in order to properly process the received signal and recover the different layers. Once again, the figure should not be interpreted such that CRS are inserted specifically for a given PDSCH transmission.

There are two operational modes for codebook-based precoding, *closed-loop operation* and *open-loop operation*. These two modes differ in terms of the exact structure of the precoder matrix and how the matrix is selected by the network and made known to the device.

6.3.3.1 Closed-Loop Precoding

In case of closed-loop precoding it is assumed that the network selects the precoder matrix based on feedback from the device. As already mentioned, closed-loop precoding is associated with transmission mode 4.

Based on measurements on the CRS, the device selects a suitable transmission rank and corresponding precoder matrix. Information about the selected rank and precoder matrix is then reported to the network in the form of a *Rank Indicator* (RI) and a *Precoder-Matrix Indicator* (PMI), as described in Chapter 10. It is important to understand though that the RI and PMI are only recommendations and the network does not need to follow the RI/PMI provided by the device when selecting the actual transmission rank and precoder matrix to be used for transmission to the device. When not following the device recommendation, the network must explicitly inform the device what precoder matrix is used for the downlink transmission. On the other hand, if the network uses the precoder matrix recommended by the device, only a confirmation that the network is using the recommended matrix is signaled.

To limit the signaling on both uplink and downlink only a limited set of precoder matrices, also referred to as the *codebook*, is defined for each transmission rank for a given number of antenna ports. Both the device (when reporting PMI) and the network (when selecting the actual precoder matrix to use for the subsequent downlink transmission to the device) should select a precoder matrix from the corresponding codebook. Thus, for device PMI reporting, as well as when the network informs the device about the actual precoder matrix used for the downlink transmission, only the index of the selected matrix needs to be signaled.

As LTE supports codebook-based precoding for two and four antenna ports, codebooks are defined for:

- Two antenna ports and one and two layers, corresponding to precoder matrices of size 2×1 and 2×2, respectively.
- Four antenna ports and one, two, three, and four layers, corresponding to precoder matrices of size 4×1, 4×2, 4×3, and 4×4, respectively.

As an example, the precoder matrices specified for the case of two antenna ports are illustrated in Table 6.3. As can be seen, there are four 2×1 precoder matrices for single-layer transmission and three 2×2 precoder matrices for two-layer transmission. In the same way, sets of 4×1, 4×2, 4×3, and 4×4 matrices are defined for the case of four antenna ports and one, two, three, and four layers, respectively. It should be pointed out that the first rank-2 (2×2) matrix in Table 6.3 is not used in closed-loop operation but only for *open-loop precoding*, as described in the next section.

Even if the network is following the precoder-matrix recommendation provided by the device, the network may, for different reasons, decide to use a lower rank for the transmission, so-called *rank override*. In that case the network will use a subset of the columns of the recommended precoder matrix. The network precoder confirmation will then include explicit information about the set of columns being used or, equivalently, about the set of layers being transmitted.

There is also a possibility to apply closed-loop precoding strictly limited to single-layer (rank 1) transmission. This kind of multi-antenna transmission is associated with transmission mode 6. The reason for defining an additional transmission mode limited to single-layer transmission rather than relying on the general closed-loop precoding associated with transmission mode 4 is that, by strictly limiting to single-layer transmission, the signaling overhead on both downlink and uplink can be reduced. Transmission mode 6 can, for example, be configured for devices with low SINR for which multi-layer transmission would not apply in order to harvest the beam-forming gain.

6.3.3.2 Open-Loop Precoding

Open-loop precoding does not rely on any detailed precoder recommendation being reported by the device and does not require any explicit network signaling of the actual precoder used

Table 6.3 Precoder Matrices for Two Antenna Ports and One and Two Layers. The First 2 × 2 Matrix is Only used for Open-Loop Precoding

One layer	$\frac{1}{\sqrt{2}}\begin{bmatrix} +1 \\ +1 \end{bmatrix}$	$\frac{1}{\sqrt{2}}\begin{bmatrix} +1 \\ -1 \end{bmatrix}$	$\frac{1}{\sqrt{2}}\begin{bmatrix} +1 \\ +j \end{bmatrix}$	$\frac{1}{\sqrt{2}}\begin{bmatrix} +1 \\ -j \end{bmatrix}$
Two layers	$\frac{1}{\sqrt{2}}\begin{bmatrix} +1 & 0 \\ 0 & +1 \end{bmatrix}$	$\frac{1}{2}\begin{bmatrix} +1 & +1 \\ +1 & -1 \end{bmatrix}$	$\frac{1}{2}\begin{bmatrix} +1 & +1 \\ +j & -j \end{bmatrix}$	

for the downlink transmission. Instead, the precoder matrix is selected in a predefined and deterministic way known to the device in advance. One use of open-loop precoding is in high-mobility scenarios where accurate feedback is difficult to achieve due to the latency in the PMI reporting. As already mentioned, open-loop precoding is associated with transmission mode 3.

The basic transmission structure for open-loop precoding is aligned with the general codebook-based precoding outlined in Figure 6.17 and only differs from closed-loop precoding in the structure of the precoding matrix $\mathbf{W}$.

In the case of open-loop precoding, the precoder matrix can be described as the product of two matrices $\mathbf{W}'$ and $\mathbf{P}$, where $\mathbf{W}'$ and $\mathbf{P}$ are of size $N_A \times N_L$ and $N_L \times N_L$, respectively:

$$\mathbf{W} = \mathbf{W}' \cdot \mathbf{P} \tag{6.1}$$

In the case of two antenna ports, the matrix $\mathbf{W}'$ is the normalized 2×2 identity matrix:[25]

$$\mathbf{W}' = \frac{1}{\sqrt{2}} \begin{bmatrix} +1 & 0 \\ 0 & +1 \end{bmatrix} \tag{6.2}$$

In the case of four antenna ports, $\mathbf{W}'$ is given by cycling through four of the defined $4 \times N_L$ precoder matrices and is different for consecutive resource elements.

The matrix $\mathbf{P}$ can be expressed as $\mathbf{P} = \mathbf{D}_i \neq \mathbf{U}$, where $\mathbf{U}$ is a constant matrix of size $N_L \times N_L$ and $\mathbf{D}_i$ is a matrix of size $N_L \times N_L$ that varies between subcarriers (indicated by the index i). As an example, the matrices $\mathbf{U}$ and $\mathbf{D}_i$ for the case of two layers ($N_L = 2$) are given by:

$$\mathbf{U} = \frac{1}{\sqrt{2}} \begin{bmatrix} 1 & 1 \\ 1 & e^{-j2\pi/2} \end{bmatrix} \quad \mathbf{D}_i = \begin{bmatrix} 1 & 0 \\ 0 & e^{-j2\pi i/2} \end{bmatrix} \tag{6.3}$$

The basic idea with the matrix $\mathbf{P}$ is to average out any differences in the channel conditions as seen by the different layers.

Similar to closed-loop precoding, the transmission rank for open-loop precoding can also vary dynamically down to a minimum of two layers. Transmission mode 3, associated with open-loop precoding, also allows for rank-1 transmission. In that case, transmit diversity as described in Section 6.3.2 is used—that is, SFBC for two antenna ports and combined SFBC/FSTD for four antenna ports.

6.3.4 NON-CODEBOOK-BASED PRECODING

Similar to codebook-based precoding, non-codebook-based precoding is only applicable to DL-SCH transmission. Non-codebook-based precoding was introduced in LTE release 9 but was then limited to a maximum of two layers. The extension to eight layers was then introduced as part of release 10. The release 9 scheme, associated with transmission mode 8, is a subset of the extended release 10 scheme (transmission mode 9, later extended to transmission mode 10).

---

[25]As non-codebook-based precoding is not used for rank-1 transmission (see later), there is no need for any matrix $\mathbf{W}'$ of size 2×1.

There is also a release 8 non-codebook-based precoding defined, associated with transmission mode 7. Transmission mode 7 relies on the release 8 DM-RS mentioned but not described in detail in Section 6.2.2 and only supports single-layer transmission. In this description we will focus on non-codebook-based precoding corresponding to transmission modes 8–10.

The basic principles for non-codebook-based precoding can be explained based on Figure 6.19 (where the precoder is intentionally shaded; see later). As can be seen, this figure is very similar to the corresponding figure illustrating codebook-based precoding (Figure 6.17), with layer mapping of modulation symbols corresponding to one or two transport blocks followed by precoding. The layer mapping also follows the same principles as that of codebook-based precoding (see Figure 6.18) but is extended to support up to eight layers. In particular, at least for an initial transmission, there are two transport blocks per TTI except for the case of a single layer, in which case there is only one transport block within the TTI. Similar to codebook-based precoding, for hybrid-ARQ retransmissions there may in some cases be a single transport block also in the case of multi-layer transmission.

The main difference in Figure 6.19 compared to Figure 6.17 (codebook-based precoding) is the presence of DM-RS before the precoding. The transmission of precoded reference signals allows for demodulation and recovery of the transmitted layers at the receiver side *without explicit receiver knowledge of the precoding applied at the transmitter side*. Put simply, channel estimation based on precoded DM-RS will reflect the channel experienced by the layers, *including the precoding*, and can thus be used directly for coherent demodulation of the different layers. There is no need to signal any precoder-matrix information to the device, which only needs to know the number of layers—that is, the transmission rank. As a consequence, the network can select an arbitrary precoder and there is no need for any explicit codebook to select from. This is the reason for the term *non-codebook-based* precoding. It should be noted though that non-codebook-based precoding may still rely on codebooks for the device feedback, as described below.

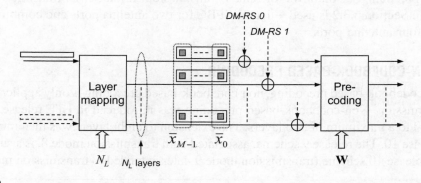

FIGURE 6.19

Basic principles for non-codebook-based antenna precoding.

The possibility to select an arbitrary precoder matrix for the transmission is also the reason why the precoder is shaded in Figure 6.19. The precoder part of Figure 6.19 is not visible in the LTE specification and, strictly speaking, in the case of non-codebook-based precoding the antenna mapping defined according to Figure 6.14 consists of only the layer mapping. This also means that the antenna ports defined in Figure 6.14 correspond to the different layers in Figure 6.19, or, expressed differently, precoding occurs *after* the antenna ports.

Still, there must be a way for the network to select a suitable precoder matrix for the transmission. There are essentially two ways by which this can be done in the case of non-codebook-based precoding.

The network may estimate the uplink channel state, for example based on transmission of uplink sounding reference signals as described in the next chapter, and rely on uplink/downlink channel reciprocity when selecting the precoder matrix to use for the downlink transmission. This is especially of interest for TDD operation for which the use of the same frequency for uplink and downlink transmission typically leads to a higher degree of downlink/uplink channel reciprocity. It should be noted though that if the device uses multiple receive antennas, it also has to transmit on multiple antennas in order for uplink measurements to fully reflect the downlink channel state.

Alternatively, the network may rely on device feedback for precoder-matrix selection. For transmission modes 8—10, this feedback is actually very similar to the corresponding feedback for closed-loop codebook-based precoding, see Chapter 10. Furthermore, for transmission mode 9, the device measurements should be based on CSI-RS, as described in Section 6.2.3, rather than the CRS.

Thus, despite the name, non-codebook-based precoding may also use defined codebooks. However, in contrast to codebook-based precoding, the codebooks are then only used for the device CSI reporting and not for the actual downlink transmission.

6.3.5 DOWNLINK MU-MIMO

Spatial multiplexing implies the transmission of multiple layers—that is, multiple parallel transmissions on the same time—frequency resource—to the same device. The presence of multiple antennas at both the transmitter and receiver sides in combination with transmitter and/or receiver signal processing is then used to suppress interference between the different layers.

Spatial multiplexing has often been referred to as MIMO transmission, reflecting the fact that the channel in the case of spatial multiplexing can be seen as having *multiple inputs*, corresponding to the multiple transmit antennas, and *multiple outputs*, corresponding to the multiple receive antennas. The more specific term *single-user MIMO* (SU-MIMO) is also often used for reasons that will be more clear below.[26]

---

[26]The term MIMO is then rather used to refer to any transmission using multiple transmit antennas and multiple receive antennas—that is, not only limited to spatial multiplexing but also, for example, single-layer precoding to get beam-forming gains.

The term MU-MIMO is, in 3GPP, used to denote transmission to *different* devices using *the same time—frequency resource*, in practice relying on multiple antennas at the transmitter (network) side to separate the two transmissions.

In principle, one could realize MU-MIMO as a direct extension to spatial multiplexing, with the different layers simply being intended for different devices. The set of devices would demodulate and decode the full set of layers in the same way as for SU-MIMO. The data on a layer not intended for a specific device would then just be discarded by that device after demodulation/decoding.

However, such an approach would imply that all devices involved in the MU-MIMO transmission would need to know about the full set of layers being transmitted. It would also imply that one would need to have exactly the same resource assignment—that is, transmission on the same set of resource blocks, for all devices involved in the MU-MIMO transmission. All devices would also need to include the full set of receive antennas necessary to receive the overall multi-layer transmission.

Instead, 3GPP has chosen an MU-MIMO approach that does not require device knowledge about the presence of the other transmissions, allows for only partly overlapping resource assignments, and, at least in principle, does not require the presence of multiple receive antennas at the mobile device.[27] There are two approaches to MU-MIMO with explicit support in the LTE specifications, one being an integrated part of transmission modes 8, 9, and 10 corresponding to non-codebook-based precoding, and one being based on codebook-based precoding but associated with a special transmission mode, *transmission mode 5*.

6.3.5.1 MU-MIMO within Transmission Modes 8/9

In principle, MU-MIMO based on transmission modes 8—10 is straightforward. Based on feedback of CSI from devices within the cell, the base station selects two or more devices to transmit to, using the same set of time-frequency resources. Non-codebook-based precoding for one, or in some cases even multiple layers, is then applied to each transmission in such way that they are spatially separated at the receiver (device) sides.

The spatial separation at the device side will typically not be perfect. To enhance channel estimation at the device side it is therefore preferred to use different DM-RS for the different transmissions in order to improve channel estimation.[28]

As discussed in Section 6.2.2.1 there are two methods by which reference-signal sequences can be assigned to downlink DM-RS:

- Cell-specific assignment supported from LTE release 9, that is, the release in which transmission mode 8 and DM-RS were introduced.
- Fully device-specific assignment supported from LTE release 11.

---

[27]Note, though, that the LTE performance requirements in general assume the presence of at least two receive antennas at the mobile device.

[28]As discussed in Section 6.3.1, transmission mode 8 and 9 are assumed to rely on DM-RS for channel estimation.

As also described in Section 6.2.2.1, in case of cell-specific assignment there is the possibility to dynamically select between two different cell-specific reference-signal sequences.

With pre-release 11 (cell-specific) DM-RS sequences there is the possibility for up to four different DM-RS supporting MU-MIMO between up to four different devices:

- In case of single-layer transmission, the network can explicitly signal on which of antenna ports 7 and 8, corresponding to the two OCC, a transmission is carried out. This, in combination with the possibility to dynamically select between two reference-signal sequences, allows for up to four different DM-RS and a corresponding possibility for single-layer MU-MIMO transmission to up to four devices in parallel.
- In case of dual-layer transmission (on antenna ports 7 and 8), the possibility to dynamically select between two reference-signal sequences allows for MU-MIMO transmission to up to two devices in parallel.
- In case of more than two layers, there is no signaling support for selecting reference-signal sequence and thus no pre-release 11 support for MU-MIMO.

Note that one can also do MU-MIMO for single-layer and dual-layer transmission in parallel. There could, for example, be one two-layer transmission using one of the two reference-signal sequences and up to two one-layer transmissions using the other reference-signal sequence, separated by means of the two different OCCs.

However, with the release 11 introduction of device-specific assignment of reference-signal sequences as described in Section 6.2.2.1, MU-MIMO can, at least in principle, be applied to an arbitrary number of devices regardless of the number of layers. In a normal, for example, macro deployment, the number of devices for which MU-MIMO can jointly be carried out is limited by the number of transmitter-side antennas and, in practice, the possibility for MU-MIMO for up to four devices in parallel as supported prior to release 11 is typically sufficient. However, the possibility for MU-MIMO between significantly more devices is important in specific scenarios, especially in so-called shared-cell heterogeneous deployments, see further Chapter 14.

6.3.5.2 MU-MIMO Based on CRS

The MU-MIMO transmission described in the preceding section is part of transmission modes 8, 9, and 10 and thus became available in LTE release 9, with further extension in subsequent releases. However, already in LTE release 8, MU-MIMO was possible by a minor modification of transmission mode 4—that is, closed-loop codebook-based beam-forming, leading to *transmission mode 5*. The only difference between transmission modes 4 and 5 is the signaling of an additional power offset between PDSCH and the CRS.

In general, for transmission modes relying on CRS (as well as when relying on DM-RS) for channel estimation the device will use the reference signal as a phase reference and also as a power/amplitude reference for the demodulation of signals transmitted by means of higher-order modulation (16QAM and 64QAM). Thus, for proper demodulation of higher-order modulation, the device needs to know the power offset between the CRS and the PDSCH.

The device is informed about this power offset by means of higher-layer signaling. However, what is then provided is the offset between CRS power and the overall PDSCH power, including all layers. In the case of spatial multiplexing, the overall PDSCH power has to be divided between the different layers, and it is the relation between the CRS power and the per-layer PDSCH power that is relevant for demodulation.

In the case of pure spatial multiplexing (no MU-MIMO)—that is, transmission modes 3 and 4—the device knows about the number of layers and thus, indirectly, about the offset between the CRS power and the per-layer PDSCH power.

In the case of MU-MIMO, the total available power will typically also be divided between the transmissions to the different devices, with less PDSCH power being available for each transmission. However, devices are not aware of the presence of parallel transmissions to other devices and are thus not aware of any per-PDSCH power reduction. For this reason, transmission mode 5 includes the explicit signaling of an *additional power offset* of −3 dB to be used by the device in addition to the CRS/PDSCH power offset signaled by higher layers.

Transmission mode 5 is limited to single-rank transmission and, in practice, limited to two users being scheduled in parallel, as there is only a single −3 dB offset defined.

Note that the power-offset signaling is not needed for MU-MIMO based on DM-RS as, in this case, each transmission has its own set of reference signals. The power per reference signal will thus scale with the number of layers and transmissions, similar to the PDSCH power, and the reference signal to per-layer PDSCH power ratio will remain constant.

6.4 DOWNLINK L1/L2 CONTROL SIGNALING

To support the transmission of downlink and uplink transport channels, there is a need for certain *associated downlink control signaling*. This control signaling is often referred to as *downlink L1/L2 control signaling*, indicating that the corresponding information partly originates from the physical layer (Layer 1) and partly from Layer 2 MAC. Downlink L1/L2 control signaling consists of downlink scheduling assignments, including information required for the device to be able to properly receive, demodulate, and decode the DL-SCH[29] on a component carrier, uplink scheduling grants informing the device about the resources and transport format to use for uplink (UL-SCH) transmission, and hybrid-ARQ acknowledgments in response to UL-SCH transmissions. In addition, the downlink control signaling can also be used for the transmission of power-control commands for power control of uplink physical channels, as well as for certain special purposes such as MBSFN notifications.

The basic time—frequency structure for transmission of L1/L2 control signaling is illustrated in Figure 6.20 with control signaling being located at the beginning of each subframe

---

[29]L1/L2 control signaling is also needed for the reception, demodulation, and decoding of the PCH transport channel.

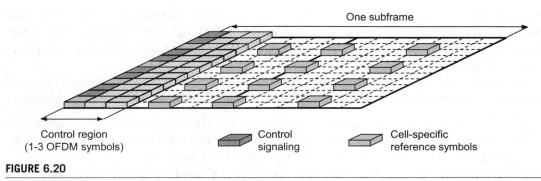

One subframe

Control region
(1-3 OFDM symbols)

Control
signaling

Cell-specific
reference symbols

FIGURE 6.20

LTE time—frequency grid illustrating the split of the subframe into (variable-sized) control and data regions.

and spanning the full downlink carrier bandwidth. Each subframe can therefore be said to be divided into a *control region* followed by a *data region*, where the control region corresponds to the part of the subframe in which the L1/L2 control signaling is transmitted. Starting from release 11, there is also a possibility to locate parts of the L1/L2 control signaling in the data region as described later. However, the split of a subframe into a control region and a data region still applies.

To simplify the overall design, the control region always occupies an integer number of OFDM symbols, more specifically one, two, or three OFDM symbols (for narrow cell bandwidths, 10 resource blocks or less, the control region consists of two, three, or four OFDM symbols to allow for a sufficient amount of control signaling).

The size of the control region expressed in number of OFDM symbols, or, equivalently, the start of the data region, can be dynamically varied on a per-subframe basis. Thus, the amount of radio resources used for control signaling can be dynamically adjusted to match the instantaneous traffic situation. For a small number of users being scheduled in a subframe, the required amount of control signaling is small and a larger part of the subframe can be used for data transmission (larger data region).

The maximum size of the control region is normally three OFDM symbols (four in the case of narrow cell bandwidths), as mentioned in the preceding paragraphs. However, there are a few exceptions to this rule. When operating in TDD mode, the control region in subframes one and six is restricted to at most two OFDM symbols since, for TDD, the primary synchronization signal (see Chapter 11) occupies the third OFDM symbol in those subframes. Similarly, for MBSFN subframes (see Chapter 5), the control region is restricted to a maximum of two OFDM symbols.

The reason for transmitting the control signaling at the beginning of the subframe is to allow for devices to decode downlink scheduling assignments as early as possible. Processing of the data region—that is, demodulation and decoding of the DL-SCH transmission—can then begin before the end of the subframe. This reduces the delay in the DL-SCH decoding and thus the overall downlink transmission delay. Furthermore, by transmitting the L1/L2 control channel at the beginning of the subframe—that is, by allowing for early decoding of the L1/L2 control information—devices that are not scheduled in the subframe may power down the receiver circuitry for a part of the subframe, allowing for reduced device power consumption.

The downlink L1/L2 control signaling consists of six different physical-channel types, all located in the control region with two exceptions located in the data region of a subframe:

- The *physical control format indicator channel* (PCFICH), informing the device about the size of the control region (one, two, or three OFDM symbols). There is one and only one PCFICH on each component carrier or, equivalently, in each cell.
- The *physical hybrid-ARQ indicator channel* (PHICH), used to signal hybrid-ARQ acknowledgments in response to uplink UL-SCH transmissions. Multiple PHICHs can exist in each cell.
- The *physical downlink control channel* (PDCCH), used to signal downlink scheduling assignments, uplink scheduling grants, or power-control commands. Each PDCCH typically carries signaling for a single device, but can also be used to address a group of devices. Multiple PDCCHs can exist in each cell.
- The *enhanced physical downlink control channel* (EPDCCH), introduced in release 11 to support DM-RS-based control signaling reception and carrying similar types of information as the PDCCH. However, in contrast to the PDCCH, the EPDCCH is located in the data region. Also, the EPDCCH can be subject to non-codebook-based precoding.
- The *MTC physical downlink control channel* (MPDCCH), introduced in release 13 as part of the improved MTC support, see Chapter 20. In essence it is a variant of the EPDCCH.
- The *relay physical downlink control channel* (R-PDCCH), introduced in release 10 to support relaying. A detailed discussion can be found in Chapter 18 in conjunction with the overall description of relays; at this stage it suffices to note that the R-PDCCH is transmitted in the data region.

In the following sections, the PCFICH, PHICH, PDCCH, and EPDCCH are described in detail while the MPDCCH and the R-PDCCH are described in Chapters 20 and 18, respectively.

6.4.1 PHYSICAL CONTROL FORMAT INDICATOR CHANNEL

The PCFICH indicates the instantaneous size of the control region in terms of the number of OFDM symbols—that is, indirectly where in the subframe the data region starts. Correct decoding of the PCFICH information is thus essential. If the PCFICH is incorrectly decoded, the device will neither know how to process the control channels nor where the data region starts for the corresponding subframe.[30] The PCFICH consists of two bits of information,

---

[30]Theoretically, the device could blindly try to decode all possible control channel formats and, from which format that was correctly decoded, deduce the starting position of the data region, but this can be a very complex procedure.

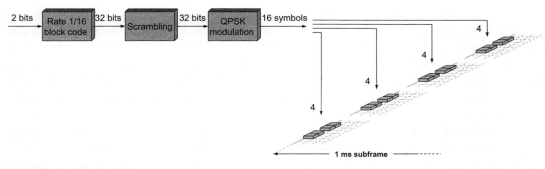

FIGURE 6.21

Overview of the PCFICH processing.

corresponding to the three[31] control-region sizes of one, two, or three OFDM symbols (two, three, or four for narrow bandwidths), which are coded into a 32-bit codeword. The coded bits are scrambled with a cell- and subframe-specific scrambling code to randomize inter-cell interference, QPSK modulated, and mapped to 16 resource elements. As the size of the control region is unknown until the PCFICH is decoded, the PCFICH is always mapped to the first OFDM symbol of each subframe.

The mapping of the PCFICH to resource elements in the first OFDM symbol in the subframe is done in groups of four resource elements, with the four groups being well separated in frequency to obtain good diversity. Furthermore, to avoid collisions between PCFICH transmissions in neighboring cells, the location of the four groups in the frequency domain depends on the physical-layer cell identity.

The transmission power of the PCFICH is under control of the eNodeB. If necessary for coverage in a certain cell, the power of the PCFICH can be set higher than for other channels by "borrowing" power from, for example, simultaneously transmitted PDCCHs. Obviously, increasing the power of the PCFICH to improve the performance in an interference-limited system depends on the neighboring cells not increasing their transmit power on the inter-fering resource elements. Otherwise, the interference would increase as much as the signal power, implying no gain in received SIR. However, as the PCFICH-to-resource-element mapping depends on the cell identity, the probability of (partial) collisions with PCFICH in neighboring cells in synchronized networks is reduced, thereby improving the performance of PCFICH power boosting as a tool to control the error rate.

The overall PCFICH processing is illustrated in Figure 6.21.

To describe the mapping of the PCFICH, and L1/L2 control signaling in general, to resource elements, some terminology is required. As previously mentioned, the mapping is specified in terms of groups of four resource elements, so-called *resource-element groups*. To

---

[31]The fourth combination is reserved for future use.

each resource-element group, a *symbol quadruplet* consisting of four (QPSK) symbols is mapped. The main motivation behind this, instead of simply mapping the symbols one by one, is the support of transmit diversity. As discussed in Section 6.3, transmit diversity with up to four antenna ports is specified for L1/L2 control signaling. Transmit diversity for four antenna ports is specified in terms of groups of four symbols (resource elements) and, consequently, the L1/L2 control-channel processing is also defined in terms of symbol quadruplets.

The definition of the resource-element groups assumes that reference symbols corresponding to two antenna ports are present in the first OFDM symbol, regardless of the actual number of antenna ports configured in the cell. This simplifies the definition and reduces the number of different structures to handle. Thus, as illustrated in Figure 6.22, in the first OFDM symbol there are two resource-element groups per resource block, as every third resource element is reserved for reference signals (or nonused resource elements corresponding to reference symbols on the other antenna port). As also illustrated in Figure 6.22, in the second OFDM symbol (if part of the control region) there are two or three resource-element groups depending on the number of antenna ports configured. Finally, in the third OFDM symbol (if part of the control region) there are always three resource-element groups per resource block. Figure 6.22 also illustrates how resource-element groups are numbered in a time-first manner within the size of the control region.

Returning to the PCFICH, four resource-element groups are used for the transmission of the 16 QPSK symbols. To obtain good frequency diversity the resource-element groups should be well spread in frequency and cover the full downlink cell bandwidth. Therefore, the four resource-element groups are separated by one-fourth of the downlink cell bandwidth in the frequency domain, with the starting position given by physical-layer cell identity. This is illustrated in Figure 6.23, where the PCFICH mapping to the first OFDM symbol in a sub-frame is shown for three different physical-layer cell identities in the case of a downlink cell bandwidth of eight resource blocks. As seen in the figure, the PCFICH mapping depends on the physical-layer cell identity to reduce the risk of inter-cell PCFICH collisions. The cell-specific shifts of the reference symbols, described in Section 6.2.1, are also seen in the figure.

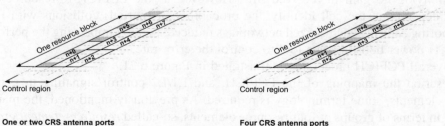

One or two CRS antenna ports   Four CRS antenna ports

FIGURE 6.22

Numbering of resource-element groups in the control region (assuming a size of three OFDM symbols).

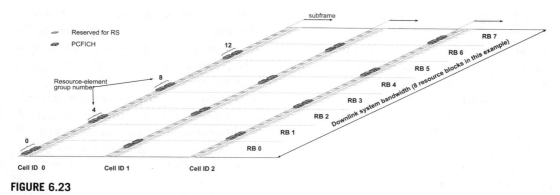

FIGURE 6.23

Example of PCFICH mapping in the first OFDM symbol for three different physical-layer cell identities.

6.4.2 PHYSICAL HYBRID-ARQ INDICATOR CHANNEL

The PHICH is used for transmission of hybrid-ARQ acknowledgments in response to UL-SCH transmission. In essence the PHICH is a one-bit scheduling grant commanding a retransmission on the UL-SCH. There is one PHICH transmitted per received transport block and TTI—that is, when uplink spatial multiplexing is used on a component carrier, two PHICHs are used to acknowledge the transmission, one per transport block.

For proper operation of the hybrid-ARQ protocol and to avoid spurious uplink transmissions, the error rate of the PHICH should be sufficiently low. The operating point of the PHICH is not specified and is up to the network operator to decide upon, but typically ACK-to-NAK and NAK-to-ACK error rates of the order of 10^{-2} and 10^{-3}–10^{-4}, respectively, are targeted. The reason for the asymmetric error rates is that an NAK-to-ACK error would imply loss of a transport block at the MAC level, a loss that has to be recovered by RLC retransmissions with the associated delays, while an ACK-to-NAK error only implies an unnecessary retransmission of an already correctly decoded transport block. To meet these error-rate targets without excessive power, it is beneficial to control the PHICH transmission power as a function of the radio-channel quality of the device to which the PHICH is directed. This has influenced the design of the PHICH structure.

In principle, a PHICH could be mapped to a set of resource elements exclusively used by this PHICH. However, taking the dynamic PHICH power setting into account, this could result in significant variations in transmission power between resource elements, which can be challenging from an RF implementation perspective. Therefore, it is preferable to spread each PHICH on multiple resource elements to reduce the power differences while at the same time providing the energy necessary for accurate reception. To fulfill this, a structure where several PHICHs are code multiplexed on to a set of resource elements is used in LTE. The hybrid-ARQ acknowledgment (one single bit of information per transport block) is repeated three times, followed by BPSK modulation on either the I or the Q branch and spreading with a length-four orthogonal sequence. A set of PHICHs transmitted on the same set of resource elements is called a PHICH group, where a PHICH group consists of eight PHICHs in the case of normal cyclic prefix. An individual PHICH can thus be uniquely represented by a single number from which the number of the PHICH group, the number of the orthogonal sequence within the group, and the branch, I or Q, can be derived.

For extended cyclic prefix, which is typically used in highly time-dispersive environments, the radio channel may not be flat over the frequency spanned by a length-four sequence. A nonflat channel would negatively impact the orthogonality between the sequences. Hence, for extended cyclic prefix, orthogonal sequences of length two are used for spreading, implying only four PHICHs per PHICH group. However, the general structure remains the same as for the normal cyclic prefix.

After forming the composite signal representing the PHICHs in a group, cell-specific scrambling is applied and the 12 scrambled symbols are mapped to three resource-element groups. Similarly to the other L1/L2 control channels, the mapping is described using resource-element groups to be compatible with the transmit diversity schemes defined for LTE.

The overall PHICH processing is illustrated in Figure 6.24.

The requirements on the mapping of PHICH groups to resource elements are similar to those for the PCFICH, namely to obtain good frequency diversity and to avoid collisions between neighboring cells in synchronized networks. Hence, each PHICH group is mapped to three resource-element groups, separated by about one-third of the downlink cell bandwidth. In the first OFDM symbol in the control region, resources are first allocated to the PCFICH, the PHICHs are mapped to resource elements not used by the PCFICH, and finally, as will be discussed later, the PDCCHs are mapped to the remaining resource elements.

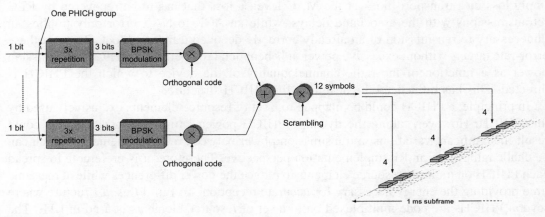

FIGURE 6.24

PHICH structure.

Typically, the PHICH is transmitted in the first OFDM symbol only, which allows the device to attempt to decode the PHICH even if it failed decoding of the PCFICH. This is advantageous as the error requirements on the PHICH typically are stricter than for PCFICH. However, in some propagation environments, having a PHICH duration of a single OFDM symbol would unnecessarily restrict the coverage. To alleviate this, it is possible to semi-statically configure a PHICH duration of three OFDM symbols. In this case, the control region is three OFDM symbols long in all subframes to fulfill the general principle of separating the control region from data in the time domain only. The value transmitted on the PCFICH will be fixed (and can be ignored) in this case. For narrow bandwidths, where the control region can be up to four OFDM symbols long, there is still a need to use the PCFICH to differentiate between a size-three and size-four control region.

The PHICH configuration is part of the system information transmitted on the PBCH; one bit indicates whether the duration is one or three OFDM symbols and two bits indicate the amount of resources in the control region reserved for PHICHs, expressed as a fraction of the downlink cell bandwidth in terms of resource blocks. Having the possibility to configure the amount of PHICH resources is useful as the PHICH capacity depends on, for example, whether the network uses MU-MIMO or not. The PHICH configuration must reside on the PBCH, as it needs to be known in order to properly process the PDCCHs for reception of the part of the system information on the DL-SCH. For TDD, the PHICH information provided on the PBCH is not sufficient for the device to know the exact set of resources used by PHICH, as there is also a dependency on the uplink–downlink allocation, provided as part of the system information transmitted on the PDSCH. In order to receive the system information on the DL-SCH, which contains the uplink–downlink allocation, the device therefore has to blindly process the PDCCHs under different PHICH configuration hypotheses.

In order to minimize the overhead and not introduce any additional signaling in the uplink grants, the PHICH that the device will expect the hybrid-ARQ acknowledgment upon is derived from the number of the first resource block upon which the corresponding uplink PUSCH transmission occurred. This principle is also compatible with semi-persistently scheduled transmission (see Chapter 9) as well as retransmissions. In addition, the resources used for a particular PHICH further depend on the reference-signal phase rotation signaled as part of the uplink grant (see Section 6.4.7). In this way, multiple devices scheduled on the same set of resources using MU-MIMO will use different PHICH resources as their reference signals are assigned different phase rotations through the corresponding field in the uplink grant. For spatial multiplexing, where two PHICH resources are needed, the second PHICH uses the same principle as the first, but to ensure that different PHICHs are used for the two transport blocks, the resource for the second PHICH is derived not from the first but from the second resource block upon which the PUSCH was transmitted.[32]

---

[32]In essence, this implies that uplink spatial multiplexing must use at least two resource blocks in the frequency domain.

6.4.3 PHYSICAL DOWNLINK CONTROL CHANNEL

The PDCCH is used to carry downlink control information (DCI) such as scheduling decisions and power-control commands. More specifically, the DCI can include:

- Downlink scheduling assignments, including PDSCH resource indication, transport format, hybrid-ARQ information, and control information related to spatial multiplexing (if applicable). A downlink scheduling assignment also includes a command for power control of the PUCCH used for transmission of hybrid-ARQ acknowledgments in response to downlink scheduling assignments.
- Uplink scheduling grants, including PUSCH resource indication, transport format, and hybrid-ARQ-related information. An uplink scheduling grant also includes a command for power control of the PUSCH.
- Power-control commands for a set of devices as a complement to the commands included in the scheduling assignments/grants.
- Control information related to sidelink operation as described in Chapter 21.
- Control information to support eMTC devices as described in Chapter 20.

The different types of control information, both between the groups and within the groups, correspond to different DCI message sizes. For example, supporting spatial multiplexing with noncontiguous allocation of resource blocks in the frequency domain requires a larger scheduling message in comparison with an uplink grant allowing for frequency-contiguous allocations only. The DCI is therefore categorized into different *DCI formats*, where a format corresponds to a certain message size and usage. The DCI formats are summarized in Table 6.4, including the size for an example of 20 MHz FDD operation with 2 Tx antennas at the base station and no carrier aggregation. The actual message size depends on, among other factors, the cell bandwidth as, for larger bandwidths, a larger number of bits is required to indicate the resource-block allocation. The number of CRS in the cell and whether cross-carrier scheduling is configured or not will also affect the absolute size of most DCI formats. Hence, a given DCI format may have different sizes depending on the overall configuration of the cell. This will be discussed later; at this stage it suffices to note that formats 0, 1A, 3, and 3A have the same message size.[33]

One PDCCH carries one DCI message with one of the formats mentioned earlier. To support different radio-channel conditions, link adaptation can be used, where the code rate (and transmission power) of the PDCCH is selected to match the radio-channel conditions. As multiple devices can be scheduled simultaneously, on both downlink and uplink, there must be a possibility to transmit multiple scheduling messages within each subframe. Each scheduling message is transmitted on a separate PDCCH, and consequently there are typically multiple simultaneous PDCCH transmissions within each cell. A device may also

---

[33]The smaller of DCI formats 0 and 1A is padded to ensure the same payload size. Which of 0 and 1A is padded depends on the uplink and downlink cell bandwidths; in the case of identical uplink and downlink bandwidths in a cell there is a single bit of padding in format 0.

Table 6.4 DCI Formats

	DCI Format	Example Size (Bits)	Usage
Uplink	0	45	Uplink scheduling grant
	4	53	Uplink scheduling grant with spatial multiplexing
	6-0A, 6-0B	46, 36	Uplink scheduling grant for eMTC devices (see Chapter 20)
Downlink	1C	31	Special purpose compact assignment
	1A	45	Contiguous allocations only
	1B	46	Codebook-based beam-forming using CRS
	1D	46	MU-MIMO using CRS
	1	55	Flexible allocations
	2A	64	Open-loop spatial multiplexing using CRS
	2B	64	Dual-layer transmission using DM-RS (TM8)
	2C	66	Multi-layer transmission using DM-RS (TM9)
	2D	68	Multi-layer transmission using DM-RS (TM10)
	2	67	Closed-loop spatial multiplexing using CRS
	6-1A, 6-1B	46, 36	Downlink scheduling grants for eMTC devices (see Chapter 20)
Special	3, 3A	45	Power control commands
	5		Sidelink operation (see Chapter 21)
	6-2		Paging/direct indication for eMTC devices (see Chapter 20)

receive *multiple* DCI messages in the same subframe (on different PDCCHs), for example if it is scheduled simultaneously in uplink and downlink.

For carrier aggregation, scheduling assignments/grants are transmitted individually per component carrier as further elaborated upon in Chapter 12.

Having introduced the concept of DCI formats, the transmission of the DCI message on a PDCCH can be described. The processing of downlink control signaling is illustrated in Figure 6.25. A CRC is attached to each DCI message payload. The identity of the device (or devices) addressed—that is, the *Radio Network Temporary Identifier* (RNTI)—is included in the CRC calculation and not explicitly transmitted. Depending on the purpose of the DCI message (unicast data transmission, power-control command, random-access response, etc.), different RNTIs are used; for normal unicast data transmission, the device-specific C-RNTI is used.

Upon reception of DCI, the device will check the CRC using its set of assigned RNTIs. If the CRC checks, the message is declared to be correctly received and intended for the device. Thus, the identity of the device that is supposed to receive the DCI message is implicitly encoded in the CRC and not explicitly transmitted. This reduces the amount of bits necessary to transmit on the PDCCH as, from a device point of view, there is no difference between a corrupt message whose CRC will not check and a message intended for another device.

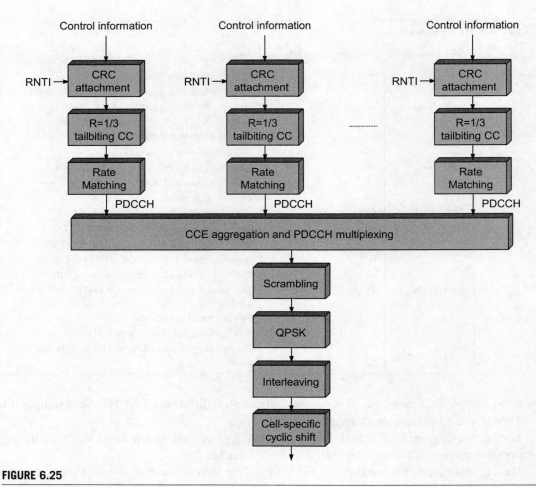

FIGURE 6.25

Processing of L1/L2 control signaling.

After CRC attachment, the bits are coded with a rate-1/3 tail-biting convolutional code and rate-matched to fit the amount of resources used for PDCCH transmission. Tail-biting convolutional coding is similar to conventional convolutional coding with the exception that no tail bits are used. Instead, the convolutional encoder is initialized with the last bits of the message prior to the encoding process. Thus, the starting and ending states in the trellis in an MLSE (Viterbi) decoder are identical.

After the PDCCHs to be transmitted in a given subframe have been allocated to the desired resource elements (the details of which are given in the following paragraphs), the sequence of bits corresponding to all the PDCCH resource elements to be transmitted in the subframe, including the unused resource elements, is scrambled by a cell- and subframe-specific scrambling sequence to randomize inter-cell interference, followed by QPSK modulation and mapping to resource elements.

To allow for simple yet efficient processing of the control channels in the device, the mapping of PDCCHs to resource elements is subject to a certain structure. This structure is based on so-called *control-channel elements* (CCEs), which in essence is a convenient name for a set of 36 useful resource elements (nine resource-element groups as defined in Section 6.4.1). The number of CCEs, one, two, four, or eight, required for a certain PDCCH depends on the payload size of the control information (DCI payload) and the channel-coding rate. This is used to realize link adaptation for the PDCCH; if the channel conditions for the device to which the PDCCH is intended are disadvantageous, a larger number of CCEs needs to be used compared to the case of advantageous channel conditions. The number of CCEs used for a PDCCH is also referred to as the aggregation level.

The number of CCEs available for PDCCHs depends on the size of the control region, the cell bandwidth, the number of downlink antenna ports, and the amount of resources occupied by PHICH. The size of the control region can vary dynamically from subframe to subframe as indicated by the PCFICH, whereas the other quantities are semi-statically configured. The CCEs available for PDCCH transmission can be numbered from zero and upward, as illustrated in Figure 6.26. A specific PDCCH can thus be identified by the numbers of the corresponding CCEs in the control region.

As the number of CCEs for each of the PDCCHs may vary and is not signaled, the device has to blindly determine the number of CCEs used for the PDCCH it is addressed upon. To reduce the complexity of this process somewhat, certain restrictions on the aggregation of contiguous CCEs have been specified. For example, an aggregation of eight CCEs can only start on CCE numbers evenly divisible by 8, as illustrated in Figure 6.26. The same principle

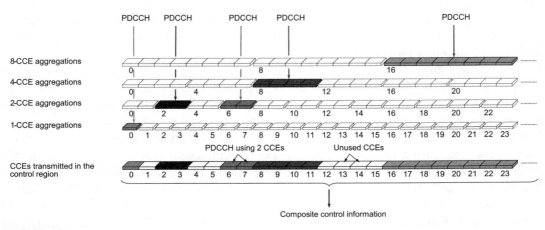

FIGURE 6.26

CCE aggregation and PDCCH multiplexing.

is applied to the other aggregation levels. Furthermore, some combinations of DCI formats and CCE aggregations that result in excessively high channel-coding rates are not supported.

The sequence of CCEs should match the amount of resources available for PDCCH transmission in a given subframe—that is, the number of CCEs varies according to the value transmitted on the PCFICH. In many cases, not all the PDCCHs that can be transmitted in the control region are used. Nevertheless, unused PDCCHs are part of the interleaving and mapping process in the same way as any other PDCCH. At the device, the CRC will not check for those "dummy" PDCCHs. Preferably, the transmission power is set to zero for those unused PDCCHs; the power can be used by other control channels.

The mapping of the modulated composite control information is, for the same reason as for PCFICH and PHICH, described in terms of symbol quadruplets being mapped to resource-element groups. Thus, the first step of the mapping stage is to group the QPSK symbols into symbol quadruplets, each consisting of four consecutive QPSK symbols. In principle, the sequence of quadruplets could be mapped directly to the resource elements in sequential order. However, this would not exploit all the frequency diversity available in the channel and diversity is important for good performance. Furthermore, if the same CCE-to-resource-element mapping is used in all neighboring cells, a given PDCCH will persistently collide with one and the same PDCCH in the neighboring cells assuming a fixed PDCCH format and inter-cell synchronization. In practice, the number of CCEs per PDCCH varies in the cell as a function of the scheduling decisions, which gives some randomization to the interference, but further randomization is desirable to obtain a robust control-channel design. Therefore, the sequence of quadruplets is first interleaved using a block interleaver to allow exploitation of the frequency diversity, followed by a cell-specific cyclic shift to randomize the interference between neighboring cells. The output from the cell-specific shift is mapped to resource-element groups in a time-first manner, as illustrated in Figure 6.22, skipping resource-element groups used for PCFICH and PHICH. Time-first mapping preserves the interleaving properties; with frequency-first over multiple OFDM symbols, resource-element groups that are spread far apart after the interleaving process may end up close in frequency, although on different OFDM symbols.

The interleaving operation described in the preceding paragraphs, in addition to enabling exploitation of the frequency diversity and randomizing the inter-cell interference, also serves the purpose of ensuring that each CCE spans virtually all the OFDM symbols in the control region. This is beneficial for coverage as it allows flexible power balancing between the PDCCHs to ensure good performance for each of the devices addressed. In principle, the energy available in the OFDM symbols in the control region can be balanced arbitrarily between the PDCCHs. The alternative of restricting each PDCCH to a single OFDM symbol would imply that power cannot be shared between PDCCHs in different OFDM symbols.

Similarly to the PCFICH, the transmission power of each PDCCH is under the control of the eNodeB. Power adjustments can therefore be used as a complementary link adaptation mechanism in addition to adjusting the code rate. Relying on power adjustments alone might seem a tempting solution but, although possible in principle, it can result in relatively large power differences between resource elements. This may have implications on the RF implementation and may violate the out-of-band emission masks specified. Hence, to keep the power differences between the resource elements reasonable, link adaptation through adjusting the channel code rate, or equivalently the number of CCEs aggregated for a PDCCH, is necessary. Furthermore, lowering the code rate is generally more efficient than increasing the transmission power. The two mechanisms for link adaptation, power adjustments, and different code rates, complement each other.

To summarize and to illustrate the mapping of PDCCHs to resource elements in the control region, consider the example shown in Figure 6.27. In this example, the size of the control region in the subframe considered equals three OFDM symbols. Two downlink antenna ports are configured (but, as explained previously, the mapping would be identical in the case of a single antenna port). One PHICH group is configured and three resource-element groups are therefore used by the PHICHs. The cell identity is assumed to be identical to zero in this case.

The mapping can then be understood as follows: First, the PCFICH is mapped to four resource-element groups, followed by allocating the resource-element groups required for the PHICHs. The resource-element groups left after the PCFICH and PHICHs are used for the different PDCCHs in the system. In this particular example, one PDCCH is using CCE numbers 0 and 1, while another PDCCH is using CCE number 4. Consequently, there is a relatively large number of unused resource-element groups in this example; either they can be used for additional PDCCHs or the power otherwise used for the unused CCEs could be allocated to the PDCCHs in use (as long as the power difference between resource elements is

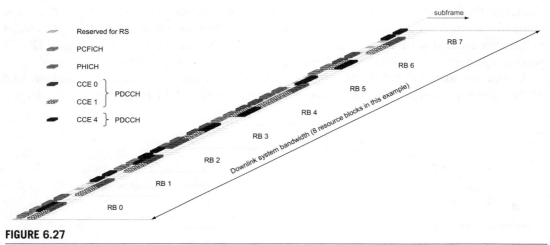

FIGURE 6.27

Example of mapping of PCFICH, PHICH, and PDCCH.

kept within the limits set by the RF requirements). Furthermore, depending on the inter-cell interference situation, fractional loading of the control region may be desirable, implying that some CCEs are left unused to reduce the average inter-cell interference.

6.4.4 ENHANCED PHYSICAL DOWNLINK CONTROL CHANNEL

In release 11, a complementary control channel was introduced, the *enhanced physical downlink control channel* (EPDCCH), primarily differing from the PDCCH in terms of precoding and mapping to physical resources. The reasons for introducing the EPDCCH are twofold:

- to enable frequency-domain scheduling and interference coordination also for control signaling;
- to enable DM-RS-based reception for the control signaling.

Unlike the PDCCH, which is transmitted in the control region and spans the full system bandwidth, the EPDCCH is transmitted in the data region and normally spans only a small portion of the overall bandwidth, see Figure 6.28. Hence, as it is possible to control in which part of the overall spectrum an EPDCCH is transmitted, it is possible not only to benefit from frequency-selective scheduling for control channels but also to implement various forms of inter-cell interference coordination schemes in the frequency domain. This can be very useful, for example, in heterogeneous deployments (see Chapter 14). Furthermore, as the EPDCCH uses DM-RS for demodulation in contrast to the CRS-based reception of the PDCCH, any precoding operations are therefore transparent to the device. The EPDCCH can therefore be seen as a prerequisite for CoMP (see Chapter 13) as well as more advanced antenna solutions in general for which it is beneficial for the network to have the freedom of changing the precoding without explicitly informing the device. Massive machine-type communication devices also use the EPDCCH for control signaling but with some smaller enhancements as described in Chapter 20. The rest of this section focuses on the EPDCCH processing in general and not the enhancements for machine-type communication.

In contrast to the PDCCH, the EPDCCH decoding result is not available until the end of the subframe, which leaves less processing time for the PDSCH. Hence, for the largest

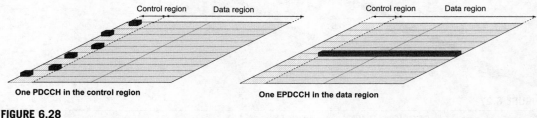

FIGURE 6.28

Illustration of principle for PDCCH mapping (left) and EPDCCH mapping (right).

DL-SCH payloads, the device may need a slightly faster PDSCH decoder compared to the PDCCH case to meet the hybrid ARQ timing.

With a few exceptions, the EPDCCH generally supports the same DCI formats as the PDCCH (see Table 6.4) and can thus be used for scheduling all the different transmission modes in both uplink and downlink. However, DCI formats 3/3A, used for uplink power control of multiple devices, and format 1C, used for scheduling system information to multiple devices, are not supported on the EPDCCH for reasons discussed in conjunction with search spaces in Section 6.4.5.

The EPDCCH processing, illustrated in Figure 6.29, is virtually identical to that of a PDCCH apart from precoding and the resource-element mapping. The identity of the device addressed—that is, the RNTI—is implicitly encoded in the CRC and not explicitly transmitted. After CRC attachment, the bits are coded with a rate-1/3 tail-biting convolutional code and rate-matched to fit the amount of resources available for the EPDCCH in question. The coded and rate-matched bits are scrambled with a device and EPDCCH-set specific

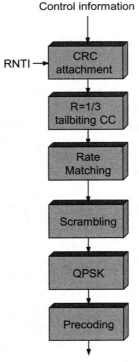

FIGURE 6.29

EPDCCH coding.

scrambling sequence (description of EPDCCH sets follows) and QPSK-modulated before being mapped to the physical resources. Link adaptation is supported by adjusting the number of control channel resources utilized by an encoded EPDCCH message, similar to the PDCCH.

To describe the mapping of EPDCCH to physical resources, it is worthwhile to view this from a device perspective. Each device using EPDCCH is configured with one or two sets of PRBs where EPDCCH transmission to that device may occur. Each set consists of two, four, or eight PRB pairs and the two sets may be of different size. There is full flexibility in the location of the resource-block pairs across the full downlink system bandwidth (the location is configured by higher layers using a combinatorial index) and the resource-block pairs do not have to be contiguous in frequency. It is important to understand that these sets are defined from a *device* perspective and only indicates where a device *may* receive EPDCCH transmissions. A resource-block pair not used for EPDCCH transmission to a particular device in a certain subframe can still be used for data transmission, either to the same device or to another device, despite being part of one of the two EPDCCH sets.

An EPDCCH set is configured as being either *localized* or *distributed*. For a localized set, a single EPDCCH is mapped to one PRB pair (or a few in case of the highest aggregation levels) to allow exploitation of frequency-domain scheduling gains. For a distributed set, on the other hand, a single EPDCCH is distributed over multiple PRB pairs with the intention of providing robustness for situations in which there is limited or no CSI available at the transmitter.

Providing two EPDCCH sets instead of just one is beneficial for several reasons. For example, one set can be configured for localized transmission in order to benefit from frequency-dependent scheduling and the other set can be configured for distributed transmission and act as a fallback in case the channel-state feedback becomes unreliable due to the device moving rapidly. It is also useful for supporting CoMP as discussed further as follows.

Inspired by the PDCCH, the EPDCCH mapping to physical resources is subject to a structure based on *enhanced control-channel elements* (ECCEs) and *enhanced resource-element groups* (EREGs). An EPDCCH is mapped to a number of ECCEs, where the number of ECCEs used for an EPDCCH is known as the *aggregation level* and is used to realize link adaptation. If the radio conditions are worse, or if the EPDCCH payload is large, a larger number of ECCEs is used than for small payloads and/or good radio conditions. An EPDCCH uses a consecutive set of ECCEs in which the number of ECCEs used for an EPDCCH can be 1, 2, 4, 8, 16, or 32 although not all these values are possible for all configurations. Note though that the ECCE numbering is in the *logical* domain; consecutive ECCEs do not necessarily imply transmission on consecutive PRB pairs.

Each ECCE in turn consists of four[34] EREGs, where an EREG in essence corresponds to nine[35] resource elements in one PRB pair. To define an EREG, number all resource elements in a PRB pair cyclically in a frequency-first manner from 0 to 15, excluding resource-

---

[34]Four holds for normal cyclic prefix; for extended cyclic prefix and some special subframe configurations in normal cyclic prefix there are eight EREGs per ECCE.

[35]Eight in case of extended cyclic prefix.

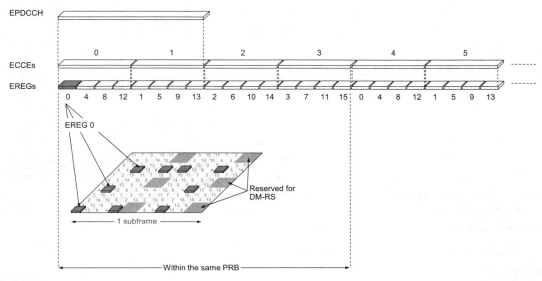

FIGURE 6.30

Example of relation between EPDCCH, ECCEs, and EREGs for localized mapping with aggregation level 2.

elements used for DM-RS. EREG number i consists of all resource elements with number i in that PRB pair and there are thus 16 EREGs in one PRB pair. However, note that not all nine resource elements in an EREG may be available for EPDCCH usage as some resource elements are occupied, for example, by the PDCCH control region, CRS, or CSI-RS. In Figure 6.30, an example with EREG 0 in one PRB pair is shown.

The mapping of ECCEs to EREGs is different for localized and distributed transmission (see Figure 6.31).

For localized transmission, the intention is to provide possibilities to select the physical resources and antenna precoding based on the instantaneous channel conditions. This is useful when exploiting channel-dependent scheduling or in conjunction with multi-antenna schemes such as CoMP (see Chapter 13). Hence, an ECCE is mapped to EREGs in the *same* PRB pair, which allows the eNodeB to select the appropriate frequency region to which to transmit the EPDCCH. Only if one PRB pair is not sufficient to to carry one EPDDCH—that is, for the highest aggregation levels—a second PRB pair is used. A single antenna port is used for transmission of the EPDCCH. The DM-RS associated with the EPDCCH transmission[36] is a function of the ECCE index and the C-RNTI. This is useful to support MU-MIMO (see Section 6.3.5.1) in which multiple EPDCCHs intended for different spatially separated devices are transmitted using the *same* time—frequency resources but with

---

[36]In the specifications, this is described as selecting *one* of *four* antenna ports for transmission of the EPDCCH, where each antenna port has an associated orthogonal reference signal.

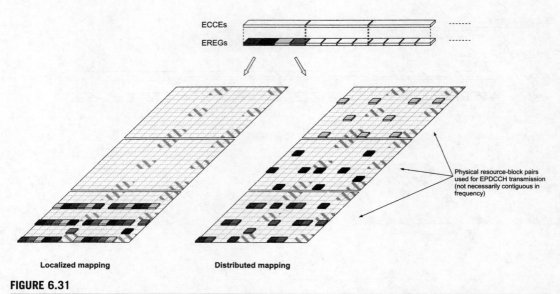

Localized mapping   **Distributed mapping**

FIGURE 6.31

Example of localized and distributed mapping of ECCEs (three PRB pairs for EPDCCH assumed in this example).

different DM-RS sequences. Up to four different orthogonal DM-RS sequences are available, implying that up to four different devices can be multiplexed in this manner.

For distributed transmission, the intention is to maximize diversity to provide diversity gain in situations in which it is not possible or desirable to exploit the instantaneous channel conditions. Consequently, an ECCE is mapped to EREGs in *different* PRB pairs, where the PRB pairs preferably are configured to be far apart in the frequency domain. To further increase the diversity order and exploit multiple transmission antennas, the resource elements in each EREG is alternating between two antenna ports to provide spatial diversity.

Quasi-colocation configuration is supported for the EPDCCH. In transmission mode 10, each EPDCCH set is associated with one of the four *PDSCH to RE mapping and quasi-colocation states* (see Section 5.3.1). This can be used to enable CoMP and dynamic point selection (see Chapter 13). Each EPDCCH set is associated with a certain transmission point and dynamic point selection is achieved by selecting from which set to transmit a DCI message. Handling different sizes of the PDCCH control region and locations of CRS from the different transmission points is possible since these pieces of information are part of the four states.

Triggering of retransmissions is done in the same manner as for the PDCCH case—that is by using the PHICH which in essence is a very compact uplink grant for retransmissions. In principle an EPHICH could be considered, but since each EPHICH would need its own set of DM-RS, the resulting structure would not be as compact as the PHICH in which multiple devices share the same set of CRS. Hence, either the PHICH is used or, if the dependency on CRS is not desirable, an EPDCCH can be used to schedule a retransmission.

6.4.5 BLIND DECODING OF PDCCHS AND EPDCCHS

As described in the previous section, each PDCCH or EPDCCH supports multiple DCI formats and the format used is a priori unknown to the device. Therefore, the device needs to blindly detect the format of the (E)PDCCHs. The CCE and ECCE structures described in the previous section help in reducing the number of blind decoding attempts, but are not sufficient. Hence, it is required to have mechanisms to limit the number of CCE/ECCE aggregations that the device is supposed to monitor. Clearly, from a scheduling point of view, restrictions in the allowed aggregations are undesirable as they may influence the scheduling flexibility and require additional processing at the transmitter side. At the same time, requiring the device to monitor all possible CCE/ECCE aggregations, also for the larger cell bandwidths and EPDCCH set sizes, is not attractive from a device-complexity point of view. To impose as few restrictions as possible on the scheduler while at the same time limit the maximum number of blind decoding attempts in the device, LTE defines so-called *search spaces*. A search space is a set of candidate control channels formed by CCEs (or ECCEs) at a given aggregation level, which the device is supposed to attempt to decode. As there are multiple aggregation levels a device has multiple search spaces. The search space concept is applied to both PDCCH and EPDCCH decoding, although there are differences between the two. In the following section, search spaces are described, starting with the case of PDCCHs on a single component carrier and later extended to EPDCCHs as well as multiple component carriers.

The PDCCH supports four different aggregation levels corresponding to one, two, four, and eight CCEs. In each subframe, the devices will attempt to decode all the PDCCHs that can be formed from the CCEs in each of its search spaces. If the CRC checks, the content of the control channel is declared as valid for this device and the device processes the information (scheduling assignment, scheduling grants, and so on). The network can only address a device if the control information is transmitted on a PDCCH formed by the CCEs in one of the device's search spaces. For example, device A in Figure 6.32 cannot be addressed on a PDCCH starting at CCE number 20, whereas device B can. Furthermore, if device A is using CCEs 16–23, device B cannot be addressed on aggregation level 4 as all CCEs in its level 4 search space are blocked by the use for the other devices. From this it can be intuitively understood that for efficient utilization of the CCEs in the system, the search spaces should differ between devices. Each device in the system therefore has a *device-specific* search space at each aggregation level.

As the device-specific search space is typically smaller than the number of PDCCHs the network could transmit at the corresponding aggregation level; there must be a mechanism determining the set of CCEs in the device-specific search space for each aggregation level. One possibility would be to let the network configure the device-specific search space in each

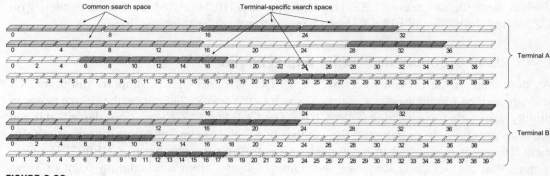

FIGURE 6.32

Illustration of principle for PDCCH search spaces in two devices.

device. However, this would require explicit signaling to each of the devices and possibly reconfiguration at handover. Instead, the device-specific search spaces for PDCCH are defined without explicit signaling through a function of the device identity and implicitly the subframe number. Dependence on the subframe number results in the device-specific search spaces being time varying, which helps resolve blocking between devices. If a given device cannot be scheduled in a subframe as all the CCEs that the device is monitoring have already been used for scheduling other devices in the same subframe, the time-varying definition of the device-specific search spaces is likely to resolve the blocking in the next subframe.

In several situations, there is a need to address a group of, or all, devices in the system. One example is dynamic scheduling of system information; another is transmission of paging messages, both described in Chapter 11. Transmission of explicit power-control commands to a group of devices is a third example. To allow multiple devices to be addressed at the same time, LTE has defined *common search spaces* for the PDCCH in addition to the device-specific search spaces. A common search space is, as the name implies, common, and all devices in the cell monitor the CCEs in the common search spaces for PDCCH control information. Although the motivation for the common search space is primarily transmission of various system messages, it can be used to schedule individual devices as well. Thus, it can be used to resolve situations where scheduling of one device is blocked due to lack of available resources in the device-specific search space. Unlike unicast transmissions, where the transmission parameters of the control signaling can be tuned to match the channel conditions of a specific device, system messages typically need to reach the cell border. Consequently, the common search spaces are only defined for aggregation levels of four and eight CCEs and only for the smallest DCI formats, 0/1A/3/3A and 1C. There is no support for DCI formats with spatial multiplexing in the common search space. This helps to reduce the number of blind decoding attempts in the device used for monitoring the common search space.

Figure 6.32 illustrates the device-specific and common search spaces for PDCCH in two devices in a certain subframe. The device-specific search spaces are different in the two devices and will, as described earlier, vary from subframe to subframe. Furthermore, the device-specific search spaces partially overlap between the two devices in this subframe (CCEs 24–31 on aggregation level 8) but, as the device-specific search space varies between subframes, the overlap in the next subframe is most likely different. There are 16 PDCCH candidates in the device-specific search spaces, mainly allocated to the smaller aggregation levels, and six PDCCH candidates in the common search spaces.

EPDCCH blind decoding in general follows the same principles as the PDCCH—that is, the device will attempt to decode all the EPDCCHs that can be formed from the ECCEs in each of its search spaces. However, only the device-specific search spaces are supported for EPDCCH. Thus, if a device has been configured to use the EPDCCH, it monitors the EPDCCH device-specific search spaces *instead of* the PDCCH device-specific search spaces. The common search spaces for PDCCH are monitored irrespective of whether an EPDCCH has been configured or not. The reason for not defining common search spaces for EPDCCH is that system information needs to be provided to all devices, including those not supporting EPDCCH, and hence the PDCCH needs to be used. The lack of a common search space for the EPDCCH is also the reason why DCI formats 3/3A and 1C are not supported on the EPDCCH.

The device-specific EPDCCH search spaces for an EPDCCH set are randomly varying within the set of PRB pairs configured for EPDCCH monitoring in a device using similar principles as the PDCCH. The two EPDCCH sets have independent pseudo-random sequences, initialized through RRC signaling. Having multiple EPDCCH sets is beneficial as it reduces the blocking probability: compare with the PDCCH, where one PDCCH on aggregation level eight may block all the PDCCH candidates on aggregation level one.

For a localized EPDCCH set, the EPDCCH candidates are spread over as many PRB pairs as possible. The intention behind this is to allow the device to be addressed over a wide frequency range in order not to limit the benefits of channel-dependent scheduling of EPDCCH.

The number of blind decoding attempts is the same irrespective of whether the device is monitoring the EPDCCH or PDCCH search spaces. Thus, there are 16 EPDCCH candidates[37] to attempt to decode, and for each candidate two different DCI formats need to be considered. The 16 EPDCCH candidates are distributed across the two EPDDCH sets, roughly following the principle that the number of candidates is proportional to the number of PRB pairs. Furthermore, for a localized set, more candidates are allocated to lower aggregation levels suitable for good channel conditions, motivated by the assumption that channel-dependent scheduling is used for localized transmission. For distributed transmission, the opposite holds—that is, more candidates are allocated to the higher aggregation levels than the lower ones to provide robustness as channel-dependent scheduling is typically not used for distributed transmission.

---

[37]For a small number of PRB pairs configured for EPDCCH there may be less than 16 EPDCCH candidates.

As mentioned, the device is monitoring *either* the device-specific PDCCH search spaces *or* the device-specific EPDCCH search spaces. The basic principle is that the device is monitoring EPDCCHs in all subframes whenever EPDCCH support is enabled, except for the special subframe in configurations where there are no DM-RS supported and hence no possibility to receive the EPDCCH. However, to provide additional control of when the EPDCCH is monitored, it is possible to provide a bitmap to the device, indicating in which subframes it should monitor EPDCCHs and in which subframes it should monitor PDCCHs. One possible use case for this is PDSCH transmission in MBSFN subframes. In MBSFN subframes where the PMCH is transmitted, it is not possible to transmit the EPDCCH and the device needs to be addressed using the PDCCH, while in MBSFN subframes not used for PMCH transmission, the EPDCCH can be used.

In the preceding paragraphs, the search spaces for PDCCH and EPDCCH in terms of control channel candidates are described. However, to determine whether a control channel candidate contains relevant downlink control information or not, the contents have to be decoded. If the CRC, which includes the device identity, checks, then the content of the control channel is declared as valid for this device and the device processes the information (scheduling assignment, scheduling grants, and so on). Thus, for each control-channel candidate, the device needs to attempt to decode the contents once for each supported DCI format. The downlink DCI formats to decode in the device-specific search spaces depend on the *transmission mode* configured for the device. Transmission modes are described in Section 6.3 and, in principle, correspond to different multi-antenna configurations. As an example, there is no need to attempt to decode DCI format 2 when the device has not been configured for spatial multiplexing, which helps to reduce the number of blind decoding attempts. The DCI formats a device addressed using the C-RNTI should monitor as a function of the transmission mode are listed in Table 6.5. Note that DCI format 1C is monitored in the common search spaces as well, although not with the C-RNTI identity. As seen from the table, there are two DCI formats to monitor in the device-specific search spaces and one in the common search spaces. In addition, a device also needs to monitor DCI format 1C in the common search space. Hence, with 16 PDCCH/EPDCCH candidates in the device-specific search spaces and six in the common search spaces, a device needs to execute $2 \cdot 16 + 2 \cdot 6 = 44$ blind decoding attempts in each subframe. With uplink spatial multiplexing, introduced in release 10, an additional uplink DCI format needs to be monitored in the device-specific search spaces, increasing the number of blind decoding attempts to $3 \cdot 16 + 2 \cdot 6 = 60$. These numbers are for a single component carrier; in case of carrier aggregation the number of blind decodes is increased further as discussed in Chapter 12. Finally, note that a certain device in some circumstances may be addressed with different RNTIs. For example, DCI format 1A in the common search spaces may be used either with the C-RNTI for normal scheduling purposes or with the SI-RNTI for scheduling of system information. This does not affect the number of blind decoding attempts, as they are related to

Table 6.5 Downlink DCI Formats Monitored in Different Search Spaces for C-RNTI. Note that DCI Format 1C is Monitored in the Common Search Space as Well, Although Not with the C-RNTI Identity

| Mode | Search Space | | | Description | Release |
	Common (PDCCH)	Device-Specific (PDCCH or EPDCCH)			
1	1A	1A	1	Single antenna transmission	8
2			1	Transmit diversity	
3			2A	Open-loop spatial multiplexing	
4			2	Closed-loop spatial multiplexing	
5			1D	MU-MIMO	
6			1B	Single-layer codebook-based precoding	
7			1	Single-layer transmission using DM-RS	
8			2B	Dual-layer transmission using DM-RS	9
9			2C	Multi-layer transmission using DM-RS	10
10			2D	Multi-layer transmission using DM-RS	11

the DCI format; checking two different RNTIs—that is, checking two different CRCs, after decoding is a very low-complexity operation.

Configuration of the transmission mode is done via RRC signaling. As the exact subframe number when this configuration takes effect in the device is not specified and may vary depending on, for example, RLC retransmissions, there is a (short) period when the network and the device may have different understandings of which transmission mode is configured. Therefore, in order not to lose the possibility of communicating with the device, it is necessary to have at least one DCI format that is decoded irrespective of the transmission mode. For downlink transmissions, DCI format 1A serves this purpose and the network can therefore always transmit data to the device using this DCI format. Another function of format 1A is to reduce overhead for transmissions when full flexibility in resource block assignment is not needed.

6.4.6 DOWNLINK SCHEDULING ASSIGNMENTS

Having described the transmission of downlink control information on PDCCH and EPDCCH, the detailed contents of the control information can be discussed, starting with the downlink scheduling assignments. Downlink scheduling assignments are valid for the same subframe in which they are transmitted. The scheduling assignments use one of the DCI formats 1, 1A, 1B, 1C, 1D, 2, 2A, 2B, 2C, or 2D and the DCI formats used depend on the transmission mode configured (see Table 6.5 for the relation between DCI formats and transmission modes). The reason for supporting multiple formats with different message sizes for the same purpose is to allow for a trade-off in control-signaling overhead and scheduling flexibility. Parts of the contents are the same for the different DCI formats, as seen in Table 6.6, but there are also differences due to the different capabilities.

DCI format 1 is the basic downlink assignment format in the absence of spatial multiplexing (transmission modes 1, 2, and 7). It supports noncontiguous allocations of resource blocks and the full range of modulation-and-coding schemes.

Table 6.6 DCI Formats used for Downlink Scheduling

Field		DCI Format									
		1	1A	1B	1C	1D	2	2A	2B	2C	2D
Resource information	Carrier indicator	•	•	•		•	•	•	•	•	•
	Resource block assignment type	0/1	2	2	2'	2	0/1	0/1	0/1	0/1	0/1
HARQ process number		•	•	•		•	•	•	•	•	•
1st transport block	MCS	•	•	•	•	•	•	•	•	•	•
	RV	•	•	•		•	•	•	•	•	•
	NDI	•	•	•		•	•	•	•	•	•
2nd transport block	MCS						•	•	•	•	•
	RV						•	•	•	•	•
	NDI						•	•	•	•	•
Multi-antenna information	PMI confirmation			•							
	Precoding information			•		•	•				
	Transport block swap flag						•	•			
	Power offset					•					
	DM-RS scrambling								•		
	#Layers/DM-RS scrambling/ antenna ports									•	•
PDSCH mapping and quasi-colocation indicator											•
Downlink assignment index		•	•	•		•	•	•	•	•	•
PUCCH power control		•	•	•		•	•	•	•	•	•
SRS request[a]			F						T	T	T
ACK/NAK offset (EPDCCH only)		•	•	•		•	•	•	•	•	•
Flag for 0/1A differentiation			•								
Padding (only if needed)		(•)	(•)	(•)		(•)	(•)	(•)	(•)	(•)	(•)
Identity		•	•	•	•	•	•	•	•	•	•

[a]*Format 1A for FDD and formats 2B, 2C, and 2D for TDD.*

DCI format 1A, also known as the "compact" downlink assignment, supports allocation of frequency-contiguous resource blocks only and can be used in all transmission modes. Contiguous allocations reduce the payload size of the control information with a somewhat reduced flexibility in resource allocations. The full range of modulation-and-coding schemes is supported. Format 1A can also be used to trigger a contention-free random access (see Chapter 11), in which case some bit fields are used to convey the necessary random-access preamble information and the remaining bit fields are set to a specific combination.

DCI format 1B is used to support codebook-based beam-forming described in Section 6.3.3, with a low control-signaling overhead (transmission mode 6). The content is similar to DCI format 1A with the addition of bits for signaling of the precoding matrix. As codebook-based beam-forming can be used to improve the data rates for cell-edge devices, it is important to keep the related DCI message size small so as not to unnecessarily limit the coverage.

DCI format 1C is used for various special purposes such as random-access response, paging, transmission of system information, MBMS-related signaling (see Chapter 19), and eIMTA support (Chapter 15). Common for these applications is simultaneous reception of a relatively small amount of information by *multiple* users. Hence, DCI format 1C supports QPSK only, has no support for hybrid-ARQ retransmissions, and does not support closed-loop spatial multiplexing. Consequently, the message size for DCI format 1C is very small, which is beneficial for coverage and efficient transmission of the type of system messages for which it is intended. Furthermore, as only a small number of resource blocks can be indicated, the size of the corresponding indication field in DCI format 1C is independent of the cell bandwidth.

DCI format 1D is used to support MU-MIMO (transmission mode 5) scheduling of one codeword with precoder information. To support dynamic sharing of the transmission power between the devices sharing the same resource block in MU-MIMO, one bit of power offset information is included in DCI format 1D, as described in Section 6.3.5.2.

DCI format 2 is an extension for DCI format 1 to support closed-loop spatial multiplexing (transmission mode 4). Thus, information about the number of transmission layers and the index of the precoder matrix used are jointly encoded in the precoding information field. Some of the fields in DCI format 1 have been duplicated to handle the two transport blocks transmitted in parallel in the case of spatial multiplexing.

DCI format 2A is similar to DCI format 2 except that it supports open-loop spatial multiplexing (transmission mode 3) instead of closed-loop spatial multiplexing. The precoder information field is used to signal the number of transmission layers only, hence the field has a smaller size than in DCI format 2. Furthermore, since DCI format 2A is used for scheduling of multi-layer transmissions only, the precoder information field is only necessary in the case of four transmit antenna ports; for two transmit antennas the number of layers is implicitly given by the number of transport blocks.

DCI format 2B was introduced in release 9 in order to support dual-layer spatial multiplexing in combination with beam-forming using DM-RS (transmission mode 8). Since scheduling with DCI format 2B relies on DM-RS, precoding/beam-forming is transparent to the device and there is no need to signal a precoder index. The number of layers can be controlled by disabling one of the transport blocks. Two different scrambling sequences for the DM-RS can be used, as described in Section 6.2.2.

DCI format 2C was introduced in release 10 and is used to support spatial multiplexing using DM-RS (transmission mode 9). To some extent, it can be seen as a generalization of format 2B to support spatial multiplexing of up to eight layers. DM-RS scrambling and the number of layers are jointly signaled by a single three-bit field.

DCI format 2D was introduced in release 11 and is used to support spatial multiplexing using DM-RS (transmission mode 10). In essence it is an extension of format 2C to support signaling of quasi-colocation of antenna ports.

Many information fields in the different DCI formats are, as already mentioned, common among several of the formats, while some types of information exist only in certain formats. Furthermore, in later releases, some of the DCI formats are extended with additional bits. One example hereof is the addition of the carrier indicator in many DCI formats in release 10, as well as the inclusion of the SRS request in DCI formats 0 and 1A in release 10. Such extensions, for which different releases have different payload sizes for the same DCI format, are possible as long the extensions are used in the device-specific search spaces only where a single device is addressed by the DCI format in question. For DCI formats used to address multiple devices at the same time, for example, to broadcast system information, payload extensions are clearly not possible as devices from previous releases are not capable of decoding these extended formats. Consequently, the extensions are allowed in the device-specific search spaces only.

The information in the DCI formats used for downlink scheduling can be organized into different groups, as shown in Table 6.6, with the fields present varying between the DCI formats. A more detailed explanation of the contents of the different DCI formats is as follows:

- Resource information, consisting of:
  - Carrier indicator (0 or 3 bit). This field is present in releases 10 and beyond if cross-carrier scheduling is enabled via RRC signaling and is used to indicate the component carrier the downlink control information relates to (see Chapter 12). The carrier indicator is not present in the common search space as this would either impact compatibility with devices not capable of carrier aggregation or require additional blind decoding attempts.
  - Resource-block allocation. This field indicates the resource blocks on one component carrier upon which the device should receive the PDSCH. The size of the field depends on the cell bandwidth and on the DCI format, more specifically on the resource indication type, as discussed in Section 6.4.6.1. Resource allocation types 0 and 1, which are the same size, support noncontiguous resource-block allocations, while resource allocation type 2 has a smaller size but supports contiguous allocations only. DCI format 1C uses a restricted version of type 2 in order to further reduce control signaling overhead.

- Hybrid-ARQ process number (3 bit for FDD, 4 bit for TDD), informing the device about the hybrid-ARQ process to use for soft combining. Not present in DCI format 1C as this DCI format is intended for scheduling of system information which does not use hybrid ARQ retransmissions.
- For the first (or only) transport block:[38]
  - Modulation-and-coding scheme (5 bit), used to provide the device with information about the modulation scheme, the code rate, and the transport-block size, as described later. DCI format 1C has a restricted size of this field as only QPSK is supported.
  - New-data indicator (1 bit), used to clear the soft buffer for initial transmissions. Not present in DCI format 1C as this format does not support hybrid ARQ.
  - Redundancy version (2 bit).
- For the second transport block (only present in DCI format supporting spatial multiplexing):
  - Modulation-and-coding scheme (5 bit).
  - New-data indicator (1 bit).
  - Redundancy version (2 bit).
- Multi-antenna information. The different DCI formats are intended for different multi-antenna schemes and which of the fields below that are included depends on the DCI format shown in Table 6.5.
  - PMI confirmation (1 bit), present in format 1B only. Indicates whether the eNodeB uses the (frequency-selective) precoding matrix recommendation from the device or if the recommendation is overridden by the information in the PMI field.
  - Precoding information, providing information about the index of the precoding matrix used for the downlink transmission and, indirectly, about the number of transmission layers. This information is present in the DCI formats used for CRS-based transmission only; for DM-RS-based transmissions the precoder used is transparent for the device and consequently there is no need to signal this information in this case.
  - Transport block swap flag (1 bit), indicating whether the two codewords should be swapped prior to being fed to the hybrid-ARQ processes. Used for averaging the channel quality between the codewords.
  - Power offset between the PDSCH and CRS used to support dynamic power sharing between multiple devices for MU-MIMO.
  - Reference-signal scrambling sequence, used to control the generation of quasi-orthogonal DM-RS sequences, as discussed in Section 6.2.2.
  - Number of layers, reference-signal scrambling sequence and the set of antenna ports used for the transmission (jointly encoded information in releases 10 and later, 3 bits, possibility for extension to 4 bits in release 13).

---

[38]A transport block can be disabled by setting the modulation-and-coding scheme to zero and the RV to 1 in the DCI.

- PDSCH resource-element mapping and quasi-colocation indicator (2 bit), informing the device which set of parameters to assume when demodulating the PDSCH. Up to four different sets of parameters can be configured by RRC to support different CoMP schemes as discussed in Chapter 13.
- Downlink assignment index (2 bit), informing the device about the number of downlink transmissions for which a single hybrid-ARQ acknowledgment should be generated according to Section 8.1.3. Present for TDD only or for aggregation of more than 5 carriers in which case 4 bits are used.
- Transmit-power control for PUCCH (2 bit). For scheduling of a secondary component carrier in the case of carrier aggregation, these bits are reused as *acknowledgment resource indicator* (ARI)—see Chapter 12.
- SRS request (1 bit). This field is used to trigger a one-shot transmission of a sounding reference signal in the uplink, a feature introduced in release 10 and an example of extending an existing DCI format in the device-specific search space with additional information fields in a later release. For FDD it is present in format 1A only, while for TDD it is present in formats 2B, 2C, and 2D only. For TDD, where short-term channel reciprocity can be used, it is motivated to include this field in the DCI formats used to support DM-RS-based multi-layer transmission in order to estimate the downlink channel conditions based on uplink channel sounding. For FDD, on the other hand, this would not be useful and the SRS request field is consequently not included. However, for DCI format 1A, including an SRS request bit comes "for free" as padding otherwise would have to be used to ensure the same payload size as for DCI format 0.
- ACK/NAK offset (2 bit). This field is present on EPDCCH only and thus supported in release 11 and later. It is used to dynamically control the PUCCH resource used for the hybrid-ARQ acknowledgment as discussed in Section 7.4.2.1.
- DCI format 0/1A indication (1 bit), used to differentiate between DCI formats 1A and 0 as the two formats have the same message size. This field is present in DCI formats 0 and 1A only. DCI formats 3 and 3A, which have the same size, are separated from DCI formats 0 and 1A through the use of a different RNTI.
- Padding. The smaller of DCI formats 0 and 1A is padded to ensure the same payload size irrespective of the uplink and downlink cell bandwidths. Padding is also used to ensure that the DCI size is different for different DCI formats that may occur simultaneously in the same search space (this is rarely required in practice as the payload sizes are different due to the different amounts of information). Finally, for PDCCH, padding is used to avoid certain DCI sizes that may cause ambiguous decoding.[39]

---

[39]For a small set of specific payload sizes, the control signaling on PDCCH may be correctly decoded at an aggregation level other than the one used by the transmitter. Since the PHICH resource is derived from the first CCE used for the PDCCH, this may result in incorrect PHICH being monitored by the device. To overcome this, padding is used if necessary to avoid the problematic payload sizes. Note that this padding applied for PDCCH only.

- Identity (RNTI) of the device for which the PDSCH transmission is intended (16 bit). As described in Sections 6.4.3 and 6.4.4, the identity is not explicitly transmitted but implicitly included in the CRC calculation. There are different RNTIs defined depending on the type of transmission (unicast data transmission, paging, power-control commands, etc.).

6.4.6.1 *Signaling of Downlink Resource-Block Allocations*

Focusing on the signaling of resource-block allocations, there are three different possibilities, types 0, 1, and 2, as indicated in Table 6.6. Resource-block allocation types 0 and 1 both support noncontiguous allocations of resource blocks in the frequency domain, whereas type 2 supports contiguous allocations only. A natural question is why multiple ways of signaling the resource-block allocations are supported, a question whose answer lies in the number of bits required for the signaling. The most flexible way of indicating the set of resource blocks the device is supposed to receive the downlink transmission upon is to include a bitmap with size equal to the number of resource blocks in the cell bandwidth. This would allow for an arbitrary combination of resource blocks to be scheduled for transmission to the device but would, unfortunately, also result in a very large bitmap for the larger cell bandwidths. For example, in the case of a downlink cell bandwidth corresponding to 100 resource blocks, the downlink PDCCH would require 100 bits for the bitmap alone, to which the other pieces of information need to be added. Not only would this result in a large control-signaling over-head, but it could also result in downlink coverage problems as more than 100 bits in one OFDM symbol correspond to a data rate exceeding 1.4 Mbit/s. Consequently, there is a need for a resource allocation scheme requiring a smaller number of bits while keeping sufficient allocation flexibility.

In resource allocation type 0, the size of the bitmap has been reduced by pointing not to individual resource blocks in the frequency domain, but to groups of contiguous resource blocks, as shown at the top of Figure 6.33. The size of such a group is determined by the downlink cell bandwidth; for the smallest bandwidths there is only a single resource block in a group, implying that an arbitrary set of resource blocks can be scheduled, whereas for the largest cell bandwidths, groups of four resource blocks are used (in the example in Figure 6.33, the cell bandwidth is 25 resource blocks, implying a group size of two resource blocks). Thus, the bitmap for the system with a downlink cell bandwidth of 100 resource blocks is reduced from 100 to 25 bits. A drawback is that the scheduling granularity is reduced; single resource blocks cannot be scheduled for the largest cell bandwidths using allocation type 0.

However, also in large cell bandwidths, frequency resolution of a single resource block is sometimes useful, for example, to support small payloads. Resource allocation type 1 address this by dividing the total number of resource blocks in the frequency domain into dispersed subsets, as shown in the middle of Figure 6.33. The number of subsets is given from the cell bandwidth with the number of subsets in type 1 being equal to the group size in type 0. Thus, in Figure 6.33, there are two subsets, whereas for a cell bandwidth of 100 resource blocks

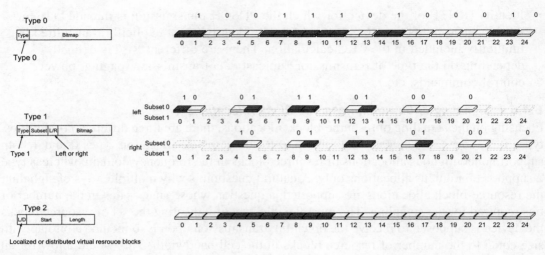

FIGURE 6.33

Illustration of resource-block allocation types (cell bandwidth corresponding to 25 resource blocks used in this example).

there would have been four different subsets. Within a subset, a bitmap indicates the resource blocks in the frequency domain upon which the downlink transmission occurs.

To inform the device whether resource allocation type 0 or 1 is used, the resource allocation field includes a flag for this purpose, denoted "type" in the leftmost part of Figure 6.33. For type 0, the only additional information is the bitmap discussed previously. For type 1, on the other hand, in addition to the bitmap itself, information about the subset for which the bitmap relates is also required. As one of the requirements in the design of resource allocation type 1 was to maintain the same number of bits in the allocation as for type 0 without adding unnecessary overhead,[40] the bitmap in resource allocation type 1 is smaller than in type 0 to allow for the signaling of the subset number. However, a consequence of a smaller bitmap is that not all resource blocks in the subset can be addressed simultaneously. To be able to address all resources with the bitmap, there is a flag indicating whether the bitmap relates to the "left" or "right" part of the resource blocks, as depicted in the middle part of Figure 6.33.

Unlike the other two types of resource-block allocation signaling, type 2 does not rely on a bitmap. Instead, it encodes the resource allocation as a start position and length of the resource-block allocation. Thus, it does not support arbitrary allocations of resource blocks but only frequency-contiguous allocations, thereby reducing the number of bits required for signaling the resource-block allocation. The number of bits required for resource-signaling type 2 compared to type 0 or 1 is shown in Figure 6.34 and, as shown, the difference is fairly large for the larger cell bandwidths.

---

[40]Allowing different sizes would result in an increase in the number of blind decoding attempts required in the device.

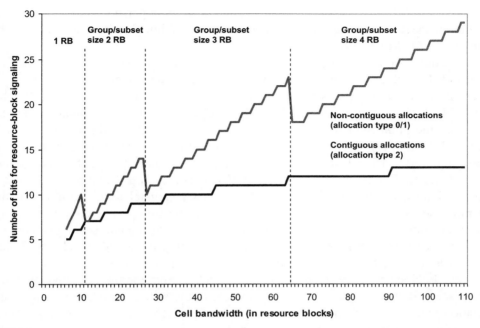

FIGURE 6.34

Number of bits used for downlink resource allocation signaling for downlink allocation types 0/1 and 2.

All three resource-allocation types refer to *VRBs* (see Section 6.1.1.8 for a discussion of resource-block types). For resource-allocation types 0 and 1, the VRBs are of localized type and the VRBs are directly mapped to PRBs. For resource-allocation type 2, on the other hand, both localized and distributed VRBs are supported. One bit in the resource allocation field indicates whether the allocation signaling refers to localized or distributed resource blocks.

6.4.6.2 Signaling of Transport-Block Sizes

Proper reception of a downlink transmission requires, in addition to the set of resource blocks, knowledge about the modulation scheme and the transport-block size, information (indirectly) provided by a 5-bit field in the different DCI formats. Of the 32 combinations, 29 are used to signal the modulation-and-coding scheme whereas 3 are reserved, the purpose of which is described later. Together, the modulation-and-coding scheme and the number of resource blocks assigned provide the transport-block size on the DL-SCH. Thus, the possible transport-block sizes can be described as a table with 29 rows and 110 columns, one column for each number of resource blocks possible to transmit upon (the number of columns follows from the maximum downlink component carrier bandwidth of 110 resource blocks). For devices configured with support for 256QAM, 4 of the 32 combinations are reserved and an alternative 28×110 table is used instead[41] to support the larger transport block sizes. The principles in the following apply to both these cases, although the numbers given below assume no 256QAM support.

Each modulation-and-coding scheme represents a particular combination of modulation scheme and channel-coding rate or, equivalently, a certain spectral efficiency measured in the number of information bits per modulation symbol. Although the 29-by-110 table of transport-block sizes in principle could be filled directly from the modulation-and-coding scheme and the number of resource blocks, this would result in arbitrary transport-block sizes, which is not desirable. First, as all the higher-layer protocol layers are byte aligned, the resulting transport-block sizes should be an integer number of bytes. Secondly, common payloads (e.g., RRC signaling messages and VoIP) should be possible to transmit without padding. Aligning with the QPP interleaver sizes is also beneficial, as this would avoid the use of filler bits (see Section 6.1.1.3). Finally, the same transport-block size should ideally appear for several different resource-block allocations, as this allows the number of resource blocks to be changed between retransmission attempts, providing increased scheduling flexibility. Therefore, a "mother table" of transport-block sizes is first defined, fulfilling the said requirements. Each entry in the 29-by-110 table is picked from the mother table such that the resulting spectral efficiency is as close as possible to the spectral efficiency of the signaled modulation-and-coding scheme. The mother table spans the full range of transport-block sizes possible, with an approximately constant worst-case padding.

From a simplicity perspective, it is desirable if the transport-block sizes do not vary with the configuration of the system. The set of transport-block sizes is therefore independent of the actual number of antenna ports and the size of the control region.[42] The design of the table assumes a control region of three OFDM symbols and two antenna ports, the "reference configuration." If the actual configuration is different, the resulting code rate for the DL-SCH will be slightly different as a result of the rate-matching procedure. However, the difference is small and of no practical concern. Also, if the actual size of the control region is smaller than the three-symbol assumption in the reference configuration, the spectral efficiencies will be somewhat smaller than the range indicated by the modulation-and-coding scheme signaled as part of the DCI. Thus, information about the modulation scheme used is obtained directly from the modulation-and-coding scheme, whereas the exact code rate and rate matching is obtained from the implicitly signaled transport-block size together with the number of resource elements used for DL-SCH transmission.

For bandwidths smaller than the maximum of 110 resource blocks, a subset of the table is used. More specifically, in case of a cell bandwidth of N resource blocks, the first N columns of the table are used. Also, in the case of spatial multiplexing, a single transport block can be

---

[41]The alternative table is only used for transmissions scheduled with the C-RNTI. Random-access response and system information cannot use 256QAM for backward-compatibility reasons.

[42]For DwPTS, the transport-block size is scaled by a factor of 0.75 compared to the values found in the table, motivated by the DwPTS having a shorter duration than a normal subframe.

mapped to up to four layers. To support the higher data rates this facilitates, the set of supported transport-block sizes needs to be extended beyond what is possible in the absence of spatial multiplexing. The additional entries are in principle obtained by multiplying the sizes with the number of layers to which a transport block is mapped and adjusting the result to match the QPP interleaver size.

The 29 combinations of modulation-and-coding schemes each represent a reference spectral efficiency in the approximate range of 0.1–5.9 bits/s/symbol (with 256QAM the upper limit is 7.4 bit/s/symbol).[43] There is some overlap in the combinations in the sense that some of the 29 combinations represent the same spectral efficiency. The reason is that the best combination for realizing a specific spectral efficiency depends on the channel properties; sometimes higher-order modulation with a low code rate is preferable over lower-order modulation with a higher code rate, and sometimes the opposite is true. With the overlap, the eNodeB can select the best combination, given the propagation scenario. As a consequence of the overlap, two of the rows in the 29-by-110 table are duplicates and result in the same spectral efficiency but with different modulation schemes, and there are only 27 unique rows of transport-block sizes.

Returning to the three reserved combinations in the modulation-and-coding field mentioned at the beginning of this section, those entries can be used for retransmissions only. In the case of a retransmission, the transport-block size is, by definition, unchanged and fundamentally there is no need to signal this piece of information. Instead, the three reserved values represent the modulation scheme, QPSK, or 16QAM or 64QAM,[44] which allows the scheduler to use an (almost) arbitrary combination of resource blocks for the retransmission. Obviously, using any of the three reserved combinations assumes that the device properly received the control signaling for the initial transmission; if this is not the case, the retransmission should explicitly indicate the transport-block size.

The derivation of the transport-block size from the modulation-and-coding scheme and the number of scheduled resource blocks is illustrated in Figure 6.35.

6.4.7 UPLINK SCHEDULING GRANTS

Uplink scheduling grants use one of DCI formats 0 or 4; DCI format 4 was added in release 10 to support uplink spatial multiplexing. The basic resource-allocation scheme for the uplink is single-cluster allocations where the resource blocks are contiguous in the frequency domain, although release 10 added support for multi-cluster transmissions of up to two clusters on a single component carrier.

DCI format 0 is used for scheduling uplink transmissions not using spatial multiplexing on one component carrier. It has the same size control-signaling message as the "compact" downlink assignment (DCI format 1A). A flag in the message is used to inform the device

---

[43]The exact values vary slightly with the number of resource blocks allocated due to rounding.
[44]Four reserved values representing QPSK, 16QAM, 64QAM, and 256QAM for the alternative table.

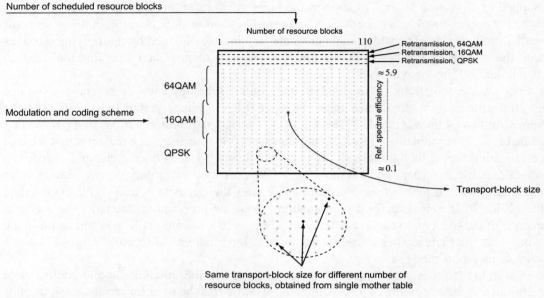

FIGURE 6.35

Computing the transport-block size (no 256QAM configured).

whether the message is an uplink scheduling grant (DCI format 0) or a downlink scheduling assignment (DCI format 1A).

DCI format 4 is used for uplink transmissions using spatial multiplexing on one component carrier. Consequently, the size of DCI format 4 is larger than that of DCI format 0, as additional information fields are required.

Many information fields are common to the two DCI formats, but there are also differences, as shown in Table 6.7. Similarly to the downlink scheduling assignments, some DCI formats have been extended with additional bits in later releases. The contents of the different DCI formats are explained in more detail in the following:

- Resource information, consisting of:
  - Carrier indicator (0 or 3 bit). This field is present in releases 10 and beyond only if cross-carrier scheduling is enabled via RRC signaling and is used to indicate the uplink component carrier the grant relates to (see Chapter 12). The carrier indicator is not present in the common search space as this would either impact compatibility with devices not capable of carrier aggregation or require additional blind decoding attempts.
  - Resource allocation type or multi-cluster flag (1 bit), indicating whether resource allocation type 0 (one cluster of resource blocks is used for uplink transmissions) or type 1 (two clusters of resource blocks are used for the uplink transmission) is used.

Table 6.7 DCI Formats for Uplink Scheduling Grants

Field		DCI Format	
		0	**4**
Resource information	Carrier indicator	•	•
	Resource allocation type	•	•
	Resource block assignment	0/(1)	0/1
1st transport block	MCS/RV	•	•
	NDI	•	•
2nd transport block	MCS/RV		•
	NDI		•
DM-RS phase rotation and OCC index		•	•
Precoding information			•
CSI request		•	•
SRS request		•	•
Uplink index/DAI (TDD only)		•	•
PUSCH power control		•	•
Flag for 0/1A differentiation		•	
Padding (only if needed)		(•)	(•)
Identity		•	•

This flag is not present in releases prior to release 10. In previous releases, the downlink bandwidth was, in practice, always at least as large as the uplink bandwidth, implying that one padding bit was used for DCI format 0 in those releases to align with the size of format 1A. The padding bit could therefore be replaced by the multi-cluster flag in release 10 without sacrificing backward compatibility. In DCI format 4, supported in releases 10 and later, the multi-cluster flag is always present.

- Resource-block allocation, including hopping indication. This field indicates the resource blocks upon which the device should transmit the PUSCH using uplink resource-allocation type 0 (DCI format 0) or type 1 (DCI format 4), as described in Section 6.4.7.1. The size of the field depends on the cell bandwidth. For single-cluster allocations, uplink frequency hopping, as described in Chapter 7, can be applied to the uplink PUSCH transmission.

- For the first (or only) transport block:
  - Modulation-and-coding scheme including redundancy version (5 bit), used to provide the device with information about the modulation scheme, the code rate, and the transport-block size. The signaling of the transport-block size uses the same transport-block table as for the downlink—that is, the modulation-and-coding scheme together with the number of scheduled resource blocks provides the transport-block size. However, as the support of 64QAM in the uplink is not mandatory for all devices,

devices not capable of 64QAM use 16QAM when 64QAM is indicated in the modulation-and-coding field. The use of the three reserved combinations is slightly different than for the downlink; the three reserved values are used for implicit signaling of the RV, as described later. A transport block can be disabled by signaling a specific combination of modulation-and-coding scheme and number of resource blocks. Disabling one transport block is used when retransmitting a single transport block only.

- New-data indicator (1 bit), used to indicate to the device whether transmission of a new transport block or retransmission of the previous transport block is granted.
- For the second transport block (DCI format 4 only):
  - Modulation-and-coding scheme including redundancy version (5 bit).
  - New-data indicator (1 bit).
- Phase rotation of the uplink demodulation reference signal (3 bit), used to support MU-MIMO, as described in Chapter 7. By assigning different reference-signal phase rotations to devices scheduled on the same time—frequency resources, the eNodeB can estimate the uplink channel response from each device and suppress the interdevice interference by the appropriate processing. In releases 10 and later, it also controls the orthogonal cover sequence, see Section 7.2.
- Precoding information, used to signal the precoder to use for the uplink transmission in releases 10 and later.
- Channel-state request flag (1, 2, or 3 bit). The network can explicitly request an aperiodic channel-state report to be transmitted on the UL-SCH by setting this bit(s) in the uplink grant. In the case of carrier aggregation of up to five carriers, 2 bits are used to indicate which downlink component carrier the CSI should be reported for (see Chapter 10), a number increased to 3 bits if more than five carriers are configured in the device.
- SRS request (2 bit), used to trigger aperiodic sounding using one of up to three preconfigured settings, as discussed in Chapter 7. The SRS request, introduced in release 10, is supported in the device-specific search space only for reasons already described.
- Uplink index/DAI (2, 4 bit for carrier aggregation of more than five carriers). This field is present only when operating in TDD or when aggregating more than five carriers. For uplink—downlink configuration 0 (uplink-heavy configuration), it is used as an uplink index to signal for which uplink subframe(s) the grant is valid, as described in Chapter 9. For other uplink—downlink configurations, it is used as downlink assignment index to indicate the number of downlink transmissions the eNodeB expects hybrid-ARQ acknowledgment for.
- Transmit-power control for PUSCH (2 bit).
- DCI format 0/1A indication (1 bit), used to differentiate between DCI formats 1A and 0 as the two formats have the same message size. This field is present in DCI formats 0 and 1A only.
- Padding; the smaller of DCI formats 0 and 1A is padded to ensure the same payload size irrespective of the uplink and downlink cell bandwidths. Padding is also used to ensure

that the DCI size is different for DCI formats 0 and 4 (this is rarely required in practice as the payload sizes are different due to the different amounts of information). Finally, for PDCCH, padding is used to avoid certain DCI sizes that may cause ambiguous decoding.

- Identity (RNTI) of the device for which the grant is intended (16 bit). As described in Sections 6.4.3 and 6.4.4, the identity is not explicitly transmitted but implicitly included in the CRC calculation.

There is no explicit signaling of the redundancy version in the uplink scheduling grants. This is motivated by the use of a synchronous hybrid-ARQ protocol in the uplink; retransmissions are normally triggered by a negative acknowledgment on the PHICH and not explicitly scheduled as for downlink data transmissions. However, as described in Chapter 8, there is a possibility to explicitly schedule retransmissions. This is useful in a situation where the network will explicitly move the retransmission in the frequency domain by using the PDCCH instead of the PHICH. Three values of the modulation-and-coding field are reserved to mean redundancy version 1, 2, and 3. If one of those values is signaled, the device should assume that the same modulation and coding as the original transmission is used. The remaining entries are used to signal the modulation-and-coding scheme to use and also imply that redundancy version zero should be used. The difference in usage of the reserved values compared to the downlink scheduling assignments implies that the modulation scheme, unlike the downlink case, cannot change between uplink (re)transmission attempts.

6.4.7.1 Signaling of Uplink Resource-Block Allocations

The basic uplink resource-allocation scheme is single-cluster allocations—that is, allocations contiguous in the frequency domain—but releases 10 and later also provide the possibility for multi-cluster uplink transmissions.

Single-cluster allocations use uplink resource-allocation type 0, which is identical to downlink resource allocation type 2 described in Section 6.4.6.1 except that the single-bit flag indicating localized/distributed transmission is replaced by a single-bit frequency hopping flag. The resource allocation field in the DCI provides the set of VRBs to use for uplink transmission. The set of PRBs to use in the two slots of a subframe is controlled by the hopping flag, as described in Chapter 7.

Multi-cluster allocations with up to two clusters were introduced in release 10, using uplink resource-allocation type 1. In resource-allocation type 1, the starting and ending positions of two clusters of frequency-contiguous resource blocks are encoded into an index. Uplink resource-allocation type 1 does not support frequency hopping (diversity is achieved through the use of two clusters instead). Indicating two clusters of resources naturally requires additional bits compared to the single-cluster case. At the same time, the total number of bits used for resource-allocation type 1 should be identical to that of type 0. This is similar to the situation for the downlink allocation types 0 and 1; without aligning the sizes a new DCI format with a corresponding negative impact on the number of blind decodings is necessary. Since frequency hopping is not supported for allocation type 1, the bit otherwise

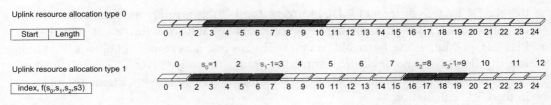

FIGURE 6.36

Illustration of uplink resource-block allocation types (uplink bandwidth corresponding to 25 resource blocks used in this example).

used for the hopping flag can be reused for extending the resource-allocation field. However, despite the extension of the resource-allocation field by one bit, the number of bits is not sufficient to provide a single-resource-block resolution in the two clusters for all bandwidths. Instead, similar to downlink resource-allocation type 0, groups of resource blocks are used and the starting and ending positions of the two clusters are given in terms of group numbers. The size of such a group is determined by the uplink carrier bandwidth in a similar way as for the downlink. For the smallest bandwidths there is only a single resource block in a group, implying that an arbitrary (as long as the limit of at most two clusters is observed) set of resource blocks can be scheduled, whereas for the largest cell bandwidths groups of four resource blocks are used. In the example in Figure 6.36, the cell bandwidth is 25 resource blocks, implying a group size of two resource blocks.

6.4.8 POWER-CONTROL COMMANDS

As a complement to the power-control commands provided as part of the downlink scheduling assignments and the uplink scheduling grants, there is the potential to transmit a power-control command using DCI formats 3 (2-bit command per device) or 3A (single-bit command per device). The main motivation for DCI format 3/3A is to support power control for semi-persistent scheduling. The power-control message is directed to a group of devices using an RNTI specific for that group. Each device can be allocated two power-control RNTIs, one for PUCCH power control and the other for PUSCH power control. Although the power-control RNTIs are common to a group of devices, each device is informed through RRC signaling which bit(s) in the DCI message it should follow. No carrier indicator is used for formats 3 and 3A.

UPLINK PHYSICAL-LAYER PROCESSING

This chapter provides a description of the basic physical-layer functionality related to the LTE uplink. It essentially follows the same outline as the corresponding downlink description provided in the previous chapter, with detailed descriptions regarding transport-channel processing (Section 7.1), the reference-signal structure (Section 7.2), multi-antenna transmission (Section 7.3), and the uplink L1/L2 control-channel structure (Section 7.4). The chapter ends with an overview of uplink power control and the uplink timing-alignment procedure in Sections 7.5 and 7.6, respectively. Physical aspects related to some specific uplink functions and procedures such as random access are provided in later chapters.

7.1 TRANSPORT-CHANNEL PROCESSING

This section describes the physical-layer processing applied to the uplink shared channel (UL-SCH) and the subsequent mapping to the uplink physical resource in the form of the basic OFDM time—frequency grid. As mentioned before, the UL-SCH is the only uplink transport-channel type in LTE[1] and is used for transmission of all uplink higher-layer information—that is, for both user data and higher-layer control signaling.

7.1.1 PROCESSING STEPS

Figure 7.1 outlines the different steps of the UL-SCH physical-layer processing in case of transmission on a single carrier. Similar to the downlink, in the case of uplink carrier aggregation the different component carriers correspond to separate transport channels with separate physical-layer processing.

The different steps of the uplink transport-channel processing are summarized below. Most of these steps are very similar to the corresponding steps for DL-SCH processing outlined in Section 6.1. For a more detailed overview of the different steps, the reader is referred to the corresponding downlink description.

---

[1]Strictly speaking, the LTE Random-Access Channel is also defined as a transport-channel type, see Chapter 4. However, RACH only includes a layer-1 preamble and carries no data in form of transport blocks.

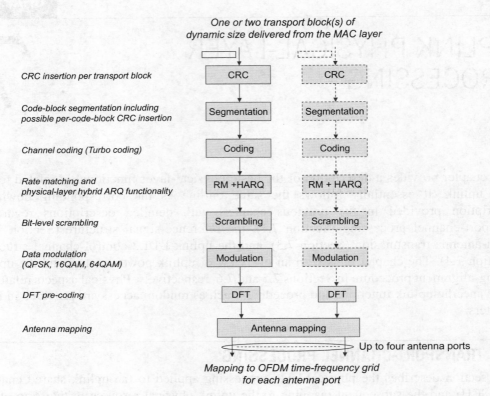

FIGURE 7.1

Physical-layer processing for the uplink shared channel (UL-SCH).

- *CRC insertion per transport block.* A 24-bit CRC is calculated for and appended to each transport block.
- *Code-block segmentation and per-code-block CRC insertion.* In the same way as for the downlink, code-block segmentation, including per-code-block CRC insertion, is applied for transport blocks larger than 6144 bits.
- *Channel coding.* Rate-1/3 Turbo coding with QPP-based inner interleaving is also used for the uplink shared channel.
- *Rate matching and physical-layer hybrid-ARQ functionality.* The physical-layer part of the uplink rate-matching and hybrid-ARQ functionality is essentially the same as the corresponding downlink functionality, with subblock interleaving and insertion into a circular buffer followed by bit selection with four redundancy versions. There are some important differences between the downlink and uplink hybrid-ARQ protocols, such as asynchronous versus synchronous operation as described in Chapter 8. However, these differences are not visible in terms of the physical-layer processing.

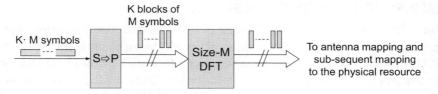

FIGURE 7.2

DFT precoding of K blocks, each consisting of M modulation symbols.

- *Bit-level scrambling.* The aim of uplink scrambling is the same as for the downlink—that is, to randomize, in this case, the *uplink* interference to ensure that the processing gain provided by the channel code can be fully utilized.
- *Data modulation.* Similar to the downlink, QPSK, 16QAM, and 64QAM modulation can also be used for uplink shared-channel transmission, while 256QAM is not supported.
- *DFT precoding.* As illustrated in Figure 7.2, the modulation symbols, in blocks of M symbols, are fed through a size-M DFT, where M corresponds to the number of subcarriers assigned for the transmission. The reason for the DFT precoding is to reduce the cubic metric [10] for the transmitted signal, thereby enabling higher power-amplifier efficiency. From an implementation complexity point of view the DFT size should preferably be constrained to a power of 2. However, such a constraint would limit the scheduler flexibility in terms of the amount of resources that can be assigned for an uplink transmission. Rather, from a flexibility point of view all possible DFT sizes should preferably be allowed. For LTE, a middle-way has been adopted where the DFT size, and thus also the size of the resource allocation, is limited to products of the integers 2, 3, and 5. Thus, for example, DFT sizes of 60, 72, and 96 are allowed but a DFT size of 84 is not allowed.[2] In this way, the DFT can be implemented as a combination of relatively low-complex radix-2, radix-3, and radix-5 FFT processing.
- *Antenna mapping.* The antenna mapping maps the output of the DFT precoder to one or several uplink antenna ports for subsequent mapping to the physical resource (the OFDM time—frequency grid). In the first releases of LTE (releases 8 and 9), only single-antenna transmission was used for the uplink.[3] However, as part of LTE release 10, support for uplink multi-antenna transmission by means of antenna precoding with up to four antennas ports was introduced. More details about LTE uplink multi-antenna transmission are provided in Section 7.3.

---

[2]As uplink resource assignments are always done in terms of resource blocks of size 12 subcarriers, the DFT size is always a multiple of 12.

[3]Uplink multi-antenna transmission in form of *antenna selection* has been part of the LTE specification since release 8. However, it is an optional device feature with limited implementation in commercially available devices.

7.1.2 MAPPING TO THE PHYSICAL RESOURCE

The scheduler assigns a set of resource-block pairs to be used for the uplink transmission, more specifically for transmission of the physical uplink shared channel (PUSCH) that carries the UL-SCH transport channel. Each such resource-block pair spans 14 OFDM symbols in time (one subframe).[4] However, as will be described in Section 7.2.1, two of these symbols are used for uplink *demodulation reference signals* (DM-RS) and are thus not available for PUSCH transmission. Furthermore, one additional symbol may be reserved for the transmission of *sounding reference signals* (SRS), see Section 7.2.2. Thus, 11 or 12 OFDM symbols are available for PUSCH transmission within each subframe.

Figure 7.3 illustrates how $K \cdot M$ DFT-precoded symbols at the output of the antenna mapping are mapped to the basic OFDM time—frequency grid, where K is the number of available OFDM symbols within a subframe (11 or 12 according to the text in the preceding paragraphs) and M is the assigned bandwidth in number of subcarriers. As there are 12 subcarriers within a resource-block pair, $M = N \cdot 12$ where N is the number of assigned resource-block pairs.

Figure 7.3 assumes that the set of assigned resource-block pairs are contiguous in the frequency domain. This is the typical assumption for DFTS-OFDM and was strictly the case for LTE releases 8 and 9. Mapping of the DFT-precoded signal to frequency-contiguous resource elements is preferred in order to retain good cubic-metric properties of the uplink transmission. At the same time, such a restriction implies an additional constraint for the uplink scheduler, something which may not always be desirable. Therefore, LTE release 10 introduced the possibility to assign partly frequency-separated resources for PUSCH transmission. More specifically, in release 10 the assigned uplink resource may consist of a maximum of two frequency-separated *clusters* as illustrated in Figure 7.4, where each cluster consists of a number of resource-block pairs (N_1 and N_2 resource-block pairs, respectively). In the case of such *multi-cluster transmission*, a single DFT

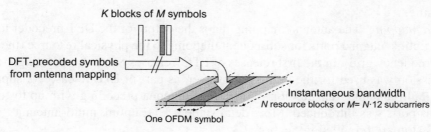

K blocks of M symbols

DFT-precoded symbols from antenna mapping

One OFDM symbol

Instantaneous bandwidth N resource blocks or $M = N \cdot 12$ subcarriers

FIGURE 7.3

Mapping to the uplink physical resource.

[4]Twelve symbols in the case of extended cyclic prefix.

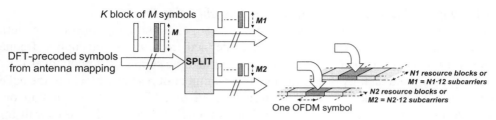

FIGURE 7.4

Uplink multi-cluster transmission.

precoding spans the overall assigned resource in the frequency domain—that is, both clusters. This means that the total assigned bandwidth in number of subcarriers ($M = M1 + M2$) should be aligned with the restrictions on available DFT sizes described in the preceding paragraphs.

7.1.3 PUSCH FREQUENCY HOPPING

In Chapter 6 it was described how the notion of *virtual resource blocks* (VRBs) in combination with the mapping from VRBs to *physical resource blocks* (PRBs) allowed for downlink distributed transmission—that is, the spreading of a downlink transmission in the frequency domain. As described, downlink distributed transmission consists of two separate steps: (1) a mapping from VRB pairs to PRB pairs such that frequency-consecutive VRB pairs are not mapped to frequency-consecutive PRB pairs and (2) a split of each resource-block pair such that the two resource blocks of the resource-block pair are transmitted with a certain frequency gap in between. This second step can be seen as frequency hopping on a slot basis.

The notion of VRBs can also be used for the LTE uplink, allowing for frequency-domain-distributed transmission for the uplink. However, in the uplink, where transmission from a device should always be over a set of consecutive subcarriers in the absence of multi-cluster transmission, distributing resource-block pairs in the frequency domain, as in the first step of downlink distributed transmission, is not possible. Rather, uplink distributed transmission is similar to the second step of downlink distributed transmission—that is, a frequency separation of the transmissions in the first and second slots of a subframe. Uplink distributed transmission for PUSCH can thus more directly be referred to as *uplink frequency hopping*.

There are two types of uplink frequency hopping defined for PUSCH:

- subband-based hopping according to cell-specific hopping/mirroring patterns;
- hopping based on explicit hopping information in the scheduling grant.

Uplink frequency hopping is not supported for multi-cluster transmission as, in that case, sufficient diversity is assumed to be obtained by proper location of the two clusters.

7.1.3.1 Hopping Based on Cell-Specific Hopping/Mirroring Patterns

To support subband-based hopping according to cell-specific hopping/mirroring patterns, a set of consecutive subbands of a certain size is defined from the overall uplink frequency band, as illustrated in Figure 7.5. It should be noted that the subbands do not cover the total uplink frequency band, mainly due to the fact that a number of resource blocks at the edges of the uplink frequency band are used for transmission of L1/L2 control signaling on the physical uplink control channel (PUCCH). For example, in Figure 7.5, the overall uplink bandwidth corresponds to 50 resource blocks and there are a total of four subbands, each consisting of 11 resource blocks. Six resource blocks are not included in the hopping bandwidth and could, for example, be used for PUCCH transmission.

In the case of subband-based hopping, the set of VRBs provided in the scheduling grant are mapped to a corresponding set of PRBs according to a cell-specific hopping pattern. The resource to use for transmission, the *PRBs*, is obtained by shifting the VRBs provided in the scheduling grant by a number of subbands according to the hopping pattern, where the hopping pattern can provide different shifts for each slot. As illustrated in Figure 7.6, a device is assigned VRBs 27, 28, and 29. In the first slot, the predefined hopping pattern takes the

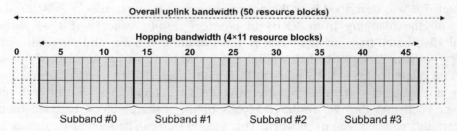

FIGURE 7.5

Definition of subbands for PUSCH hopping. The figure assumes a total of four subbands, each consisting of eleven resource blocks.

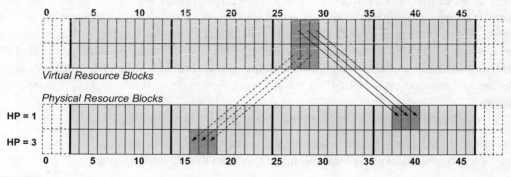

FIGURE 7.6

Hopping according to predefined hopping pattern.

value 1, implying transmission using PRBs one subband to the right—that is, PRBs 38, 39, and 40. In the second slot, the predefined hopping pattern takes the value 3, implying a shift of three subbands to the right in the figure and, consequently, transmission using resource blocks 16, 17, and 18. Note that the shifting "wraps-around"—that is, in the case of four subbands, a shift of three subbands is the same as a negative shift of one subband. As the hopping pattern is cell specific—that is, the same for all devices within a cell—different devices will transmit on nonoverlapping physical resources as long as they are assigned nonoverlapping virtual resources.

In addition to the hopping pattern, there is also a cell-specific *mirroring pattern* defined in a cell. The mirroring pattern controls, on a slot basis, whether or not mirroring within each subband should be applied to the assigned resource. In essence, mirroring implies that the resource blocks within each subband are numbered right to left instead of left to right. Figure 7.7 illustrates mirroring in combination with hopping. Here, the mirroring pattern is such that mirroring is not applied to the first slot while mirroring is applied to the second slot.

Both the hopping pattern and the mirroring pattern depend on the physical-layer cell identity and are thus typically different in neighboring cells. Furthermore, the period of the hopping/mirroring patterns corresponds to one frame.

7.1.3.2 Hopping Based on Explicit Hopping Information

As an alternative to hopping/mirroring according to cell-specific hopping/mirroring patterns as described in the preceding section, uplink slot-based frequency hopping for PUSCH can also be controlled by *explicit hopping information* provided in the scheduling grant. In such a case the scheduling grant includes

- information about the resource to use for uplink transmission in the first slot, exactly as in the nonhopping case;
- additional information about the offset of the resource to use for uplink transmission in the second slot, relative to the resource of the first slot.

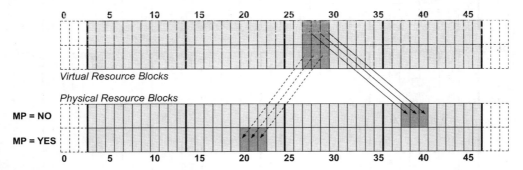

FIGURE 7.7

Hopping/mirroring according to predefined hopping/mirroring patterns. Same hopping pattern as in Figure 7.6.

Selection between hopping according to cell-specific hopping/mirroring patterns as discussed or hopping according to explicit information in the scheduling grant can be done dynamically. More specifically, for cell bandwidths less than 50 resource blocks, there is a single bit in the scheduling grant indicating if hopping should be according to the cell-specific hopping/mirroring patterns or should be according to information in the scheduling grant. In the latter case, the hop is always half of the hopping bandwidth. In the case of larger bandwidths (50 resource blocks and beyond), there are two bits in the scheduling grant. One of the combinations indicates that hopping should be according to the cell-specific hopping/mirroring patterns while the three remaining alternatives indicate hopping of 1/2, +1/4, and −1/4 of the hopping bandwidth. Hopping according to information in the scheduling grant for the case of a cell bandwidth corresponding to 50 resource blocks is illustrated in Figure 7.8. In the first subframe, the scheduling grant indicates a hop of one-half the hopping bandwidth. In the second subframe, the grant indicates a hop of one-fourth the hopping bandwidth (equivalent to a negative hop of three-fourths of the hopping bandwidth). Finally, in the third subframe, the grant indicates a negative hop of one-fourth the hopping bandwidth.

7.2 UPLINK REFERENCE SIGNALS

Similar to the downlink, reference signals are also transmitted on the LTE uplink. There are two types of reference signals defined for the LTE uplink:

- Uplink DM-RS are intended to be used by the base station for channel estimation for coherent demodulation of the uplink physical channels (PUSCH and PUCCH). A DM-RS is thus only transmitted together with PUSCH or PUCCH and is then spanning the same frequency range as the corresponding physical channel.

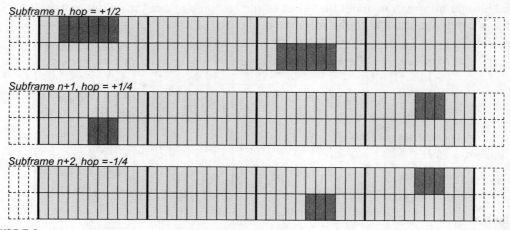

FIGURE 7.8

Frequency hopping according to explicit hopping information.

- Uplink SRS are intended to be used by the base station for channel-state estimation to support uplink channel-dependent scheduling and link adaptation. The SRS can also be used in other cases when uplink transmission is needed although there is no data to transmit. One example is when uplink transmission is needed for the network to be able to estimate the uplink receive timing as part of the *uplink-timing-alignment procedure*, see Section 7.6.

7.2.1 DEMODULATION REFERENCE SIGNALS

Uplink DM-RS are intended to be used for channel estimation for coherent demodulation of the PUSCH to which the UL-SCH transport channel is mapped, as well as for the PUCCH which carries different types of uplink L1/L2 control signaling. The basic principles for uplink DM-RS are the same for PUSCH and PUCCH transmission although there are some differences—for example, in terms of the exact set of OFDM symbols in which the reference signals are transmitted. The discussion in the following primarily focuses on PUSCH DM-RM. Some additional details on the PUCCH DM-RS structure is provided in Section 7.4 as part of the more general description of uplink L1/L2 control signaling.

7.2.1.1 Time–Frequency Structure

Due to the importance of low cubic metric and corresponding high power-amplifier efficiency for uplink transmissions, the principles for uplink reference-signal transmission are different compared to the downlink. In essence, transmitting reference signals frequency multiplexed with other uplink transmissions from the same device is not suitable for the uplink as that would negatively impact the device power-amplifier efficiency due to increased cubic metric. Instead, certain OFDM symbols within a subframe are used exclusively for DM-RS transmission—that is, the reference signals are *time multiplexed* with other uplink transmissions (PUSCH and PUCCH) from the same device. The structure of the reference signal itself then ensures a low cubic metric within these symbols as described in the following.

More specifically, in case of PUSCH transmission DM-RS is transmitted within the fourth symbol of each uplink slot[5] (Figure 7.9). Within each subframe, there are thus two reference-signal transmissions, one in each slot.

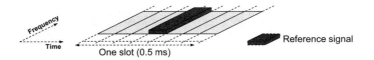

Frequency
Time
One slot (0.5 ms)
Reference signal

FIGURE 7.9

Transmission of uplink DM-RS within a slot in case of PUSCH transmission.

---

[5]The third symbol in the case of extended cyclic prefix.

In case of PUCCH transmission, the number of OFDM symbols used for DM-RS transmission in a slot, as well as the exact position of these symbols, differs between different *PUCCH formats* as further described in Section 7.4.

In general, there is no reason to estimate the channel outside the frequency band of the corresponding PUSCH/PUCCH transmission that is to be coherently demodulated. The frequency range spanned by an uplink DM-RS is therefore equal to the instantaneous frequency range spanned by the corresponding PUSCH/PUCCH transmission. This means that, for PUSCH transmission, it should be possible to generate reference signals of different bandwidths, corresponding to the possible bandwidths of a PUSCH transmission. More specifically, it should be possible to generate reference signals of a bandwidth corresponding to $12 \cdot N$ sub carriers, where N corresponds to the bandwidth of the PUSCH transmission measured in number of resource blocks.[6]

Regardless of the kind of uplink transmission (PUSCH or PUCCH), the basic structure of each reference-signal transmission is the same. As illustrated in Figure 7.10, an uplink DM-RS can be defined as a *frequency-domain reference-signal sequence* applied to consecutive inputs of an OFDM modulator—that is, to consecutive subcarriers. Referring to the preceding discussion, in case of PUSCH transmission, the frequency-domain reference-signal sequence should have a length $M = 12 \cdot N$ where N corresponds to the PUSCH bandwidth measured in number of resource blocks. In case of PUCCH transmission, the length of the reference-signal sequence should always be equal to 12.

7.2.1.2 Base Sequences

Uplink reference signals should preferably have the following properties:

- Small power variations in the frequency domain to allow for similar channel-estimation quality for all frequencies spanned by the reference signal. Note that this is equivalent to a well-focused time-domain auto-correlation of the transmitted reference signal.

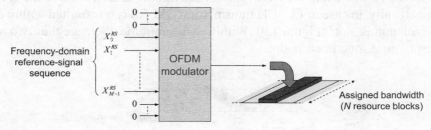

FIGURE 7.10

Generation of uplink DM-RS from a frequency-domain reference-signal sequence.

---

[6]Due to the imposed limitations on supported DFT sizes as described in Section 7.1.1 there will be some additional constraints on N.

- Limited power variations in the time domain, leading to low cubic metric of the transmitted signal.

Furthermore, a sufficient number of reference-signal sequences of a given length, corresponding to a certain reference-signal bandwidth, should be available in order to avoid an unreasonable planning effort when assigning reference-signal sequences to devices and cells.

So-called *Zadoff–Chu* sequences [34] have the property of constant power in both the frequency and time domains. The M_{ZC} elements of the q:th Zadoff–Chu sequence within the set of Zadoff–Chu sequences of (odd) length M_{ZC} can be expressed as:

$$Z_k^q = e^{-j\pi q \frac{k \cdot (k+1)}{M_{ZC}}} \quad 0 \le k < M_{ZC} \tag{7.1}$$

From the point of view of small power variations in both the frequency and time domains, Zadoff–Chu sequences would thus be excellent as uplink reference-signal sequences. However, there are two reasons why Zadoff–Chu sequences are not suitable for direct use as uplink reference-signal sequences in LTE:

- The number of available Zadoff–Chu sequences of a certain length, corresponding to the number of possible values for the parameter q in Eq. (7.1), equals the number of integers that are relative prime to the sequence length M_{ZC}. To maximize the number of Zadoff–Chu sequences and thus, in the end, to maximize the number of available uplink reference signals, prime-length Zadoff–Chu sequences would therefore be preferred. At the same time, according to above, the length of the uplink reference-signal sequences should be a multiple of 12, which is not a prime number.
- For short sequence lengths, corresponding to narrow uplink transmission bandwidths, relatively few reference-signal sequences would be available even if they were based on prime-length Zadoff–Chu sequences.

Instead, for sequence lengths larger than or equal to 36, corresponding to transmission bandwidths larger than or equal to three resource blocks, basic reference-signal sequences, in the LTE specification referred to as *base sequences*, are defined as *cyclic extensions* of Zadoff–Chu sequences of length M_{ZC} (Figure 7.11), where M_{ZC} is the largest prime number smaller than or equal to the length of the reference-signal sequence. For example, the largest prime number less than or equal to 36 is 31, implying that reference-signal sequences of length 36 are defined as cyclic extensions of Zadoff–Chu sequences of length 31. The number of available sequences is then equal to 30—that is, one less than the length of the Zadoff–Chu sequence. For larger sequence lengths, more sequences are available. For example, for a sequence length equal to 72, there are 70 sequences available.[7]

For sequence lengths equal to 12 and 24, corresponding to transmission bandwidths of one and two resource blocks, respectively, special QPSK-based sequences have instead been

---

[7]The largest prime number smaller than or equal to 72 is 71. The number of sequences is then one less than the length of the Zadoff–Chu sequence, that is, 70.

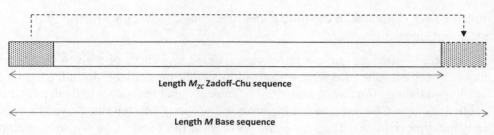

FIGURE 7.11

Length-M basic reference-signal sequence derived from cyclic extension of a length-M_{ZC} Zadoff–Chu sequence.

found from computer searches and are explicitly listed in the LTE specifications. For each of the two sequence lengths, 30 sequences have been defined.

Thus there are at least 30 sequences available for each sequence length. However, not all of these sequences are actually being used as base sequences:

- for sequence lengths less than 72, corresponding to reference-signal bandwidths less than six resource blocks, 30 sequences are being used;
- for sequence lengths equal to 72 and beyond, corresponding to reference-signal bandwidths of six resource blocks and beyond, 60 sequences are being used.

These sequences are divided into 30 *sequence groups* where each group consists of one base sequence for each sequence length less than 72 and two base sequences for each sequence length equal to 72 and beyond. A base sequence of a given length is thus fully specified by a *group index* ranging from 0 to 29 together with, in case of sequence lengths equal to 72 and beyond, a *sequence index* taking the values 0 and 1.

7.2.1.3 Phase-Rotation and Orthogonal Cover Codes

From the base sequences previously described, additional reference-signal sequences can be generated by applying different linear phase rotations in the frequency domain, as illustrated in Figure 7.12.

Applying a linear phase rotation in the frequency domain is equivalent to applying a cyclic shift in the time domain. Thus, although being *defined* as different frequency-domain phase rotations in line with Figure 7.12, the LTE specification actually refers to this as applying different *cyclic shifts*. Here the term "phase rotation" will be used. However, it should be borne in mind that what is here referred to as phase rotation is referred to as cyclic shift in the LTE specifications.

DM-RS derived from different base sequences typically have relatively low but still nonzero cross correlation. In contrast, reference signals defined from different phase rotations of the same base sequence are, at least in theory, completely orthogonal if the parameter α in

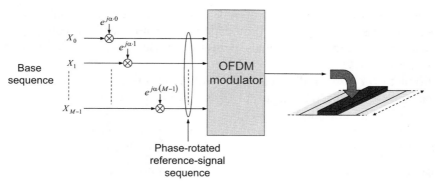

FIGURE 7.12

Generation of uplink reference-signal sequence from phase-rotated base sequence.

Figure 7.12 takes a value $m\pi/6$, where the integer m ranges from 0 to 11.[8] Up to 12 orthogonal reference signals can thus be derived from each base sequence by using different values of the parameter m.

However, to retain the orthogonality between these reference signals at the receiver side, the frequency response of the channel should be essentially constant over the span of one resource block—that is, over 12 subcarriers. Alternatively expressed, the main part of the channel time dispersion should not extend beyond the length of the cyclic shift mentioned in the preceding paragraphs. If that is not the case, a subset of the available values for α may be used—for example, only the values $\{0, 2\pi/6, 4\pi/6, \ldots, 10\pi/6\}$ or perhaps even fewer values. Limiting the set of possible values for α implies orthogonality over a smaller number of subcarriers and, as a consequence, less sensitivity to channel frequency selectivity. In other words, there is a trade-off between the number of orthogonal reference signals that can be generated from different phase rotations and the amount of channel frequency-selectivity that should be possible to cope with.

Another prerequisite for receiver-side orthogonality between reference signals defined from different phase rotations of the same base sequence is that the reference signals are transmitted well aligned in time. Thus the main uses of phase rotation include:

- to provide multiple simultaneous reference signals from the same device in case of uplink multi-layer transmission (uplink spatial multiplexing, see also Section 6.3.1);
- to provide the possibility for orthogonal reference signals between multiple devices being scheduled for PUSCH transmission on the same resource—that is, same set of resource blocks, within a cell (uplink MU-MIMO, see Section 6.3.2).

---

[8]The orthogonality is due to the fact that, for $\alpha = m\pi/6$, there will be an integer number of full-circle rotations over 12 subcarriers—that is, over one resource block.

Phase rotations may also be used to provide orthogonal reference signals between devices in neighbor cells, assuming tight synchronization between the cells. Finally, phase rotations are also used to separate reference signals of different devices in case of PUCCH transmission (see further Section 7.4).

In addition to the use of different phase rotations, orthogonal reference-signal transmissions can also be achieved by means of *Orthogonal Cover Codes* (OCC). As illustrated in Figure 7.13, two different length-two OCCs ([+1 +1] and [+1 −1], respectively) can be applied to the two PUSCH reference-signal transmissions within a subframe. This allows for overall DM-RS orthogonality over the subframe assuming that

- the channel does not change substantially over the subframe;
- the reference signals of the two slots are the same.[9]

Similar to phase rotations, receiver-side orthogonality between reference-signal transmissions based on different OCC requires that the transmissions are well-aligned in time at the receiver side. Thus the use of OCC is essentially the same as for phase rotations as described earlier:

- to provide multiple reference signals from the same device in case of uplink spatial multiplexing;
- to provide orthogonal reference signals between devices being scheduled on the same resource within a cell (uplink MU-MIMO);
- to allow for reference-signal orthogonality between uplink transmissions within neighbor cells in case of tight synchronization and time alignment between the cells.

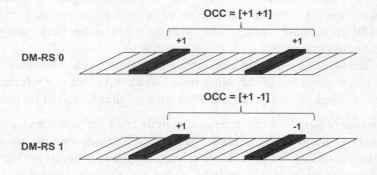

FIGURE 7.13

Generation of multiple DM-RS from orthogonal cover codes.

---

[9]Strictly speaking, the only thing required is that the correlation between the reference signals of DM-RS 0 and DM-RS 1 is the same for the two slots. If the reference signals are the same for the two slots, this is obviously the case.

It should be noted that, in contrast to phase rotations, orthogonality by means of OCC does not require that the same base sequence is used for the two DM-RS (DM-RS 0 and DM-RS 1 in Figure 7.13). Actually, the two reference signals do not even need to have the same bandwidth; having the same cross-correlation between the reference signals of DM-RS 0 and DM-RS 1 for the two slots is sufficient. Thus OCC can be used to achieve reference-signal orthogonality also for PUSCH transmissions of different bandwidths.

Similar to phase rotations, orthogonal codes can also be applied to DM-RS in case of PUCCH transmission, although in a somewhat different way compared to PUSCH due to the different time-domain structure of PUCCH DM-RS, see further Section 7.4.

7.2.1.4 Base-Sequence Assignment

According to the preceding discussion, each base sequence of a given length corresponds to a unique combination of a group index ranging from 0 to 29 and a sequence index taking the values 0 or 1. Base-sequence assignment—that is, determining which base sequence should be used by a specific device—is thus equivalent to assigning a corresponding group and sequence index.[10]

Prior to release 11, base-sequence assignment was *cell specific*—that is, for a given slot the group and sequence indices were the same for all devices having the same serving cell.

In the case of a *fixed* (*nonhopping*) group assignment, the sequence group to use for PUCCH transmission does not change between slots and was prior to release 11 directly given by the physical-layer cell identity. More specifically, the group index was equal to the cell identity modulo 30, where the cell identity may take values in the range 0 to 503 as described in Chapter 11. Thus, cell identities 0, 30, 60, …, 480 corresponded to sequence group 0, cell identities 1, 31, 61, …, 481 to sequence group 1, and so on.

In contrast, what sequence group to use for PUSCH transmission could be explicitly configured on a cell basis by adding an offset provided as part of the cell system information, to the PUCCH group index. The reason for providing the possibility for explicitly indicating what sequence group to use for PUSCH transmission in a cell was that it should be possible to use the same sequence group for PUSCH transmission in neighboring cells, despite the fact that such cells typically have different cell identities. In this case, the reference signals for PUSCH transmissions within the two cells would instead be distinguished by different phase rotations and/or OCC as discussed in Section 7.2.1.3, allowing for reference-signal orthogonality also *between* cells.[11]

In the case of *group hopping*, an additional cell-specific *group-hopping pattern* is added to the group index allowing for the group index of a cell to change on a slot basis. Prior to release 11, the group-hopping pattern was also derived from the cell identity and identical group-hopping patterns were used for PUSCH and PUCCH within a cell.

---

[10]For base-sequence lengths less than 72, the sequence index is always equal to zero.

[11]This assumes tight synchronization and time alignment between the cells.

In addition to the group index, for sequence lengths equal to or larger than 72, the reference-signal sequence also depends on the *sequence index*. The sequence index can either be fixed (nonhopping), in which case it always equals 0, or vary between slots (hopping) according to a *sequence-hopping pattern*. Similar to group hopping, prior to release 11 also the sequence-hopping pattern was cell-specific and given by the physical-layer cell identity.

In LTE release 11, the possibility for *device specific* base sequence assignment was introduced—that is, the group and sequence indices to use for PUSCH and PUCCH can, with release 11, be explicitly configured for a specific device regardless of the identity of the serving cell. The introduction of device-specific base-sequence assignment was done very similarly to the introduction of device-specific downlink DM-RS as described in Section 6.2.2—that is, by introducing the possibility for explicitly configuring a device with something that can be seen as a "virtual cell identity" which, if configured, replaces that actual physical-cell identity when deriving the group and sequence index. Similar to downlink DM-RS, if no virtual cell identity is configured the device should assume cell-specific base-sequence assignment according to the preceding discussion.

It should be pointed out that the device is not configured with a single "virtual-cell identity" to be used for deriving the base sequences for PUSCH and PUCCH. Rather, the device-specific configuration is done separately for PUSCH and PUCCH using two different "virtual cell identities."

The main reason for introducing the possibility for full device-specific assignment of uplink reference signals was to enhance the support for uplink *multi-point reception* (CoMP). In case of uplink multi-point reception, an uplink transmission may be received at a reception point not corresponding to the serving cell of the device but at a reception point corresponding to another cell. To enhance the reception quality it is beneficial to have orthogonal reference signals for different devices received at this reception point despite the fact that they, strictly speaking, have different serving cells. To allow for this, the reference-signal sequences should not be cell-specific but rather possible to assign on a device basis. Multi-point reception is discussed in somewhat more detail in Chapter 13 as part of a more general discussion on multi-point coordination and transmission.

7.2.1.5 *Assignment of Phase Rotation and OCC*

As discussed previously, the main use for phase rotations and OCC in case of PUSCH DM-RS is to provide a possibility for orthogonal reference signals for different layers in case of spatial multiplexing and for different devices scheduled on the same resource either within one cell (MU-MIMO) or in neighboring tightly synchronized cells.

In order not to limit the flexibility in terms of what devices can be jointly scheduled on the same resource, the assignment of phase rotation and OCC can be done dynamically. Thus, the exact phase rotation, given by the phase parameter m in Figure 7.12 and the OCC is provided jointly as a single parameter included as part of the uplink scheduling grant provided by the network. Each value of this parameter corresponds to a certain combination of phase rotation and OCC for each layer to be transmitted by the device. In case of spatial multiplexing, the different layers will then inherently be assigned different phase shifts and, possibly, different OCC. By providing different parameter values to different devices, the devices will be assigned different phase-shifts/OCC combinations allowing for orthogonal reference signals and thus providing enhanced MU-MIMO performance either within a cell or between cells.

7.2.2 SOUNDING REFERENCE SIGNALS

The DM-RS discussed in Section 7.2.1 are intended to be used by the base station for channel estimation to allow for coherent demodulation of uplink physical channels (PUSCH or PUCCH). A DM-RS is always transmitted together with and spanning the same frequency range as the corresponding physical channel.

In contrast, SRS are transmitted on the uplink to allow for the base station to estimate the uplink *channel state* at different frequencies. The channel-state estimates can then, for example, be used by the base-station scheduler to assign resource blocks of instantaneously good quality for uplink PUSCH transmission from the specific device (uplink channel-dependent scheduling). They can also be used to select different transmission parameters such as the instantaneous data rate and different parameters related to uplink multi-antenna transmission. The channel information obtained from the SRS can also be used for downlink transmission purposes exploiting channel reciprocity, for example downlink channel-dependent scheduling in TDD systems. As mentioned earlier, SRS transmission can also be used in other situations when uplink transmission is needed although there is no data to transmit, for example, for uplink timing estimation as part of the uplink-timing-alignment procedure. Thus, an SRS is not necessarily transmitted together with any physical channel and if transmitted together with, for example, PUSCH, the SRS may span a different, typically larger, frequency range.

There are two types of SRS transmission defined for the LTE uplink: *periodic* SRS transmission which has been available from the first release of LTE (release 8) and *aperiodic* SRS transmission introduced in LTE release 10.

7.2.2.1 Periodic SRS Transmission

Periodic SRS transmission from a device occurs at regular time intervals, from as often as once every 2 ms (every second subframe) to as infrequently as once every 160 ms (every 16th frame). When SRS is transmitted in a subframe, it occupies the last symbol of the subframe as illustrated in Figure 7.14. As an alternative, in the case of TDD operation, SRS can also be transmitted within the UpPTS.

In the frequency domain, SRS transmissions should span the frequency range of interest for the scheduler. This can be achieved in two ways:

- By means of a sufficiently wideband SRS transmission that allows for sounding of the entire frequency range of interest with a single SRS transmission as illustrated in the upper part of Figure 7.15.

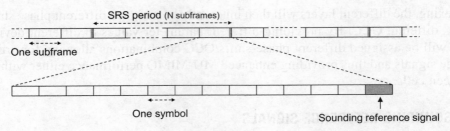

FIGURE 7.14

Transmission of SRS.

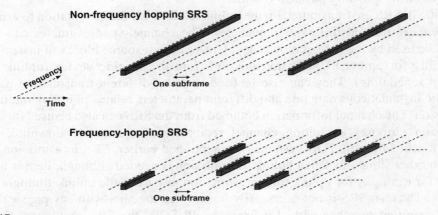

FIGURE 7.15

Nonfrequency-hopping (wideband) versus frequency-hopping SRS.

- By means of a more narrowband SRS transmission that is hopping in the frequency domain in such a way that a sequence of SRS transmissions jointly spans the frequency range of interest, as illustrated in the lower part of Figure 7.15.

The instantaneous SRS bandwidth is always a multiple of four resource blocks. Different bandwidths of the SRS transmission can simultaneously be available within a cell. A narrow SRS bandwidth, corresponding to four resource blocks, is always available, regardless of the uplink cell bandwidth. Up to three additional, more wideband SRS bandwidths may also be configured within a cell. A specific device within the cell is then explicitly configured to use one of the up to four SRS bandwidths available in the cell.

If a device is transmitting SRS in a certain subframe, the SRS transmission may very well overlap, in the frequency domain, with PUSCH transmissions from other devices within the cell. To avoid collision between SRS and PUSCH transmissions from different devices, devices should in general avoid PUSCH transmission in the OFDM symbols in which SRS transmission may occur. To achieve this, all devices within a cell are aware of the set of subframes within which SRS *may* be transmitted by *any* device within the cell. All devices should then avoid PUSCH transmission in the last OFDM symbol of those subframes. Information about the set of subframes in which SRS may be transmitted within a cell is provided as part of the cell system information.[12]

On a more detailed level, the structure for SRS is similar to that of uplink DM-RS described in Section 7.2.1. More specifically, a SRS is also defined as a frequency-domain reference-signal sequence derived in the same way as for DM-RS—that is, cyclic extension of prime-length Zadoff—Chu sequences for sequence lengths equal to 30 and above and special sequences for sequence lengths less than 30. However, in the case of SRS, the reference-signal sequence is mapped to *every second* subcarrier, creating a "comb"-like spectrum, as illustrated in Figure 7.16. Taking into account that the bandwidth of the SRS transmission is always a multiple of four resource blocks, the lengths of the reference-signal sequences for SRS are thus always multiples of 24.[13] The reference-signal sequence to use for SRS transmission within the cell is derived in the same way as the PUCCH DM-RS within the cell, assuming cell-specific reference-signal sequence assignment. Device-specific reference-signal sequences are not supported for SRS.

Starting from release 13, up to four different combs can be used instead of two as in previous releases, subject to higher-layer configuration. In case of four different combs, every fourth subcarrier is used instead of every second. The purpose of this is to increase the SRS multiplexing capacity to handle the increased number of antennas supported with the introduction of FD-MIMO, see Chapter 10.

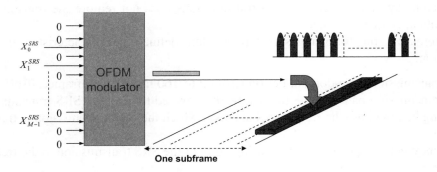

FIGURE 7.16

Generation of SRS from a frequency-domain reference-signal sequence.

---

[12]What is provided as part of the system information is a periodicity (2 to 160 ms) and a subframe offset, compare the following bullet list.

[13]Four resource blocks, each spanning 12 subcarriers but only every second subcarrier used for a certain SRS transmission.

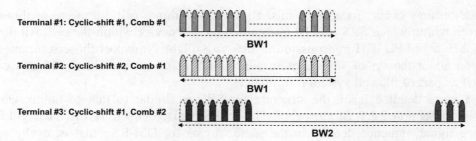

FIGURE 7.17

Multiplexing of SRS transmissions from different devices.

Similar to DM-RS, different phase rotations, also for SRS referred to as "cyclic shifts" in the LTE specifications, can be used to generate different SRS that are orthogonal to each other. By assigning different phase rotations to different devices, multiple SRS can thus be transmitted in parallel in the same subframe, as illustrated by devices #1 and #2 in the upper part of Figure 7.17. However, it is then required that the reference signals span the same frequency range.

Another way to allow for SRS to be simultaneously transmitted from different devices is to rely on the fact that each SRS only occupies every second (or every fourth) subcarrier. Thus, SRS transmissions from two devices can be *frequency multiplexed* by assigning them to different frequency shifts or "combs," as illustrated by device #3 in the lower part of Figure 7.17. In contrast to the multiplexing of SRS transmission by means of different "cyclic shifts," frequency multiplexing of SRS transmissions does not require the transmissions to cover identical frequency ranges.

To summarize, the following set of parameters defines the characteristics of an SRS transmission:

- SRS transmission time-domain period (from 2 to 160 ms) and subframe offset.
- SRS transmission bandwidth—the bandwidth covered by a single SRS transmission.
- Hopping bandwidth—the frequency band over which the SRS transmission is frequency hopping.
- Frequency-domain position—the starting point of the SRS transmission in the frequency domain.
- Transmission comb as illustrated in Figure 7.17.
- Phase rotation (or equivalently cyclic shift) of the reference-signal sequence.
- Number of combs (introduced in release 13).

A device that is to transmit SRS is configured with these parameters by means of higher layer (RRC) signaling.

7.2.2.2 Aperiodic SRS Transmission

In contrast to periodic SRS, aperiodic SRS are *one-shot* transmissions triggered by signaling on PDCCH as part of the scheduling grant. The frequency-domain structure of an aperiodic SRS transmission is identical to that of periodic SRS. Also, in the same way as for periodic SRS transmission, aperiodic SRS are transmitted within the last symbol of a subframe. Furthermore, the time instants when aperiodic SRS may be transmitted are configured per device using higher-layer signaling.

The frequency-domain parameters for aperiodic SRS (bandwidth, odd or even "comb," and so on) are configured by higher-layer (RRC) signaling. However, no SRS transmission will actually be carried out until the device is explicitly triggered to do so by an explicit *SRS trigger* on PDCCH/EPDCCH. When such a trigger is received, a single SRS is transmitted in the next available aperiodic SRS instant configured for the device using the configured frequency-domain parameters. Additional SRS transmissions can then be carried out if additional triggers are received.

Three different parameter sets can be configured for aperiodic SRS, for example differing in the frequency position of the SRS transmission and/or the transmission comb. Information on what parameters to use when the SRS is actually transmitted is included in the PDCCH/EPDCCH L1/L2 control signaling information, which consists of two bits, three combinations of which indicate the specific SRS parameter set. The fourth combination simply dictates that no SRS should be transmitted.

7.3 UPLINK MULTI-ANTENNA TRANSMISSION

Downlink multi-antenna transmission was supported by the LTE specification from its first release (release 8). With LTE release 10, support for uplink multi-antenna transmission—that is, uplink transmission relying on multiple transmit antennas at the device side—was also introduced for LTE. Uplink multi-antenna transmission can be used to improve the uplink link performance in different ways:

- to improve the achievable data rates and spectral efficiency for uplink data transmission by allowing for antenna precoding supporting uplink beam-forming as well as spatial multiplexing with up to four layers for the uplink physical data channel PUSCH;
- to improve the uplink control-channel performance by allowing for transmit diversity for the PUCCH.

7.3.1 PRECODER-BASED MULTI-ANTENNA TRANSMISSION FOR PUSCH

As illustrated in Figure 7.18, the structure of the uplink antenna precoding is very similar to that of downlink antenna precoding (Section 6.3), including the presence of precoded DM-RS (one per layer) similar to downlink non-codebook-based precoding (Figure 6.18). Uplink

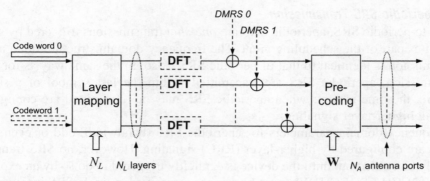

FIGURE 7.18

Precoder-based multi-antenna transmission for LTE uplink.

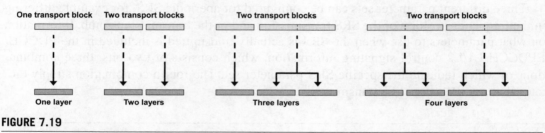

FIGURE 7.19

Uplink transport-channel-to-layer mapping (initial transmission).

antenna precoding supports transmission using up to four antenna ports, allowing for spatial multiplexing with up to four layers.

The principles for mapping of the modulation symbols to layers are also the same as for the downlink. For an initial transmission, there is one transport block in the case of a single layer and two transport blocks for more than one layer, as illustrated in Figure 7.19. Similar to the downlink, in the case of a hybrid-ARQ retransmission, a single transport block may also be transmitted on multiple layers in some cases.

As can be seen in Figure 7.18, the DFT precoding is actually taking place after layer mapping—that is, each layer is separately DFT precoded. To simplify the description this was not really visible in Figure 7.1 outlining the overall physical-layer transport-channel processing.

It can also be noted that, in contrast to Figure 6.19, the precoder in Figure 7.18 is not shaded. As discussed in Section 6.3.3, for downlink non-codebook-based precoding, the precoder part of the antenna mapping is not visible in the specification and the network can, in essence, apply an arbitrary precoding for the downlink transmission. Due to the use of

Table 7.1 Uplink Precoder Matrices for Two Antenna Ports

Transmission Rank	Codebook Index					
	0	**1**	**2**	**3**	**4**	**5**
1	$\frac{1}{\sqrt{2}}\begin{bmatrix}1\\1\end{bmatrix}$	$\frac{1}{\sqrt{2}}\begin{bmatrix}1\\-1\end{bmatrix}$	$\frac{1}{\sqrt{2}}\begin{bmatrix}1\\j\end{bmatrix}$	$\frac{1}{\sqrt{2}}\begin{bmatrix}1\\-j\end{bmatrix}$	$\frac{1}{\sqrt{2}}\begin{bmatrix}1\\0\end{bmatrix}$	$\frac{1}{\sqrt{2}}\begin{bmatrix}0\\1\end{bmatrix}$
2	$\frac{1}{\sqrt{2}}\begin{bmatrix}1&0\\0&1\end{bmatrix}$	—	—	—	—	—

precoded DM-RS, the device can recover the different layers without knowledge of exactly what precoding has been applied at the transmitted side.

The same is also true for the uplink—that is, the presence of precoded DM-RS would allow for the base station to demodulate the uplink multi-antenna transmission and recover the different layers without knowledge of the precoding taking place at the transmitter side. However, for LTE the uplink precoder matrix is selected by the network and conveyed to the device as part of the scheduling grant. The device should then follow the precoder matrix selected by the network. Thus, in the uplink, the precoder is visible in the specification and, in order to limit the downlink signaling, there is a limited set of precoder matrices specified for each transmission rank.

More specifically, for each combination of transmission rank N_L and number of antennas ports N_A, a set of precoder matrices of size $N_A \times N_L$ is defined, as illustrated in Tables 7.1 and 7.2 for two and four antenna ports, respectively. For full-rank transmission—that is, when the transmission rank or number of layers equals the number of transmit antennas—only a single precoder matrix is defined, namely the identity matrix of size $N_A \times N_A$ (not shown in the tables). Note that, for the case of four antenna ports, only a subset of the defined matrices is shown. In total there are 24 rank-1 matrices, 16 rank-2 matrices, and 12 rank-3 matrices defined for four antenna ports, in addition to the single rank-4 matrix.

As can be seen, all the precoder matrices in Table 7.1 contain one and only one nonzero element in each row, and this is generally true for all precoder matrices defined for the uplink. As a consequence, the signal transmitted on a certain antenna port (corresponding to a certain row of the precoder matrix) always depends on one and only one specific layer (corresponding to a specific column of the precoder matrix). Expressed alternatively, the precoder matrix maps the layers to the antenna ports *with at most one layer being mapped to each antenna port*. Due to this, the good cubic-metric properties of the transmitted signal are also preserved for each antenna port when antenna precoding is applied. The precoder matrices of Tables 7.1 and 7.2 are therefore also referred to as *cubic-metric-preserving precoder matrices*.

In order to select a suitable precoder, the network needs information about the uplink channel. Such information can, for example, be based on measurements on the uplink SRS

Table 7.2 Subset of Uplink Precoder Matrices for Four Antenna Ports and Different Transmission Ranks

Transmission Rank	Codebook Index 0	1	2	3	...
1	$\frac{1}{2}\begin{bmatrix}1\\1\\1\\-1\end{bmatrix}$	$\frac{1}{2}\begin{bmatrix}1\\1\\j\\j\end{bmatrix}$	$\frac{1}{2}\begin{bmatrix}1\\1\\-1\\1\end{bmatrix}$	$\frac{1}{2}\begin{bmatrix}1\\1\\-j\\-j\end{bmatrix}$	...
2	$\frac{1}{2}\begin{bmatrix}1&0\\1&0\\0&1\\0&-j\end{bmatrix}$	$\frac{1}{2}\begin{bmatrix}1&0\\1&0\\0&1\\0&j\end{bmatrix}$	$\frac{1}{2}\begin{bmatrix}1&0\\-j&0\\0&1\\0&1\end{bmatrix}$	$\frac{1}{2}\begin{bmatrix}1&0\\-j&0\\0&1\\0&-1\end{bmatrix}$	...
3	$\frac{1}{2}\begin{bmatrix}1&0&0\\1&0&0\\0&1&0\\0&0&1\end{bmatrix}$	$\frac{1}{2}\begin{bmatrix}1&0&0\\-1&0&0\\0&1&0\\0&0&1\end{bmatrix}$	$\frac{1}{2}\begin{bmatrix}1&0&0\\0&1&0\\1&0&0\\0&0&1\end{bmatrix}$	$\frac{1}{2}\begin{bmatrix}1&0&0\\0&1&0\\-1&0&0\\0&0&1\end{bmatrix}$	...
4	$\frac{1}{2}\begin{bmatrix}1&0&0&0\\0&1&0&0\\0&0&1&0\\0&0&0&1\end{bmatrix}$	—	—	—	...

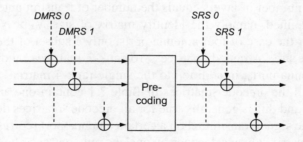

FIGURE 7.20

Illustration of SRS transmitted *after* uplink antenna precoding.

(Section 7.2.2). As indicated in Figure 7.20 SRS are transmitted non-precoded—that is, directly on the different antenna ports. The received SRS thus reflect the channel of each antenna port, not including any precoding. Based on the received SRS, the network can thus decide on a suitable uplink transmission rank and corresponding uplink precoder matrix, and provide information about the selected rank and precoder matrix as part of the scheduling grant.

The previous paragraph assumed the same number of antenna ports for PUSCH as for SRS. This is a relevant situation and the SRS is, in this case, used to aid the selection of the precoding matrix, as discussed in the preceding paragraphs. However, there are also situations when SRS and PUSCH use *different* numbers of antenna ports. One example is uplink transmission of two layers (two antenna ports), where the eNodeB would like to use SRS to probe the channel for potential four-layer transmission. In this case the SRS is transmitted on a *different* set of antenna ports than the PUSCH to aid the eNodeB in assessing the benefits, if any, of switching to four-layer transmission.

7.3.2 UPLINK MULTI-USER MIMO

As described in Section 6.3.5, downlink multi-user MIMO (MU-MIMO) implies downlink transmission to different devices using *the same time—frequency resource* and relying on the availability of multiple antennas, at least on the network side, to suppress interference between the transmissions. The term MU-MIMO originated from the resemblance to SU-MIMO (spatial multiplexing).

Uplink MU-MIMO is essentially the same thing but for the uplink transmission direction—that is, uplink MU-MIMO implies uplink transmissions from multiple devices using *the same uplink time—frequency resource* and relying on the availability of multiple receive antennas at the base station to separate the two or more transmissions. Thus, MU-MIMO is really just another term for uplink *space-division multiple access* (SDMA).

Actually, on the uplink, the relation between MU-MIMO and SU-MIMO (spatial multiplexing) is even closer. Uplink spatial multiplexing, for example with two antenna ports and two layers, implies that the device transmits two transport blocks with one transport block transmitted on each layer and thus on each antenna port,[14] as illustrated in the left part of Figure 7.21. As illustrated in the right part of the figure, MU-MIMO is essentially equivalent to separating the two antennas into two different devices and transmitting one transport block from each device. The base-station processing to separate the two

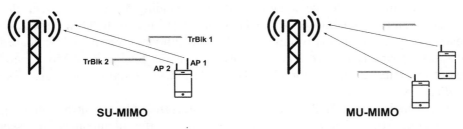

FIGURE 7.21

SU-MIMO and MU-MIMO.

---

[14]Note that the 2 × 2 precoder matrix is the identity matrix, see Table 7.1.

transmissions could essentially be identical to the processing used to separate the two layers in the case of spatial multiplexing. It should be noted that the separation of the two transmissions at the receiver side could be simplified, or at least the possible means to achieve this separation are extended, if the two devices are well separated in space, something which is not the case for two antennas attached to the same device. As an example, for sufficiently separated devices, classical beam-forming relying on correlated receiver antennas can be used to separate the uplink transmissions. Alternatively, uncorrelated receiver antennas can be used, and the separation means are then essentially the same as for SU-MIMO.

One important benefit of uplink MU-MIMO is that one, in many cases, can get similar gains in *system throughput* as SU-MIMO (spatial multiplexing) without the need for multiple transmit antennas at the device side, allowing for less complex device implementation. It should be noted though that spatial multiplexing could still provide substantial gains in terms of *user throughput* and peak data rates that can be provided from a single device. Furthermore, the potential system gains of uplink MU-MIMO rely on the fact that more than one device is actually available for transmission to in a subframe. The process of "pairing" devices that should share the time—frequency resources is also nontrivial and requires suitable radio-channel conditions.

Essentially, support for uplink MU-MIMO only requires the possibility to explicitly assign a specific orthogonal reference signal for the uplink transmission, thereby ensuring orthogonality between reference-signal transmissions from the different devices involved in the MU-MIMO transmission. As described in Section 7.2.1.5, this is supported by means of the dynamic assignment of DM-RS phase rotation and OCC as part of the uplink scheduling grant.

7.3.3 PUCCH TRANSMIT DIVERSITY

Precoder-based multi-layer transmission is only used for the uplink data transmission on PUSCH. However, in the case of a device with multiple transmit antennas, one obviously wants to use the full set of device antennas and corresponding device power amplifiers also for the L1/L2 control signaling on PUCCH in order to be able to utilize the full power resource and achieve maximum diversity. To achieve additional diversity, LTE release 10 also introduced the possibility for two-antenna *transmit diversity* for PUCCH. More specifically, the transmit diversity supported for PUCCH is referred to as *spatial orthogonal-resource transmit diversity* (SORTD).

The basic principle of SORTD is simply to transmit the uplink control signaling using different resources (time, frequency, and/or code) on the different antennas. In essence, the PUCCH transmissions from the two antennas will be identical to PUCCH transmissions from two different devices using different resources. Thus, SORTD creates additional diversity but achieves this by using twice as many PUCCH resources, compared to non-SORTD transmission.

For four physical antennas at the device, implementation-specific *antenna virtualization* is used. In essence, a transparent scheme is used to map the two-antenna-port signal to four physical antennas.

7.4 UPLINK L1/L2 CONTROL SIGNALING

Similar to the LTE downlink, there is also a need for uplink L1/L2 control signaling to support data transmission on downlink and uplink transport channels. Uplink L1/L2 control signaling consists of:

- hybrid-ARQ acknowledgments for received DL-SCH transport blocks;
- channel-state information (CSI) related to the downlink channel conditions, used to assist downlink scheduling; and
- scheduling requests, indicating that a device needs uplink resources for UL-SCH transmission.

There is no information indicating the UL-SCH transport format signaled on the uplink. As mentioned in Chapter 4, the eNodeB is in complete control of the uplink UL-SCH transmissions and the device always follows the scheduling grants received from the network, including the UL-SCH transport format specified in those grants. Thus, the network knows the transport format used for the UL-SCH transmission in advance and there is no need for any explicit transport-format signaling on the uplink.

Uplink L1/L2 control signaling needs to be transmitted on the uplink regardless of whether or not the device has any uplink transport-channel data to transmit and thus regardless of whether or not the device has been assigned any uplink resources for UL-SCH transmission. Hence, two different methods are supported for the transmission of the uplink L1/L2 control signaling, depending on whether or not the device has been assigned an uplink resource for UL-SCH transmission:

- *Nonsimultaneous transmission of UL-SCH and L1/L2 control.* If the device does not have a valid scheduling grant—that is, no resources have been assigned for the UL-SCH in the current subframe—a separate physical channel, the PUCCH, is used for transmission of uplink L1/L2 control signaling.
- *Simultaneous transmission of UL-SCH and L1/L2 control.* If the device has a valid scheduling grant—that is, resources have been assigned for the UL-SCH in the current subframe—the uplink L1/L2 control signaling is time multiplexed with the coded UL-SCH on to the PUSCH prior to DFT precoding and OFDM modulation. As the device has been assigned UL-SCH resources, there is no need to support transmission of the scheduling request in this case. Instead, scheduling information can be included in the MAC headers, as described in Chapter 13.

The reason to differentiate between the two cases is to minimize the cubic metric for the uplink power amplifier in order to maximize coverage. However, in situations when there is sufficient power available in the device, simultaneous transmission of PUSCH and PUCCH can be used with no impact on the coverage. The possibility for simultaneous PUSCH and PUCCH transmission was therefore introduced in release 10 as one part of several features[15] adding flexibility at the cost of a somewhat higher cubic metric. In situations where this cost is not acceptable, simultaneous PUSCH and PUCCH can always be avoided by using the basic mechanism introduced in the first version of LTE.

In the following section, the basic PUCCH structure and the principles for PUCCH control signaling are described, followed by control signaling on PUSCH.

7.4.1 BASIC PUCCH STRUCTURE

If the device has not been assigned an uplink resource for UL-SCH transmission, the L1/L2 control information (CSI reports, hybrid-ARQ acknowledgments, and scheduling requests) is transmitted on uplink resources (resource blocks) specifically assigned for uplink L1/L2 control on PUCCH. Transmission of control signaling on PUCCH is characterized by the PUCCH format used.

The first LTE releases, release 8 and release 9, defined two main PUCCH formats:[16]

- PUCCH format 1, carrying 0, 1, or 2 bits of information and used for hybrid-ARQ acknowledgments and scheduling requests;
- PUCCH format 2, carrying up to 11 bits of control information and used for reporting CSI.

There is one PUCCH per device. Given the relatively small payload size of PUCCH formats 1 and 2, the bandwidth of one resource block during one subframe is too large for the control signaling needs of a single device. Therefore, to efficiently exploit the resources set aside for control signaling, multiple devices can share the same resource-block pair.

With the introduction of carrier aggregation in release 10, where up to five component carriers can be aggregated, the number of hybrid-ARQ acknowledgments increased. To address this situation, an additional PUCCH format was introduced,

- PUCCH format 3, carrying up to 22 bits of control information.

Carrier aggregation was extended to handle up to 32 component carriers in release 13, calling for an even higher PUCCH capacity which was solved by

- PUCCH format 4, capable of a large number of hybrid-ARQ acknowledgments by using multiple resource-block pairs, and
- PUCCH format 5, capable of a payload between PUCCH formats 3 and 4.

---

[15]Other examples of such features are simultaneous transmission on multiple uplink component carriers and uplink multi-cluster transmission.

[16]There are actually three subtypes each of format 1 and 2, see further the detailed description.

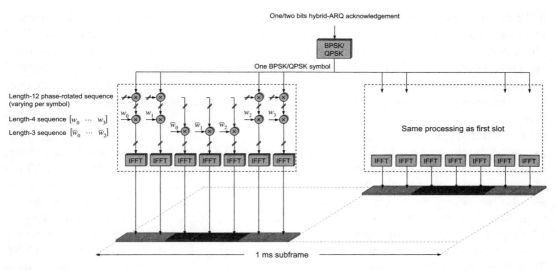

FIGURE 7.22

PUCCH format 1 (normal cyclic prefix).

The detailed structure of each of the different PUCCH formats is discussed in the following section, followed by an overview on how the formats are used.

7.4.1.1 PUCCH Format 1

PUCCH format 1,[17] for transmission of hybrid-ARQ acknowledgments and scheduling requests, is capable of carrying up to two bits of information. The same structure is used in the two slots of a subframe as illustrated in Figure 7.22. For transmission of a hybrid-ARQ acknowledgment, the one or two hybrid-ARQ acknowledgment bits are used to generate a BPSK or QPSK symbol, respectively. For a scheduling request, the same constellation point as for a negative acknowledgment is used. The modulation symbol is then used to generate the signal to be transmitted in each of the two PUCCH slots.

Different devices sharing the same resource-block pair in a subframe are separated by different orthogonal phase rotations of a length-12 frequency-domain sequence, where the sequence is identical to a length-12 reference-signal sequence. Furthermore, as described in conjunction with the reference signals in Section 7.2, a linear phase rotation in the frequency domain is equivalent to applying a cyclic shift in the time domain. Thus, although the term "phase rotation" is used here, the term cyclic shift is sometimes used with an implicit reference to the time domain. Similarly to the case of reference signals, there are up

---

[17]There are actually three variants in the LTE specifications, formats 1, 1a, and 1b, used for transmission of scheduling requests and one or two hybrid-ARQ acknowledgments, respectively. However, for simplicity, they are all referred to as format 1 herein.

to 12 different phase rotations specified, providing up to 12 different orthogonal sequences from each base sequence.[18] However, in the case of frequency-selective channels, not all 12 phase rotations can be used if orthogonality is to be retained. Typically, up to six rotations are considered usable in a cell from a radio-propagation perspective, although inter-cell interference may result in a smaller number being useful from an overall system perspective. Higher-layer signaling is used to configure the number of rotations that are used in a cell.

There are seven OFDM symbols per slot for a normal cyclic prefix (six in the case of an extended cyclic prefix). In each of those seven OFDM symbols, a length-12 sequence, obtained by phase rotation of the base sequence as described earlier, is transmitted. Three of the symbols are used as reference signals to enable channel estimation by the eNodeB and the remaining four[19] are modulated by the BPSK/QPSK symbols described earlier. In principle, the BPSK/QPSK modulation symbol could directly modulate the rotated length-12 sequence used to differentiate devices transmitting on the same time—frequency resource. However, this would result in unnecessarily low capacity on the PUCCH. Therefore, the BPSK/QPSK symbol is multiplied by a length-4 orthogonal cover sequence.[20] Multiple devices may transmit on the same time—frequency resource using the same phase-rotated sequence and be separated through different orthogonal covers. To be able to estimate the channels for the respective devices, the reference signals also employ an orthogonal cover sequence, with the only difference being the length of the sequence—three for the case of a normal cyclic prefix. Thus, since each base sequence can be used for up to $3 \cdot 12 = 36$ different devices (assuming all 12 rotations are available; typically at most six of them are used), there is a threefold improvement in the PUCCH capacity compared to the case of no cover sequence. The cover sequences are three Walsh sequences of length 4 for the data part and three DFT sequences of length 3 for the reference signals.

A PUCCH format 1 resource, used for either a hybrid-ARQ acknowledgment or a scheduling request, is represented by a single scalar resource index. From the index, the phase rotation and the orthogonal cover sequence are derived.

The use of a phase rotation of a base sequence together with orthogonal sequences as described earlier provides orthogonality between different devices in the same cell transmitting PUCCH on the same set of resource blocks. Hence, in the ideal case, there will be no intra-cell interference, which helps improve the performance. However, there will typically

---

[18]In releases 8 to 10, the base sequence is cell-specific, while release 11 adds the possibility of configuring a "virtual cell identity" from which the base sequence is derived. See further the discussion in conjunction with uplink reference signals in Section 7.2.

[19]The number of symbols used for reference signals and the acknowledgment is a trade-off between channel-estimation accuracy and energy in the information part; three symbols for reference symbols and four symbols for the acknowledgment have been found to be a good compromise.

[20]In the case of simultaneous SRS and PUCCH transmissions in the same subframe, a length-3 sequence is used, thereby making the last OFDM symbol in the subframe available for the SRS.

be inter-cell interference for the PUCCH as the different sequences used in neighboring cells are nonorthogonal. To randomize the inter-cell interference, the phase rotation of the sequence used in a cell varies on a symbol-by-symbol basis in a slot according to a hopping pattern derived from the physical-layer cell identity of the primary carrier. In release 11, the randomization can be configured to use the virtual cell identity instead of the physical one. On top of this, slot-level hopping is applied to the orthogonal cover and phase rotation to further randomize the interference. This is exemplified in Figure 7.23 assuming normal cyclic prefix and six of 12 rotations used for each cover sequence. To the phase rotation given by the cell-specific hopping a slot-specific offset is added. In cell A, a device is transmitting on PUCCH resource number 3, which in this example corresponds to using the (phase rotation, cover sequence) combination (6, 0) in the first slot and (11, 1) in the second slot of this particular subframe. PUCCH resource number 11, used by another device in cell A transmitting in the same subframe, corresponds to (11, 1) and (8, 2) in the first and second slots, respectively, of the subframe. In another cell the PUCCH resource numbers are mapped to different sets (rotation, cover sequence) in the slots. This helps to randomize the inter-cell interference.

For an extended cyclic prefix, the same structure as in Figure 7.22 is used with the difference being the number of reference symbols in each slot. In this case, the six OFDM symbols in each slot are divided such that the two middle symbols are used for reference signals and the remaining four symbols used for the information. Thus, the length of the orthogonal sequence used to spread the reference symbols is reduced from 3 to 2 and the multiplexing capacity is lower. However, the general principles described in the preceding paragraphs still apply.

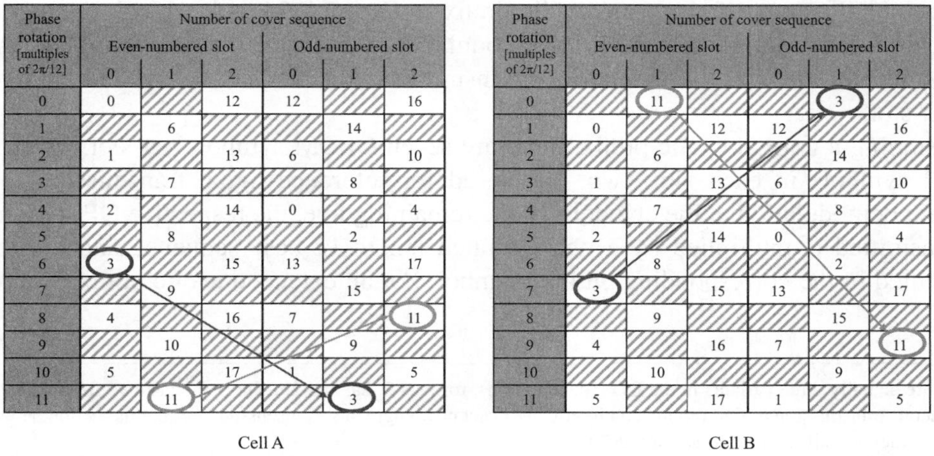

Cell A

Phase rotation [multiples of 2π/12]	Number of cover sequence					
	Even-numbered slot			Odd-numbered slot		
	0	1	2	0	1	2
0	0		12	12		16
1		6			14	
2	1		13	6		10
3		7			8	
4	2		14	0		4
5		8			2	
6	(3)		15	13		17
7		9			15	
8	4		16	7		(11)
9		10			9	
10	5		17	1		5
11		(11)			(3)	

Cell B

Phase rotation [multiples of 2π/12]	Number of cover sequence					
	Even-numbered slot			Odd-numbered slot		
	0	1	2	0	1	2
0		(11)			(3)	
1	0		12	12		16
2		6			14	
3	1		13	6		10
4		7			8	
5	2		14	0		4
6		8			2	
7	(3)		15	13		17
8		9			15	
9	4		16	7		(11)
10		10			9	
11	5		17	1		5

FIGURE 7.23

Example of phase rotation and cover hopping for two PUCCH resource indices in two different cells.

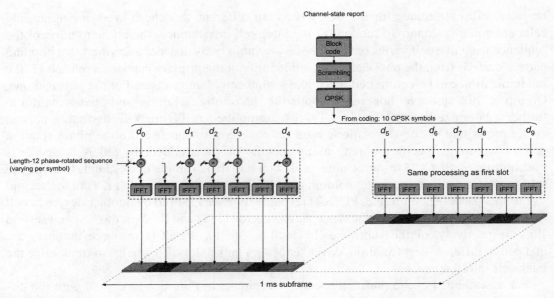

FIGURE 7.24

PUCCH format 2 (normal cyclic prefix).

7.4.1.2 PUCCH Format 2

PUCCH format 2, used for CSI reporting, is capable of handling up to 11 information bits per subframe.[21] Similarly to PUCCH format 1, multiple devices, using the same resource-block pair in a subframe, are separated through different orthogonal phase rotations of a length-12 sequence as illustrated for normal cyclic prefix in Figure 7.24. After block coding using a punctured Reed—Müller code and QPSK modulation, there are 10 QPSK symbols to transmit in the subframe: the first five symbols are transmitted in the first slot and the remaining five in the last slot.

Assuming a normal cyclic prefix, there are seven OFDM symbols per slot. Of the seven OFDM symbols in each slot, two[22] are used for reference-signal transmission to allow coherent demodulation at the eNodeB. In the remaining five, the respective QPSK symbol to be transmitted is multiplied by a phase-rotated length-12 base sequence and the result is transmitted in the corresponding OFDM symbol. For an extended cyclic prefix, where there

---

[21]There are actually three variants in the LTE specifications, formats 2, 2a, and 2b, where the last two formats are used for simultaneous transmission of CSI reports and hybrid-ARQ acknowledgments as discussed later in this section. However, for simplicity, they are all referred to as format 2 here.

[22]Similarly to format 1, the number of symbols used for reference signals and the coded channel-quality information is a trade-off between channel-estimation accuracy and energy in the information part. Two symbols for reference symbols and five symbols for the coded information part in each slot were found to be the best compromise.

are six OFDM symbols per slot, the same structure is used but with one reference-signal symbol per slot instead of two.

Basing the format 2 structure on phase rotations of the same base sequence as format 1 is beneficial as it allows the two formats to be transmitted in the same resource block. As phase-rotated sequences are orthogonal, one rotated sequence in the cell can be used either for one PUCCH instance using format 2 or three PUCCH instances using format 1. Thus, the "resource consumption" of one CSI report is equivalent to three hybrid-ARQ acknowledgments (assuming normal cyclic prefix). Note that no orthogonal cover sequences are used for format 2.

The phase rotations to use in the different symbols for PUCCH format 2 are hopping in a similar way as for format 1, motivated by interference randomization. Resources for PUCCH format 2 can, similar to format 1, be represented by a scalar index, which can be seen as a "channel number."

7.4.1.3 PUCCH Format 3

For downlink carrier aggregation, see Chapter 12, multiple hybrid-ARQ acknowledgment bits need to be fed back in the case of simultaneous transmission on multiple component carriers. Although PUCCH format 1 with resource selection can be used to handle the case of two downlink component carriers, this is not sufficient as a general solution as carrier aggregation of up to five component carriers is part of release 10. PUCCH format 3 was therefore introduced in release 10 to enable the possibility of transmitting up to 22 bits of control information on PUCCH in an efficient way. A device capable of more than two downlink component carriers—that is, capable of more than four bits for hybrid-ARQ acknowledgments—needs to support PUCCH format 3. For such a device, PUCCH format 3 can also be used for less than four bits of feedback relating to simultaneous transmission on multiple component carriers if configured by higher-layer signaling not to use PUCCH format 1 with resource selection.

The basis for PUCCH format 3, illustrated in Figure 7.25, is DFT-precoded OFDM—that is, the same transmission scheme as used for UL-SCH. The acknowledgment bits, one or two per downlink component carrier depending on the transmission mode configured for that particular component carrier, are concatenated with a bit reserved for scheduling request into a sequence of bits where bits corresponding to unscheduled transport blocks are set to zero. Block coding is applied,[23] followed by scrambling to randomize inter-cell interference. The resulting 48 bits are QPSK-modulated and divided into two groups, one per slot, of 12 QPSK symbols each.

Assuming a normal cyclic prefix, there are seven OFDM symbols per slot. Similarly to PUCCH format 2, two OFDM symbols (one in the case of an extended cyclic prefix) in each slot are used for reference-signal transmission, leaving five symbols for data transmission. In each slot, the block of 12 DFT-precoded QPSK symbols is transmitted in the

---

[23]A (32,k) Reed—Müller code is used, but for twelve or more bits in TDD two Reed—Müller codes are used in combination.

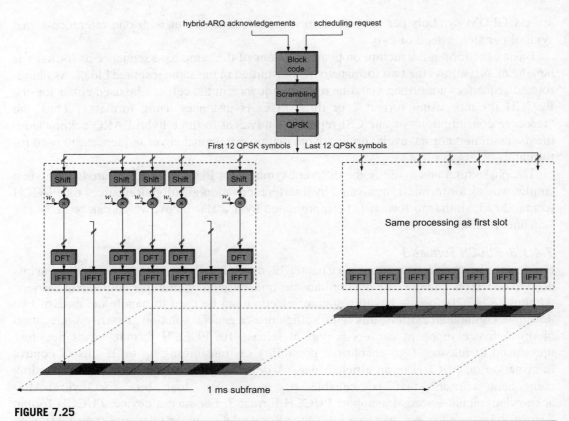

FIGURE 7.25

PUCCH format 3 (normal cyclic prefix).

five available DFTS-OFDM symbols. To further randomize the inter-cell interference, a cyclic shift of the 12 inputs to the DFT, varying between OFDM symbols in a cell-specific manner, is applied to the block of 12 QPSK symbols prior to DFT precoding (in releases 11 and later, the cyclic shift can be based on the virtual cell identity instead of the physical one).

To increase the multiplexing capacity, a length-5 orthogonal sequence is used with each of the five OFDM symbols carrying data in a slot being multiplied by one element of the sequence. Thus, up to five devices may share the same resource-block pair for PUCCH format 3. Different length-5 sequences are used in the two slots to improve the performance in high-Doppler scenarios. To facilitate channel estimation for the different transmissions sharing the same resource block, different reference-signal sequences are used.

The length-5 orthogonal cover sequences are obtained as five DFT sequences. There is also the possibility to use a length-4 Walsh sequence for the second slot in order to leave the last OFDM symbol unused for the case when sounding is configured in the subframe.

In the same manner as for the other two PUCCH formats, a resource can be represented by a single index from which the orthogonal sequence and the resource-block number can be derived.

Note that, due to the differences in the underlying structure of PUCCH format 3 compared to the two previous formats, resource blocks cannot be shared between format 3 and formats 1 and 2.

7.4.1.4 PUCCH Format 4

With the extension of carrier aggregation to handle up to 32 component carriers, the payload capacity of PUCCH format 3 is not sufficient to handle the resulting number of hybrid-ARQ acknowledgments. PUCCH formats 4 and 5 were introduced in release 13 to address this problem.

PUCCH format 4, illustrated in Figure 7.26, is to a large extent modeled after the PUSCH processing with a single DFT precoder covering multiple resource-block pairs. An 8-bit CRC is added to the payload, followed by tailbiting convolutional coding and rate matching to match the number of coded bits to the number of available resource elements. Scrambling,

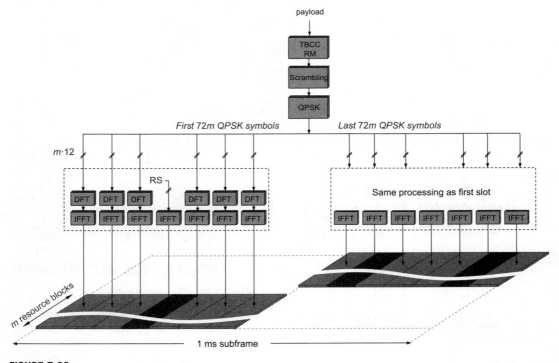

FIGURE 7.26

PUCCH format 4 (normal cyclic prefix).

QPSK modulation, DFT-precoding, and mapping to resource elements follow the same structure as the PUSCH—that is each DFT-spread OFDM symbol carries separate sets of coded bits. Multiple resource blocks in the frequency domain, 1, 2, 3, 4, 5, 6, or 8, can be used for PUCCH format 4, allowing for a very large payload. Inter-slot frequency hopping is used, similar to the other PUCCH formats and both normal and extended cyclic prefix is supported. There is also a possibility for a shortened format, leaving the last OFDM symbol in the subframe unused, for the case when sounding is configured in the subframe.

7.4.1.5 PUCCH Format 5

PUCCH format 5 is, similar to format 4, used to handle a large number of hybrid-ARQ feedback bits. However, as it uses a single resource-block pair, the payload capacity is smaller than PUCCH format 4. In addition, format 5 multiplexes two users onto a single resource-block pair as shown in Figure 7.27, making it suitable for efficient handling of payloads larger than format 3 but smaller than format 4.

The channel coding including CRC attachment and rate matching, and the QPSK modulation is identical to PUCCH format 4. However, the resource-block mapping and the usage of spreading is different. Each DFT-spread OFDM symbol carries six QPSK symbols. Prior to

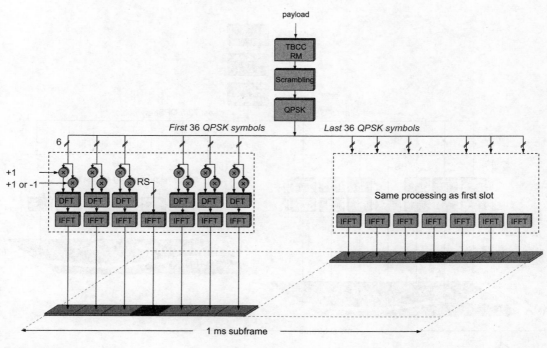

FIGURE 7.27

PUCCH format 5 (cyclic prefix).

DFT precoding, the six QPSK symbols are block-wise repeated where the second block is multiplied with $+1$ or -1 depending on which of the two orthogonal sequences are used. Hence, by using different orthogonal sequences, two users may share the same resource-block pair and transmit 144 coded bits each.[24]

The two users sharing the same resource-block pair use mutually orthogonal reference-signal sequences.

7.4.1.6 Resource-Block Mapping for PUCCH

The signals described for all of the PUCCH formats are, as already explained, transmitted on a (set of) resource-block pair. The resource-block pair to use is determined from the PUCCH resource index. Multiple resource-block pairs can be used to increase the control-signaling capacity in the cell; when one resource-block pair is full, the next PUCCH resource index is mapped to the next resource-block pair in sequence.

The resource-block pair(s) where a PUCCH is transmitted is located at the edges of the bandwidth allocated to the primary component carrier[25] as illustrated in Figure 7.28. To provide frequency diversity, frequency hopping on the slot boundary is used—that is, one "frequency resource" consists of 12 (or more in case of PUCCH format 4) subcarriers at the upper part of the spectrum within the first slot of a subframe and an equally sized resource at the lower part of the spectrum during the second slot of the subframe (or vice versa).

The reason for locating the PUCCH resources at the edges of the overall available spectrum is twofold:

- Together with the frequency hopping described previously, this maximizes the frequency diversity experienced by the control signaling.
- Assigning uplink resources for the PUCCH at other positions within the spectrum—that is, not at the edges—would have fragmented the uplink spectrum, making it impossible

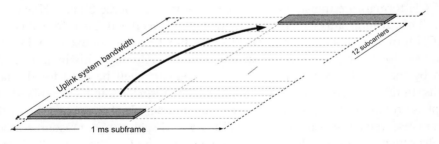

FIGURE 7.28

Uplink L1/L2 control signaling transmission on PUCCH.

---

[24]There are 12 OFDM symbols, each carrying 6 QPSK symbols for one user, resulting in $12 \cdot 6 \cdot 2 = 144$ bits for normal CP.
[25]Note that the primary component carrier in the uplink is specified on a per-device basis. Hence, different devices may view different carriers as their primary component carrier.

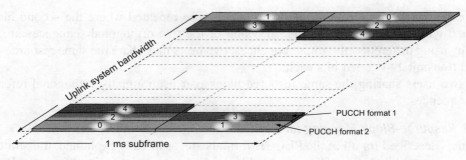

FIGURE 7.29

Allocation of resource blocks for PUCCH.

to assign very wide transmission bandwidths to a single device and still preserve the low-cubic-metric properties of the uplink transmission.

The resource-block mapping is in principle done such that PUCCH format 2 (CSI reports) is transmitted closest to the edges of the uplink cell bandwidth with PUCCH format 1 (hybrid-ARQ acknowledgments, scheduling requests) next as illustrated in Figure 7.29. The locations of PUCCH formats 3, 4, and 5 are configurable and can, for example, be located between formats 1 and 2. A semi-static parameter, provided as part of the system information, controls on which resource-block pair the mapping of PUCCH format 1 starts. Furthermore, the semi-statically configured scheduling requests are located at the outermost parts of the format 1 resources, leaving dynamic acknowledgments closest to the data. As the amount of resources necessary for hybrid-ARQ acknowledgments varies dynamically, this maximizes the amount of contiguous spectrum available for PUSCH.

In many scenarios, the configuration of the PUCCH resources can be done such that the three PUCCH formats are transmitted on separate sets of resource blocks. However, for the smallest cell bandwidths, this would result in too high an overhead. Therefore, it is possible to mix PUCCH formats 1 and 2 in one of the resource-block pairs—for example, in Figure 7.29 this is the case for the resource-block pair denoted "2." Although this mixture is primarily motivated by the smaller cell bandwidths, it can equally well be used for the larger cell bandwidths. In the resource-block pair where PUCCH formats 1 and 2 are mixed, the set of possible phase rotations are split between the two formats. Furthermore, some of the phase rotations are reserved as "guard"; hence the efficiency of such a mixed resource-block pair is slightly lower than a resource-block pair carrying only one of the first two PUCCH formats.

7.4.2 UPLINK CONTROL SIGNALING ON PUCCH

Having described the three PUCCH formats, the details on how these different formats are used to convey uplink control information can be discussed. As already mentioned, uplink control signaling on PUCCH can in principle be any combination of hybrid-ARQ

Table 7.3 Usage of Different PUCCH Formats for Different Pieces of Information (the Superscripts 10, 11, and 13 Denote the First Release Supporting this Combination)

Information		PUCCH Format					
		Format 1					
		Selection	Bundling	Format 2	Format 3	Format 4	Format 5
ACK	•	•10			•10	•13	•13
SR	•						
SR + ACK	•		•10		•10	•13	•13
CSI				•		•13	•13
CSI + ACK				•	•11	•13	•13
CSI + SR						•13	•13
CSI + SR + ACK					•11	•13	•13

acknowledgments (ACK), CSI, and scheduling requests (SR). Depending on whether these pieces of information are transmitted alone or in combination, different PUCCH formats and mechanisms are used as summarized in Table 7.3. In principle, simultaneous transmission of multiple control signaling messages from a single device could use multiple PUCCHs. However, this would increase the cubic metric and a single PUCCH structure supporting simultaneous transmission of multiple feedback signals is used instead.

7.4.2.1 Hybrid-ARQ Acknowledgments

Hybrid-ARQ acknowledgments are used to acknowledge receipt of one (or two in the case of spatial multiplexing) transport blocks on the DL-SCH. PUCCH format 1 is used in absence of carrier aggregation but can also support carrier aggregation of up to two downlink carriers—that is, up to four acknowledgment bits—as discussed in the following. PUCCH formats 3, 4, or 5 are used for more than four acknowledgments bits.

The hybrid-ARQ acknowledgment is only transmitted when the device correctly received control signaling related to DL-SCH transmission intended for this device on an PDCCH or EPDCCH. If no valid DL-SCH-related control signaling is detected, then nothing is transmitted on the PUCCH (that is, DTX). Apart from not unnecessarily occupying PUCCH resources that can be used for other purposes, this allows the eNodeB to perform three-state detection, ACK, NAK, or DTX, on the PUCCH received when using PUCCH format 1. Three-state detection is useful as NAK and DTX may need to be treated differently. In the case of NAK, retransmission of additional parity bits is useful for incremental redundancy, while for DTX the device has most likely missed the initial transmission of systematic bits and a better alternative than transmitting additional parity bits is to retransmit the systematic bits.

Transmission of one or two hybrid-ARQ acknowledgment bits uses PUCCH format 1. As mentioned in Section 7.4.1.1, a PUCCH resource can be represented by an index. How to determine this index depends on the type of information and whether the PDCCH or the EPDCCH was used to schedule the downlink data transmission.

For PDCCH-scheduled downlink transmissions, the resource index to use for a hybrid-ARQ acknowledgment is given as a function of the first CCE in the PDCCH used to schedule the downlink transmission to the device. In this way, there is no need to explicitly include information about the PUCCH resources in the downlink scheduling assignment, which of course reduces overhead. Furthermore, as described in Chapter 8, hybrid-ARQ acknowledgments are transmitted a fixed time after the reception of a DL-SCH transport block and when to expect a hybrid ARQ on the PUCCH is therefore known to the eNodeB.

For EPDCCH-scheduled transmissions, the index of the first ECCE in the EPDCCH cannot be used alone. Since the ECCE numbering is configured per device and therefore is device-specific, two different devices with control signaling on different resource blocks may have the same number of the first ECCE in the EPDCCH. Therefore, the ACK/NAK resource offset (ARO) being part of the EPDCCH information (see Section 6.4.6) is used in addition to the index of the first ECCE to determine the PUCCH resource. In this way, PUCCH collisions between two devices scheduled with EPDCCH can be avoided.

In addition to dynamic scheduling by using the (E)PDCCH, there is also, as described in Chapter 9, the possibility to semi-persistently schedule a device according to a specific pattern. In this case there is no PDCCH or EPDCCH to derive the PUCCH resource index from. Instead, the configuration of the semi-persistent scheduling pattern includes information on the PUCCH index to use for the hybrid-ARQ acknowledgment. In either of these cases, a device is using PUCCH resources only when it has been scheduled in the downlink. Thus, the amount of PUCCH resources required for hybrid-ARQ acknowledgments does not necessarily increase with an increasing number of devices in the cell, but, for dynamic scheduling, is rather related to the number of CCEs in the downlink control signaling.

The description in the preceding paragraphs addressed the case of downlink carrier aggregation not being used. Extensions to handle carrier aggregation are covered in Chapter 12.

7.4.2.2 Scheduling Request

Scheduling requests are used to request uplink resources for data transmission. Obviously, a scheduling request should only be transmitted when the device is requesting resources, otherwise the device should be silent to save battery resources and not create unnecessary interference.

Unlike the hybrid-ARQ acknowledgment, whose occurrence is known to the eNodeB from the downlink scheduling decision, the need for uplink resources for a certain device is in principle unpredictable by the eNodeB. One way to handle this would be to have a contention-based mechanism for requesting uplink resources. The random-access mechanism is based on this principle and can, to some extent, also be used for scheduling requests, as discussed in Chapter 9. Contention-based mechanisms typically work well for low intensities, but for higher scheduling-request intensities, the collision rate between different devices simultaneously requesting resources becomes too large. Therefore, LTE provides a contention-free scheduling-request mechanism on the PUCCH, where each device in the cell is given a reserved resource on which it can transmit a request for uplink resources. Unlike hybrid-ARQ acknowledgments, no explicit information bit is transmitted by the scheduling request; instead the information is conveyed by the presence (or absence) of energy on the corresponding PUCCH. However, the scheduling request, although used for a completely different purpose, shares the same PUCCH format as the hybrid-ARQ acknowledgment, namely PUCCH format 1.

The contention-free scheduling-request resource is represented by a PUCCH format 1 resource index as described earlier, occurring at every nth subframe. The more frequently these time instants occur, the lower the scheduling-request delay at the cost of higher PUCCH resource consumption. As the eNodeB configures all the devices in the cell, when and on which resources a device can request resources is known to the eNodeB. A single scheduling request resource is also sufficient for the case of carrier aggregation, as it only represents a request for uplink resources, which is independent of whether carrier aggregation is used or not.

7.4.2.3 Hybrid-ARQ Acknowledgments and Scheduling Request

The discussion in the two previous sections concerned transmission of *either* a hybrid-ARQ acknowledgment *or* a scheduling request. However, there are situations when the device needs to transmit *both* of them.

If PUCCH format 1 is used for the acknowledgments, simultaneous transmission of the acknowledgments and scheduling request is handled by transmitting the hybrid-ARQ acknowledgment on the scheduling-request resource (see Figure 7.30). This is possible as the same PUCCH structure is used for both of them and the scheduling request carries no explicit information. By comparing the amount of energy detected on the acknowledgment resource and the scheduling-request resource for a specific device, the eNodeB can determine whether or not the device is requesting uplink data resources. Once the PUCCH resource used for transmission of the acknowledgment is detected, the hybrid-ARQ acknowledgment can be decoded. Other, more advanced methods jointly decoding hybrid-ARQ and scheduling request can also be envisioned.

Channel selection, which is a way to transmit up to four acknowledgments on PUCCH in absence of a simultaneous scheduling request, cannot be used for joint transmission of

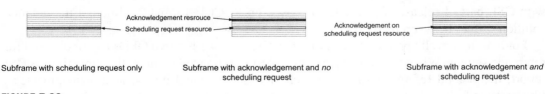

Acknowledgement resrouce
Scheduling request resource
Acknowledgement on scheduling request resource

Subframe with scheduling request only   Subframe with acknowledgement and *no* scheduling request   Subframe with acknowledgement *and* scheduling request

FIGURE 7.30

Multiplexing of scheduling request and hybrid-ARQ acknowledgment from a single device.

acknowledgments and scheduling request. Instead, up to four acknowledgment bits are bundled (combined) into two bits that are transmitted as described in the previous paragraph. Bundling implies that two or more acknowledgment bits are combined into a smaller number of bits. In essence, one acknowledgment bit represents the decoding outcome of multiple transport blocks and all these transport blocks need to be retransmitted as soon as one of them is incorrectly received.

PUCCH formats 3 to 5 support joint coding of acknowledgments and scheduling request in conjunction with carrier aggregation as described in Chapter 12.

7.4.2.4 Channel-State Information

CSI reports, the contents of which are discussed in Chapter 10, are used to provide the eNodeB with an estimate of the downlink radio-channel properties as seen from the device to aid channel-dependent scheduling. A CSI report consists of multiple bits transmitted in one subframe. There are *two* types of CSI reports, namely

- *periodic* reports, occurring at regular time instants;
- *aperiodic* reports, triggered by downlink control signaling on the PDCCH (or EPDCCH).

Aperiodic reports can only be transmitted on PUSCH as described later in Section 7.4.3, while periodic reports can be transmitted on PUCCH using PUCCH format 2.

A PUCCH format 2 resource is represented by an index and higher-layer signaling is used to configure each device with a resource to transmit its CSI report on, as well as when those reports should be transmitted. Hence, the eNodeB has full knowledge of when and on which resources each of the devices will transmit CSI on PUCCH.

7.4.2.5 Hybrid-ARQ Acknowledgments and CSI

Transmission of data in the downlink implies transmission of hybrid-ARQ acknowledgments in the uplink. At the same time, since data is transmitted in the downlink, up-to-date CSI is beneficial to optimize the downlink transmissions. Hence, simultaneous transmission of hybrid-ARQ acknowledgments and CSI is supported by LTE.

The handling of simultaneous transmission of acknowledgments and a CSI report depends on the number of acknowledgment bits as well as higher-layer configuration. There is also the possibility to configure the device to drop the CSI report and only transmit the acknowledgments.

The basic way of supporting transmission of one or two acknowledgments simultaneously with CSI, part of releases 8 and later, is based on PUCCH format 2, although the detailed solution differs between the two.

For a normal cyclic prefix, each slot in PUCCH format 2 has two OFDM symbols used for reference signals. When transmitting a hybrid-ARQ acknowledgment at the same time as CSI report, the second reference signal in each slot is modulated by the acknowledgment, as illustrated in Figure 7.31(a). Either BPSK or QPSK is used, depending on whether one or two acknowledgment bits are to be fed back. The fact that the acknowledgment is superimposed

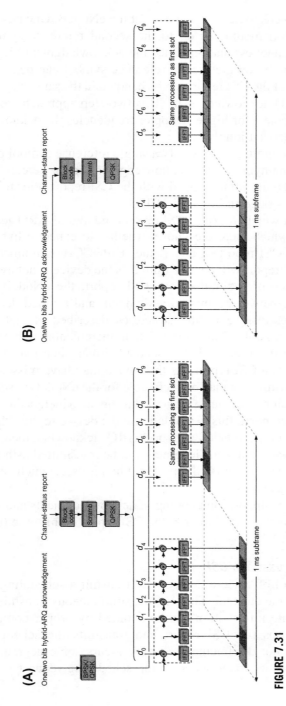

FIGURE 7.31

Simultaneous transmission of CSI and hybrid-ARQ acknowledgments: (A) normal cyclic prefix and (B) extended cyclic prefix.

on the reference signal needs to be accounted for at the eNodeB. One possibility is to decode the acknowledgment bit(s) modulated on to the second reference symbol using the first reference symbol for channel estimation. Once the acknowledgment bit(s) have been decoded, the modulation imposed on the second reference symbol can be removed and channel estimation and decoding of the CSI report can be handled in the same way as in the absence of simultaneous hybrid-ARQ acknowledgment. This two-step approach works well for low to medium Doppler frequencies; for higher Doppler frequencies the acknowledgment and CSI reports are preferably decoded jointly.

For an extended cyclic prefix, there is only a single reference symbol per slot. Hence, it is not possible to overlay the hybrid-ARQ acknowledgment on the reference symbol. Instead, the acknowledgment bit(s) are jointly coded with the CSI report prior to transmission using PUCCH format 2, as illustrated in Figure 7.31(b).

The time instances for which to expect CSI reports and hybrid-ARQ acknowledgments are known to the eNodeB, which therefore knows whether to expect a hybrid-ARQ acknowledgment along with the CSI report or not. If the (E)PDCCH assignment is missed by the device, then only the CSI report will be transmitted as the device is not aware that it has been scheduled. In the absence of a simultaneous CSI report, the eNodeB can employ DTX detection to discriminate between a missed assignment and a failed decoding of downlink data. However, one consequence of the structures described is that DTX detection is cumbersome, if not impossible. This implies that incremental redundancy needs to be operated with some care if the eNodeB has scheduled data such that the acknowledgment occurs at the same time as a CSI report. As the device may have missed the original transmission attempt in the downlink, it may be preferable for the eNodeB to select the redundancy version of the retransmission such that systematic bits are also included in the retransmission.

One possibility to circumvent this is to configure the device to drop the CSI report in the case of simultaneous transmission of a hybrid-ARQ acknowledgment. In this case, the eNodeB can detect DTX as the acknowledgment can be transmitted using PUCCH format 1, as described earlier. There will be no CSI report sent in this case, which needs to be taken into account in the scheduling process.

PUCCH formats 3 to 5 support joint coding of acknowledgments and scheduling request, see Chapter 10, and there is no need to drop the CSI report in this case (unless the device is configured to do so).

7.4.2.6 Scheduling Request and CSI

The eNodeB is in control of when a device may transmit a scheduling request and when the device should report the channel state. Hence, simultaneous transmission of scheduling requests and channel-state information can be avoided by proper configuration. If this is not done the device drops the CSI report and transmits the scheduling request only. Missing a CSI report is not detrimental and only incurs some degradation in the scheduling and rate-adaptation accuracy, whereas the scheduling request is critical for uplink transmissions.

7.4.2.7 *Hybrid-ARQ Acknowledgments, CSI and Scheduling Request*

For devices not supporting or configured to use PUCCH format 3, 4, or 5, simultaneous transmission of acknowledgments, CSI, and scheduling request are handled similarly to the description in the previous section; the CSI report is dropped and the acknowledgments and scheduling request are multiplexed as previously described in Section 7.4.2.3. However, devices that are using PUCCH format 3 or higher for multiple acknowledgments support simultaneous transmission of all three pieces of information. Since there is a bit reserved for scheduling requests in this case, the transmission structure is no different from the case of simultaneous transmission of acknowledgments and CSI reports described in Section 7.4.2.5.

7.4.3 UPLINK L1/L2 CONTROL SIGNALING ON PUSCH

If the device is transmitting data on PUSCH—that is, has a valid scheduling grant in the subframe—control signaling is time multiplexed[26] with data on the PUSCH instead of using the PUCCH (in release 10 and later simultaneous PUSCH and PUCCH can be used, avoiding the need for control signaling on PUSCH for most cases at the cost of a somewhat worse cubic metric). Only hybrid-ARQ acknowledgments and CSI reports are transmitted on the PUSCH. There is no need to request a scheduling grant when the device is already scheduled; instead, in-band buffer-status reports are sent as part of the MAC headers, as described in Chapter 9.

Time multiplexing of CSI reports and hybrid-ARQ acknowledgments is illustrated in Figure 7.32. However, although they both use time multiplexing there are some differences in the details for the two types of uplink L1/L2 control signaling motivated by their different properties.

The hybrid-ARQ acknowledgment is important for proper operation of the downlink. For one and two acknowledgments, robust QPSK modulation is used, regardless of the modulation scheme used for the data, while for a larger number of bits the same modulation scheme as for the data is used. Channel coding for more than two bits is done in the same way as for the PUCCH and bundling is applied if the number of bits exceeds a limit—that is, two transport blocks on the same component carrier share a single bit instead of having independent bits. Furthermore, the hybrid-ARQ acknowledgment is transmitted near to the reference symbols as the channel estimates are of better quality close to the reference symbols. This is especially important at high Doppler frequencies, where the channel may vary during a slot. Unlike the data part, the hybrid-ARQ acknowledgment cannot rely on retransmissions and strong channel coding to handle these variations.

In principle, the eNodeB knows when to expect a hybrid-ARQ acknowledgment from the device and can therefore perform the appropriate demultiplexing of the acknowledgment and the data parts. However, there is a certain probability that the device has missed the scheduling assignment on the downlink control channels (PDCCH or EPDCCH), in which case the

---

[26]In the case of spatial multiplexing, the CQI/PMI is time multiplexed with one of the code words, implying that it is spatially multiplexed with the other code word.

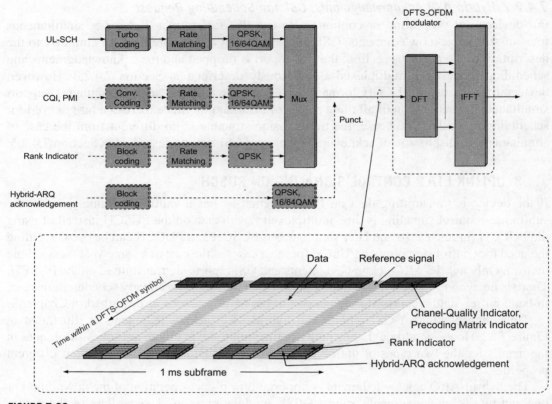

FIGURE 7.32

Multiplexing of control and data onto PUSCH.

eNodeB will expect a hybrid-ARQ acknowledgment while the device will not transmit one. If the rate-matching pattern was to depend on whether an acknowledgment is transmitted or not, all the coded bits transmitted in the data part could be affected by a missed assignment, which is likely to cause the UL-SCH decoding to fail. To avoid this error, the hybrid-ARQ acknowledgments are therefore punctured into the coded UL-SCH bit stream. Thus, the non-punctured bits are not affected by the presence/absence of hybrid-ARQ acknowledgments and the problem of a mismatch between the rate matching in the device and the eNodeB is avoided.

The contents of the CSI reports are described in Chapter 9; at this stage it suffices to note that a CSI report consists of *channel-quality indicator* (CQI), *precoding matrix indicator* (PMI), and *rank indicator* (RI). The CQI and PMI are time multiplexed with the coded data bits from PUSCH and transmitted using the same modulation as the data part. CSI reports are mainly useful for low-to-medium Doppler frequencies for which the radio channel is relatively constant, hence the need for special mapping is less pronounced. The RI, however, is mapped differently than the CQI and PMI; as illustrated in Figure 7.32, the RI is located near to the reference symbols using a similar mapping as the hybrid-ARQ acknowledgments. The more robust mapping of the RI is motivated by the fact that the RI is required in order to correctly interpret the CQI/PMI. The CQI/PMI, on the other hand, is simply mapped across the full subframe duration. Modulation-wise, the RI uses QPSK.

For uplink spatial multiplexing, in which case two transport blocks are transmitted simultaneously on the PUSCH, the CQI and PMI are multiplexed with the coded transport block using the highest modulation-and-coding scheme (MCS), followed by applying the previously described multiplexing scheme per layer (Figure 7.33). The intention behind this approach is to transmit the CQI and PMI on the (one or two) layers with the best quality.[27]

The hybrid-ARQ acknowledgments and the rank indicator are replicated across all transmission layers and multiplexed with the coded data in each layer in the same way as the single layer case described in the preceding paragraphs. The bits may, though, have been scrambled differently on the different layers. In essence, as the same information is transmitted on multiple layers with different scrambling, this provides diversity.

The basis for CSI reporting on the PUSCH is *aperiodic reports*, where the eNodeB requests a report from the device by setting the CSI request bit in the scheduling grant, as mentioned in Chapter 6. UL-SCH rate matching takes the presence of the CSI reports into account; by using a higher code rate a suitable number of resource elements is made available for transmission of the CSI report. Since the reports are explicitly requested by the eNodeB, their presence is known and the appropriate rate de-matching can be done at the receiver. If one of the configured transmission instances for a periodic report coincides with the

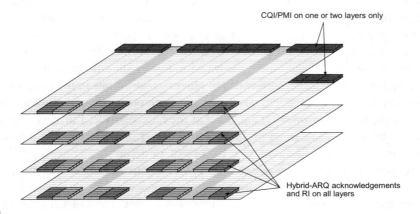

FIGURE 7.33

Multiplexing of CQI/PMI, RI and hybrid-ARQ acknowledgments in case of uplink spatial multiplexing.

[27]Assuming the MCS follows the channel quality, this holds for one, two, and four layers but not necessarily for three layers.

device being scheduled on the PUSCH, the periodic report is "rerouted" and transmitted on the PUSCH resources. Also, in this case there is no risk of mismatch in rate matching; the transmission instants for periodic reports are configured by robust RRC signaling and the eNodeB knows in which subframes such reports will be transmitted.

The channel coding of the CSI reports depends on the report size. For the smaller sizes such as a periodic report that otherwise would have been transmitted on the PUCCH, the same block coding as used for the PUCCH reports is used. For the larger reports, a tail-biting convolutional code is used for CQI/PMI, whereas the RI uses a (3, 2) block code for a single component carrier.

Unlike the data part, which relies on rate adaptation to handle different radio conditions, this cannot be used for the L1/L2 control-signaling part. Power control could, in principle, be used as an alternative, but this would imply rapid power variations in the time domain, which negatively impact the RF properties. Therefore, the transmission power is kept constant over the subframe and the amount of resource elements allocated to L1/L2 control signaling—that is, the code rate of the control signaling—is varied according to the scheduling decision for the data of the data part. High data rates are typically scheduled when the radio conditions are advantageous and hence a smaller amount of resource needs to be used by the L1/L2 control signaling compared to the case of poor radio conditions. To account for different hybrid-ARQ operating points, an offset between the code rate for the control-signaling part and the MCS used for the data part can be configured via higher-layer signaling.

7.5 UPLINK POWER CONTROL

Uplink power control for LTE is the set of algorithms and tools by which the transmit power for different uplink physical channels and signals are controlled to ensure that they, to the extent possible, are received with the appropriate power. This means that the transmission should be received with sufficient power to allow for proper demodulation of the corresponding information. At the same time, the transmit power should not be unnecessarily high as that would cause unnecessary interference to other transmissions in the same or other cells. The transmit power will thus depend on the channel properties, including the channel attenuation and the noise and interference level at the receiver side. Furthermore, in the case of UL-SCH transmission on PUSCH, if the received power is too low one can either increase the transmit power or reduce the data rate by use of rate control. Thus, for PUSCH transmission there is an intimate relation between power control and the link adaptation (rate control).

How to set the transmit power for random access is discussed in Chapter 11. Here we mainly discuss the power-control mechanism for the PUCCH and PUSCH physical channels. We also briefly discuss the power setting for SRS. Uplink DM-RS are always transmitted together and time-multiplexed with PUSCH or PUCCH. The DM-RS are then transmitted with the same power as the corresponding physical channel. This is also true in the case of uplink spatial multiplexing if the reference-signal power is defined as the total power of all DM-RS transmitted by the device. Expressed differently, the power of a single DM-RS is equal to the corresponding *per-layer* PUSCH power.

Fundamentally, LTE uplink power control is a combination of an *open-loop* mechanism, implying that the device transmit power depends on estimates of the downlink path loss, and a *closed-loop* mechanism, implying that the network can, in addition, directly adjust the device transmit power by means of explicit *power-control commands* transmitted on the downlink. In practice, these power-control commands are determined based on prior network measurements of the received uplink power, thus the term "*closed loop.*"

7.5.1 UPLINK POWER CONTROL: SOME BASIC RULES

Before going into the details of the power-control algorithms for PUSCH and PUCCH, some basic rules for the power assignment to different physical channels will be discussed. These rules mainly deal with the presence of different transmit-power limitations and how these limitations impact the transmit-power setting for different physical channels. This is especially of interest in the case of the simultaneous transmission of multiple physical channels from the same device, a situation that may occur for LTE releases 10 and beyond:

- Release 10 introduced the possibility for carrier aggregation, implying that multiple PUSCH may be transmitted in parallel on different component carriers.
- Release 10 also introduced the possibility for simultaneous PUSCH/PUCCH transmission on the same or different component carriers.

In principle, each physical channel is separately and independently power controlled. However, in the case of multiple physical channels to be transmitted in parallel from the same device, the total power to be transmitted for all physical channels may, in some cases, exceed the maximum device output power P_{TMAX} corresponding to the device power class. As will be seen below, the basic strategy is then to first ensure that transmission of any L1/L2 control signaling is assigned the power assumed to be needed for reliable transmission. The remaining available power is then assigned to the remaining physical channels.

For each uplink component carrier configured for a device there is also an associated and explicitly configured *maximum per-carrier transmit power* $P_{CMAX,c}$, which may be different for different component carriers (indicated by the index c). Furthermore, although it obviously does not make sense for $P_{CMAX,c}$ to exceed the maximum device output power P_{TMAX}, the sum of $P_{CMAX,c}$ for all configured component carriers may very well, and typically will, exceed P_{TMAX}. The reason is that, in many cases, the device will not be scheduled for uplink transmission on all its configured component carriers and the device should also in that case be able to transmit with its maximum output power.

As will be seen in the next sections, the power control of each physical channel explicitly ensures that the total transmit power for a given component carrier does not exceed $P_{CMAX,c}$ for that carrier. However, the separate power-control algorithms do not ensure that the total transmit power for all component carriers to be transmitted by the device does not exceed the maximum device output power P_{TMAX}. Rather, this is ensured by a subsequent *power scaling* applied to the physical channels to be transmitted. This power scaling is carried out in such a way that any L1/L2 control signaling has higher priority, compared to data (UL-SCH) transmission.

If PUCCH is to be transmitted in the subframe, it is first assigned the power determined by its corresponding power-control algorithm, before any power is assigned to any PUSCH to be transmitted in parallel PUCCH. This ensures that L1/L2 control signaling on PUCCH is assigned the power assumed to be needed for reliable transmission before any power is assigned for data transmission.

If PUCCH is not transmitted in the subframe but L1/L2 control signaling is multiplexed on to PUSCH, the PUSCH carrying the L1/L2 control signaling is first assigned the power determined by its corresponding power-control algorithm, before any power is assigned to any other PUSCH to be transmitted in parallel. Once again, this ensures that L1/L2 control signaling is assigned the power assumed to be needed before any power is assigned for other PUSCH transmissions only carrying UL-SCH. Note that, in the case of transmission of multiple PUSCH in parallel (carrier aggregation), at most one PUSCH may include L1/L2 control signaling. Also, there cannot be PUCCH transmission and L1/L2 control signaling multiplexed on to PUSCH in the same subframe. Thus, there will never be any conflict between the said rules.

If the remaining available transmit power is not sufficient to fulfill the power requirements of any remaining PUSCH to be transmitted, the powers of these remaining physical channels, which only carry UL-SCH, are scaled so that the total power for all physical channels to be transmitted does not exceed the maximum device output power.

Overall, the PUSCH power scaling, including the priority for PUSCH with L1/L2 control signaling, can thus be expressed as:

$$\sum_{c} w_c \cdot P_{PUSCH,c} \leq P_{TMAX} - P_{PUCCH} \tag{7.2}$$

where $P_{PUSCH,c}$ is the transmit power for PUSCH on carrier c as determined by the power-control algorithm (before power scaling but including the per-carrier limitation $P_{CMAX,c}$), P_{PUCCH} is the transmit power for PUCCH (which is zero if there is no PUCCH transmission in the subframe), and w_c is the power-scaling factor for PUSCH on carrier c ($w_c \leq 1$). For any PUSCH carrying L1/L2 control signaling the scaling factor w_c should be set to 1. For the remaining PUSCH, some scaling factors may be set to zero by decision of the device, in practice implying that the PUSCH, as well as the corresponding UL-SCH mapped to the PUSCH, are not transmitted. For the remaining PUSCH the scaling factors w_c are set to the same value less than or equal to 1 to ensure that the above inequality is fulfilled. Thus, all PUSCH that are actually transmitted are power scaled by the same factor.

After this overview of some general rules for the power setting of different devices, especially for the case of multiple physical channels transmitted in parallel from the same device, the power control carried out separately for each physical channel will be described in more detail.

7.5.2 POWER CONTROL FOR PUCCH

For PUCCH, the appropriate received power is simply the power needed to achieve a desired—that is, a sufficiently low—error rate in the decoding of the L1/L2 control information transmitted on the PUCCH. However, it is then important to bear the following in mind:

- In general, decoding performance is not determined by the *received signal strength* but rather by the *received signal-to-interference-plus-noise ratio* (SINR). What is an appropriate received power thus depends on the interference level at the receiver side, an interference level that may differ between different deployments and which may also vary in time as, for example, the load of the network varies.
- As previously described, there are different PUCCH formats which are used to carry different types of uplink L1/L2 control information (hybrid-ARQ acknowledgments, scheduling requests, CSI, or combinations thereof). The different PUCCH formats thus carry different numbers of information bits per subframe and the information they carry may also have different error-rate requirements. The required received SINR may therefore differ between the different PUCCH formats, something that needs to be taken into account when setting the PUCCH transmit power in a given subframe.

Overall, power control for PUCCH can be described by the following expression:

$$P_{\text{PUCCH}} = \min\{P_{\text{CMAX},c}, P_{0,\text{PUCCH}} + \text{PL}_{\text{DL}} + \Delta_{\text{Format}} + \delta\} \tag{7.3}$$

where P_{PUCCH} is the PUCCH transmit power to use in a given subframe and PL_{DL} is the downlink path loss as estimated by the device. The "min $\{P_{\text{CMAX},c}, \ldots\}$" term ensures that the PUCCH transmit power as determined by the power control will not exceed the per-carrier maximum power $P_{\text{CMAX},c}$.

The parameter $P_{0,\text{PUCCH}}$ in expression (7.3) is a cell-specific parameter that is broadcast as part of the cell system information. Considering only the part $P_{0,\text{PUCCH}} + \text{PL}_{\text{DL}}$ in the PUCCH power-control expression and assuming that the (estimated) downlink path loss accurately reflects the true uplink path loss, it is obvious that $P_{0,\text{PUCCH}}$ can be seen as the *desired* or *target* received power. As discussed earlier, the required received power will depend on the uplink noise/interference level. From this point of view, the value of $P_{0,\text{PUCCH}}$ should take the interference level into account and thus vary in time as the interference level varies. However, in practice it is not feasible to have $P_{0,\text{PUCCH}}$ varying with the instantaneous interference level. One simple reason is that the device does not read the system information continuously and thus the device would anyway not have access to a fully up-to-date $P_{0,\text{PUCCH}}$ value. Another reason is that the uplink path-loss estimates derived from downlink measurements will anyway not be fully accurate, for example due to differences between the instantaneous downlink and uplink path loss and measurement inaccuracies.

Thus, in practice, $P_{0,PUCCH}$ may reflect the average interference level, or perhaps only the relatively constant noise level. More rapid interference variations can then be taken care of by closed-loop power control, see below.

For the transmit power to reflect the typically different SINR requirements for different PUCCH formats, the PUCCH power-control expression includes the term Δ_{Format}, which adds a format-dependent power offset to the transmit power. The power offsets are defined such that a baseline PUCCH format, more exactly the format corresponding to the transmission of a single hybrid-ARQ acknowledgment (format 1 with BPSK modulation, as described in Section 7.4.1.1), has an offset equal to 0 dB, while the offsets for the remaining formats can be explicitly configured by the network. For example, PUCCH format 1 with QPSK modulation, carrying two simultaneous acknowledgments and used in the case of downlink spatial multiplexing, should have a power offset of roughly 3 dB, reflecting the fact that twice as much power is needed to communicate two acknowledgments instead of just a single acknowledgment.

Finally, it is possible for the network to directly adjust the PUCCH transmit power by providing the device with explicit power-control commands that adjust the term δ in the power-control expression (7.3). These power-control commands are *accumulative*—that is, each received power-control command increases or decreases the term δ by a certain amount. The power-control commands for PUCCH can be provided to the device by two different means:

- As mentioned in Section 6.4, a power-control command is included in each downlink scheduling assignment—that is, the device receives a power-control command every time it is explicitly scheduled on the downlink. One reason for uplink PUCCH transmissions is the transmission of hybrid-ARQ acknowledgments as a response to downlink DL-SCH transmissions. Such downlink transmissions are typically associated with downlink scheduling assignments on PDCCH and the corresponding power-control commands could thus be used to adjust the PUCCH transmit power prior to the transmission of the hybrid-ARQ acknowledgments.
- Power-control commands can also be provided on a special PDCCH that simultaneously provides power-control commands to multiple devices (PDCCH using DCI format 3/3A; see Section 6.4.8). In practice, such power-control commands are then typically transmitted on a regular basis and can be used to adjust the PUCCH transmit power, for example prior to (periodic) uplink CSI reports. They can also be used in the case of semi-persistent scheduling (see Chapter 9), in which case there may be uplink transmission of both PUSCH (UL-SCH) and PUCCH (L1/L2 control) without any explicit scheduling assignments/grants.

The power-control command carried within the uplink scheduling grant consists of two bits, corresponding to the four different update steps -1, 0, $+1$, or $+3$ dB. The same is true for the power-control command carried on the special PDCCH assigned for power control when this is configured to DCI format 3A. On the other hand, when the PDCCH is configured to use DCI format 3, each power-control command consists of a single bit, corresponding to the update steps -1 and $+1$ dB. In the latter case, twice as many devices can be power controlled by a single PDCCH. One reason for including the possibility for 0 dB (no change of power) as one power-control step is that a power-control command is included in *every* downlink scheduling assignment and it is desirable not to have to update the PUCCH transmit power for each assignment.

7.5.3 **POWER CONTROL FOR PUSCH**

Power control for PUSCH transmission can be described by the following expression:

$$P_{\text{PUSCH},c} = \min\{P_{\text{CMAX},c} - P_{\text{PUCCH}}, P_{0,PUSCH} + \alpha \cdot \text{PL}_{\text{DL}} + 10 \cdot \log_{10}(M) + \Delta_{\text{MCS}} + \delta\} \qquad (7.4)$$

where M indicates the instantaneous PUSCH bandwidth measured in number of resource blocks and the term Δ_{MCS} is similar to the term Δ_{Format} in the expression for PUCCH power control—that is, it reflects the fact that different SINR is required for different modulation schemes and coding rates used for the PUSCH transmission.

Equation (7.4) is similar to the power-control expression for PUCCH transmission, with some key differences:

- The use of "$P_{\text{CMAX},c} - P_{\text{PUCCH}}$" reflects the fact that the transmit power available for PUSCH on a carrier is the maximum allowed per-carrier transmit power *after power has been assigned to any PUCCH transmission* on that carrier. This ensures priority of L1/L2 signaling on PUCCH over data transmission on PUSCH in the power assignment, as described in Section 7.5.1.
- The term $10 \cdot \log_{10}(M)$ reflects the fact that what is fundamentally controlled by the parameter $P_{0,\text{PUSCH}}$ is the power *per resource block*. For a larger resource assignment, a correspondingly higher received power and thus a correspondingly higher transmit power is needed.[28]
- The parameter α, which can take a value smaller than or equal to 1, allows for so-called *partial path-loss compensation*, as described in the following.

In general, the parameters $P_{0,\text{PUSCH}}$, α, and Δ_{MCS} can be different for the different component carriers configured for a device.

In the case of PUSCH transmission, the explicit power-control commands controlling the term δ are included in the uplink scheduling grants, rather than in the downlink scheduling assignments. This makes sense as PUSCH transmissions are preceded by an uplink scheduling grant except for the case of semi-persistent scheduling. Similar to the power-control commands for PUCCH in the downlink scheduling assignment, the power-control

---

[28]One could also have included a corresponding term in the equation for PUCCH power control. However, as the PUCCH bandwidth always corresponds to one resource block, the term would always equal zero.

commands for PUSCH are multi-level. Furthermore, also in the same way as for PUCCH power control, explicit power-control commands for PUSCH can be provided on the special PDCCH that simultaneously provides power-control commands to multiple devices. These power-control commands can, for example, be used for the case of PUSCH transmission using semi-persistent scheduling.

Assuming α equal to 1, also referred to as *full path-loss compensation*, the PUSCH power-control expression becomes very similar to the corresponding expression for PUCCH. Thus, the network can select a MCS and the power-control mechanism, including the term Δ_{MCS}, will ensure that the received SINR will match the SINR required for that MCS, *assuming that the device transmit power does not reach its maximum value.*

In the case of PUSCH transmission, it is also possible to "turn off" the Δ_{MCS} function by setting all Δ_{MCS} values to zero. In that case, the PUSCH received power will be matched to a certain MCS given by the selected value of $P_{0,PUSCH}$.

With the parameter α less than 1, the PUSCH power control operates with so-called *partial path-loss compensation*—that is, an increased path loss is not fully compensated for by a corresponding increase in the uplink transmit power. In that case, the received power, and thus the received SINR per resource block, will vary with the path loss and, consequently, the scheduled MCS should vary accordingly. Clearly, in the case of fractional path-loss compensation, the Δ_{MCS} function should be disabled. Otherwise, the device transmit power would be further reduced when the MCS is reduced to match the partial path-loss compensation.

Figure 7.34 illustrates the differences between full path-loss compensation ($\alpha = 1$) and partial path-loss compensation ($\alpha < 1$). As can be seen, with partial path-loss compensation, the device transmit power increases more slowly than the increase in path loss (left in the figure) and, consequently, the received power, and thus also the received SINR, is reduced as

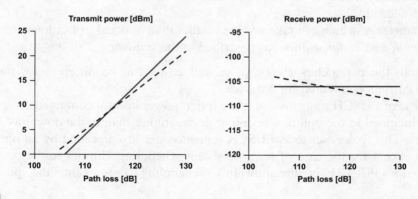

FIGURE 7.34

Full versus partial path-loss compensation. Solid curve: full compensation ($\alpha = 1$). Dashed curve: partial compensation ($\alpha = 0.8$).

the path loss increases (right in the figure). To compensate for this, the MCS—that is, the PUSCH data rate—should be reduced as the path loss increases.

The potential benefit of partial path-loss compensation is a relatively lower transmit power for devices closer to the cell border, implying less interference to other cells. At the same time, this also leads to a reduced data rate for these devices. It should also be noted that a similar effect can be achieved with full path-loss compensation by having the scheduled MCS depend on the estimated downlink path loss, which can be derived from the power headroom report, and rely on Δ_{MCS} to reduce the relative device transmit power for devices with higher path loss. However, an even better approach would then be to not only base the MCS selection on the path loss to the current cell, but also on the path loss to the neighboring interfered cells.

7.5.4 POWER CONTROL FOR SRS

The SRS transmit power basically follows that of the PUSCH, compensating for the exact bandwidth of the SRS transmission and with an additional power offset. Thus, the power control for SRS transmission can be described according to the equation:

$$P_{SRS} = \min\left\{P_{CMAX,c}, P_{0,PUSCH} + \alpha \cdot PL_{DL} + 10 \cdot \log_{10}(M_{SRS}) + \delta + P_{SRS}\right\} \qquad (7.5)$$

where the parameters $P_{0,PUSCH}$, α, and δ are the same as for PUSCH power control, as discussed in Section 7.5.3. Furthermore, M_{SRS} is the bandwidth, expressed as number of resource blocks, of the SRS transmission and P_{SRS} is a configurable offset.

7.6 UPLINK TIMING ALIGNMENT

The LTE uplink allows for uplink intra-cell orthogonality, implying that uplink transmissions received from different devices within a cell do not cause interference to each other. A requirement for this *uplink orthogonality* to hold is that the signals transmitted from different devices within the same subframe but within different frequency resources (different resource blocks) arrive approximately time aligned at the base station. More specifically, any timing misalignment between received signals should fall within the cyclic prefix. To ensure such receiver-side time alignment, LTE includes a mechanism for *transmit-timing advance*.

In essence, timing advance is a negative offset, at the device, between the start of a received downlink subframe and a transmitted uplink subframe. By controlling the offset appropriately for each device, the network can control the timing of the signals received at the base station from the devices. Devices far from the base station encounter a larger propagation delay and therefore need to start their uplink transmissions somewhat in advance, compared to devices closer to the base station, as illustrated in Figure 7.35. In this specific example, the first device is located close to the base station and experiences a small propagation delay, $T_{P,1}$. Thus, for this device, a small value of the timing advance offset $T_{A,1}$ is sufficient to compensate for the propagation delay and to ensure the correct timing at the base

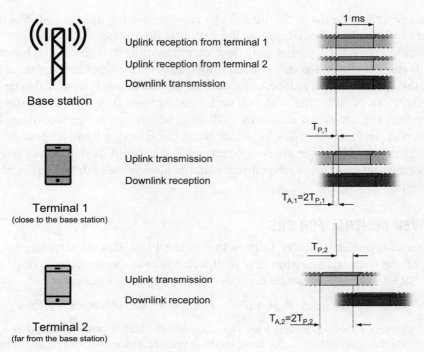

FIGURE 7.35

Uplink timing advance.

station. However, a larger value of the timing advance is required for the second device, which is located at a larger distance from the base station and thus experiences a larger propagation delay.

The timing-advance value for each device is determined by the network based on measurements on the respective uplink transmissions. Hence, as long as a device carries out uplink data transmission, this can be used by the receiving base station to estimate the uplink receive timing and thus be a source for the timing-advance commands. SRS can be used as a regular signal to measure upon, but in principle the base station can use any signal transmitted from the devices.

Based on the uplink measurements, the network determines the required timing correction for each device. If the timing of a specific device needs correction, the network issues a timing-advance command for this specific device, instructing it to retard or advance its timing relative to the current uplink timing. The user-specific timing-advance command is transmitted as a MAC control element (see Chapter 4 for a description of MAC control elements) on the DL-SCH. The maximum value possible for timing advance is 0.67 ms, corresponding to a device-to-base-station distance of slightly more than 100 km. This is also the value assumed when determining the processing time for decoding, as discussed in Section 8.1.

Typically, timing-advance commands to a device are transmitted relatively infrequently—for example, one or a few times per second.

If the device has not received a timing-advance command during a (configurable) period, the device assumes it has lost the uplink synchronization. In this case, the device must re-establish uplink timing using the random-access procedure prior to any PUSCH or PUCCH transmission in the uplink.

RETRANSMISSION PROTOCOLS

Transmissions over wireless channels are subject to errors, for example due to variations in the received signal quality. To some degree, such variations can be counteracted through link adaptation as is discussed in the next chapter. However, receiver noise and unpredictable interference variations cannot be counteracted. Therefore, virtually all wireless communications systems employ some form of *forward error correction* (FEC), adding redundancy to the transmitted signal allowing the receiver to correct errors, based on the concept whose roots can be traced to the pioneering work of Claude Shannon in 1948 [7]. In LTE, Turbo coding is used as discussed in Chapter 6.

Despite the error-correcting code, there will be data units received in error. *Hybrid automatic repeat request* (ARQ), first proposed in [18] and relying on a combination of error-correcting coding and retransmission of erroneous data units, is therefore commonly used in many modern communication systems. Data units in error despite the error-correcting coding are detected by the receiver, which requests a retransmission from the transmitter.

In LTE, there are *two* mechanisms responsible for retransmission handling, namely the MAC and RLC sublayers. Retransmissions of missing or erroneous data units are handled primarily by the hybrid-ARQ mechanism in the MAC layer, complemented by the retransmission functionality of the RLC protocol. The reasons for having a two-level retransmission structure can be found in the trade-off between fast and reliable feedback of the status reports. The hybrid-ARQ mechanism targets very fast retransmissions and, consequently, feedback on success or failure of the decoding attempt is provided to the transmitter after each received transport block. Although it is in principle possible to attain a very low error probability of the hybrid-ARQ feedback, it comes at a cost in transmission power. Keeping the cost reasonable typically results in a feedback error rate of around 1%, which results in a hybrid-ARQ residual error rate of a similar order. Such an error rate is in many cases far too high; high data rates with TCP may require virtually error-free delivery of packets to the TCP protocol layer. As an example, for sustainable data rates exceeding 100 Mbit/s, a packet-loss probability less than 10^{-5} is required [35]. The reason is that TCP assumes packet errors to be due to congestion in the network. Any packet error therefore triggers the TCP congestion-avoidance mechanism with a corresponding decrease in data rate.

Compared to the hybrid-ARQ acknowledgments, the RLC status reports are transmitted relatively infrequently and thus the cost of obtaining a reliability of 10^{-5} or better is relatively

227

small. Hence, the combination of hybrid-ARQ and RLC attains a good combination of small round-trip time and a modest feedback overhead where the two components complement each other—fast retransmissions due to the hybrid-ARQ mechanism and reliable packet delivery due to the RLC. As the MAC and RLC protocol layers are located in the same network node, a tight interaction between the two protocols is possible. Hence, to some extent, the combination of the two can be viewed as *one* retransmission mechanism with *two* feedback channels. However, note that, as discussed in Chapter 4 and illustrated in Figure 8.1, the RLC operates per logical channel, while the hybrid-ARQ operates per transport channel (that is, per component carrier). One hybrid-ARQ entity may therefore retransmit data belonging to multiple logical channels.

In the following section, the principles behind the hybrid-ARQ and RLC protocols are discussed in more detail.

8.1 HYBRID ARQ WITH SOFT COMBINING

The hybrid-ARQ operation described in the preceding section discards erroneously received packets and requests retransmission. However, despite it was not possible to decode the packet, the received signal still contains information, which is lost by discarding erroneously received packets. This shortcoming is addressed by *hybrid ARQ with soft combining*. In hybrid ARQ with soft combining, the erroneously received packet is stored in a buffer

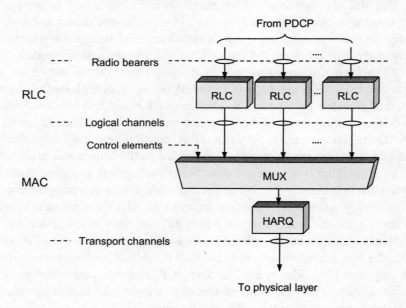

FIGURE 8.1

RLC and hybrid-ARQ retransmission mechanisms in LTE.

memory and later combined with the retransmission to obtain a single, combined packet that is more reliable than its constituents. Decoding of the error-correction code operates on the combined signal.

The hybrid-ARQ functionality spans both the physical layer and the MAC layer; generation of different redundancy versions at the transmitter as well as the soft combining at the receiver are handled by the physical layer, while the hybrid-ARQ protocol is part of the MAC layer.

The basis for the LTE hybrid-ARQ mechanism is a structure with multiple stop-and-wait protocols, each operating on a single transport block. In a stop-and-wait protocol, the transmitter stops and waits for an acknowledgment after each transmitted transport block. This is a simple scheme; the only feedback required is a single bit indicating positive or negative acknowledgment of the transport block. However, since the transmitter stops after each transmission, the throughput is also low. LTE therefore applies *multiple* stop-and-wait processes operating in parallel such that, while waiting for acknowledgment from one process, the transmitter can transmit data to another hybrid-ARQ process. This is illustrated in Figure 8.3; while processing the data received in the first hybrid-ARQ process the receiver can continue to receive using the second process and so on. This structure, multiple hybrid-ARQ processes operating in parallel to form one hybrid-ARQ entity, combines the simplicity of a stop-and-wait protocol while still allowing continuous transmission of data.

There is one hybrid-ARQ entity per device. Spatial multiplexing, where two transport blocks can be transmitted in parallel on the same transport channel as described in Chapter 6, is supported by one hybrid-ARQ entity having two sets of hybrid-ARQ processes with independent hybrid-ARQ acknowledgments. The details for the physical-layer transmission of the downlink and uplink hybrid-ARQ acknowledgments are described in Chapters 6 and 7.

Upon receiving a transport block for a certain hybrid-ARQ process, the receiver makes an attempt to decode the transport block and informs the transmitter about the outcome through a hybrid-ARQ acknowledgment, indicating whether the transport block was correctly decoded or not. The time from reception of data until transmission of the hybrid-ARQ acknowledgment is fixed, hence the transmitter knows from the timing relation which hybrid-ARQ process a received acknowledgment relates to. This is beneficial from an overhead perspective as there is no need to signal the process number along with the acknowledgment.

An important part of the hybrid-ARQ mechanism is the use of *soft combining*, which implies that the receiver combines the received signal from multiple transmission attempts.

By definition, retransmission in any hybrid-ARQ scheme must represent the same set of information bits as the original transmission. However, the set of coded bits transmitted in each retransmission may be selected differently as long as they represent the same set of information bits. Depending on whether the retransmitted bits are required to be identical to the original transmission or not, the soft combining scheme is often referred to as *chase combining*, first proposed in [19], or *incremental redundancy* (IR), which is used in LTE. With IR, each retransmission does not have to be identical to the original transmission. Instead, *multiple sets* of coded bits are generated, each representing the same set of information bits [20,21]. The rate matching functionality of LTE, described in Chapter 6, is used to generate different sets of coded bits as a function of the redundancy version as illustrated in Figure 8.2. In addition to a gain in accumulated received E_b/N_0, IR also results in a coding gain for each retransmission. The gain with IR compared to pure energy accumulation (chase combining) is larger for high initial code rates [22]. Furthermore, as shown in [23], the performance gain of IR compared to chase combining can also depend on the relative power difference between the transmission attempts.

In the discussion so far, it has been assumed that the receiver has received all the previously transmitted redundancy versions. If all redundancy versions provide the same amount of information about the data packet, the order of the redundancy versions is not critical. However, for some code structures, not all redundancy versions are of equal importance. One example here is Turbo codes, where the systematic bits are of higher importance than the parity bits. Hence, the initial transmission should at least include the systematic bits and some parity bits. In the retransmission(s), parity bits not in the initial transmission can be included.

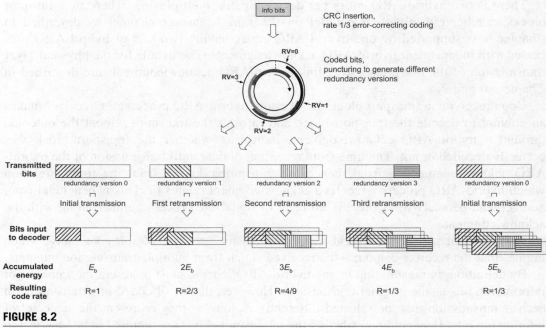

FIGURE 8.2

Example of incremental redundancy.

This is the background to why systematic bits are inserted first in the circular buffer discussed in Chapter 6.

Hybrid ARQ with soft combining, regardless of whether chase combining or IR is used, leads to an implicit reduction of the data rate by means of retransmissions and can thus be seen as implicit link adaptation. However, in contrast to link adaptation based on explicit estimates of the instantaneous channel conditions, hybrid ARQ with soft combining implicitly adjusts the coding rate based on the result of the decoding. In terms of overall throughput this kind of implicit link adaptation can be superior to explicit link adaptation, as additional redundancy is only added *when needed*—that is, when previous higher-rate transmissions were not possible to decode correctly. Furthermore, as it does not try to predict any channel variations, it works equally well, regardless of the speed at which the terminal is moving. Since implicit link adaptation can provide a gain in system throughput, a valid question is why explicit link adaptation is necessary at all. One major reason for having explicit link adaptation is the reduced delay. Although relying on implicit link adaptation alone is sufficient from a system throughput perspective, the end-user service quality may not be acceptable from a delay perspective.

For proper operation of soft combining, the receiver needs to know when to perform soft combining prior to decoding and when to clear the soft buffer—that is, the receiver needs to differentiate between the reception of an initial transmission (prior to which the soft buffer should be cleared) and the reception of a retransmission. Similarly, the transmitter must know whether to retransmit erroneously received data or to transmit new data. Therefore, an explicit *new-data indicator* is included for each of the one or two scheduled transport blocks as part of the scheduling information transmitted in the downlink. The new-data indicator is present in both downlink assignments and uplink grants, although the meaning is slightly different for the two.

For downlink data transmission, the new-data indicator is toggled for a new transport block—that is, it is essentially a single-bit sequence number. Upon reception of a downlink scheduling assignment, the device checks the new-data indicator to determine whether the current transmission should be soft combined with the received data currently in the soft buffer for the hybrid-ARQ process in question, or if the soft buffer should be cleared.

For uplink data transmission, there is also a new-data indicator transmitted on the downlink PDCCH. In this case, toggling the new-data indicator requests transmission of a new transport block, otherwise the previous transport block for this hybrid-ARQ process should be retransmitted (in which case the eNodeB should perform soft combining).

The use of multiple hybrid-ARQ processes operating in parallel can result in data being delivered from the hybrid-ARQ mechanism out of sequence. For example, transport block 5 in Figure 8.3 was successfully decoded before transport block 1, which required two retransmissions. Out-of-sequence delivery can also occur in the case of carrier aggregation, where transmission of a transport block on one component carrier could be successful while a retransmission is required on another component carrier. To handle out-of-sequence delivery

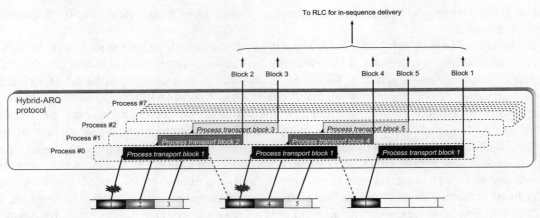

FIGURE 8.3

Multiple parallel hybrid-ARQ processes forming one hybrid-ARQ entity.

from the hybrid-ARQ protocol, the RLC protocol includes an in-sequence-delivery mechanism, as described in Section 8.2.

Hybrid-ARQ protocols can be characterized as synchronous versus asynchronous, related to the flexibility in the time domain, as well as adaptive versus nonadaptive, related to the flexibility in the frequency domain:

- An *asynchronous* hybrid-ARQ protocol implies that retransmissions can occur at any time, whereas a *synchronous* protocol implies that retransmissions occur at a fixed time after the previous transmission (see Figure 8.4). The benefit of a synchronous protocol is

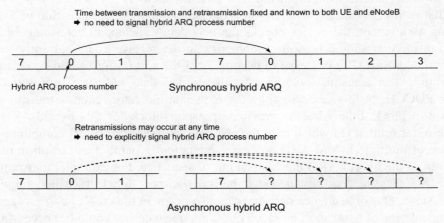

FIGURE 8.4

Synchronous and asynchronous hybrid ARQ.

that there is no need to explicitly signal the hybrid-ARQ process number as this information can be derived from the subframe number. On the other hand, an asynchronous protocol allows for more flexibility in the scheduling of retransmissions.

- An *adaptive* hybrid-ARQ protocol implies that the frequency location and possibly also the more detailed transmission format can be changed between retransmissions. A *nonadaptive protocol*, in contrast, implies that the retransmission must occur at the same frequency resources and with the same transmission format as the initial transmission.

In the case of LTE, asynchronous adaptive hybrid ARQ is used for the downlink. For the uplink, synchronous hybrid ARQ is used. Typically, the retransmissions are nonadaptive, but there is also the possibility to use adaptive retransmissions as a complement.

8.1.1 DOWNLINK HYBRID ARQ

In the downlink, retransmissions are scheduled in the same way as new data—that is, they may occur at any time and at an arbitrary frequency location within the downlink cell bandwidth. Hence, the downlink hybrid-ARQ protocol is *asynchronous* and *adaptive*. The support for asynchronous and adaptive hybrid ARQ for the LTE downlink is motivated by the need to avoid collisions with, for example, transmission of system information and MBSFN subframes. Instead of dropping a retransmission that otherwise would collide with MBSFN subframes or transmission of system information, the eNodeB can move the retransmission in time and/or frequency to avoid the overlap in resources.

Support for soft combining is, as described in the introduction, provided through an explicit new-data indicator, toggled for each new transport block. In addition to the new-data indicator, hybrid-ARQ-related downlink control signaling consists of the hybrid-ARQ process number (three bits for FDD, four bits for TDD) and the redundancy version (two bits), both explicitly signaled in the scheduling assignment for each downlink transmission.

Downlink spatial multiplexing implies, as already mentioned, transmission of two transport blocks in parallel on a component carrier. To provide the possibility to retransmit only one of the transport blocks, which is beneficial as error events for the two transport blocks can be fairly uncorrelated, each transport block has its own separate new-data indicator and redundancy-version indication. However, there is no need to signal the process number separately as each process consists of two sub-processes in the case of spatial multiplexing or, expressed differently, once the process number for the first transport block is known, the process number for the second transport block is given implicitly.

Transmissions on downlink component carriers are acknowledged independently. On each component carrier, transport blocks are acknowledged by transmitting one or two bits on the uplink, as described in Chapter 7. In the absence of spatial multiplexing, there is only a single transport block within a TTI and consequently only a single acknowledgment bit is required in response. However, if the downlink transmission used spatial multiplexing, there are two transport blocks per TTI, each requiring its own hybrid-ARQ acknowledgment bit. The total number of bits required for hybrid-ARQ acknowledgments thus depends on the number of component carriers and the transmission mode for each of the component carriers. As each downlink component carrier is scheduled separately from its own PDCCH, the hybrid-ARQ process numbers are signaled independently for each component carrier.

The device should not transmit a hybrid-ARQ acknowledgment in response to reception of system information, paging messages, and other broadcast traffic. Hence, hybrid-ARQ acknowledgments are only sent in the uplink for "normal" unicast transmission.

8.1.2 UPLINK HYBRID ARQ

Shifting the focus to the uplink, a difference compared to the downlink case is the use of synchronous, nonadaptive operation as the basic principle of the hybrid-ARQ protocol, motivated by the lower overhead compared to an asynchronous, adaptive structure. Hence, uplink retransmissions always occur at an a priori known subframe; in the case of FDD operation uplink retransmissions occur eight subframes after the prior transmission attempt for the same hybrid-ARQ process. The set of resource blocks used for the retransmission on a component carrier is identical to the initial transmission. Thus, the only control signaling required in the downlink for a retransmission is a retransmission command, transmitted on the PHICH as described in Chapter 6. In the case of a negative acknowledgment on the PHICH, the data are retransmitted.

Spatial multiplexing is handled in the same way as in the downlink—two transport blocks transmitted in parallel on the uplink, each with its own modulation-and-coding scheme and new-data indicator but sharing the same hybrid-ARQ process number. The two transport blocks are individually acknowledged and hence two bits of information are needed in the downlink to acknowledge an uplink transmission using spatial multiplexing. TDD is another example, as is discussed later, where uplink transmissions in different subframes may need to be acknowledged in the same downlink subframe.[1] Thus, multiple PHICHs may be needed as each PHICH is capable of transmitting a single bit only.

Despite the fact that the basic mode of operation for the uplink is synchronous, nonadaptive hybrid ARQ, there is also the possibility to operate the uplink hybrid ARQ in a synchronous, *adaptive* manner, where the resource-block set and modulation-and-coding scheme for the retransmissions are changed. Although nonadaptive retransmissions are typically used due to the very low overhead in terms of downlink control signaling, adaptive retransmissions are sometimes useful to avoid fragmenting the uplink frequency resource or to avoid collisions with random-access resources. This is illustrated in Figure 8.5. A device is scheduled for an initial transmission in subframe n; a transmission that is not correctly received and consequently a retransmission is required in subframe $n + 8$ (assuming FDD; for TDD the timing depends on the uplink–downlink allocation, as discussed later). With nonadaptive hybrid ARQ, the retransmissions occupy the same part of the uplink spectrum as

---

[1]A third example is carrier aggregation, see Chapter 12.

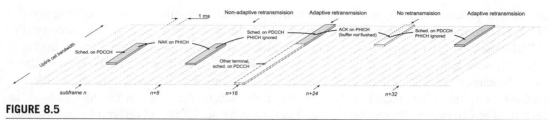

FIGURE 8.5

Nonadaptive and adaptive hybrid-ARQ operation.

the initial transmission. Hence, in this example the spectrum is fragmented, which limits the bandwidth available to another device (unless the other device is capable of multi-cluster transmission). In subframe $n + 16$, an example of an adaptive retransmission is found; to make room for another device to be granted a large part of the uplink spectrum, the retransmission is moved in the frequency domain. It should be noted that the uplink hybrid-ARQ protocol is still synchronous—that is, a retransmission should always occur eight subframes after the previous transmission.

The support for both adaptive and nonadaptive hybrid ARQ is realized by *not* flushing the transmission buffer when receiving a positive hybrid-ARQ acknowledgment on PHICH for a given hybrid-ARQ process. Instead, the actual control of whether data should be retransmitted or not is done by the new-data indicator included in the uplink scheduling grant sent on the PDCCH. The new-data indicator is toggled for each new transport block. If the new-data indicator is toggled, the device flushes the transmission buffer and transmits a new-data packet. However, if the new-data indicator does not request transmission of a new transport block, the previous transport block is retransmitted. Hence, clearing of the transmission buffer is not handled by the PHICH but by the PDCCH as part of the uplink grant. The negative hybrid-ARQ acknowledgment on the PHICH should preferably be seen as a single-bit scheduling grant for retransmissions where the set of bits to transmit and all the resource information are known from the previous transmission attempt. In Figure 8.5 an example of postponing a transmission is seen in subframe $n + 24$. The device has received a positive acknowledgment and therefore does *not* retransmit the data. However, the transmission buffer is not flushed, which later is exploited by an uplink grant requesting retransmission in subframe $n + 32$.

A consequence of the above-mentioned method of supporting both adaptive and nonadaptive hybrid ARQ is that the PHICH and PDCCH related to the same uplink subframe have the same timing. If this were not the case, the complexity would increase as the device would not know whether to obey the PHICH or wait for a PDCCH overriding the PHICH.

As explained earlier, the new-data indicator is explicitly transmitted in the uplink grant. However, unlike the downlink case, the redundancy version is *not* explicitly signaled for each retransmission. With a single-bit acknowledgment on the PHICH, this is not possible. Instead, as the uplink hybrid-ARQ protocol is synchronous, the redundancy version follows a predefined pattern, starting with zero when the initial transmission is scheduled by the PDCCH. Whenever a retransmission is requested by a negative acknowledgment on the PHICH, the next redundancy version in the sequence is used. However, if a retransmission is explicitly scheduled by the PDCCH overriding the PHICH, then there is the potential to affect the redundancy version to use. Grants for retransmissions use the same format as ordinary grants (for initial transmissions). One of the information fields in an uplink grant is, as described in Chapter 6, the modulation-and-coding scheme. Of the 32 different combinations this five-bit field can take, three of them are reserved. Those three combinations represent different redundancy versions; hence, if one of these combinations is signaled as part of an uplink grant indicating a retransmission, the corresponding redundancy version is used for the transmission. The transport-block size is already known from the initial transmission as it, by definition, cannot change between retransmission attempts. In Figure 8.5, the initial transmission in subframe n uses the first redundancy version in sequence as the transport-block size must be indicated for the initial transmission. The retransmission in subframe $n + 8$ uses the next redundancy version in the sequence, while the explicitly scheduled retransmission in subframe $n + 16$ can use any redundancy scheme as indicated on the PDCCH.

Whether to exploit the possibility to signal an arbitrary redundancy version in a retransmission scheduled by the PDCCH is a trade-off between incremental-redundancy gain and robustness. From an incremental-redundancy perspective, changing the redundancy value between retransmissions is typically beneficial to fully exploit the gain from IR. However, as the modulation-and-coding scheme is normally not indicated for uplink retransmissions, as either the single-bit PHICH is used or the modulation-and-coding field is used to explicitly indicate a new redundancy version, it is implicitly assumed that the device did not miss the initial scheduling grant. If this is the case, it is necessary to explicitly indicate the modulation-and-coding scheme, which also implies that the first redundancy version in the sequence is used.

8.1.3 HYBRID-ARQ TIMING

The receiver must know to which hybrid-ARQ process a received acknowledgment is associated. This is handled by the timing of the hybrid-ARQ acknowledgment being used to associate the acknowledgment with a certain hybrid-ARQ process; the timing relation between the reception of data in the downlink and transmission of the hybrid-ARQ acknowledgment in the uplink (and vice versa) is fixed. From a latency perspective, the time between the reception of downlink data at the device and transmission of the hybrid-ARQ acknowledgment in the uplink should be as short as possible. At the same time, an unnecessarily short time would increase the demand on the device processing capacity and a trade-off between latency and implementation complexity is required. The situation is similar for uplink data transmissions. For LTE, this trade-off led to the decision to have eight hybrid-ARQ processes per component carrier in both uplink and downlink for FDD. For TDD, the number of processes depends on the uplink–downlink allocation, as discussed later.

8.1.3.1 Hybrid-ARQ Timing for FDD

Starting with the FDD case, transmission of acknowledgments in the uplink in response to downlink data transmission is illustrated in Figure 8.6. Downlink data on the DL-SCH is transmitted to the device in subframe n and received by the device, after the propagation delay T_p, in subframe n. The device attempts to decode the received signal, possibly after soft combining with a previous transmission attempt, and transmits the hybrid-ARQ acknowledgment in uplink subframe $n + 4$ (note that the start of an uplink subframe at the device is offset by T_{TA} relative to the start of the corresponding downlink subframe at the device as a result of the timing-advance procedure described in Section 7.6). Upon reception of the hybrid-ARQ acknowledgment, the eNodeB can, if needed, retransmit the downlink data in subframe $n + 8$. Thus, eight hybrid-ARQ processes are used—that is, the hybrid-ARQ round-trip time is 8 ms.

The description in the previous paragraph, as well as the illustration in Figure 8.6, describe the timing for downlink data transmission—that is, data on DL-SCH and acknowledgments on PUCCH (or PUSCH). However, the timing of uplink data transmission—that is, data on PUSCH and acknowledgments on PHICH—is identical, namely uplink data transmission in subframe n results in a PHICH transmission in subframe $n + 4$. This is also the same timing relation as for uplink scheduling grants in general (as described in Section 9.3) and allows a scheduling grant on the PDCCH to override the PHICH, as shown in Figure 8.5.

In Figure 8.6 it is seen that the processing time available to the device, T_{UE}, depends on the value of the timing advance or, equivalently, on the device-to-base-station distance. As a device must operate at any distance up to the maximum size supported by the specifications, the device must be designed such that it can handle the worst-case scenario. LTE is designed to handle at least 100 km, corresponding to a maximum timing advance of 0.67 ms. Hence, there is approximately 2.3 ms left for the device processing, which is considered a reasonable

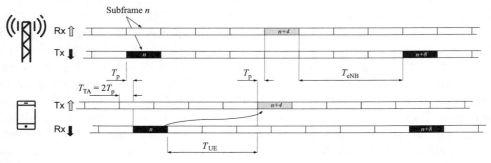

FIGURE 8.6

Timing relation between downlink data in subframe n and uplink hybrid-ARQ acknowledgment in subframe $n + 4$ for FDD.

trade-off between the processing requirements imposed on a device and the associated delays.

For the eNodeB, the processing time available, denoted as T_{eNB}, is 3 ms and thus of the same order as for the device. In the case of downlink data transmission, the eNodeB performs scheduling of any retransmissions during this time and, for uplink data transmission, the time is used for decoding of the received signal. The timing budget is thus similar for the device and the eNodeB, which is motivated by the fact that, although the eNodeB typically has more processing power than a device, it also has to serve multiple devices and to perform the scheduling operation.

Having the same number of hybrid-ARQ processes in uplink and downlink is beneficial for half-duplex FDD operation, discussed in Chapter 9. By proper scheduling, the uplink hybrid-ARQ transmission from the device will coincide with transmission of uplink data and the acknowledgments related to the reception of uplink data will be transmitted in the same subframe as the downlink data. Thus, using the same number of hybrid-ARQ processes in uplink and downlink results in a 50:50 split between transmission and reception for a half-duplex device.

8.1.3.2 Hybrid-ARQ Timing for TDD

For TDD operation, the time relation between the reception of data in a certain hybrid-ARQ process and the transmission of the hybrid-ARQ acknowledgment depends on the uplink–downlink allocation. An uplink hybrid-ARQ acknowledgment can only be transmitted in an uplink subframe and a downlink acknowledgment only in a downlink subframe. Furthermore, starting from release 11, different uplink–downlink allocations can be used on component carriers operating in different frequency bands. Therefore, the hybrid-ARQ timing relations for TDD are more intricate than their FDD counterparts.

For simplicity, consider the situation when the same uplink–downlink allocation is used across all component carriers (the case of different allocations is described later). Since the minimum processing time required in the device and the eNodeB remains the same in FDD and TDD as the same turbo decoders are used and the scheduling decisions are similar, acknowledgments cannot be transmitted earlier than in FDD. Therefore, for TDD the acknowledgment of a transport block in subframe n is transmitted in subframe $n + k$, where $k \geq 4$ and is selected such that $n + k$ is an uplink subframe when the acknowledgment is to be transmitted from the device (on PUCCH or PUSCH) and a downlink subframe when the acknowledgment is transmitted from the eNodeB (on PHICH). The value of k depends on the uplink–downlink configuration, as shown in Table 8.1 for both downlink and uplink transmissions. From the table it is seen that, as a consequence of the uplink–downlink configuration, the value of k is sometimes larger than the FDD value, $k = 4$. For example, assuming configuration 2, an uplink transmission on PUSCH received in subframe 2 should be acknowledged on PHICH in subframe $2 + 6 = 8$. Similarly, for the same configuration, downlink transmission on PDSCH in subframe 0 should be acknowledged on PUCCH (or PUSCH) in subframe $0 + 7 = 7$.

Table 8.1 Number of Hybrid-ARQ Processes and Acknowledgment Timing k for Different TDD Configurations

Configuration (DL:UL)	Downlink											Uplink										
	Proc	PDSCH Reception in Subframe n										Proc	PUSCH Reception in Subframe n									
		0	1	2	3	4	5	6	7	8	9		0	1	2	3	4	5	6	7	8	9
0 (2:3)	4	4	6	—	—	—	4	6	—	—	—	6	—	—	4	7	6	—	—	4	7	6
1 (3:2)	7	7	6	—	—	4	7	6	—	—	4	4	—	—	4	6	—	—	—	4	6	—
2 (4:1)	10	7	6	—	4	8	7	6	—	4	8	2	—	—	6	—	—	—	—	6	—	—
3 (7:3)	9	4	11	—	—	—	7	6	6	5	5	3	—	—	6	6	6	—	—	—	—	—
4 (8:2)	12	12	11	—	—	8	7	7	6	5	4	2	—	—	6	6	—	—	—	—	—	—
5 (9:1)	15	12	11	—	9	8	7	6	5	4	13	1	—	—	6	—	—	—	—	—	—	—
6 (5:5)	6	7	7	—	—	—	7	7	—	—	5	6	—	—	4	6	6	—	—	4	7	—

From the table it is also seen that the number of hybrid-ARQ processes used for TDD depends on the uplink—downlink configuration, implying that the hybrid-ARQ round-trip time is configuration dependent for TDD (actually, it may even vary between subframes, as seen in Figure 8.7). For the downlink-heavy configurations 2, 3, 4, and 5, the number of downlink hybrid-ARQ processes is larger than for FDD. The reason is the limited number of uplink subframes available, resulting in k-values well beyond 4 for some subframes (Table 8.1).

The PHICH timing in TDD is identical to the timing when receiving an uplink grant, as described in Section 9.3. The reason is the same as in FDD, namely to allow a PDCCH uplink grant to override the PHICH in order to implement adaptive retransmissions, as illustrated in Figure 8.5.

The uplink—downlink allocation for TDD has implications on the number of transport blocks to acknowledge in a single subframe. For FDD there is always a one-to-one relation between uplink and downlink subframes. Hence, in absence of carrier aggregation, a subframe only needs to carry acknowledgments for one subframe in the other direction. For TDD, in contrast, there is not necessarily a one-to-one relation between uplink and downlink subframes. This can be seen in the possible uplink—downlink allocations described in Chapter 5.

For uplink transmissions using UL-SCH, each uplink transport block is acknowledged individually by using the PHICH. Hence, in uplink-heavy asymmetries (uplink—downlink configuration 0), the device may need to receive two hybrid-ARQ acknowledgments in downlink subframes 0 and 5 in the absence of uplink spatial multiplexing; for spatial multiplexing up to four acknowledgments per component carrier are needed. There are also some subframes where no PHICH will be transmitted. The amount of PHICH groups may therefore vary between subframes for TDD.

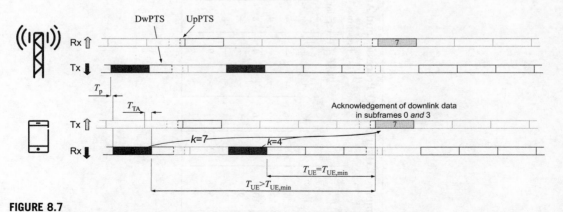

FIGURE 8.7

Example of timing relation between downlink data and uplink hybrid-ARQ acknowledgment for TDD (configuration 2).

For downlink transmissions, there are some configurations where DL-SCH receipt in multiple downlink subframes needs to be acknowledged in a single uplink subframe, as illustrated in Figure 8.7 (the illustration corresponds to the two entries shown in bold in Table 8.1). Two different mechanisms to handle this are provided in TDD: multiplexing and bundling.

Multiplexing implies that independent acknowledgments for each of the received transport blocks are fed back to the eNodeB. This allows independent retransmission of erroneous transport blocks. However, it also implies that multiple bits need to be transmitted from the device, which may limit the uplink coverage. This is the motivation for the bundling mechanism.

Bundling of acknowledgments implies that the outcome of the decoding of downlink transport blocks from multiple downlink subframes can be combined into a single hybrid-ARQ acknowledgment transmitted in the uplink. Only if both of the downlink transmissions in subframes 0 and 3 in the example in Figure 8.7 are correctly decoded will a positive acknowledgment be transmitted in uplink subframe 7.

Combining acknowledgments related to multiple downlink transmissions into a single uplink message assumes that the device has not missed any of the scheduling assignments upon which the acknowledgment is based. Assume, as an example, that the eNodeB scheduled the device in two (subsequent) subframes, but the device missed the PDCCH transmission in the first of the two subframes and successfully decoded the data transmitted in the second subframe. Without any additional mechanism, the device will transmit an acknowledgment based on the assumption that it was scheduled in the second subframe only, while the eNodeB will interpret acknowledgment as the device successfully received both transmissions. To avoid such errors, the *downlink assignment index* (see Section 6.4.6) in the scheduling assignment is used. The downlink assignment index in essence informs the device about the number of transmissions it should base the combined acknowledgment upon. If there is a mismatch between the assignment index and the number of transmissions the device received, the device concludes at least one assignment was missed and transmits no hybrid-ARQ acknowledgment, thereby avoiding acknowledging transmissions not received.

8.2 RADIO-LINK CONTROL

The *radio-link control* (RLC) protocol takes data in the form of RLC SDUs from PDCP and delivers them to the corresponding RLC entity in the receiver by using functionality in MAC and physical layers. The relation between RLC and MAC, including multiplexing of multiple logical channels into a single transport channel, is illustrated in Figure 8.8. Multiplexing of several logical channels into a single transport channel is mainly used for priority handling, as described in Section 9.2 in conjunction with downlink and uplink scheduling.

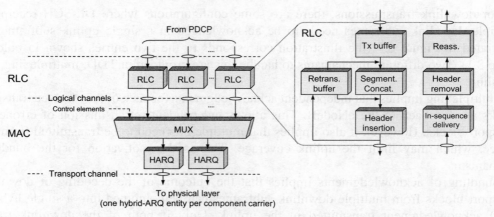

FIGURE 8.8

MAC and RLC structure (single-device view).

There is one RLC entity per logical channel configured for a device, where each RLC entity is responsible for:

- segmentation, concatenation, and reassembly of RLC SDUs;
- RLC retransmission;
- in-sequence delivery and duplicate detection for the corresponding logical channel.

Other noteworthy features of the RLC are: (1) the handling of varying PDU sizes; and (2) the possibility for close interaction between the hybrid-ARQ and RLC protocols. Finally, the fact that there is one RLC entity per logical channel and one hybrid-ARQ entity per component carrier implies that one RLC entity may interact with multiple hybrid-ARQ entities in case of carrier aggregation.

8.2.1 SEGMENTATION, CONCATENATION, AND REASSEMBLY OF RLC SDUS

The purpose of the segmentation and concatenation mechanism is to generate RLC PDUs of appropriate size from the incoming RLC SDUs. One possibility would be to define a fixed PDU size, a size that would result in a compromise. If the size were too large, it would not be possible to support the lowest data rates. Also, excessive padding would be required in some scenarios. A single small PDU size, however, would result in a high overhead from the header included with each PDU. To avoid these drawbacks, which is especially important given the very large dynamic range of data rates supported by LTE, the RLC PDU size varies dynamically.

Segmentation and concatenation of RLC SDUs into RLC PDUs are illustrated in Figure 8.9. The header includes, among other fields, a sequence number, which is used by the

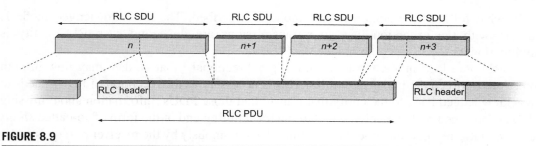

FIGURE 8.9

Generation of RLC PDUs from RLC SDUs.

reordering and retransmission mechanisms. The reassembly function at the receiver side performs the reverse operation to reassemble the SDUs from the received PDUs.

8.2.2 RLC RETRANSMISSION

Retransmission of missing PDUs is one of the main functionalities of the RLC. Although most of the errors can be handled by the hybrid-ARQ protocol, there are, as discussed at the beginning of the chapter, benefits of having a second-level retransmission mechanism as a complement. By inspecting the sequence numbers of the received PDUs, missing PDUs can be detected and a retransmission requested from the transmitting side.

Different services have different requirements; for some services (e.g., transfer of a large file), error-free delivery of data is important, whereas for other applications (e.g., streaming services), a small amount of missing packets is not a problem. The RLC can therefore operate in three different modes, depending on the requirements from the application:

- *Transparent mode* (TM), where the RLC is completely transparent and is essentially bypassed. No retransmissions, no segmentation/reassembly, and no in-sequence delivery take place. This configuration is used for control-plane broadcast channels such as BCCH, CCCH, and PCCH, where the information should reach multiple users. The size of these messages are selected such that all intended devices are reached with a high probability and hence there is neither need for segmentation to handle varying channel conditions, nor retransmissions to provide error-free data transmission. Furthermore, retransmissions are not feasible for these channels as there is no possibility for the device to feed back status reports as no uplink has been established.
- *Unacknowledged mode* (UM) supports segmentation/reassembly and in-sequence delivery, but not retransmissions. This mode is used when error-free delivery is not required, for example voice-over IP, or when retransmissions cannot be requested, for example broadcast transmissions on MTCH and MCCH using MBSFN.
- *Acknowledged mode* (AM) is the main mode of operation for TCP/IP packet data transmission on the DL-SCH. Segmentation/reassembly, in-sequence delivery, and retransmissions of erroneous data are all supported.

In what follows, the acknowledged mode is described. The unacknowledged mode is similar, with the exception that no retransmissions are done and each RLC entity is unidirectional.

In acknowledged mode, the RLC entity is bidirectional—that is, data may flow in both directions between the two peer entities. This is necessary as the reception of PDUs needs to be acknowledged back to the entity that transmitted those PDUs. Information about missing PDUs is provided by the receiving end to the transmitting end in the form of so-called *status reports*. Status reports can either be transmitted autonomously by the receiver or requested by the transmitter. To keep track of the PDUs in transit, the transmitter attaches an RLC header to each PDU, including, among other fields, a sequence number.

Both RLC entities maintain two windows, the transmission and reception windows, respectively. Only PDUs in the transmission window are eligible for transmission; PDUs with sequence number below the start of the window have already been acknowledged by the receiving RLC. Similarly, the receiver only accepts PDUs with sequence numbers within the reception window. The receiver also discards any duplicate PDUs as each PDU should be assembled into an SDU only once.

8.2.3 IN-SEQUENCE DELIVERY

In-sequence delivery implies that data blocks are delivered by the receiver in the same order as they were transmitted. This is an essential part of RLC; the hybrid-ARQ processes operate independently and transport blocks may therefore be delivered out of sequence, as seen in Figure 8.3. In-sequence delivery implies that SDU n should be delivered prior to SDU $n + 1$. This is an important aspect as several applications require the data to be received in the same order as it was transmitted. TCP can, to some extent, handle IP packets arriving out of sequence, although with some performance impact, while for some streaming applications in-sequence delivery is essential. The basic idea behind in-sequence delivery is to store the received PDUs in a buffer until all PDUs with lower sequence number have been delivered. Only when all PDUs with lower sequence number have been used for assembling SDUs is the next PDU used. RLC retransmission, provided when operating in acknowledged mode only, operates on the same buffer as the in-sequence delivery mechanism.

8.2.4 RLC OPERATION

The operation of the RLC with respect to retransmissions and in-sequence delivery is perhaps best understood by the simple example in Figure 8.10, where two RLC entities are illustrated, one in the transmitting node and the other in the receiving node. When operating in acknowledged mode, as assumed below, each RLC entity has both transmitter and receiver functionality, but in this example only one of the directions is discussed as the other direction is identical. In the example, PDUs numbered from n to $n + 4$ are awaiting transmission in the transmission buffer. At time t_0, PDUs with sequence number up to and including n have been

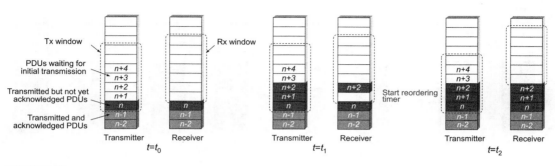

FIGURE 8.10

In-sequence delivery.

transmitted and correctly received, but only PDUs up to and including $n - 1$ have been acknowledged by the receiver. As seen in the figure, the transmission window starts from n, the first not-yet-acknowledged PDU, while the reception window starts from $n + 1$, the next PDU expected to be received. Upon reception of PDU n, the PDU is forwarded to the SDU reassembly functionality for further processing.

The transmission of PDUs continues and, at time t_1, PDUs $n + 1$ and $n + 2$ have been transmitted but, at the receiving end, only PDU $n + 2$ has arrived. One reason for this could be that the missing PDU, $n + 1$, is being retransmitted by the hybrid-ARQ protocol and therefore has not yet been delivered from the hybrid ARQ to the RLC. The transmission window remains unchanged compared to the previous figure, as none of the PDUs n and higher have been acknowledged by the receiver. Hence, any of these PDUs may need to be retransmitted as the transmitter is not aware of whether they have been received correctly. The reception window is not updated when PDU $n + 2$ arrives. The reason is that PDU $n + 2$ cannot be forwarded to SDU assembly as PDU $n + 1$ is missing. Instead, the receiver waits for the missing PDU $n + 1$. Waiting for the missing PDU for an infinite time would stall the queue. Hence, the receiver starts a timer, the *reordering timer*, for the missing PDU. If the PDU is not received before the timer expires, a retransmission is requested. Fortunately, in this example, the missing PDU arrives from the hybrid-ARQ protocol at time t_2, before the timer expires. The reception window is advanced and the reordering timer is stopped as the missing PDU has arrived. PDUs $n + 1$ and $n + 2$ are delivered for reassembly into SDUs.

Duplicate detection is also the responsibility of the RLC, using the same sequence number as used for reordering. If PDU $n + 2$ arrives again (and is within the reception window), despite it having already been received, it is discarded.

This example illustrates the basic principle behind in-sequence delivery, which is supported by both acknowledged and unacknowledged modes. However, the acknowledged mode of operation also provides retransmission functionality. To illustrate the principles behind this, consider Figure 8.11, which is a continuation of the example mentioned earlier. At time t_3, PDUs up to $n + 5$ have been transmitted. Only PDU $n + 5$ has arrived and PDUs

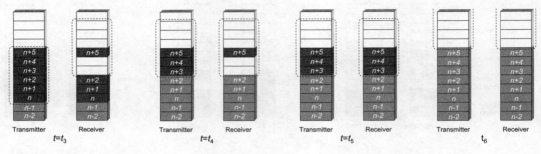

FIGURE 8.11

Retransmission of missing PDUs.

$n + 3$ and $n + 4$ are missing. Similar to the preceding case, this causes the reordering timer to start. However, in this example no PDUs arrive prior to the expiration of the timer. The expiration of the timer at time t_4 triggers the receiver to send a control PDU containing a status report, indicating the missing PDUs, to its peer entity. Control PDUs have higher priority than data PDUs to avoid the status reports being unnecessarily delayed and negatively impacting the retransmission delay. Upon reception of the status report at time t_5, the transmitter knows that PDUs up to $n + 2$ have been received correctly and the transmission window is advanced. The missing PDUs $n + 3$ and $n + 4$ are retransmitted and, this time, correctly received.

The retransmission was triggered by the reception of a status report in this example. However, as the hybrid-ARQ and RLC protocols are located in the same node, tight interaction between the two is possible. The hybrid-ARQ protocol at the transmitting end could therefore inform the RLC at the transmitting end in case the transport block(s) containing PDUs $n + 3$ and $n + 4$ have failed. The RLC can use this to trigger retransmission of missing PDUs without waiting for an explicit RLC status report, thereby reducing the delays associated with RLC retransmissions.

Finally, at time t_6, all PDUs, including the retransmissions, have been delivered by the transmitter and successfully received. As $n + 5$ was the last PDU in the transmission buffer, the transmitter requests a status report from the receiver by setting a flag in the header of the last RLC data PDU. Upon reception of the PDU with the flag set, the receiver will respond by transmitting the requested status report, acknowledging all PDUs up to and including $n + 5$. Reception of the status report by the transmitter causes all the PDUs to be declared as correctly received and the transmission window is advanced.

Status reports can, as mentioned earlier, be triggered for multiple reasons. However, to control the amount of status reports and to avoid flooding the return link with an excessive number of status reports, it is possible to use a status prohibit timer. With such a timer, status reports cannot be transmitted more often than once per time interval as determined by the timer.

For the initial transmission, it is relatively straightforward to rely on a dynamic PDU size as a means to handle the varying data rates. However, the channel conditions and the amount of resources may also change between RLC retransmissions. To handle these variations, already transmitted PDUs can be (re)segmented for retransmissions. The reordering and retransmission mechanisms described in the preceding paragraphs still apply; a PDU is assumed to be received when all the segments have been received. Status reports and retransmissions operate on individual segments; only the missing segment of a PDU needs to be retransmitted.

SCHEDULING AND RATE ADAPTATION

Scheduling is a central part of an LTE system. For each time instant, the scheduler determines to which user(s) the shared time—frequency resource should be assigned and determines the data rate to use for the transmission. The scheduler is a key element and to a large degree determines the overall behavior of the system. Both uplink and downlink transmissions are scheduled and, consequently, there is a downlink and an uplink scheduler in the eNodeB.

The *downlink scheduler* is responsible for dynamically controlling the device(s) to transmit to. Each of the scheduled devices is provided with a *scheduling assignment* consisting of the set of resource blocks upon which the device's DL-SCH[1] is transmitted, and the associated transport-format. The basic mode of operation is the so-called *dynamic* scheduling, where the eNodeB for each 1 ms TTI conveys scheduling assignments to the selected devices using the (E)PDCCHs as described in Chapter 6, but there is also a possibility for *semi-persistent* scheduling to reduce the control-signaling overhead. Downlink scheduling assignments and logical channel multiplexing are controlled by the eNodeB as illustrated in the left part of Figure 9.1.

The *uplink scheduler* serves a similar purpose, namely to dynamically control which devices are to transmit on their UL-SCH. Similarly to the downlink case, each scheduled device is provided with a *scheduling grant* consisting of the set of resource blocks upon which the device should transmit its UL-SCH and the associated transport-format. Also in this case, either dynamic or semi-persistent scheduling can be used. The uplink scheduler is in complete control of the transport format the device shall use but, unlike the downlink case, not the logical-channel multiplexing. Instead, the logical-channel multiplexing is controlled by the device according to a set of rules. Thus, uplink scheduling is *per device* and not per radio bearer. This is illustrated in the right part of Figure 9.1, where the scheduler controls the transport format and the device controls the logical channel multiplexing.

The following sections describe the details in the LTE scheduling framework after a brief review of basic scheduling principles.

---

[1]In case of carrier aggregation there is one DL-SCH (or UL-SCH) per component carrier.

4G, LTE-Advanced Pro and The Road to 5G. http://dx.doi.org/10.1016/B978-0-12-804575-6.00009-1

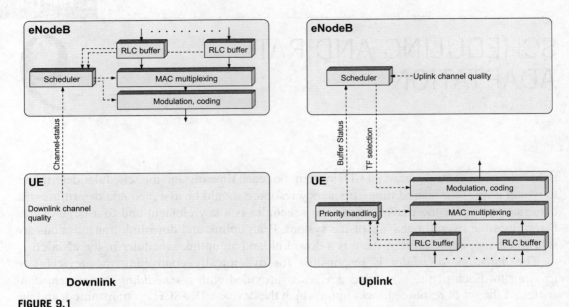

FIGURE 9.1

Transport format selection in downlink (left) and uplink (right).

9.1 SCHEDULING STRATEGIES

The scheduling strategy in LTE is not standardized but is a base-station-implementation issue—and an important one as the scheduler is a key element in LTE and to a large extent defines the overall behavior. Different vendors may choose different strategies in various scenarios to match the user needs. What is standardized is the supporting functions for scheduling such as transmission of scheduling grants, quality-of-service mechanisms, and various feedback information, for example channel-state reports and buffer-status reports. However, there are some basic scheduling strategies in the literature, useful to illustrate the principles.

For the purpose of illustrating the principles, consider time-domain-only scheduling with a single user being scheduled at a time and all users having an infinite amount of data to transmit. In this case, the utilization of the radio resources is maximized if, at each time instant, all resources are assigned to the user with the best instantaneous channel condition. Together with rate control, this implies that the highest data rate is achieved for a given transmit power or, in other words, for a given interference to other cells, the highest link utilization is achieved. Rate control is more efficient compared to power control, which adjusts the transmission power to follow the channel variations while keeping the data rate constant [11,12]. This scheduling strategy is an example of channel-dependent scheduling known as *max*-C/I (or *maximum rate*) scheduling. Since the radio conditions for the different radio links within a cell typically vary independently, at each point in time there is almost always a radio link whose channel quality is near its peak and supporting a correspondingly high data rate. This translates into a high system capacity. The gain obtained by transmitting to users with favorable radio-link conditions is commonly known as multi-user diversity; the multi-user diversity gains are larger, the larger the channel variations and the larger the number of users in a cell. Hence, in contrast to the traditional view that rapid variations in the radio-link quality is an undesirable effect that has to be combated, the possibility of channel-dependent scheduling implies that *rapid variations are in fact potentially beneficial and should be exploited.*

Mathematically, the max-C/I scheduler can be expressed as scheduling user k given by

$$k = \arg \max_i R_i$$

where R_i is the instantaneous data rate for user i. Although, from a system capacity perspective, max-C/I scheduling is beneficial, this scheduling principle will not be fair in all situations. If all devices are, on average, experiencing similar channel conditions and the variations in the instantaneous channel conditions are only due to, for example, fast multi-path fading, all users will experience the same average data rate. Any variations in the instantaneous data rate are rapid and often not even noticeable by the user. However, in practice different devices will also experience differences in the (short-term) average channel conditions—for example, due to differences in the distance between the base station and the device. In this case, the channel conditions experienced by one device may, for a relatively long time, be worse than the channel conditions experienced by other devices. A pure max-C/I-scheduling strategy would then "starve" the device with the bad channel conditions, and the device with bad channel conditions will never be scheduled. This is illustrated in Figure 9.2(a), where a max-C/I

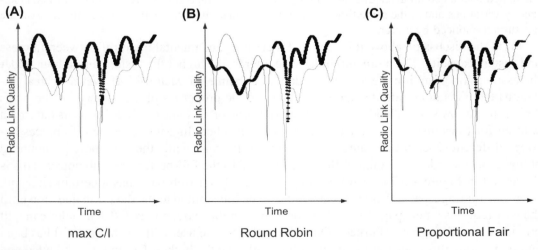

FIGURE 9.2

Example of three different scheduling behaviors for two users with different average channel quality: (a) max C/I, (b) round robin, and (c) proportional fair. The selected user is shown with bold lines.

scheduler is used to schedule between two different users with different average channel quality. Although resulting in the highest system capacity, this situation is often not acceptable from a quality-of-service point of view.

An alternative to the max-C/I scheduling strategy is so-called *round-robin* scheduling, illustrated in Figure 9.2(b). This scheduling strategy lets users take turns in using the shared resources, without taking the instantaneous channel conditions into account. Round-robin scheduling can be seen as fair scheduling in the sense that the same amount of radio resources (the same amount of time) is given to each communication link. However, round-robin scheduling is not fair in the sense of providing the same service quality to all communication links. In that case more radio resources (more time) must be given to communication links with bad channel conditions. Furthermore, as round-robin scheduling does not take the instantaneous channel conditions into account in the scheduling process, it will lead to lower overall system performance but more equal service quality between different communication links, compared to max-C/I scheduling.

A third possibility is the so-called *proportional fair* scheduler, see Figure 9.2(c), which tries to exploit rapid channel variations while suppressing the effect of differences in the average channel gain. In this strategy, the shared resources are assigned to the user with the *relatively* best radio-link conditions—that is, at each time instant, user k is selected for transmission according to

$$k = \arg \max_i \frac{R_i}{\overline{R_i}}$$

where R_i is the instantaneous data rate for user i and $\overline{R_i}$ is the average data rate for user i. The average is calculated over a certain averaging period long enough to average out differences in the fast channel-quality variations and at the same time short enough so that quality variations within the interval are not strongly noticed by a user.

From the discussion above it is seen that there is a fundamental trade-off between fairness and system capacity. The more unfair the scheduler is, the higher the system throughput under the assumption of an infinite amount of data to transmit for each user. However, in real situations there is not an infinite amount of data and the properties of the (bursty) traffic plays large role. At low system load—that is, when only one or, in some cases, a few users have data waiting for transmission at the base station at each scheduling instant—the differences between different scheduling strategies above are fairly small while they are more pronounced at higher loads and can be quite different compared to the full-buffer scenario above. This is illustrated in Figure 9.3 for web-browsing scenario. Each web page has a certain size and, after transmitting a page, there is no more data to be transmitted to the device in question until the user requests a new page by clicking on a link. In this case, a max-C/I scheduler can still provide a certain degree of fairness. Once the buffer for the user with the highest C/I has been emptied, another user with non-empty buffers will have the highest C/I and be scheduled and so on. The proportional fair scheduler has similar performance in both scenarios.

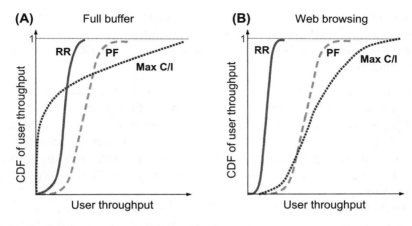

FIGURE 9.3

Illustration of the principle behavior of different scheduling strategies: (a) for full buffers and (b) for web browsing traffic model.

Clearly, the degree of fairness introduced by the traffic properties depends heavily on the actual traffic; a design made with certain assumptions may be less desirable in an actual network where the traffic pattern may be different from the assumptions made during the design. Therefore, relying solely on the traffic properties to achieve fairness is not a good strategy, but the discussion above also emphasizes the need to design the scheduler not only for the full buffer case. Traffic priorities, for example prioritizing a latency-critical services over a delay-tolerant service despite the channel quality for the latter being superior, is another example where the full-buffer discussion above was simplified to illustrate the basic principles. Other examples of scheduling input are DRX cycles, retransmissions, device capabilities, and device power consumption, all impacting the overall scheduling behavior.

The general discussion above is applicable to both downlink and uplink transmissions. However, there are some differences between the two. Fundamentally, the uplink power resource is *distributed* among the users, while in the downlink the power resource is *centralized* within the base station. Furthermore, the maximum uplink transmission power of a single device is typically significantly lower than the output power of a base station. This has a significant impact on the scheduling strategy. Unlike the downlink, where pure time-domain scheduling often can be used [15,16] and can be shown to be preferable from a theoretical point of view [13,14], uplink scheduling typically has to rely on sharing in the frequency domain in addition, as a single device may not have sufficient power for efficiently utilizing the link capacity. There are also other reasons to complement the time domain with the frequency domain, both in uplink and downlink, for example,

- in the case of insufficient payload—that is, the amount of data to transfer to a user is not sufficiently large to utilize the full channel bandwidth, or
- in the case where not only time-domain variations but also frequency-domain variations are to be exploited.

The scheduling strategies in these cases can be seen as generalizations of the schemes discussed for the time-domain-only cases in the preceding paragraphs. For example, to handle small payloads, a greedy filling approach can be used, where the scheduled user is selected according to max-C/I (or any other scheduling scheme). Once this user has been assigned resources matching the amount of data awaiting transmission, the second best user according to the scheduling strategy is selected and assigned (a fraction of) the residual resources and so on.

In the following sections, dynamic downlink and uplink scheduling, as well as related functionality such as uplink priority handling, scheduling requests, buffer status and power headroom reporting, semi-persistent scheduling, half-duplex FDD operation, and DRX functionality are described. Channel-state reporting, an important input to any channel-dependent scheduling, is discussed in Chapter 10. Remember that in LTE, only the general framework, and not the scheduling strategy, is standardized.

9.2 DOWNLINK SCHEDULING

The task of the downlink scheduler is to dynamically determine the device(s) to transmit to and, for each of these devices, the set of resource blocks upon which the device's DL-SCH should be transmitted. As discussed in the previous section, the amount of data in the transmission buffers as well as the desire to exploit channel-variations in the frequency domain implies that transmissions to multiple users in different parts of the spectrum is needed. Therefore, multiple devices can be scheduled in parallel in a subframe, in which case there is one DL-SCH per scheduled device (and component carrier), each dynamically mapped to a (unique) set of frequency resources.

The scheduler is in control of the instantaneous data rate used, and the RLC segmentation and MAC multiplexing will therefore be affected by the scheduling decision. Although formally part of the MAC layer but to some extent better viewed as a separate entity, the scheduler is thus controlling most of the functions in the eNodeB associated with downlink data transmission:

- *RLC*: Segmentation/concatenation of RLC SDUs is directly related to the instantaneous data rate. For low data rates, it may only be possible to deliver a part of an RLC SDU in a TTI, in which case segmentation is needed. Similarly, for high data rates, multiple RLC SDUs may need to be concatenated to form a sufficiently large transport block.
- *MAC*: Multiplexing of logical channels depends on the priorities between different streams. For example, radio resource control signaling, such as handover commands, typically has a higher priority than streaming data, which in turn has higher priority than a background file transfer. Thus, depending on the data rate and the amount of traffic of

different priorities, the multiplexing of different logical channels is affected. Hybrid-ARQ retransmissions also need to be accounted for.

- *L1*: Coding, modulation, and, if applicable, the number of transmission layers and the associated precoding matrix are affected by the scheduling decision. The choices of these parameters are mainly determined by the radio conditions and the selected data rate, that is, the transport block size.

The scheduling strategy is implementation-specific and not a part of the 3GPP specifications; in principle any strategy can be applied as discussed in Section 9.1. However, the overall goal of most schedulers is to take advantage of the channel variations between devices and preferably to schedule transmissions to a device when the channel conditions are advantageous. Most scheduling strategies therefore at least need information about:

- channel conditions at the device;
- buffer status and priorities of the different data flows;
- the interference situation in neighboring cells (if some form of interference coordination is implemented).

Information about the channel conditions at the device can be obtained in several ways. In principle, the eNodeB can use any information available, but typically the CSI reports from the device are used. The details of the transmission-mode-dependent CSI reports are found in Chapter 10. Other sources of channel knowledge, for example, exploiting channel reciprocity to estimate the downlink quality from uplink channel estimates in the case of TDD, can also be exploited by a particular scheduler implementation, either alone or in combination with CSI reports.

In addition to the channel-state information, the scheduler should take buffer status and priority levels into account. For example, it does not make sense to schedule a device with empty transmission buffers. Priorities of the different types of traffic may also vary; RRC signaling may be prioritized over user data. Furthermore, RLC and hybrid-ARQ retransmissions, which are in no way different from other types of data from a scheduler perspective, are typically also given priority over initial transmissions.

Downlink inter-cell interference coordination is also part of the implementation-specific scheduler strategy. A cell may signal to its neighboring cells the intention to transmit with a lower transmission power in the downlink on a set of resource blocks. This information can then be exploited by neighboring cells as a region of low interference where it is advantageous to schedule devices at the cell edge, devices that otherwise could not attain high data rates due to the interference level. Inter-cell interference handling is further discussed in Chapter 13.

9.3 UPLINK SCHEDULING

The basic function of the *uplink scheduler* is similar to its downlink counterpart, namely to dynamically determine, for each 1 ms interval, which devices are to transmit and on which uplink resources. As discussed before, the LTE uplink is primarily based on maintaining orthogonality between different uplink transmissions and the shared resource controlled by the eNodeB scheduler is time—frequency resource units. In addition to assigning the time—frequency resources to the device, the eNodeB scheduler is also responsible for controlling the transport format the device will use for each of the uplink component carriers. This allows the scheduler to tightly control the uplink activity to maximize the resource usage compared to schemes where the device autonomously selects the data rate, as autonomous schemes typically require some margin in the scheduling decisions. A consequence of the scheduler being responsible for selection of the transport format is that accurate and detailed knowledge in the eNodeB about the device situation with respect to buffer status and power availability is more accentuated in LTE compared to systems where the device autonomously controls the transmission parameters.

The basis for uplink scheduling is *scheduling grants*, containing the scheduling decision and providing the device information about the resources and the associated transport format to use for transmission of the UL-SCH on one component carrier. Only if the device has a valid grant is it allowed to transmit on the corresponding UL-SCH; autonomous transmissions are not possible. Dynamic grants are valid for one subframe, that is, for each subframe in which the device is to transmit on the UL-SCH, the scheduler issues a new grant.

Similarly to the downlink case, the uplink scheduler can exploit information about channel conditions, and, if some form of interference coordination is employed, the interference situation in neighboring cells. Information about the buffer status in the device, and its available transmission power, is also beneficial to the scheduler. This calls for the reporting mechanisms described in the following, unlike the downlink case where the scheduler, power amplifier, and transmission buffers all are in the same node. Uplink priority handling is, as already touched upon, another area where uplink and downlink scheduling differs.

Channel-dependent scheduling, which typically is used for the downlink, can be used for the uplink as well. In the uplink, estimates of the channel quality can be obtained from the use of uplink channel sounding, as described in Chapter 7. For scenarios where the overhead from channel sounding is too costly, or when the variations in the channel are too rapid to be tracked, for example at high device speeds, uplink diversity can be used instead. The use of frequency hopping as discussed in Chapter 7 is one example of obtaining diversity in the uplink.

Finally, inter-cell interference coordination can be used in the uplink for similar reasons as in the downlink by exchanging information between neighboring cells, as discussed in Chapter 13.

9.3.1 UPLINK PRIORITY HANDLING

Multiple logical channels of different priorities can be multiplexed into the same transport block using the similar MAC multiplexing functionality as in the downlink (described in Chapter 4). However, unlike the downlink case, where the prioritization is under control of the scheduler and up to the implementation, the uplink multiplexing is done according to a set of well-defined rules in the device with parameters set by the network as a scheduling grant applies to a specific uplink carrier of a device, not to a specific radio bearer within the device. Using radio-bearer-specific scheduling grants would increase the control signaling overhead in the downlink and hence per-device scheduling is used in LTE.

The simplest multiplexing rule would be to serve logical channels in strict priority order. However, this may result in starvation of lower-priority channels; all resources would be given to the high-priority channel until its transmission buffer is empty. Typically, an operator would instead like to provide at least some throughput for low-priority services as well. Therefore, for each logical channel in an LTE device, a *prioritized data rate* is configured in addition to the priority value. The logical channels are then served in decreasing priority order up to their prioritized data rate, which avoids starvation as long as the scheduled data rate is at least as large as the sum of the prioritized data rates. Beyond the prioritized data rates, channels are served in strict priority order until the grant is fully exploited or the buffer is empty. This is illustrated in Figure 9.4.

9.3.2 SCHEDULING REQUESTS

The scheduler needs knowledge about devices having data to transmit and therefore need to be scheduled uplink resources. There is no need to provide uplink resources to a device with no data to transmit as this would only result in the device performing padding to fill up the granted resources. Hence, as a minimum, the scheduler needs to know whether the device has data to transmit and should be given a grant. This is known as a *scheduling request*. Scheduling requests are used for devices not having a valid scheduling grant; devices that have a valid grant and are transmitting in the uplink provide more detailed information to the eNodeB as discussed in the next section.

A scheduling request is a simple flag, raised by the device to request uplink resources from the uplink scheduler. Since the device requesting resources by definition has no PUSCH resource, the scheduling request is transmitted on the PUCCH. Each device can be assigned a dedicated PUCCH scheduling request resource, occurring every nth subframe, as described in Chapter 7. With a dedicated scheduling-request mechanism, there is no need to provide the identity of the device requesting to be scheduled as the identity of the device is implicitly known from the resources upon which the request is transmitted. When data with higher

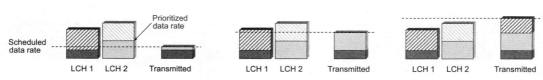

FIGURE 9.4

Prioritization of two logical channels for three different uplink grants.

priority than already existing in the transmit buffers arrives at the device and the device has no grant and hence cannot transmit the data, the device transmits a scheduling request at the next possible instant, as illustrated in Figure 9.5. Upon reception of the request, the scheduler can assign a grant to the device. If the device does not receive a scheduling grant until the next possible scheduling-request instant, the scheduling request is repeated up to a configurable limit after which the device resorts to random access to request resources form the eNodeB.

The use of a single bit for the scheduling request is motivated by the desire to keep the uplink overhead small, as a multi-bit scheduling request would come at a higher cost. A consequence of the single-bit scheduling request is the limited knowledge at the eNodeB about the buffer situation at the device when receiving such a request. Different scheduler implementations handle this differently. One possibility is to assign a small amount of resources to ensure that the device can exploit them efficiently without becoming power limited. Once the device has started to transmit on the UL-SCH, more detailed information about the buffer status and power headroom can be provided through the inband MAC control message, as discussed in the following text. Knowledge of the service type may also be used—for example, in the case of voice the uplink resource to grant is preferably the size of a typical voice-over-IP package. The scheduler may also exploit, for example, path-loss measurements used for mobility and handover decisions to estimate the amount of resources the device may efficiently utilize.

An alternative to a dedicated scheduling-request mechanism would be a contention-based design. In such a design, multiple devices share a common resource and provide their identity as part of the request. This is similar to the design of the random access. The number of bits transmitted from a device as part of a request would, in this case, be larger with the correspondingly larger need for resources. In contrast, the resources are shared by multiple users. Basically, contention-based designs are suitable for a situation where there are a large number of devices in the cell and the traffic intensity, and hence the scheduling intensity, is low. In situations with higher intensities, the collision rate between different devices simultaneously requesting resources would be too high and lead to an inefficient design.

Since the scheduling-request design for LTE relies on dedicated resources, a device that has not been allocated such resources cannot transmit a scheduling request. Instead, devices

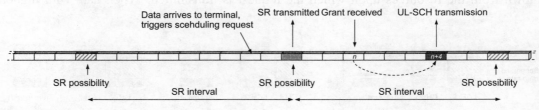

FIGURE 9.5

Scheduling request transmission.

without scheduling-request resources configured rely on the random-access mechanism described in Chapter 11. In principle, an LTE device can therefore be configured to rely on a contention-based mechanism if this is advantageous in a specific deployment.

9.3.3 BUFFER STATUS REPORTS

Devices that already have a valid grant do not need to request uplink resources. However, to allow the scheduler to determine the amount of resources to grant to each device in future subframes, information about the buffer situation, discussed in this section, and the power availability, discussed in the next section, is useful. This information is provided to the scheduler as part of the uplink transmission through MAC control elements (see Chapter 4 for a discussion on MAC control elements and the general structure of a MAC header). The LCID field in one of the MAC subheaders is set to a reserved value indicating the presence of a buffer status report, as illustrated in Figure 9.6.

From a scheduling perspective, buffer information for each logical channel is beneficial, although this could result in a significant overhead. Logical channels are therefore grouped into logical-channel groups and the reporting is done per group. The buffer-size field in a buffer-status report indicates the amount of data awaiting transmission across all logical channels in a logical-channel group. A buffer-status report can be triggered for the following reasons:

- Arrival of data with higher priority than currently in the transmission buffer—that is, data in a logical-channel group with higher priority than the one currently being transmitted—as this may impact the scheduling decision.
- Change of serving cell, in which case a buffer-status report is useful to provide the new serving cell with information about the situation in the device.
- Periodic reporting, as controlled by a timer.

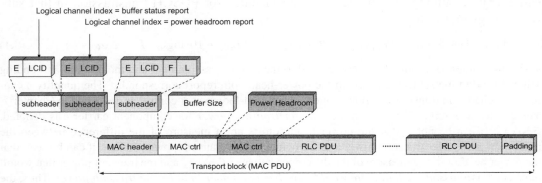

FIGURE 9.6

Signaling of buffer status and power headroom reports.

- To reduce padding. If the amount of padding required to match the scheduled transport block size is larger than a buffer-status report, a buffer-status report is inserted as it is better to exploit the available payload for useful scheduling information instead of padding, if possible.

9.3.4 POWER HEADROOM REPORTS

In addition to buffer status, the amount of transmission power available in each device is also relevant for the uplink scheduler. There is little reason to schedule a higher data rate than the available transmission power can support. In the downlink, the available power is immediately known to the scheduler as the power amplifier is located in the same node as the scheduler. For the uplink, the power availability, or *power headroom*, needs to be provided to the eNodeB. Power headroom reports are therefore transmitted from the device to the eNodeB in a similar way as the buffer-status reports, that is, only when the device is scheduled to transmit on the UL-SCH. A power headroom report can be triggered for the following reasons:

- Periodic reporting as controlled by a timer.
- Change in path loss (when the difference between the current power headroom and the last report is larger than a configurable threshold).
- To reduce padding (for the same reason as buffer-status reports).

It is also possible to configure a prohibit timer to control the minimum time between two power-headroom reports and thereby the signaling load on the uplink.

There are two different types of power-headroom reports defined in LTE, *Type 1* and *Type 2*. Type 1 reporting reflects the power headroom assuming PUSCH-only transmission on the carrier, while the Type 2 report, introduced in release 10, assumes combined PUSCH and PUCCH transmission.

The Type 1 power headroom, valid for a certain subframe (and a certain component carrier) and assuming that the device was scheduled for PUSCH transmission in that subframe, is given (in dB) by

$$\text{Power Headroom} = P_{\text{CMAX},c} - \left(P_{0,\text{PUSCH}} + \alpha \cdot PL_{\text{DL}} + 10 \cdot \log_{10}(M) + \Delta_{\text{MCS}} + \delta \right) \tag{9.1}$$

where the values for M and Δ_{MCS} correspond to the resource assignment and modulation-and-coding scheme used in the subframe to which the power-headroom report corresponds.[2] The quantity δ captures the change in transmission power due to the closed-loop power control as described in Chapter 7. The explicitly configured *maximum per-carrier transmit power* for component carrier c is denoted, $P_{\text{CMAX},c}$. It can be noted that the power headroom is not a measure of the difference between the maximum per-carrier transmit power and the actual carrier transmit power. Rather, it can be seen that the power headroom is a measure of the difference between $P_{\text{CMAX},c}$ and the transmit power that would have been used *assuming that there would have been no upper limit on the transmit power*. Thus, the

---

[2]In case of carrier aggregation, type 1 reports are supported for each of the component carriers.

power headroom can very well be negative, indicating that the per-carrier transmit power was limited by $P_{CMAX,c}$ at the time of the power headroom reporting, that is, the network has scheduled a higher data rate than the device can support given the available transmission power. As the network knows what modulation-and-coding scheme and resource size the device used for transmission in the subframe to which the power-headroom report corresponds, it can determine the valid combinations of modulation-and-coding scheme and resource size M, assuming that the downlink path loss PL_{DL} and the term δ have not changed substantially.

Type 1 power headroom can also be reported for subframes where there is no actual PUSCH transmission. In such cases, $10 \cdot \log_{10}(M)$ and Δ_{MCS} in the expression above are set to zero. This can be seen as the power headroom assuming a default transmission configuration corresponding to the minimum possible resource assignment ($M = 1$) and the modulation-and-coding scheme associated with $\Delta_{MCS} = 0$ dB.

Similarly, Type 2 power headroom reporting is defined as the difference between the maximum per-carrier transmit power and the sum of the PUSCH and PUCCH transmit power (Eqs. (7.4) and (7.3), respectively), once again not taking into account any maximum per-carrier power when calculating the PUSCH and PUCCH transmit power.[3]

Along the lines of Type 1 power headroom reporting, the Type 2 power headroom can also be reported for subframes in which no PUSCH and/or PUCCH is transmitted. In that case a virtual PUSCH and PUCCH transmit power is calculated, assuming the smallest possible resource assignment ($M = 1$) and $\Delta_{MCS} = 0$ dB for PUSCH and $\Delta_{Format} = 0$ for PUCCH.

9.4 TIMING OF SCHEDULING ASSIGNMENTS/GRANTS

The scheduling decisions, downlink scheduling assignments and uplink scheduling grants, are communicated to each of the scheduled devices through the downlink L1/L2 control signaling as described in Chapter 6, using one (E)PDCCH per downlink assignment. Each device monitors a set of (E)PDCCHs for valid scheduling assignment or grants and, upon detection of a valid assignment or grant, receives PDSCH or transmits PUSCH, respectively. The device needs to know which subframe the scheduling command relates to.

9.4.1 DOWNLINK SCHEDULING TIMING

For downlink data transmission, the scheduling assignment is transmitted in the same subframe as the data. Having the scheduling assignment in the same subframe as the corresponding data minimizes the latency in the scheduling process. Also, note that there is no possibility for dynamically scheduling future subframes—data and the scheduling assignment are always in the same subframe. This holds for FDD as well as TDD.

---

[3]In case of carrier aggregation, type 2 reports are supported for the primary component carrier only, as PUCCH cannot be transmitted on a secondary component carrier (prior to release 13).

9.4.2 UPLINK SCHEDULING TIMING

The timing for uplink scheduling grants is more intricate than the corresponding downlink scheduling assignments, especially for TDD. The grant cannot relate to the same subframe it was received in as the uplink subframe has already started when the device has decoded the grant. The device also needs some time to prepare the data to transmit. Therefore, a grant received in subframe n affects the uplink transmission in a later subframe.

For FDD, the grant timing is straightforward. An uplink grant received in a subframe n triggers an uplink transmission in subframe $n + 4$, as illustrated in Figure 9.7. This is the same timing relation as used for uplink retransmission triggered by the PHICH, motivated by the possibility to override the PHICH by a dynamic scheduling grant, as described in Chapter 8.

For TDD, the situation is more complicated. A grant received in subframe n in TDD may not necessarily trigger an uplink transmission in subframe $n + 4$—the timing relation used in FDD—as subframe $n + 4$ may not be an uplink subframe. Hence, for TDD configurations 1−6 the timing relation is modified such that the uplink transmission occurs in subframe $n + k$, where k is the smallest value larger than or equal to 4 such that subframe $n + k$ is an uplink subframe. This provides at least the same processing time for the device as in the FDD

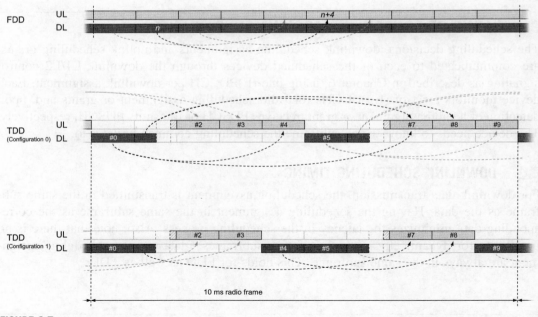

FIGURE 9.7

Timing relation for uplink grants in FDD and TDD configurations 0 and 1.

case while minimizing the delay from receipt of the uplink grant to the actual transmission. Note that this implies that the time between grant reception and uplink transmission may differ between different subframes. Furthermore, for the downlink-heavy configurations 1–5, another property is that uplink scheduling grants can only be received in some of the downlink subframes.

For TDD configuration 0 there are more uplink subframes than downlink subframes, which calls for the possibility to schedule transmissions in multiple uplink subframes from a single downlink subframe. The same timing relation as for the other TDD configurations is used but with slight modifications. Recall from Section 6.4.7 that the grant transmitted in the downlink contains an uplink index consisting of two bits. For uplink-downlink configuration 0, the index field specifies which uplink subframe(s) a grant received in a downlink subframe applies to. For example, as illustrated in Figure 9.7, an uplink scheduling grant received in downlink subframe 0 applies to one or both of the uplink subframes 4 and 7, depending on which of the bits in the uplink index are set.

9.5 SEMI-PERSISTENT SCHEDULING

The basis for uplink and downlink scheduling is dynamic scheduling, as described in Sections 9.2 and 9.3. Dynamic scheduling with a new scheduling decision taken in each subframe allows for full flexibility in terms of the resources used and can handle large variations in the amount of data to transmit at the cost of the scheduling decision being sent on an (E)PDCCH in each subframe. In many situations, the overhead in terms of control signaling on the (E) PDCCH is well motivated and relatively small compared to the payload on DL-SCH/UL-SCH. However, some services, most notably voice-over IP, are characterized by regularly occurring transmission of relatively small payloads. To reduce the control signaling overhead for those services, LTE provides semi-persistent scheduling in addition to dynamic scheduling.

With semi-persistent scheduling, the device is provided with the scheduling decision on the (E)PDCCH, together with an indication that this applies to every nth subframe until further notice. Hence, control signaling is only used once and the overhead is reduced, as illustrated in Figure 9.8. The periodicity of semi-persistently scheduled transmissions, that is, the value of n, is configured by RRC signaling in advance, while activation (and deactivation)

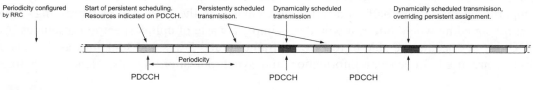

FIGURE 9.8

Example of semi-persistent scheduling.

is done using the (E)PDCCH with the semi-persistent C-RNTI.[4] For example, for voice-over IP the scheduler can configure a periodicity of 20 ms for semi-persistent scheduling and, once a talk spurt starts, the semi-persistent pattern is triggered by the (E)PDCCH.

After enabling semi-persistent scheduling, the device continues to monitor the set of candidate (E)PDCCHs for uplink and downlink scheduling commands. When a dynamic scheduling command is detected, it takes precedence over the semi-persistent scheduling in that particular subframe, which is useful if the semi-persistently allocated resources occasionally need to be increased. For example, for voice-over IP in parallel with web browsing it may be useful to override the semi-persistent resource allocation with a larger transport block when downloading the web page.

For the downlink, only initial transmissions use semi-persistent scheduling. Retransmissions are explicitly scheduled using an (E)PDCCH assignment. This follows directly from the use of an asynchronous hybrid-ARQ protocol in the downlink. Uplink retransmissions, in contrast, can either follow the semi-persistently allocated subframes or be dynamically scheduled.

9.6 SCHEDULING FOR HALF-DUPLEX FDD

Half-duplex FDD implies that a single device cannot receive and transmit at the same time while the eNodeB still operates in full duplex. In LTE, half-duplex FDD is implemented as a scheduler constraint, implying it is up to the scheduler to ensure that a single device is not simultaneously scheduled in uplink and downlink. Hence, from a device perspective, subframes are dynamically used for uplink or downlink. Briefly, the basic principle for half-duplex FDD is that a device is receiving in the downlink unless it has been explicitly instructed to transmit in the uplink (either UL-SCH transmission or hybrid-ARQ acknowledgements triggered by a downlink transmission). The timing and structure for control signaling are identical between half- and full-duplex FDD devices. Note that, as the eNodeB is operating in full duplex, regardless of the duplex capability of the devices, the cell capacity is hardly affected by the presence of half-duplex devices as, given a sufficient number of devices with data to transmit/receive, the scheduler can with a high likelihood find a set of devices to schedule in the uplink and another set to schedule in the downlink in a given subframe.

An alternative to a dynamic half-duplex FDD based on scheduling restrictions would be to base half-duplex FDD on the TDD control signaling structure and timing, with a semi-static configuration of subframes to either downlink or uplink. However, this would complicate supporting a mixture of half- and full duplex devices in the same cell as the timing of the control signaling would differ. It would also imply a waste of uplink spectrum resources. All FDD devices need to be able to receive subframes 0 and 5 in some situations as those subframes are used for system information and synchronization signals. Hence, if a fixed

---

[4]Each device has two identities, the "normal" C-RNTI for dynamic scheduling and the semi-persistent C-RNTI for semi-persistent scheduling.

uplink—downlink allocation were to be used, no uplink transmissions could take place in those two subframes, resulting in a loss in uplink spectral efficiency of 20%. This is not attractive and led to the choice of implementing half-duplex FDD as a scheduling strategy instead.

Support for half-duplex FDD has been part of LTE since the beginning but has so far seen limited use in practice. However, with the increased interest in massive machine-type communication in LTE release 12 and later, there is a renewed interest in half-duplex FDD as part of reducing the device cost for these devices. Hence, there are two ways of handling half-duplex FDD in LTE, differing in the optimization criterion and how the necessary guard time between reception and transmission is created:

- type A, part of LTE from the beginning and aiming at high performance by minimizing the guard time between reception and transmission, and
- type B, introduced in LTE release 12 and providing a long guard time to facilitate simple low-cost implementation for massive MT devices.

As stated, half-duplex type A has been part of LTE from the beginning, focusing on minimizing the guard time between reception and transmission. In this mode-of-operation, guard time for the downlink-to-uplink switch is created by allowing the device to skip reception of the last OFDM symbols in a downlink subframe immediately preceding an uplink subframe, as described in Chapter 5. Note that skipping reception of the last symbol in the downlink is only required if there is an uplink transmission immediately after the downlink subframe, otherwise the full downlink subframe is received. Guard time for the uplink-to-downlink switch is handled by setting the appropriate amount of timing advance in the devices. Compared to type B, the guard time is fairly short, resulting in high performance.

An example of half-duplex type A operation as seen from a device perspective is shown in Figure 9.9. In the leftmost part of the figure, the device is explicitly scheduled in the uplink and, consequently, cannot receive data in the downlink in the same subframe. The uplink transmission implies the receipt of an acknowledgement on the PHICH four subframes later, as mentioned in Chapter 8, and therefore the device cannot be scheduled in the uplink in this subframe. Similarly, when the device is scheduled to receive data in the downlink in subframe n, the corresponding hybrid-ARQ acknowledgement needs to be transmitted in the uplink

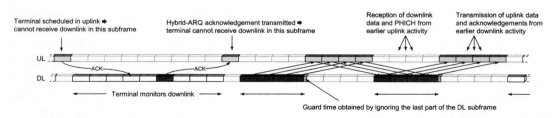

FIGURE 9.9

Example of type A half-duplex FDD device operation.

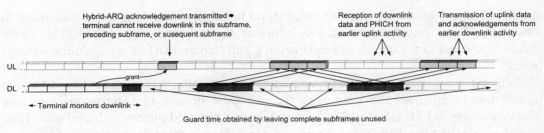

FIGURE 9.10

Example of type B half-duplex FDD device operation.

subframe $n + 4$, preventing downlink reception in subframe $n + 4$. The scheduler can exploit this by scheduling downlink data in four consecutive subframes and uplink transmission in the four next subframes when the device needs to transmit hybrid-ARQ acknowledgements in the uplink anyway, and so on. Hence, at most half of the time can be used in the downlink and half in the uplink or, in other words, the asymmetry in half-duplex FDD type A is 4:4. Efficient support of half-duplex FDD is one of the reasons why the same number of hybrid-ARQ processes was selected in uplink and downlink.

Half-duplex FDD type B was introduced in LTE release 12 as part of the overall work on enhancements for massive machine-type communication. In type B, a complete subframe is used as guard time between reception and transmission as well as transmission and reception as illustrated in Figure 9.10. The motivation behind half-duplex type B, as well as a thorough description of the enhancements for massive machine-type communication can be found in Chapter 20.

9.7 DISCONTINUOUS RECEPTION

Packet-data traffic is often highly bursty, with occasional periods of transmission activity followed by longer periods of silence. From a delay perspective, it is beneficial to monitor the downlink control signaling in each subframe to receive uplink grants or downlink data transmissions and instantaneously react on changes in the traffic behavior. At the same time this comes at a cost in terms of power consumption at the device; the receiver circuitry in a typical device represents a non-negligible amount of power consumption. To reduce the device power consumption, LTE includes mechanisms for *discontinuous reception* (DRX).

The basic mechanism for DRX is a configurable DRX cycle in the device. With a DRX cycle configured, the device monitors the downlink control signaling only in one subframe per DRX cycle, sleeping with the receiver circuitry switched off in the remaining subframes. This allows for a significant reduction in power consumption: the longer the cycle, the lower the power consumption. Naturally, this implies restrictions to the scheduler as the device can be addressed only in the active subframes.

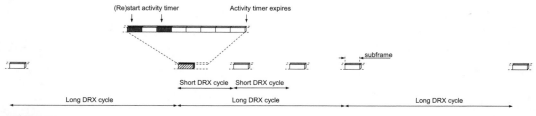

FIGURE 9.11

Illustration of DRX operation.

In many situations, if the device has been scheduled and active with receiving or transmitting data in one subframe, it is highly likely it will be scheduled again in the near future. One reason could be that it was not possible to transmit all the data in the transmission buffer in one subframe and additional subframes are required. Waiting until the next active subframe according to the DRX cycle, although possible, would result in additional delays. Hence, to reduce the delays, the device remains in the active state for a certain configurable time after being scheduled. This is implemented by the device (re)starting an inactivity timer every time it is scheduled and remaining awake until the time expires, as illustrated at the top of Figure 9.11.

Retransmissions take place regardless of the DRX cycle. Thus, the device receives and transmits hybrid-ARQ acknowledgements as normal in response to data transmission. In the uplink, this also includes retransmissions in the subframes given by the synchronous hybrid-ARQ timing relation. In the downlink, where asynchronous hybrid ARQ is used, the retransmission time is not fixed in the specifications. To handle this, the device monitors the downlink for retransmissions in a configurable time window after the previous transmission.

The above mechanism, a (long) DRX cycle in combination with the device remaining awake for some period after being scheduled, is sufficient for most scenarios. However, some services, most notably voice-over IP, are characterized by periods of regular transmission, followed by periods of no or very little activity. To handle these services, a second short DRX cycle can optionally be used in addition to the long cycle described above. Normally, the device follows the long DRX cycle, but if it has recently been scheduled, it follows a shorter DRX cycle for some time. Handling voice-over IP in this scenario can be done by setting the short DRX cycle to 20 ms, as the voice codec typically delivers a voice-over-IP packet per 20 ms. The long DRX cycle is then used to handle longer periods of silence between talk spurts.

CHANNEL-STATE INFORMATION AND FULL-DIMENSION MIMO

The possibility for downlink channel-dependent scheduling—that is, selecting the downlink transmission configuration and related parameters depending on the instantaneous downlink channel condition, including the interference situation—is a key feature of LTE. As already mentioned, an important part of the support for downlink channel-dependent scheduling is *channel-state information* (CSI) provided by devices to the network, information on which the latter can base its scheduling decisions. CSI reports has been part of LTE since the first release but has evolved over the releases to support more advanced antenna configurations with enhancements for full-dimension MIMO (FD-MIMO) in release 13 being the latest example. This chapter describes the basics of CSI reporting and some of these enhancements, including the support for FD-MIMO.

10.1 CSI REPORTS

CSI reports provide the network with information about the current channel conditions. The CSI consists of one or several pieces of information:

- *Rank indicator* (RI), providing a recommendation on the transmission rank to use or, expressed differently, the number of layers that should preferably be used for DL-SCH transmission to the device. The RI is further discussed in Section 10.5.
- *Precoder matrix indicator* (PMI), indicating a preferred precoder to use for DL-SCH transmission, conditioned on the number of layers indicated by the RI. The precoder recommended by the device is not explicitly signaled but provided as an index into a set of predefined matrices, a so-called *codebook*. The PMI is discussed in Section 10.5 with enhancements for FD-MIMO in Section 10.6.
- *Channel-quality indicator* (CQI), representing the highest modulation-and-coding scheme that, if used, would mean a DL-SCH transmission using the recommended RI and PMI would be received with a block-error probability of at most 10%.
- *CSI-RS resource indicator* (CRI), used in conjunction with beam-formed CSI reference signals introduced in release 13. The CRI indicates the beam the device prefers in case the device is configured to monitor multiple beams, see Section 10.6.

4G, LTE-Advanced Pro and The Road to 5G. http://dx.doi.org/10.1016/B978-0-12-804575-6.00010-8

Together, a combination of the RI, PMI, CQI, and CRI forms a CSI report. Exactly what is included in a CSI report depends on the reporting mode the device is configured to be in. For example, RI and PMI do not need to be reported unless the device is in a spatial-multiplexing transmission mode. However, also given the transmission mode, there are different reporting configurations that typically differ as to what set of resource blocks the report is valid for and whether precoding information is reported or not. The type of information useful to the network also depends on the particular implementation and antenna deployment.

Although referred to as CSI, what a device delivers to the network are not explicit reports of the downlink channel state in terms of the radio channel. Rather, what the device delivers are *suggestions* on the transmission rank and precoding matrix to use, together with an indication of the highest possible modulation-and-coding scheme that the network should not exceed if it intends to keep the block-error probability below 10%.

Information about the actual modulation scheme and coding rate used for DL-SCH transmission as well as the set of resource blocks used for the transmission is always included in the downlink scheduling assignment. Hence, the eNodeB is free to follow the CSI report or to select transmission parameters on its own.

10.2 PERIODIC AND APERIODIC CSI REPORTS

There are two types of CSI reports in LTE, *aperiodic* and *periodic* reports, which are different in terms of how a report is triggered:

- Aperiodic CSI reports are delivered when explicitly requested by the network by means of the channel-state-request flag included in uplink scheduling grants (see Section 6.4.7). An aperiodic CSI report is always delivered using the PUSCH—that is, on a dynamically assigned resource.
- Periodic CSI reports are configured by the network to be delivered with a certain periodicity, possibly as often as once every 2 ms, on a semi-statically configured PUCCH resource. However, similar to hybrid-ARQ acknowledgments normally delivered on PUCCH, channel-state reports are "re-routed" to the PUSCH[1] if the device has a valid uplink grant.

Aperiodic and periodic reports, despite both providing estimates of the channel and interference conditions at the device, are quite different in terms of their detailed contents and the usage. In general, aperiodic reports are larger and more detailed than their periodic counterparts. There are several reasons for this. First, the PUSCH, upon which the aperiodic report is transmitted, is capable of a larger payload, and hence a more detailed report, than the PUCCH used for the periodic reports. Furthermore, as aperiodic reports are transmitted

---

[1]In release 10 and later, a device can be configured for simultaneous PUSCH and PUCCH transmission, in which case the periodic channel-state reports can remain on the PUCCH.

on a per-need basis only, the overhead from these reports is less of an issue compared to periodic reports. Finally, if the network requests a report it is likely that it will transmit a large amount of data to the device, which makes the overhead from the report less of an issue compared to a periodic report that is transmitted irrespective of whether the device in question is scheduled in the near future or not. Hence, as the structure and usage of aperiodic and periodic reports are different, they are described separately in the following, starting with aperiodic reports.

10.2.1 APERIODIC CSI REPORTS

Aperiodic CSI reports are transmitted on the PUSCH upon request by the network. Three aperiodic reporting modes, where each mode has several sub-modes depending on the configuration, are supported in LTE:

- Wideband reports, reflecting the average channel quality across the entire cell bandwidth with a single CQI value. Despite a single average CQI value being provided for the whole bandwidth, the PMI reporting is frequency selective. Frequency-selective reporting is obtained, for reporting purposes only, by dividing the overall downlink bandwidth (of each component carrier) into a number of equally sized *sub-bands*, where each sub-band consists of a set of consecutive resource blocks. The size of a sub-band, ranging from four to eight resource blocks, depends on the cell bandwidth. The PMI is then reported for each sub-band. For transmission modes supporting spatial multiplexing, the CQI and the PMI are calculated assuming the channel rank indicated by the RI, otherwise rank-1 is assumed. Wideband reports are smaller than their frequency-selective counterparts, but do not provide any information about the frequency domain.
- UE-selected reports, where the device selects the best M sub-bands and reports, in addition to the indices of the selected sub-bands, one CQI reflecting the average channel quality over the selected M sub-bands together with one wideband CQI reflecting the channel quality across the full downlink carrier bandwidth. This type of report thus provides frequency-domain information about the channel conditions. The sub-band size, ranging from two to four resource blocks, and the value of M, ranging from 1 to 6, depends on the downlink carrier bandwidth. Depending on the sub-mode configured, the PMI and RI are also provided as part of this type of report.
- Configured reports, where the device reports one wideband CQI reflecting the channel quality across the full downlink carrier bandwidth and one CQI per sub-band. The sub-band size depends on the downlink carrier bandwidth and is in the range of four to eight resource blocks. Depending on the sub-mode configured, the PMI and RI are also provided as part of this type of report.

The different aperiodic reporting modes are summarized in Table 10.1.

Table 10.1 Possible Aperiodic Reporting Modes for Different Transmission Modes

Transmission Mode	Reporting Mode								
	Wideband CQI			Frequency-Selective CQI — UE-Selected Sub-bands			Frequency-Selective CQI — Conf. Sub-bands		
	1-0: No PMI	1-1: Wideband PMI	1-2: Selective PMI	2-0: No PMI	2-1: Wideband PMI	2-2: Selective PMI	3-0: No PMI	3-1: Wideband PMI	3-2: Selective PMI
1 Single antenna, CRS	•[13]			•			•		
2 Transmit diversity	•[13]			•			•		
3 Open-loop spatial mux	•[13]			•			•		
4 Closed-loop spatial mux		•[13]	•			•		•	•[12]
5 Multi-user MIMO								•	
6 Codebook-based beam-form		•[13]	•			•		•	•[12]
7 Single-layer trans., DM-RS	•[13]			•			•		
8 Dual-layer trans., DM-RS			•[10]	•[10]		•[10]	•[10]	•[10]	
9 Multi-layer trans., DM-RS			•[10]	•[10]		•[10]	•[10]	•[10]	•[12]
10 Multi-layer trans., DM-RS			•[11]	•[11]		•[11]	•[11]	•[11]	•[12]

The superscript indicates in which release a certain reporting mode was introduced.

10.2.2 PERIODIC CSI REPORTS

Periodic reports are configured by the network to be delivered with a certain periodicity. The periodic reports are transmitted on the PUCCH (unless the device has a simultaneous grant for PUSCH transmission and is not using simultaneous PUSCH and PUCCH transmission). The limited payload possible on PUCCH, compared to PUSCH, implies that the different types of information in a periodic report may not be possible to transmit in a single subframe. Therefore, some of the reporting modes will transmit one or several of the wideband CQI, including PMI, the RI, and the CQI for the UE-selected sub-bands at different time instants. Furthermore, the RI can typically be reported less often, compared to the reporting of PMI and CQI, reflecting the fact that the suitable number of layers typically varies on a slower basis, compared to the channel variations that impact the choice of precoder matrix and modulation-and-coding scheme.

Two periodic reporting modes, again with different sub-modes possible, are supported in LTE:

- *Wideband reports.* Reflect the average channel quality across the entire cell bandwidth with a single CQI value. If PMI reporting is enabled, a single PMI valid across the full bandwidth is reported.
- *UE-selected reports.* Although named in the same way as for aperiodic reports, the principle for UE-selected periodic reports is different. The total bandwidth (of a component carrier) is divided into one to four *bandwidth parts*, with the number of bandwidth parts obtained from the cell bandwidth. For each bandwidth part, the device selects the best sub-band within that part. The sub-band size ranges from four to eight resource blocks. Since the supported payload size of the PUCCH is limited, the reporting cycles through the bandwidth parts and in one subframe report the wideband CQI and PMI (if enabled) for that bandwidth part, as well as the best sub-band and the CQI for that sub-band. The RI (if enabled) is reported in a separate subframe.

The different periodic reporting modes are summarized in Table 10.2. Note that all PMI reporting, if enabled, is of wideband type. There is no support for frequency-selective PMI in periodic reporting, as the amount of bits would result in a too large overhead.

A typical use of periodic and aperiodic reporting could be to configure lightweight periodic CSI reporting on PUCCH, for example, to provide feedback of the wideband CQI and no PMI information (mode 1-0). Upon arrival of data to transmit in the downlink to a specific device, aperiodic reports could be requested as needed, for example, with frequency-selective CQI and PMI (mode 3-1).

10.3 INTERFERENCE ESTIMATION

Channel-state reports, irrespective of whether they are aperiodic or periodic, need measurements of the channel properties as well as the interference level.

Table 10.2 Possible Periodic Reporting Modes for Different Transmission Modes

		Reporting Mode								
		Wideband CQI			Frequency-Selective CQI					
					UE-Selected Sub-bands			Conf. Sub-bands		
	Transmission Mode	1-0: No PMI	1-1: Wideband PMI	1-2: Selective PMI	2-0: No PMI	2-1: Wideband PMI	2-2: Selective PMI	3-0: No PMI	3-1: Wideband PMI	3-2: Selective PMI
1	Single antenna, CRS	•			•					
2	Transmit diversity	•			•					
3	Open-loop spatial mux	•			•					
4	Closed-loop spatial mux		•			•				
5	Multi-user MIMO		•			•				
6	Codebook-based beam-form		•			•				
7	Single-layer trans., DM-RS	•			•					
8	Dual-layer trans., DM-RS	•[10]	•[10]		•[10]	•[10]				
9	Multi-layer trans., DM-RS	•[10]	•[10]		•[10]	•[10]				
10	Multi-layer trans., DM-RS	•[11]	•[11]		•[11]	•[11]				

The superscript indicates in which release a certain reporting mode was introduced.

Measuring the channel gain is relatively straightforward and from the first release, it is well specified which subframe the CSI report relates to. The reference signal which the channel-gain estimate is based upon, CRS or CSI-RS, depends on the transmission mode. For transmission modes already supported in release 8/9, the cell-specific reference signals are used, while for transmission modes 9 and 10, introduced in releases 10 and 11, respectively, the CSI-RS is used.

Measuring the interference level, which is required in order to form a relevant CSI, is more cumbersome and the measurement is greatly affected by the transmission activity in neighboring cells. LTE release 10 and earlier does not specify how to measure the interference level and leave the details for the implementation. However, in practice, interference is measured as the noise on the cell-specific reference signals—that is, the residual after subtracting the reference signal from the received signal in the appropriate resource elements is used as an estimate of the interference level. At low loads, this approach unfortunately often results in overestimating the interference level as the measurements are dominated by CRS transmissions in neighboring cells (assuming the same CRS positions in the neighboring cells), irrespective of the actual load in those cells. Furthermore, the device may also choose to average the interference level across multiple subframes[2], further adding to the uncertainty on how the interference is measured by the device.

To address these shortcomings and to better support various CoMP schemes, transmission mode 10, introduced in release 11, provides tools for the network to control on which resource elements the interference is measured. The basis is a so-called *CSI interference measurement* (CSI-IM) configuration, where a CSI-IM configuration is the set of resource elements in one subframe the device should use for measuring interference. The received power in the resource elements corresponding to the CSI-IM configuration is used as an estimate of the interference (and noise). The single subframe in which the interference should be measured is also specified, thereby avoiding device-specific, and to the network unknown, interference averaging across subframes.

Configuring a CSI-IM is done in a similar manner as a CSI-RS and the same set of configurations is available, see Chapter 6. In practice a CSI-IM resource would typically correspond to a CSI-RS resource in which nothing is transmitted from the cell or, in the general case, from a certain transmission point. Thus, in practice the CSI-IM resource will typically be covered by the set of zero-power CSI-RS resources configured for the device. However, CSI-IM and zero-power CSI-RS serve different purposes. CSI-IM is defined in order to specify a set of resource elements on which a device should measure the interference level while zero-power CSI-RS is defined in order to specify a set of resource elements avoided by the PDSCH mapping.

Since the CSI-IM does not collide with the CRS in neighboring cells but rather the PDSCH (assuming a synchronized network), the interference measurement better reflects the

---

[2]In releases 10 and later, there is a possibility to limit the interference averaging to different subsets of subframes in order to improve the support for heterogeneous deployments, see Chapter 14.

transmission activity in neighboring cells, leading to a more accurate interference estimate at low loads. Hence, with the channel conditions estimated from the CSI-RS and the interference situation estimated from the CSI-IM, the network has detailed control of the interference situation the CSI report reflects.

In some scenarios, the eNodeB benefits from *multiple* CSI reports, derived under different interference hypotheses. Therefore, release 11 provides support for up to four *CSI processes* in a device, where a CSI process is defined by one CSI-RS configuration and one CSI-IM configuration. CSI is then reported separately for each process[3]. A more in depth discussion on the usage of CSI processes to support CoMP is found in Chapter 13.

10.4 CHANNEL-QUALITY INDICATOR

Having discussed CSI reporting in general it is appropriate to discuss the different information fields in a CSI report. The CQI provides an estimate of the highest modulation-and-coding scheme that, if used with the recommended RI and PMI, would result in a block-error probability for the DL-SCH transmissions of at most 10%. The reason to use CQI as a feedback quantity instead of, for example, the signal-to-noise ratio, is to account for different receiver implementations in the device. Also, basing the feedback reports on CQI instead of signal-to-noise ratio simplifies the testing of devices; a device delivering data with more than 10% block-error probability when using the modulation-and-coding scheme indicated by the CQI would fail the test. As is discussed later, multiple CQI reports, each representing the channel quality in a certain part of the downlink spectrum, can be part of the CSI.

The modulation-and-coding scheme used for DL-SCH transmission can, and often will, differ from the reported CQI as the scheduler needs to account for additional information not available to the device when recommending a certain CQI. For example, the set of resource blocks used for the DL-SCH transmission also needs to account for other users. Furthermore, the amount of data awaiting transmission in the eNodeB also needs to be accounted for. There is no need to select a very high data rate, even if the channel conditions would permit this, if there is only a small amount of data to transmit and a small number of resource blocks with robust modulation is sufficient.

10.5 RANK INDICATOR AND PRECODER MATRIX INDICATOR

The multi-antenna-related part of the CSI consists of the RI and the PMI. This section will describe the basic RI and PMI reporting and the associated codebooks and in Section 10.6 the extension added in release 13 to handle FD-MIMO, including the additional codebooks and the CSI-RS resource indicator (CRI).

---

[3]It is possible to configure *rank inheritance*, in which case the rank reported by one CSI process is inherited from another CSI process.

The RI is only reported by devices that are configured in one of the spatial-multiplexing transmission modes. There is at most one RI reported (for a particular CSI process on a particular component carrier, see later for details), valid across the full bandwidth—that is, the RI is frequency nonselective. Note that frequency-selective transmission rank is not possible in LTE as all layers are transmitted on the same set of resource blocks.

The PMI provides an indication to the eNodeB of the preferred precoder to use conditioned on the number of layers indicated by the RI. The precoder recommendation may be frequency selective, implying that the device may recommend different precoders for different parts of the downlink spectrum, or frequency nonselective.

With regard to the precoder-related recommendations, the network has two choices:

- The network may follow the latest device recommendation, in which case the eNodeB only has to confirm (a one-bit indicator in the downlink scheduling assignment) that the precoder configuration recommended by the device is used for the downlink transmission. On receiving such a confirmation, the device will use its recommended configuration when demodulating and decoding the corresponding DL-SCH transmission. Since the PMI computed in the device can be frequency selective, an eNodeB following the precoding matrix recommended by the device may have to apply different precoding matrices for different (sets of) resource blocks.
- The network may select a different precoder, information about which then needs to be explicitly included in the downlink scheduling assignment. The device then uses this configuration when demodulating and decoding the DL-SCH. To reduce the amount of downlink signaling, only a single precoding matrix can be signaled in the scheduling assignment, implying that, if the network overrides the recommendation provided by the device, the precoding is frequency nonselective. The network may also choose to override the suggested transmission rank only.

The precoder recommended by the device is not explicitly signaled but provided as an index into a set of predefined matrices, a so-called *codebook*. The set of matrices from which the device selects the preferred precoder depends on the number of antenna ports. Although the specifications do not mandate any particular antenna arrangement, the assumptions on antenna arrangements affect the codebook design. These assumptions have varied across the different releases, impacting the characteristics of the resulting codebook as illustrated in Figure 10.1.

For two or four antenna ports (irrespective of whether CRS or DM-RS is used for demodulation), the codebook is given by the set of precoders for codebook-based precoding (see Section 10.3.2). This is a natural choice for the multi-antenna transmission modes based on cell-specific reference signals (transmission modes 4, 5, and 6) as the network in these modes must use a codebook from this set. For simplicity the same codebook is used for CSI reporting for transmission modes 8, 9, and 10 using demodulation-specific reference signals, even though the network in this case can use any precoders as the precoder used is transparent

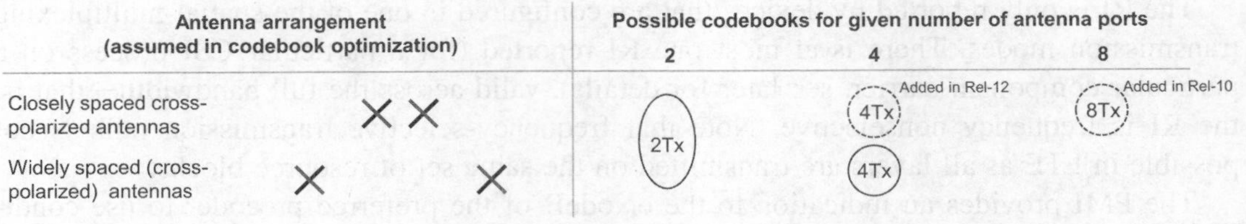

Antenna arrangement (assumed in codebook optimization)		Possible codebooks for given number of antenna ports		
		2	4	8
Closely spaced cross-polarized antennas	✕ ✕	2Tx	4Tx *Added in Rel-12*	8Tx *Added in Rel-10*
Widely spaced (cross-polarized) antennas	✕ ✕		4Tx	

FIGURE 10.1

Assumptions on antenna arrangements for different codebooks.

to the device. The four-antenna codebook is derived under the assumption of uncorrelated fading across the antennas, or, expressed differently, widely-spaced antennas.

For eight antenna ports, introduced in release 10 and supported in transmission modes 9 and 10 relying on demodulation-specific reference signals, a somewhat different approach for the codebook design than the two and four antenna cases in previous releases was taken. The assumption on uncorrelated fading across the antennas made in previous releases is less realistic as a more typical antenna arrangement is closely spaced cross-polarized antennas. Therefore, the codebook is tailored for closely spaced cross-polarized antenna elements (see Figure 10.2) and also covers the case of eight closely spaced linearly polarized antennas. All precoders in the codebook can be factorized as $W = W_1 \cdot W_2$, where the possible entries for W_1 model long-term/wideband aspects such as beam-forming while the possible entries of W_2 address short-term/frequency-selective properties such as polarization properties. One reason for factorizing the codebook is to simplify implementation as the device may apply different (time-domain) filtering to the two parts when forming the PMI. In release 12, an alternative

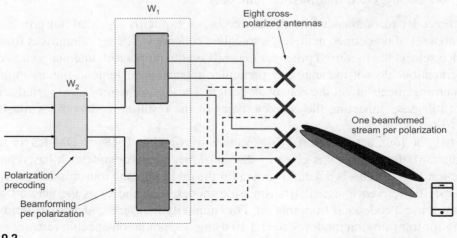

FIGURE 10.2

Illustration of the eight-antenna codebook.

four-antenna codebook following the same design principle as the eight-antenna one was introduced.

In release 13, the CSI reporting in general and the codebook design in particular was further enhanced to handle a larger number of antenna ports. This is described in Section 10.6.

Finally, for two, four, and eight antenna ports, the network can restrict the set of matrices from which the device can select the recommended precoder, so-called codebook subset restriction, to avoid reporting precoders that are not useful in the antenna setup actually used.

10.6 FULL-DIMENSION MIMO

The CSI reporting described in the previous sections covered up to eight antenna ports. Furthermore, although not explicitly stated in the specifications, the focus from a beam-forming perspective has primarily been on one-dimensional beam-forming in the azimuth domain. However, with tight integration of antenna elements and RF components, such as power amplifiers and transceivers, a significantly larger number of controllable antenna elements than eight is feasible. This can enable new advanced antenna solutions, for example, massive multi-user MIMO, two-dimensional beam-forming, and dynamic device-specific down-tilt. In principle, as the specifications described antenna ports but not the antenna elements as such, implementation-specific realization of many of these advanced antenna solutions is possible. However, to better exploit a larger number of antenna elements, release 13 increased the number of antenna ports to 16 and improved the CSI feedback taking the larger number of antenna ports into account. The number of sounding reference signals was also increased as mentioned in Chapter 6 to better support a larger number of users.

10.6.1 CSI FEEDBACK FOR MASSIVE ANTENNA ARRANGEMENTS

CSI feedback for a large number of antenna elements can be solved in many ways, grouped into two main categories: per-element reporting and per-beam reporting. Both of these approaches are supported in release 13 with the reporting class configured through RRC signaling.

Per-element reporting is the approach used in previous releases. Each CSI reference signal is mapped to (a fixed group of) antenna elements and the feedback consists of high-dimensional precoder matrices based on measurements on all the different CSI reference signals. The amount of CSI reference signals would in this case be determined by the number of antenna elements, and not by the number of simultaneously served devices, making this approach primarily suitable for a modest number of antennas. In LTE per-element reporting is referred to as CSI reporting class A.

Per-beam reporting implies that each CSI reference signal is beam-formed (or precoded) using all the antenna elements. The device measures on the beam-formed CSI reference signals it has been configured to measure upon and, among those beams, recommends a suitable beam for subsequent data transmission. The amount of CSI reference signals, the device channel estimator complexity, and the associated feedback would in this case be determined by the number of simultaneous beams rather than the number of antenna elements, making this approach primarily suitable for a very large number of antenna elements. Since each CSI reference signal is beam-formed, this approach can have coverage benefit compared to the per-element approach but potentially also be less robust as the device does not estimate the full channel. In LTE per-beam reporting is known as CSI reporting class B.

The two approaches are illustrated in Figure 10.3 and described in more detail in the following section.

10.6.2 CSI REPORTING CLASS A

CSI reporting class A is a direct extension of the CSI reporting framework in earlier releases to a larger number of antenna ports. Each antenna element transmits a unique CSI reference signal per polarization. Based on measurements on these reference signals the devices computes a recommended precoder as well as RI and CQI conditioned on the precoder.

The precoders available in earlier releases were developed assuming a one-dimensional array as illustrated in Figure 10.1, although this is not explicitly stated in the specifications. For a two-dimensional array a given number of antenna elements can be arranged in many different ways with different arrangements being preferred in different deployments. To handle this, a codebook parametrized by the number of horizontal and vertical antenna ports was selected for release 13, capable of handling the antenna arrangements shown in Figure 10.4. The device is informed about the codebook to use through RRC signaling.

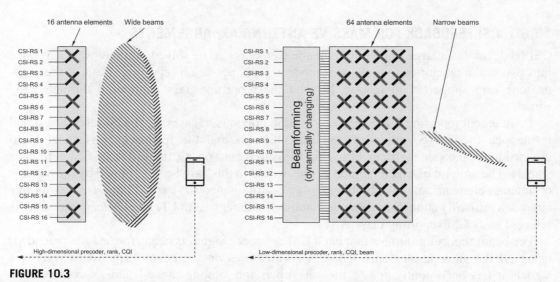

FIGURE 10.3

Illustration of per-element (left) and per-beam (right) CSI-RS transmission.

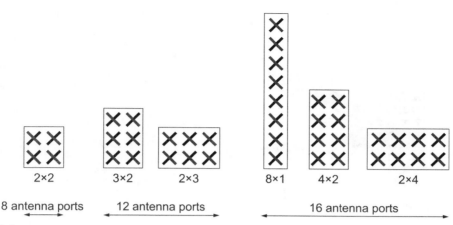

FIGURE 10.4

Supported two-dimensional antenna configurations.

The same codebook structure is used for 8, 12, and 16 antenna ports, following the principles in release 10 with all precoders in the codebook factorized as $W = W_1 \cdot W_2$, where W_1 model long-term/wideband aspects and W_2 address short-term/frequency-selective properties. The details of the codebook structure can be found in [26], but in essence the DFT structure used results in a grid-of-beams codebook. It is also prepared for extensions to a larger number of antenna elements in future releases.

Transmitting one CSI reference signal per element and polarization (or small group of elements) may lead to coverage problems as there is no array gain, calling for the possibility of power boosting to compensate. The possibility for length-four orthogonal cover code, in addition to the length-two part of earlier release, was introduced with this in mind. The longer the cover code, the larger the possibility to "borrow" power between CSI reference signals.

10.6.3 CSI REPORTING CLASS B

CSI reporting class B is a new framework in release 13, different from the per-element approach used in previous releases. In this approach the device measures on a relatively small number of beam-formed CSI reference signals. In principle, arbitrarily large antenna arrays can be used, unlike class A reporting which is limited to the 16 antenna pots.

Up to eight CSI-RS resources can be configured in the device per CSI process where each CSI-RS resource consists of 1, 2, 4, or 8 antenna ports in one beam, see Figure 10.5 for an illustration. The device will measure on each of the CSI-RS resources and report back the recommended beam in the form of a CRI (CSI-RS resource index), along with CQI, PMI, and RI under the assumption of the preferred beam being used for transmission. The same beam is used across the whole bandwidth—that is, the CRI is frequency nonselective unlike CQI and PMI, which can vary across the frequency range. As an example, if the topmost red beam is the

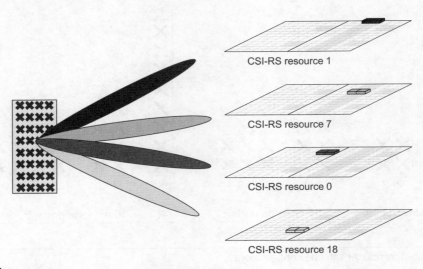

FIGURE 10.5

Example of CSI-RS resources for beam measurements.

preferred one from a device perspective, the device will report CRI equal to one, along with preferred CQI, PMI, and RI conditioned on the topmost beam being used for transmission. For PMI reporting, the 2, 4, or 8 antenna codebooks defined in previous releases is used.

In case a single CSI-RS resource of 1, 2, 4, or 8 antenna ports is configured, a codebook selecting pairs of antenna ports is used. Each pair corresponds to a beam and the device selects a preferred beam, possibly per sub-band (unlike the CRI case described earlier where the same beam is used across all sub-bands). This approach provides a faster feedback of the preferred beam than the CRI approach described earlier as the PMI, which points in to the codebook, can be reported more frequently than the CRI.

Beam-formed CSI reference signals are beneficial from a coverage perspective and can be used with very large antenna arrays. However, it also poses the challenges of beam finding and beam tracking. Upon entering a new cell, the direction in which to transmit the CSI-RS needs to be determined and updated as the device moves within the cell. There are multiple ways of addressing this challenge.

One possibility is to transmit discovery reference signals in a large set of fixed beams and use discovery signal measurement reports from the device to determine a relevant direction. Since the discovery reference signals are transmitted periodically, the eNodeB would obtain continuous updates on the "best" beam. However, as the discovery signal periodicity is fairly large, the method is mainly suitable for slowly moving devices. A similar behavior can be obtained by configuring (multiple) resources to measure upon in the device and vary the directions in which to transmit the beam-formed CSI reference signals. If inter-subframe averaging of the interference measurements is turned off, which is possible in release 13, an instantaneous CSI measurement is achieved and the eNodeB can determine which directions resulted in "high" CSI values and hence correspond to promising transmission directions.

Uplink sounding and channel reciprocity can also be used for beam finding, either alone or in combination with measurements from the device. Reciprocity is often associated with TDD systems, but the long-term statistics, for example, angle-of-arrival, is typically reciprocal also in the case of FDD which is sufficient for beam finding.[4] The additional sounding capacity available in release 13, mentioned in Chapter 7, was partially added with this in mind.

Tracking of beams could, for example, be done by configuring multiple CSI resources in the device, including the "best" beam and a couple of surrounding beams. If the set of resources in the device updated when needed such that it always includes the "best" beam and the surrounding beams, it is possible to track beam changes. As the CSI reporting framework is fairly flexible, there are other possibilities as well. Also in this case uplink sounding can be exploited.

---

[4]Short-term properties such as the instantaneous channel impulse response are not reciprocal in FDD systems, but this does not matter in this case.

ACCESS PROCEDURES

11

The previous chapters have described the LTE uplink and downlink transmission schemes. However, prior to transmission of data, the device needs to connect to the network. This chapter describes the procedures necessary for a device to be able to access an LTE-based network.

11.1 ACQUISITION AND CELL SEARCH

Before an LTE device can communicate with an LTE network it has to do the following:

- Find and acquire synchronization to a cell within the network.
- Receive and decode the information, also referred to as the *cell system information*, needed to communicate with and operate properly within the cell.

The first of these steps, often simply referred to as *cell search*, is discussed in this section. The next section then discusses, in more detail, the means by which the network provides the cell system information.

Once the system information has been correctly decoded, the device can access the cell by means of the random-access procedure as described in Section 11.3.

11.1.1 OVERVIEW OF LTE CELL SEARCH

A device does not only need to carry out cell search at power-up—that is, when initially accessing the system, but, to support mobility, it also needs to continuously search for, synchronize to, and estimate the reception quality of neighboring cells. The reception quality of the neighboring cells, in relation to the reception quality of the current cell, is then evaluated to conclude if a handover (for devices in RRC_CONNECTED) or cell reselection (for devices in RRC_IDLE) should be carried out.

LTE cell search consists of the following basic steps:

- Acquisition of frequency and symbol synchronization to a cell.
- Acquisition of frame timing of the cell—that is, determination of the start of the downlink frame.
- Determination of the physical-layer cell identity of the cell.

4G, LTE-Advanced Pro and The Road to 5G. http://dx.doi.org/10.1016/B978-0-12-804575-6.00011-X

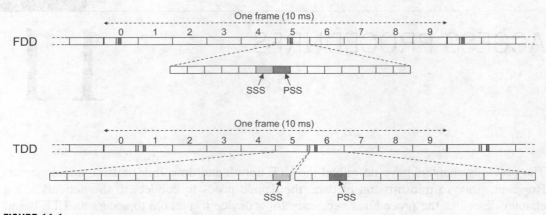

FIGURE 11.1

Time-domain positions of PSS and SSS in case of FDD and TDD.

As already mentioned, for example, in Chapter 6, there are 504 different physical-layer cell identities defined for LTE. The set of physical-layer cell identities is further divided into 168 cell-identity groups, with three cell identities within each group.

To assist the cell search, two special signals are transmitted on each downlink component carrier, the *primary synchronization signal* (PSS) and the *secondary synchronization signal* (SSS). Although having the same detailed structure, the time-domain positions of the synchronization signals within the frame differ somewhat depending on whether the cell is operating in FDD or TDD:

- In the case of FDD (upper part of Figure 11.1), the PSS is transmitted within the last symbol of the first slot of subframes 0 and 5, while the SSS is transmitted within the second last symbol of the same slot—that is, just prior to the PSS.
- In the case of TDD (lower part of Figure 11.1), the PSS is transmitted within the third symbol of subframes 1 and 6—that is, within the DwPTS—while the SSS is transmitted in the last symbol of subframes 0 and 5—that is, three symbols ahead of the PSS.

It should be noted that the difference in PSS/SSS time-domain structure between FDD and TDD allows for the device to detect the duplex mode of the acquired carrier if this is not known in advance.

Within one cell, the two PSSs within a frame are identical. Furthermore, the PSS of a cell can take three different values depending on the physical-layer cell identity of the cell. More specifically, the three cell identities within a cell-identity group always correspond to different PSS. Thus, once the device has detected and identified the PSS of the cell, it has found the following:

- Five-millisecond timing of the cell and thus also the position of the SSS which has a fixed offset relative to the PSS.[1]
- The cell identity within the cell-identity group. However, the device has not yet determined the cell-identity group itself—that is, the number of possible cell identities has been reduced from 504 to 168.

Thus, from the SSS, the position of which is known once the PSS has been detected, the device should find the following:

- Frame timing (two different alternatives given the found position of the PSS)
- The cell-identity group (168 alternatives)

Furthermore, it should be possible for a device to do this by the reception of one single SSS. The reason is that, for example, in the case when the device is searching for cells on other carriers, the search window may not be sufficiently large to cover more than one SSS.

To enable this, each SSS can take 168 different values corresponding to the 168 different cell-identity groups. Furthermore, the set of values valid for the two SSSs within a frame (SSS1 in subframe 0 and SSS2 in subframe 5) are different, implying that, from the detection of a single SSS, the device can determine whether SSS1 or SSS2 has been detected and thus determine frame timing.

Once the device has acquired frame timing and the physical-layer cell identity, it has identified the cell-specific reference signal. The behavior is slightly different depending on whether it is an initial cell search or cell search for the purpose of neighboring cell measurements:

- In the case of initial cell search—that is, the device state is in RRC_IDLE mode—the reference signal will be used for channel estimation and subsequent decoding of the BCH transport channel to obtain the most basic set of system information.
- In the case of mobility measurements—that is, the device is in RRC_CONNECTED mode—the device will measure the received power of the reference signal. If the measurement fulfills a configurable condition, it will trigger sending of a reference signal received power (RSRP) measurement report to the network. Based on the measurement report, the network will conclude whether a handover should take place. The RSRP reports can also be used for component carrier management,[2] for example, whether an

---

[1]This assumes that the device knows if it has acquired an FDD or a TDD carrier. Otherwise the device needs to try two different hypotheses regarding the SSS position relative to the PSS, thereby also indirectly detecting the duplex mode of the acquired carrier.

[2]As discussed in Chapter 5, the specifications use the terms primary and secondary "cells" instead of primary and secondary "component carriers."

additional component carrier should be configured or if the primary component carrier should be reconfigured.

11.1.2 PSS STRUCTURE

On a more detailed level, the three PSSs are three length-63 Zadoff–Chu (ZC) sequences (see Section 11.2) extended with five zeros at the edges and mapped to the center 73 subcarriers (center six resource blocks) as illustrated in Figure 11.2. It should be noted that the center subcarrier is actually not transmitted as it coincides with the DC subcarrier. Thus, only 62 elements of the length-63 ZC sequences are actually transmitted (element X_{32}^{PSS} is not transmitted).

The PSS thus occupies 72 resource elements (not including the DC carrier) in subframes 0 and 5 (FDD) and subframes 1 and 6 (TDD). These resource elements are then not available for transmission of DL-SCH.

11.1.3 SSS STRUCTURE

Similar to PSS, the SSS occupies the center 72 resource elements (not including the DC carrier) in subframes 0 and 5 (for both FDD and TDD). As described earlier, the SSS should be designed so that:

- The two SSSs (SSS1 in subframe 0 and SSS2 in subframe 5) take their values from sets of 168 possible values corresponding to the 168 different cell-identity groups.
- The set of values applicable for SSS2 is different from the set of values applicable for SSS1 to allow for frame-timing detection from the reception of a single SSS.

The structure of the two SSSs is illustrated in Figure 11.3. SSS1 is based on the frequency interleaving of two length-31-m sequences X and Y, each of which can take 31 different values (actually 31 different shifts of the same m-sequence). Within a cell, SSS2 is based on exactly the same two sequences as SSS1. However, the two sequences have been swapped in the frequency domain, as outlined in Figure 11.3. The set of valid combinations of X and Y for SSS1 has then been selected so that a swapping of the two sequences in the frequency

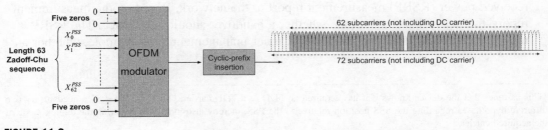

FIGURE 11.2

Definition and structure of PSS.

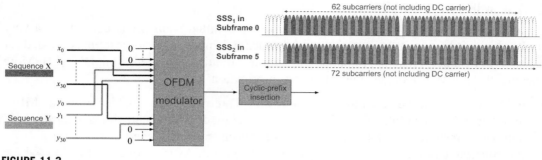

FIGURE 11.3

Definition and structure of SSS.

domain is not a valid combination for SSS1. Thus, the earlier described requirements are fulfilled:

- The set of valid combinations of X and Y for SSS1 (as well as for SSS2) are 168, allowing for detection of the physical-layer cell identity.
- As the sequences X and Y are swapped between SSS1 and SSS2, frame timing can be found.

11.2 SYSTEM INFORMATION

By means of the basic cell-search procedure described in Section 11.1, a device synchronizes to a cell, acquires the physical-layer identity of the cell, and detects the cell frame timing. Once this has been achieved, the device has to acquire the cell system information. This is information that is repeatedly broadcast by the network and which needs to be acquired by devices in order for them to be able to access and, in general, operate properly within the network and within a specific cell. The system information includes, among other things, information about the downlink and uplink cell bandwidths, the uplink/downlink configuration in the case of TDD, detailed parameters related to random-access transmission, and so on.

In LTE, system information is delivered by two different mechanisms relying on two different transport channels:

- A limited amount of system information, corresponding to the so-called *master-information lock* (MIB), is transmitted using the BCH.
- The main part of the system information, corresponding to different so-called *system-information blocks* (SIBs), is transmitted using the downlink shared channel (DL-SCH).

It should be noted that system information in both the MIB and the SIBs corresponds to the BCCH logical channel. Thus, as also illustrated in Figure 8.7, BCCH can be mapped to both BCH and DL-SCH depending on the exact BCCH information.

11.2.1 MIB AND BCH TRANSMISSION

As mentioned in the previous section, the MIB transmitted using BCH consists of a very limited amount of system information, mainly such information that is absolutely needed for a device to be able to read the remaining system information provided using DL-SCH. More specifically, the MIB includes the following information:

- Information about the downlink cell bandwidth. Three bits are available within the MIB to indicate the downlink bandwidth. Thus, up to eight different bandwidths, measured in number of resource blocks, can be defined for each frequency band.
- Information about the PHICH configuration of the cell. As mentioned in Section 6.4.2, the device must know the PHICH configuration to be able to receive the L1/L2 control signaling on PDCCH. The PDCCH information, in turn, is needed to acquire the remaining part of the system information which is carried on the DL-SCH, see later. Thus, information about the PHICH configuration (three bits) is included in the MIB—that is, transmitted using BCH, which can be received and decoded without first receiving any PDCCH.
- The system frame number (SFN) or, more exactly, all bits except the two least significant bits of the SFN are included in the MIB. As described in the following, the device can indirectly acquire the two least significant bits of the SFN from the BCH decoding.

The MIB also includes ten unused or "spare" information bits. The intention with such bits is to be able to include more information in the MIB in later releases while retaining backwards compatibility. As an example, LTE release 13 uses five of the ten spare bits to include additional eMTC-related information in the MIB, see Chapter 20 for more details.

BCH physical-layer processing, such as channel coding and resource mapping, differs quite substantially from the corresponding processing and mapping for DL-SCH outlined in Chapter 6.

As can be seen in Figure 11.4, one BCH transport block, corresponding to the MIB, is transmitted every 40 ms. The BCH transmission time interval (TTI) thus equals 40 ms.

The BCH relies on a 16-bit CRC, in contrast to a 24-bit CRC used for all other downlink transport channels. The reason for the shorter BCH CRC is to reduce the relative CRC overhead, having the very small BCH transport-block size in mind.

BCH channel coding is based on the same rate-1/3 tail-biting convolutional code as is used for the PDCCH control channel. The reason for using convolutional coding for BCH, rather than the Turbo code used for all other transport channels, is the small size of the BCH transport block. With such small blocks, tail-biting convolutional coding actually outperforms Turbo coding. The channel coding is followed by rate matching, in practice, repetition of the coded bits, and bit-level scrambling. QPSK modulation is then applied to the coded and scrambled BCH transport block.

BCH multi-antenna transmission is limited to transmit diversity—that is, SFBC in the case of two antenna ports and combined SFBC/FSTD in the case of four antenna ports. Actually, as

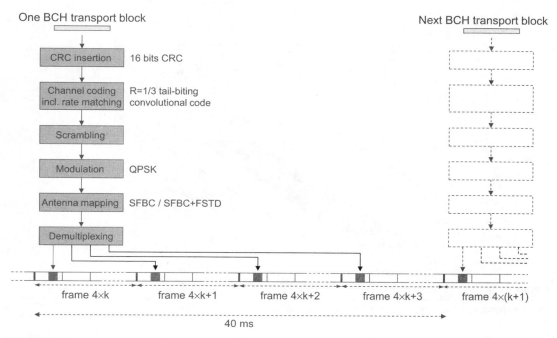

FIGURE 11.4

Channel coding and subframe mapping for the BCH transport channel.

mentioned in Chapter 6, if two antenna ports are available within the cell, SFBC must be used for BCH. Similarly, if four antenna ports are available, combined SFBC/FSTD must be used. Thus, by blindly detecting what transmit-diversity scheme is used for BCH, a device can indirectly determine the number of cell-specific antenna ports within the cell and also the transmit-diversity scheme used for the L1/L2 control signaling.

As can also be seen from Figure 11.4, the coded BCH transport block is mapped to the first subframe of each frame in four consecutive frames. However, as can be seen in Figure 11.5 and in contrast to other downlink transport channels, the BCH is not mapped on a resource-block basis. Instead, the BCH is transmitted within the first four OFDM symbols of the second slot of subframe 0 and only over the 72 center subcarriers.[3] Thus, in the case of FDD, BCH follows immediately after the PSS and SSS in subframe 0. The corresponding resource elements are then not available for DL-SCH transmission.

The reason for limiting the BCH transmission to the 72 center subcarriers, regardless of the cell bandwidth, is that a device may not know the downlink cell bandwidth when receiving BCH. Thus, when first receiving BCH of a cell, the device can assume a cell

---

[3]Not including the DC carrier.

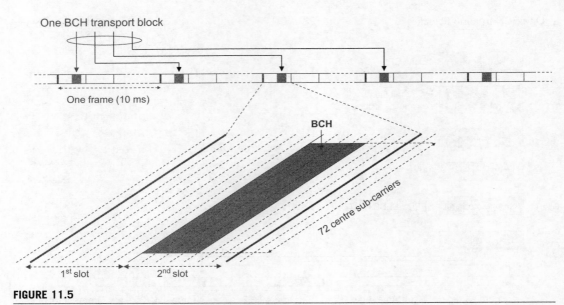

One BCH transport block

One frame (10 ms)

BCH

72 centre sub-carriers

1ˢᵗ slot   2ⁿᵈ slot

FIGURE 11.5

Detailed resource mapping for the BCH transport channel.

bandwidth equal to the minimum possible downlink bandwidth—that is, six resource blocks corresponding to 72 subcarriers. From the decoded MIB, the device is then informed about the actual downlink cell bandwidth and can adjust the receiver bandwidth accordingly.

The total number of resource elements to which the coded BCH is mapped is very large compared to the size of the BCH transport block, implying extensive repetition coding or, equivalently, massive processing gain for the BCH transmission. Such large processing gain is needed, as it should be possible to receive and correctly decode the BCH also by devices in neighboring cells, implying potentially very low receiver signal-to-interference-and-noise ratio (SINR) when decoding the BCH. At the same time, many devices will receive BCH in much better channel conditions. Such devices then do not need to receive the full set of four subframes over which a BCH transport block is transmitted to acquire sufficient energy for correct decoding of the transport block. Instead, already by receiving only a few or perhaps only a single subframe, the BCH transport block may be decodable.

From the initial cell search, the device has found only the cell frame timing. Thus, when receiving BCH, the device does not know to what set of four subframes a certain BCH transport block is mapped. Instead, a device must try to decode the BCH at four possible timing positions. Depending on which decoding is successful, indicated by a correct CRC check, the device can implicitly determine 40 ms timing or, equivalently, the two least significant bits of the SFN.[4] This is the reason why these bits do not need to be explicitly included in the MIB.

11.2.2 SYSTEM-INFORMATION BLOCKS

As already mentioned, the MIB on the BCH only includes a very limited part of the system information. The main part of the system information is instead included in different SIBs that are transmitted using the DL-SCH. The presence of system information on DL-SCH in a subframe is indicated by the transmission of a corresponding PDCCH marked with a special system-information RNTI (SI-RNTI). Similar to the PDCCH providing the scheduling assignment for "normal" DL-SCH transmission, this PDCCH also indicates the transport format and physical resource (set of resource blocks) used for the system-information transmission.

LTE defines a number of different SIBs characterized by the type of information that is included within them:

- SIB1 includes information mainly related to whether a device is allowed to camp on the cell. This includes information about the operator/operators of the cell, if there are restrictions with regard to which users may access the cell, and so on. SIB1 also includes information about the allocation of subframes to uplink/downlink and configuration of the special subframe in the case of TDD. Finally, SIB1 includes information about the time-domain scheduling of the remaining SIBs (SIB2 and beyond).
- SIB2 includes information that devices need in order to be able to access the cell. This includes information about the uplink cell bandwidth, random-access parameters, and parameters related to uplink power control.
- SIB3 mainly includes information related to cell reselection.
- SIB4—SIB8 include neighboring-cell-related information, including information related to neighboring cells on the same carrier, neighboring cells on different carriers, and neighboring non-LTE cells, such as WCDMA/HSPA, GSM, and CDMA2000 cells.
- SIB9 contains the name of the home-eNodeB.
- SIB10—SIB12 contain public warning messages, for example, earthquake information.
- SIB13 contains information necessary for MBMS reception (see also Chapter 19).
- SIB14 is used to provide enhanced access barring information, controlling the possibilities for devices to access the cell.
- SIB15 contains information necessary for MBMS reception on neighboring carrier frequencies.
- SIB16 contains information related to GPS time and coordinated universal time (UTC).
- SIB17 contains information related to interworking between LTE and WLAN

---

[4]BCH scrambling is defined with 40 ms periodicity, hence even if the device successfully decodes the BCH after observing only a single transmission instant, it can determine the 40 ms timing.

- SIB18 and SIB19 contain information related to sidelink (direct device-to-device) connectivity, see Chapter 21.
- SIB20 contains information related to single-cell point to multipoint, see Chapter 19.

Not all the SIBs need to be present. For example, SIB9 is not relevant for an operator-deployed node and SIB13 is not necessary if MBMS is not provided in the cell.

Similar to the MIB, the SIBs are broadcasted repeatedly. How often a certain SIB needs to be transmitted depends on how quickly devices need to acquire the corresponding system information when entering the cell. In general, a lower-order SIB is more time critical and is thus transmitted more often compared to a higher-order SIB. SIB1 is transmitted every 80 ms, whereas the transmission period for the higher-order SIBs is flexible and can be different for different networks.

The SIBs represent the basic system information to be transmitted. The different SIBs are then mapped to different system-information messages (SIs), which correspond to the actual transport blocks to be transmitted on DL-SCH. SIB1 is always mapped, by itself, on to the first system-information message SI-1,[5] whereas the remaining SIBs may be group-wise multiplexed on to the same SI subject to the following constraints:

- The SIBs mapped to the same SI must have the same transmission period. Thus, as an example, two SIBs with a transmission period of 320 ms can be mapped to the same SI, whereas an SIB with a transmission period of 160 ms must be mapped to a different SI.
- The total number of information bits that are mapped to a single SI must not exceed what is possible to transmit within a transport block.

It should be noted that the transmission period for a given SIB might be different in different networks. For example, different operators may have different requirements concerning the period when different types of neighboring-cell information needs to be transmitted. Furthermore, the amount of information that can fit into a single transport block very much depends on the exact deployment situation, such as cell bandwidth, cell size, and so on.

Thus, in general, the SIB-to-SI mapping for SIBs beyond SIB1 is flexible and may be different for different networks or even within a network. An example of SIB-to-SI mapping is illustrated in Figure 11.6. In this case, SIB2 is mapped to SI-2 with a transmission period of 160 ms. SIB3 and SIB4 are multiplexed into SI-3 with a transmission period of 320 ms, whereas SIB5, which also requires a transmission period of 320 ms, is mapped to a separate SI (SI-4). Finally, SIB6, SIB7, and SIB8 are multiplexed into SI-5 with a transmission period of 640 ms. Information about the detailed SIB-to-SI mapping, as well as the transmission period of the different SIs, is provided in SIB1.

Regarding the more detailed transmission of the different SIs there is a difference between the transmission of SI-1, corresponding to SIB1, and the transmission of the remaining SIs.

---

[5]Strictly speaking, as SIB1 is not multiplexed with any other SIBs, it is not even said to be mapped to an SI. Rather, SIB1 in itself directly corresponds to the transport block.

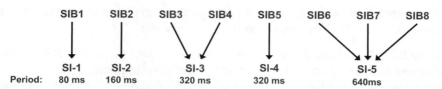

FIGURE 11.6

Example of mapping of SIBs to SIs.

The transmission of SI-1 has only a limited flexibility. More specifically, SI-1 is always transmitted within subframe 5. However, the bandwidth or, in general, the set of resource blocks over which SI-1 is transmitted, as well as other aspects of the transport format, may vary and is signaled on the associated PDCCH.

For the remaining SIs, the scheduling on DL-SCH is more flexible in the sense that each SI can, in principle, be transmitted in any subframe within time windows with well-defined starting points and durations. The starting point and duration of the time window of each SI are provided in SIB-1. It should be noted that an SI does not need to be transmitted on consecutive subframes within the time window, as is illustrated in Figure 11.7. Within the time window, the presence of system information in a subframe is indicated by the SI-RNTI on PDCCH, which also provides the frequency-domain scheduling as well as other parameters related to the system-information transmission.

Different SIs have different nonoverlapping time windows. Thus, a device knows what SI is being received without the need for any specific identifier for each SI.

In case of a relatively small SI and a relatively large system bandwidth, a single subframe may be sufficient for the transmission of the SI. In other cases, multiple subframes may be needed for the transmission of a single SI. In the latter case, instead of segmenting each SI into sufficiently small blocks that are separately channel coded and transmitted in separate subframes, the complete SI is channel coded and mapped to multiple, not necessarily consecutive, subframes.

Similar to the case of the BCH, devices that are experiencing good channel conditions may then be able to decode the complete SI after receiving only a subset of the subframes to which

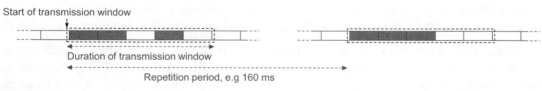

FIGURE 11.7

Transmission window for the transmission of an SI.

the coded SI is mapped, while devices in bad positions need to receive more subframes for proper decoding of the SI. This approach has two benefits:

- Similar to BCH decoding, devices in good positions need to receive fewer subframes, implying the possibility for reduced device power consumption.
- The use of larger code blocks in combination with Turbo coding leads to improved channel-coding gain.

Strictly speaking the single transport block containing the SI is not transmitted over multiple subframes. Rather, the subsequent SI transmissions are seen as autonomous hybrid-ARQ retransmissions of the first SI transmission—that is, retransmissions taking place without any explicit feedback signaling provided on the uplink.

For devices capable of carrier aggregation, the system information for the primary component carrier is obtained as discussed. For secondary component carriers, the device does not need to read the SIBs but assumes that the information obtained for the primary component carrier also holds for the secondary component carriers. System information specific for the secondary component carrier is provided through dedicated RRC signaling as part of the procedure to configure an additional secondary component carrier. Using dedicated signaling instead of reading the system information on the secondary component carrier enables faster activation of secondary component carriers as the device otherwise would have to wait until the relevant system information had been transmitted.

11.3 RANDOM ACCESS

A fundamental requirement for any cellular system is the possibility for the device to request a connection setup, commonly referred to as random access. In LTE, random access is used for several purposes, including:

- for initial access when establishing a radio link (moving from RRC_IDLE to RRC_CONNECTED; see Chapter 4 for a discussion on different device states);
- to reestablish a radio link after radio-link failure;
- for handover when uplink synchronization needs to be established to the new cell;
- to establish uplink synchronization if uplink or downlink data arrives when the device is in RRC_CONNECTED and the uplink is not synchronized;
- for the purpose of positioning using positioning methods based on uplink measurements;
- as a scheduling request if no dedicated scheduling-request resources have been configured on PUCCH (see Chapter 9 for a discussion on uplink scheduling procedures).

Acquisition of uplink timing is a main objective for all these cases; when establishing an initial radio link (i.e., when moving from RRC_IDLE to RRC_CONNECTED), the random-access procedure also serves the purpose of assigning a unique identity, the C-RNTI, to the device.

Either a contention-based or a contention-free scheme can be used, depending on the purpose. Contention-based random access can be used for all previously discussed purposes, while contention-free random access can only be used for reestablishing uplink synchronization upon downlink data arrival, uplink synchronization of secondary component carriers, handover, and positioning. The basis for the random access is the four-step procedure illustrated in Figure 11.8, with the following steps:

1. The device transmission of a random-access preamble, allowing the network to estimate the transmission timing of the device. Uplink synchronization is necessary as the device otherwise cannot transmit any uplink data.
2. The network transmission of a timing advance command to adjust the device transmit timing, based on the timing estimate obtained in the first step. In addition to establishing uplink synchronization, the second step also assigns uplink resources to the device to be used in the third step in the random-access procedure.

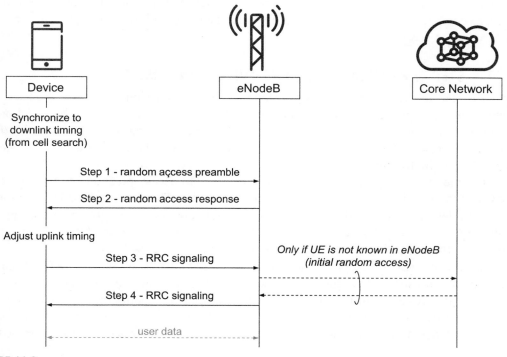

FIGURE 11.8

Overview of the random-access procedure.

3. The device transmission of the mobile-device identity to the network using the UL-SCH similar to normal scheduled data. The exact content of this signaling depends on the state of the device, in particular whether it is previously known to the network or not.

4. The network transmissions of a contention-resolution message from the network to the device on the DL-SCH. This step also resolves any contention due to multiple devices trying to access the system using the same random-access resource.

Only the first step uses physical-layer processing specifically designed for random access. The subsequent three steps utilize the same physical-layer processing as used for normal uplink and downlink data transmission. In the following, each of these steps is described in more detail. Only the first two steps of the preceding procedure are used for contention-free random access as there is no need for contention resolution in a contention-free scheme.

Both the device and the network can initiate a random-access attempt. In the latter case, RRC signaling or a so-called PDCCH order is used. A PDCCH order is a specific message transmitted on the PDCCH, containing information about when to initiate the random-access procedure and, in case of a contention-free random access, the preamble to use. PDCCH orders are primarily intended as a tool for the network to reestablish uplink synchronization but can be used for other purposes as well. A device may perform random access on its primary component carrier only[6] except for establishing uplink timing alignment for a secondary timing advance group.

11.3.1 STEP 1: RANDOM-ACCESS PREAMBLE TRANSMISSION

The first step in the random-access procedure is the transmission of a random-access preamble. The main purpose of the preamble transmission is to indicate to the base station the presence of a random-access attempt and to allow the base station to estimate the delay between the eNodeB and the device. The delay estimate will be used in the second step to adjust the uplink timing.

The time—frequency resource on which the random-access preamble is transmitted is known as the physical random-access channel (PRACH). The network broadcasts information to all devices in which time—frequency resource random-access preamble transmission is allowed (i.e., the PRACH resources), in SIB-2. As part of the first step of the random-access procedure, the device selects one preamble to transmit on the PRACH.

In each cell, there are 64 preamble sequences available. Two subsets of the 64 sequences are defined as illustrated in Figure 11.9, where the set of sequences in each subset is signaled as part of the system information. When performing a (contention-based) random-access attempt, the device selects at random one sequence in one of the subsets. As long as no other device is performing a random-access attempt using the same sequence at the same time

---

[6]The primary component carrier is device-specific as already discussed; hence, from an eNodeB perspective, random access may occur on multiple component carriers.

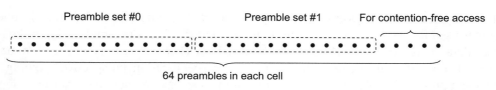

FIGURE 11.9

Preamble subsets.

instant, no collisions will occur and the attempt will, with a high likelihood, be detected by the eNodeB.

The subset to select the preamble sequence from is given by the amount of data the device would like to (and from a power perspective can) transmit on the UL-SCH in the third random-access step. Hence, from the preamble the device used, the eNodeB, will get some guidance on the amount of uplink resources to be granted to the device.

If the device has been requested to perform a contention-free random access, for example, for handover to a new cell, the preamble to use is explicitly indicated from the eNodeB. To avoid collisions, the eNodeB should preferably select the contention-free preamble from sequences outside the two subsets used for contention-based random access.

11.3.1.1 PRACH Time–Frequency Resources

In the frequency domain, the PRACH resource, illustrated in Figure 11.10, has a bandwidth corresponding to six resource blocks (1.08 MHz). This nicely matches the smallest uplink cell bandwidth of six resource blocks in which LTE can operate. Hence, the same random-access preamble structure can be used, regardless of the transmission bandwidth in the cell.

In the time domain, the length of the preamble region depends on configured preamble, as will be discussed later. The basic random-access resource is 1 ms in duration, but it is also possible to configure longer preambles. Also, note that the eNodeB uplink scheduler in principle can reserve an arbitrary long-random-access region by simply avoiding scheduling devices in multiple subsequent subframes.

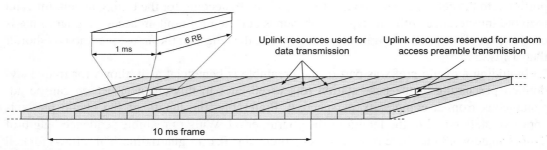

FIGURE 11.10

Principal illustration of random-access-preamble transmission.

Typically, the eNodeB avoids scheduling any uplink transmissions in the time—frequency resources used for random access, resulting in the random-access preamble being orthogonal to user data. This avoids interference between UL-SCH transmissions and random-access attempts from different devices. However, from a specification perspective, nothing prevents the uplink scheduler from scheduling transmissions in the random-access region. Hybrid-ARQ retransmissions are examples of this; synchronous nonadaptive hybrid-ARQ retransmissions may overlap with the random-access region and it is up to the implementation to handle this, either by moving the retransmissions in the frequency domain as discussed in Chapter 8 or by handling the interference at the eNodeB receiver.

For FDD, there is at most one random-access region per subframe—that is, multiple random-access attempts are not multiplexed in the frequency domain. From a delay perspective, it is better to spread out the random-access opportunities in the time domain to minimize the average waiting time before a random-access attempt can be initialized.

For TDD, multiple random-access regions can be configured in a single subframe. The reason is the smaller number of uplink subframes per radio frame in TDD. To maintain the same random-access capacity as in FDD, frequency-domain multiplexing is sometimes necessary. The number of random-access regions is configurable and can vary from one per 20 ms to one per 1 ms for FDD; for TDD up to six attempts per 10 ms radio frame can be configured.

11.3.1.2 Preamble Structure and Sequence Selection

The preamble consists of two parts:

- A preamble sequence
- A cyclic prefix

Furthermore, the preamble transmission uses a guard period to handle the timing uncertainty. Prior to starting the random-access procedure, the device has obtained downlink synchronization from the cell-search procedure. However, as uplink synchronization has not yet been established prior to random access, there is an uncertainty in the uplink timing[7] as the location of the device in the cell is not known. The uplink timing uncertainty is proportional to the cell size and amounts to 6.7 µs/km. To account for the timing uncertainty and to avoid interference with subsequent subframes not used for random access, a guard time is used as part of the preamble transmission—that is, the length of the actual preamble is shorter than 1 ms.

Including a cyclic prefix as part of the preamble is beneficial as it allows for frequency-domain processing at the base station (discussed later in this chapter), which can be advantageous from a complexity perspective. Preferably, the length of the cyclic prefix is approximately equal to the length of the guard period. With a preamble sequence length of approximately 0.8 ms, there is 0.1 ms cyclic prefix and 0.1 ms guard time. This allows for cell

---

[7]The start of an uplink frame at the device is defined relative to the start of a downlink frame received at the device.

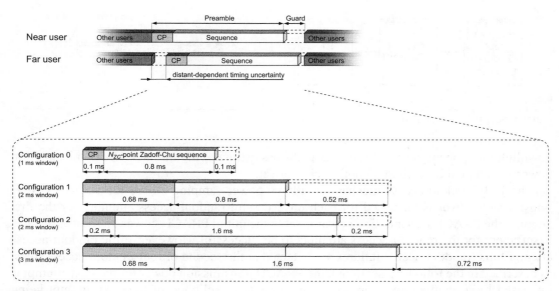

FIGURE 11.11

Different preamble formats.

sizes up to 15 km and is the typical random-access configuration, configuration 0 in Figure 11.11. To handle larger cells, where the timing uncertainty is larger, preamble configurations 1—3 can be used. Some of these configurations also support a longer preamble sequence to increase the preamble energy at the detector, which can be beneficial in larger cells. The preamble configuration used in a cell is signaled as part of the system information. Finally, note that guard times larger than those in Figure 11.11 can easily be created by not scheduling any uplink transmissions in the subframe following the random-access resource.

The preamble formats in Figure 11.11 are applicable to both FDD and TDD. However, for TDD, there is an additional fourth preamble configuration for random access. In this configuration, the random-access preamble is transmitted in the UpPTS field of the special subframe instead of in a normal subframe. Since this field is at most two OFDM symbols long, the preamble and the possible guard time are substantially shorter than the preamble formats described earlier. Hence, format 4 is applicable to very small cells only. The location of the UpPTS, next to the downlink-to-uplink switch for TDD, also implies that the interference from distant base stations may interfere with this short random-access format, which limits its usage to small cells and certain deployment scenarios.

11.3.1.3 PRACH Power Setting

The basis for setting the transmission power of the random-access preamble is a downlink path-loss estimate obtained from measuring the cell-specific reference signals on the primary

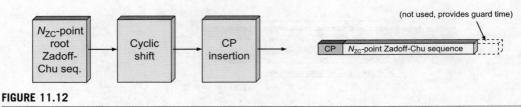

FIGURE 11.12

Random-access preamble generation.

downlink component carrier. From this path-loss estimate, the initial PRACH transmission power is obtained by adding a configurable offset.

The LTE random-access mechanism allows power ramping where the actual PRACH transmission power is increased for each unsuccessful random-access attempt. For the first attempt, the PRACH transmission power is set to the initial PRACH power. In most cases, this is sufficient for the random-access attempts to be successful. However, if the random-access attempt fails (random-access failures are detected at the second of four random-access steps, as described in the following sections), the PRACH transmission power for the next attempt is increased by a configurable step size to increase the likelihood of the next attempt being successful.

Since the random-access preamble is typically orthogonal to the user data, the need for power ramping to control intra-cell interference is smaller than in other systems with non-orthogonal random access and in many cases the transmission power is set such that the first random-access attempt with a high likelihood is successful. This is beneficial from a delay perspective.

11.3.1.4 Preamble Sequence Generation

The preamble sequences are generated from cyclic shifts of root Zadoff–Chu sequences [34]. Zadoff–Chu sequences are also used for creating the uplink reference signals as described in Chapter 7, where the structure of those sequences is described. From each root Zadoff–Chu sequence $X_{ZC}^{(u)}(k)$, $\lfloor N_{ZC}/N_{CS} \rfloor$ cyclically shifted[8] sequences are obtained by cyclic shifts of N_{CS} each, where N_{ZC} is the length of the root Zadoff–Chu sequence. The generation of the random-access preamble is illustrated in Figure 11.12. Although the figure illustrates generation in the time domain, frequency-domain generation can equally well be used in an implementation.

Cyclically shifted Zadoff–Chu sequences possess several attractive properties. The amplitude of the sequences is constant, which ensures efficient power amplifier utilization and maintains the low PAR properties of the single-carrier uplink. The sequences also have ideal cyclic auto-correlation, which is important for obtaining an accurate timing estimation at the eNodeB. Finally, the cross-correlation between different preambles based on cyclic

---

[8]The cyclic shift is in the time domain. Similar to the uplink reference signals and control signaling, this can equivalently be described as a phase rotation in the frequency domain.

shifts of the same Zadoff–Chu root sequence is zero at the receiver as long as the cyclic shift N_{CS} used when generating the preambles is larger than the maximum round-trip propagation time in the cell plus the maximum delay spread of the channel. Therefore, due to the ideal cross-correlation property, there is no intra-cell interference from multiple random-access attempts using preambles derived from the same Zadoff–Chu root sequence.

To handle different cell sizes, the cyclic shift N_{CS} is signaled as part of the system information. Thus, in smaller cells, a small cyclic shift can be configured, resulting in a larger number of cyclically shifted sequences being generated from each root sequence. For cell sizes below 1.5 km, all 64 preambles can be generated from a single root sequence. In larger cells, a larger cyclic shift needs to be configured and to generate the 64 preamble sequences, multiple root Zadoff–Chu sequences must be used in the cell. Although the larger number of root sequences is not a problem in itself, the zero cross-correlation property only holds between shifts of the same root sequence and from an interference perspective it is therefore beneficial to use as few root sequences as possible.

Reception of the random-access preamble is discussed later in this chapter. In principle, it is based on correlation of the received signal with the root Zadoff–Chu sequences. One disadvantage of Zadoff–Chu sequences is the difficulties in separating a frequency offset from the distance-dependent delay. A frequency offset results in an additional correlation peak in the time domain—a correlation peak that corresponds to a spurious device-to-base-station distance. In addition, the true correlation peak is attenuated. At low-frequency offsets, this effect is small and the detection performance is hardly affected. However, at high Doppler frequencies, the spurious correlation peak can be larger than the true peak. This results in erroneous detection; the correct preamble may not be detected or the delay estimate may be incorrect.

To avoid the ambiguities from spurious correlation peaks, the set of preamble sequences generated from each root sequence can be restricted. Such restrictions imply that only some of the sequences that can be generated from a root sequence are used to define random-access preambles. Whether such restrictions should be applied or not to the preamble generation is signaled as part of the system information. The location of the spurious correlation peak relative to the "true" peak depends on the root sequence and hence different restrictions have to be applied to different root sequences.

11.3.1.5 Preamble Detection

The base-station processing is implementation specific, but due to the cyclic prefix included in the preamble, low-complexity frequency-domain processing is possible. An example hereof is shown in Figure 11.13. Samples taken in a time-domain window are collected and converted into the frequency-domain representation using an FFT. The window length is 0.8 ms, which is equal to the length of the Zadoff–Chu sequence without a cyclic prefix. This allows handling timing uncertainties up to 0.1 ms and matches the guard time defined for the basic preamble configuration.

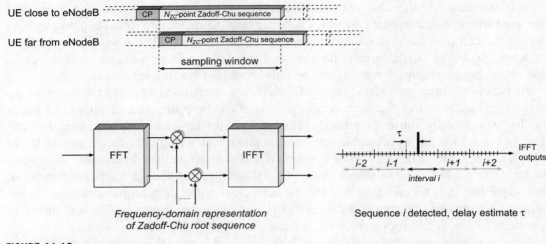

FIGURE 11.13

Random-access preamble detection in the frequency domain.

The output of the FFT, representing the received signal in the frequency domain, is multiplied by the complex-conjugate frequency-domain representation of the root Zadoff–Chu sequence and the result is fed through an IFFT. By observing the IFFT outputs, it is possible to detect which of the shifts of the root Zadoff–Chu sequence has been transmitted and its delay. Basically, a peak of the IFFT output in interval i corresponds to the ith cyclically shifted sequence and the delay is given by the position of the peak within the interval. This frequency-domain implementation is computationally efficient and allows simultaneous detection of multiple random-access attempts using different cyclically shifted sequences generated from the same root Zadoff–Chu sequence; in the case of multiple attempts there will simply be a peak in each of the corresponding intervals.

11.3.2 STEP 2: RANDOM-ACCESS RESPONSE

In response to the detected random-access attempt, the network will, as the second step of the random-access procedure, transmit a message on the DL-SCH, containing:

- The index of the random-access preamble sequences the network detected and for which the response is valid.
- The timing correction calculated by the random-access-preamble receiver.
- A scheduling grant, indicating what resources the device should use for the transmission of the message in the third step.
- A temporary identity, the TC-RNTI, used for further communication between the device and the network.

If the network detects multiple random-access attempts (from different devices), the individual response messages of multiple devices can be combined in a single transmission. Therefore, the response message is scheduled on the DL-SCH and indicated on a PDCCH using an identity reserved for random-access response, the RA-RNTI.[9] The usage of the RA-RNTI is also necessary as a device may not have a unique identity in the form of a C-RNTI allocated at this stage. All devices that have transmitted a preamble monitor the L1/L2 control channels for random-access response within a configurable time window. The timing of the response message is not fixed in the specification in order to be able to respond to many simultaneous accesses. It also provides some flexibility in the base-station implementation. If the device does not detect a random-access response within the time window, the attempt will be declared as failed and the procedure will repeat from the first step again, possibly with an increased preamble transmission power.

As long as the devices that performed random access in the same resource used different preambles, no collision will occur and from the downlink signaling it is clear to which device(s) the information is related. However, there is a certain probability of contention—that is, multiple devices using the same random-access preamble at the same time. In this case, multiple devices will react upon the same downlink response message and a collision occurs. Resolving these collisions is part of the subsequent steps, as discussed in the following. Contention is also one of the reasons why hybrid ARQ is not used for transmission of the random-access response. A device receiving a random-access response intended for another device will have incorrect uplink timing. If hybrid ARQ were used, the timing of the hybrid-ARQ acknowledgment for such a device would be incorrect and may disturb uplink control signaling from other users.

Upon reception of the random-access response in the second step, the device will adjust its uplink transmission timing and continue to the third step. If contention-free random access using a dedicated preamble is used, this is the last step of the random-access procedure as there is no need to handle contention in this case. Furthermore, the device already has a unique identity allocated in the form of a C-RNTI.

11.3.3 STEP 3: DEVICE IDENTIFICATION

After the second step, the uplink of the device is time synchronized. However, before user data can be transmitted to/from the device, a unique identity within the cell, the C-RNTI, must be assigned to the device (unless the device already has a C-RNTI assigned). Depending on the device state, there may also be a need for additional message exchange for setting up the connection.

In the third step, the device transmits the necessary messages to the eNodeB using the UL-SCH resources assigned in the random-access response in the second step. Transmitting the

---

[9]There are actually several RA-RNTIs defined. Which RA-RNTI a device is listening to is given by the time and frequency resource upon which the random-access preamble was transmitted.

uplink message in the same manner as scheduled uplink data instead of attaching it to the preamble in the first step is beneficial for several reasons. First, the amount of information transmitted in the absence of uplink synchronization should be minimized, as the need for a large guard time makes such transmissions relatively costly. Secondly, the use of the "normal" uplink transmission scheme for message transmission allows the grant size and modulation scheme to be adjusted to, for example, different radio conditions. Finally, it allows for hybrid ARQ with soft combining for the uplink message. The latter is an important aspect, especially in coverage-limited scenarios, as it allows for the use of one or several retransmissions to collect sufficient energy for the uplink signaling to ensure a sufficiently high probability of successful transmission. Note that RLC retransmissions are not used for the uplink RRC signaling in step 3.

An important part of the uplink message is the inclusion of a device identity, as this identity is used as part of the contention-resolution mechanism in the fourth step. If the device is in the RRC_CONNECTED state—that is, connected to a known cell and therefore has a C-RNTI assigned—this C-RNTI is used as the device identity in the uplink message.[10] Otherwise, a core-network device identifier is used and the eNodeB needs to involve the core network prior to responding to the uplink message in step 3.

Device-specific scrambling is used for transmission on UL-SCH, as described in Chapter 7. However, as the device may not yet have been allocated its final identity, the scrambling cannot be based on the C-RNTI. Instead, a temporary identity is used (TC-RNTI).

11.3.4 STEP 4: CONTENTION RESOLUTION

The last step in the random-access procedure consists of a downlink message for contention resolution. Note that, from the second step, multiple devices performing simultaneous random-access attempts using the same preamble sequence in the first step listen to the same response message in the second step and therefore have the same temporary identifier. Hence, the fourth step in the random-access procedure is a contention-resolution step to ensure that a device does not incorrectly use the identity of another device. The contention resolution mechanism differs somewhat depending on whether the device already has a valid identity in the form of a C-RNTI or not. Note that the network knows from the uplink message received in step 3 whether or not the device has a valid C-RNTI.

If the device already had a C-RNTI assigned, contention resolution is handled by addressing the device on the PDCCH using the C-RNTI. Upon detection of its C-RNTI on the PDCCH the device will declare the random-access attempt successful and there is no need for contention-resolution-related information on the DL-SCH. Since the C-RNTI is unique to one device, unintended devices will ignore this PDCCH transmission.

If the device does not have a valid C-RNTI, the contention resolution message is addressed using the TC-RNTI and the associated DL-SCH contains the contention-resolution message.

---

[10]The device identity is included as a MAC control element on the UL-SCH.

The device will compare the identity in the message with the identity transmitted in the third step. Only a device which observes a match between the identity received in the fourth step and the identity transmitted as part of the third step will declare the random-access procedure successful and promote the TC-RNTI from the second step to the C-RNTI. Since uplink synchronization has already been established, hybrid ARQ is applied to the downlink signaling in this step and devices with a match between the identity they transmitted in the third step and the message received in the fourth step will transmit a hybrid-ARQ acknowledgment in the uplink.

Devices that do not detect PDCCH transmission with their C-RNTI or do not find a match between the identity received in the fourth step and the respective identity transmitted as part of the third step are considered to have failed the random-access procedure and need to restart the procedure from the first step. No hybrid-ARQ feedback is transmitted from these devices. Furthermore, a device that has not received the downlink message in step 4 within a certain time from the transmission of the uplink message in step 3 will declare the random-access procedure as failed and need to restart from the first step.

11.4 PAGING

Paging is used for network-initiated connection setup when the device is in RRC_IDLE. In LTE, the same mechanism as for "normal" downlink data transmission on the DL-SCH is used and the mobile device monitors the L1/L2 control signaling for downlink scheduling assignments related to paging. Since the location of the device typically is not known on a cell level, the paging message is typically transmitted across multiple cells in the so-called tracking area (the tracking area is controlled by the MME; see [5] for a discussion on tracking areas).

An efficient paging procedure should allow the device to sleep with no receiver processing most of the time and to briefly wake up at predefined time intervals to monitor paging information from the network. Therefore, a paging cycle is defined, allowing the device to sleep most of the time and only briefly wake up to monitor the L1/L2 control signaling. If the device detects a group identity used for paging (the P-RNTI) when it wakes up, it will process the corresponding downlink paging message transmitted on the PCH. The paging message includes the identity of the device(s) being paged, and a device not finding its identity will discard the received information and sleep according to the DRX cycle. As the uplink timing is unknown during the DRX cycles, no hybrid-ARQ acknowledgments can be transmitted and consequently hybrid ARQ with soft combining is not used for paging messages.

The network configures in which subframes a device should wake up and listen for paging. Typically, the configuration is cell specific, although there is a possibility to complement the setting by device-specific configuration. In which frame a given device should wake up and search for the P-RNTI on a PDCCH is determined by an equation taking as input the identity of the device as well as a cell-specific and (optionally) a device-specific paging cycle.

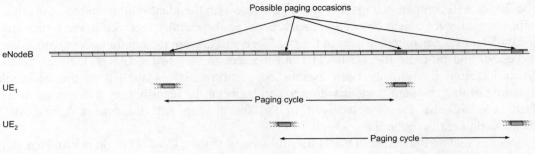

FIGURE 11.14

Illustration of paging cycles.

Table 11.1 Paging Cycles and Paging Subframes									
		Number of Paging Subframes per Paging Cycle							
		1/32	1/16	1/8	1/4	1/2	1	2	4
Paging subframes in a paging frame	FDD	9	9	9	9	9	9	4, 9	0, 4, 5, 9
	TDD	0	0	0	0	0	0	0, 5	0, 1, 5, 6

The identity used is the so-called IMSI, an identity coupled to the subscription, as an idle mode device does not have a C-RNTI allocated, and the paging cycle for a device can range from once per 256 up to once per 32 frames. The subframe within a frame to monitor for paging is also derived from the IMSI. Since different devices have different IMSI, they will compute different paging instances. Hence, from a network perspective, paging may be transmitted more often than once per 32 frames, although not all devices can be paged at all paging occasions as they are distributed across the possible paging instances as shown in Figure 11.14.

Paging messages can only be transmitted in some subframes, ranging from one subframe per 32 frames up to a very high paging capacity with paging in four subframes in every frame. The configurations are shown in Table 11.1. Note that, from a network perspective, the cost of a short paging cycle is minimal as resources not used for paging can be used for normal data transmission and are not wasted. However, from a device perspective, a short paging cycle increases the power consumption as the device needs to wake up frequently to monitor the paging instants.

In addition to initiating connection to devices being in RRC_IDLE, paging can also be used to inform devices in RRC_IDLE as well as RRC_CONNECTED about changes of system information. A device being paged for this reason knows that the system information will change and therefore needs to acquire the update system information as described in Section 11.2.

CARRIER AGGREGATION

12

The possibility for *carrier aggregation* (CA) was introduced in LTE release 10 with enhancements in the following releases. In the case of carrier aggregation, multiple LTE carriers, possibly of different bandwidths up to 20 MHz, can be transmitted in parallel to/from the same device, thereby allowing for an overall wider bandwidth and correspondingly higher per-link data rates. In the context of carrier aggregation, each carrier is referred to as a *component carrier*[1] as, from a radio-frequency (RF) point of view, the entire set of aggregated carriers can be seen as a single (RF) carrier.

Initially, up to five component carriers could be aggregated allowing for overall transmission bandwidths up to 100 MHz. In release 13 this was extended to 32 carriers allowing for an overall transmission bandwidth of 640 MHz, primarily motivated by the possibility for large bandwidths in unlicensed spectrum. A device capable of carrier aggregation may receive or transmit simultaneously on multiple component carriers. Each component carrier can also be accessed by an LTE device from earlier releases, that is, component carriers are *backward compatible*. Thus, in most respects and unless otherwise mentioned, the physical-layer description in the previous chapters applies to each component carrier separately in the case of carrier aggregation.

It should be noted that aggregated component carriers do not need to be contiguous in the frequency domain. Rather, with respect to the frequency location of the different component carriers, three different cases can be identified (see also Figure 12.1):

- Intraband aggregation with frequency-contiguous component carriers.
- Intraband aggregation with noncontiguous component carriers.
- Interband aggregation with noncontiguous component carriers.

The possibility to aggregate nonadjacent component carriers allows for exploitation of a fragmented spectrum; operators with a fragmented spectrum can provide high-data-rate services based on the availability of a wide overall bandwidth, even though they do not possess a single wideband spectrum allocation. Except from an RF point of view there is no difference between the three different cases outlined in Figure 12.1 and they are all supported

---

[1]In the specifications, the term "cell" is used instead of component carrier, but as the term "cell" is something of a misnomer in the uplink case, the term "component carrier" is used here.

4G, LTE-Advanced Pro and The Road to 5G. http://dx.doi.org/10.1016/B978-0-12-804575-6.00012-1

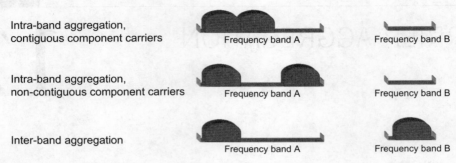

Intra-band aggregation,
contiguous component carriers

Frequency band A   Frequency band B

Intra-band aggregation,
non-contiguous component carriers

Frequency band A   Frequency band B

Inter-band aggregation

Frequency band A   Frequency band B

FIGURE 12.1

Different types of carrier aggregation.

by the LTE release 10 specification. However, the complexity of RF implementation is vastly different as discussed in Chapter 22, with the first case being the least complex. Thus, although spectrum aggregation is supported by the physical-layer and protocol specifications, the actual implementation will be strongly constrained, including specification of only a limited number of aggregation scenarios and aggregation over a dispersed spectrum only being supported by the most advanced devices.

Although carrier aggregation can be utilized to achieve very high overall bandwidths, up to 100 MHz or even 640 MHz, very few operators, if any, have such large spectrum allocations. Rather, the main usage of carrier aggregation, at least initially, was to handle fragmented spectrum allocations in which an operator may have 5–10 MHz of spectrum allocation in several bands and, despite this, would like to provide end-user performance on par with an operator having a larger amount of contiguous spectrum.

In releases 10 and 11, only downlink-heavy asymmetries are supported from an RF perspective; that is, the number of uplink component carriers configured for a device is always equal to or smaller than the number of configured downlink component carriers. Uplink-heavy asymmetries are less likely to be of practical interest and would also complicate the overall control signaling structure, as in such a case multiple uplink component carriers would need to be associated with the same downlink component carrier.

Carrier aggregation is supported for all frame structures. In release 10, all aggregated component carriers need to have the same duplex scheme and, in case of TDD, the same uplink–downlink configuration across the component carriers.

For carrier aggregation capable TDD devices in release 11, different uplink–downlink configurations can be used for component carrier in *different* frequency band. The main motivation is improved support for coexistence with other systems. One example is two independent legacy systems operating in two different frequency bands. Clearly, if LTE is to exploit spectrum in both of these bands through carrier aggregation, the uplink–downlink allocation of the component carriers in the respective band is basically determined by the legacy systems and may very well be different.

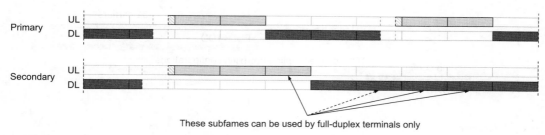

These subfames can be used by full-duplex terminals only

FIGURE 12.2

Examples of TDD interband CA with different uplink—downlink configurations.

It is worth noting that interband carrier aggregation with *different* uplink—downlink configurations may imply *simultaneous* uplink transmission and downlink reception in the device. Devices capable of simultaneous transmission and reception need, similarly to FDD devices and unlike most TDD devices, a duplex filter. Whether a TDD device is equipped with a duplex filter and capable of simultaneous transmission and reception is therefore a matter of device capability. Devices not having this capability follow the uplink—downlink configuration on one of the component carriers, the primary carrier (primary and secondary component carriers are described in the next section). These devices cannot transmit in the uplink on a secondary component carrier whenever there is a downlink subframe on the primary component carrier (and vice versa). In essence this implies certain subframes on some of the carriers being unusable by devices not capable of simultaneous reception and transmission, see Figure 12.2.[2]

The special subframe configuration can be different for the different components carriers although devices not capable of simultaneous transmission and reception require that the resulting downlink—uplink switch time is sufficiently large.

In release 12, carrier aggregation was further enhanced by allowing aggregation between FDD and TDD to enable efficient utilization of an operator's overall spectrum assets. The primary component carrier can use either FDD or TDD. Aggregation across the duplex schemes can also be used to improve the uplink coverage of TDD by relying on the possibility for continuous uplink transmission on the FDD carrier.

Release 13 increased the number of carriers possible to aggregate from 5 to 32, resulting in a maximum bandwidth of 640 MHz and a corresponding theoretical peak data rate of approximately 25 Gbit/s in the downlink. The main motivation for increasing the number of subcarriers is to allow for very large bandwidths in unlicensed spectrum as is further discussed in conjunction with license-assisted access later.

The evolution of carrier aggregation across different releases is illustrated in Figure 12.3.

---

[2]The downlink subframe on the secondary component carrier overlapping with the special subframe on the primary component carrier can only be used in the part overlapping with the DwPTS.

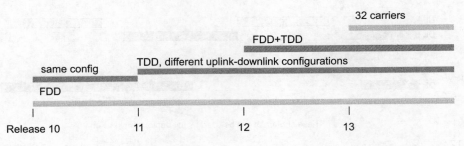

FIGURE 12.3

Evolution of carrier aggregation.

12.1 OVERALL PROTOCOL STRUCTURE

Carrier aggregation is essentially duplicating the MAC and PHY processing for each component carrier while keeping radio-link control (RLC) and above identical to the non-aggregation case (Figure 12.4). Hence, one RLC entity may handle data transmitted across multiple component carriers in the presence of carrier aggregation. The MAC entity is responsible for distributing data from each flow across the component carriers, a decision that is part of the implementation-specific scheduling approach in the downlink. Each component carrier has its own hybrid-ARQ entity, implying that hybrid-ARQ retransmissions must occur on the same component carrier as the original transmission. There is no possibility to move hybrid-ARQ retransmissions between carriers. RLC retransmissions, on the other hand, are not tied to a specific component carrier as, in essence, CA is invisible above the MAC layer. RLC retransmissions can therefore use a different component carrier than the original transmission. The RLC also handles reordering across component carriers to ensure in-sequence in

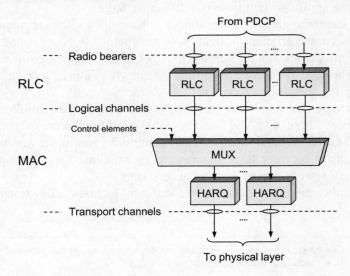

FIGURE 12.4

RLC and hybrid-ARQ retransmission mechanisms in LTE.

case a radio bearer is transmitted on multiple component carriers. Since hybrid-ARQ retransmissions are handled independently per component carrier, out-of-sequence delivery from the MAC layer may occur not only on one component carrier as described in Chapter 8 but also between component carriers.

Each component carrier is independently scheduled with individual scheduling assignments/grants per component carrier in case of carrier aggregation. The scheduling assignment/grant can either be transmitted on the same carrier as data (self-scheduling) or on another component carrier (cross-carrier scheduling) as is described in more detail in Section 12.3.

12.2 PRIMARY AND SECONDARY COMPONENT CARRIERS

A device capable of carrier aggregation has one downlink primary component carrier and an associated uplink primary component carrier. In addition, it may have one or several secondary component carriers in each direction. Different devices may have different carriers as their primary component carrier—that is, the configuration of the primary component carrier—is device specific. The association between the downlink primary carrier and the corresponding uplink primary carrier is signaled as part of the system information. This is similar to the case without carrier aggregation, although in the latter case the association is trivial. The reason for such an association is, for example, to determine to which uplink component carrier a certain scheduling grant transmitted on the downlink relates without having to explicitly signal the component-carrier number. In the uplink, the primary carrier is of particular interest as it, in many cases, carries all the L1/L2 uplink control signaling as described later in this chapter.

All idle mode procedures apply to the primary component carrier only or, expressed differently, carrier aggregation with additional secondary carriers configured only applies to devices in the RRC_CONNECTED state. Upon connection to the network, the device performs the related procedures such as cell search and random access (see Chapter 11 for a detailed description of these procedures) following the same steps as in the absence of carrier aggregation. Once the communication between the network and the device is established, additional secondary component carriers can be configured.

The fact that carrier aggregation is device specific—that is, different devices may be configured to use different sets of component carriers—is useful not only from a network perspective to balance the load across component carriers, but also to handle different capabilities between devices. Some devices may be able to transmit/receive on multiple component carriers, while other devices may do so on only a single carrier. This is a consequence of being able to serve devices from earlier releases at the same time as a carrier aggregation-capable device, but it also allows for different capabilities in terms of carrier aggregation for different devices as well as a differentiation between downlink and uplink carrier aggregation capability. For example, a device may be capable of two component carriers in the downlink but of only a single component carrier—that is, no carrier aggregation—in the uplink, as is the case for device C in Figure 12.5. Note also that the primary component-carrier configuration can differ between devices. Asymmetric carrier aggregation can also be useful to handle different spectrum allocations, for example if an

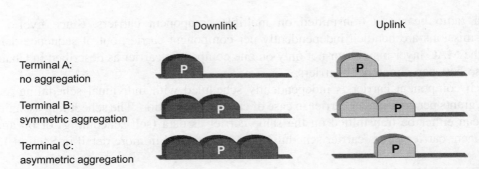

FIGURE 12.5

Examples of carrier aggregation ("P" denotes the primary component carrier).

operator has more spectrum available for downlink transmissions than uplink transmissions. Semipersistent scheduling is only supported on the primary component carrier, motivated by the fact that the main usage is for small payloads not requiring multiple component carriers.

12.3 SELF-SCHEDULING AND CROSS-CARRIER SCHEDULING

Each component carrier is, as already mentioned, individually scheduled with the scheduling assignment/grant transmitted on either the same (associated) component carrier as the data (self-scheduling) or on a different component carrier than the data (cross-carrier scheduling). The two possibilities are illustrated in Figure 12.6.

For self-scheduling, downlink scheduling assignments are valid for the component carrier upon which they are transmitted. Similarly, for uplink grants, there is an association between downlink and uplink component carriers such that each uplink component carrier has an associated downlink component carrier. The association is provided as part of the system information. Thus, from the uplink–downlink association, the device will know to which uplink component carrier the downlink control information relates to.

For cross-carrier scheduling, where downlink PDSCH or uplink PUSCH is transmitted on an (associated) component carrier other than that which (E)PDCCH is transmitted upon, the

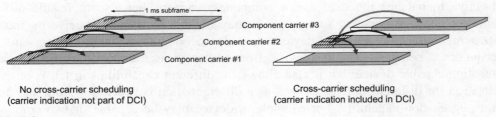

FIGURE 12.6

Self-scheduling (left) and cross-carrier scheduling (right).

carrier indicator in the PDCCH provides information about the component carrier used for the PDSCH or PUSCH.

Whether cross-carrier scheduling is used or not is configured using higher-layer signaling. Irrespective of whether self-scheduling or cross-carrier scheduling is used the hybrid-ARQ feedback is sent on the primary uplink carrier.[3] This structure was chosen to handle asymmetric carrier aggregation with more downlink carriers than uplink carriers, a common scenario. Two questions related to timing across the component carriers arise as a result of this structure, namely

- in which uplink subframe on the primary component carrier should the hybrid-ARQ acknowledgment related to data transmission in subframe n on a secondary component carrier be transmitted, and
- to which subframe does a scheduling grant received in subframe n relate to?

These timing relations are relatively straight forward for FDD but more complicated for TDD, in particular, in case of cross-carrier scheduling. The different timing scenarios are discussed in the following sections.

12.3.1 SCHEDULING TIMING FOR AGGREGATION OF FDD CARRIERS

The baseline for aggregation of multiple FDD component carriers is to follow the same timing relations for scheduling and retransmissions as in absence of carrier aggregation, that is, downlink scheduling assignments and the associated data are both transmitted in subframe n and the resulting acknowledgments in uplink subframe $n + 4$. Note that, depending on the number of component carriers scheduled, there may be a relatively large number of acknowledgment bits transmitted in a single subframe. Similarly, uplink scheduling grants received in subframe n result in data transmission in subframe $n + 4$ and PHICH being monitored in subframe $n + 8$. Multiple PHICHs are used with each PHICH being transmitted on the same component carrier as the grant initiating the uplink data transmission. These timing relations hold for both self-scheduling and cross-carrier scheduling.

12.3.2 SCHEDULING TIMING FOR AGGREGATION OF TDD CARRIERS

In case of self-scheduling, timing of scheduling assignments and grants, including the PHICH, is straight forward and the same timing relations as in absence of carrier aggregation is used. Downlink scheduling assignments are transmitted on the same carrier as the corresponding data. Uplink data transmissions occur on the uplink component carrier associated with the downlink carrier used for transmission of the scheduling grant.

Downlink transmissions imply that the device needs to respond with hybrid-ARQ acknowledgments on the PUCCH (or PUSCH) on the primary component carrier. Obviously, an

---

[3]With aggregation of up to 32 carriers in release 13 this is extended to two uplink carriers, see further Section 12.6.

acknowledgment can only be transmitted in an uplink subframe on the primary component. In case of identical uplink–downlink allocations across all component carriers, the timing of the acknowledgments is the same as in absence of CA. However, release 11 introduced the possibility of having *different* allocations on different component carriers and release 12 introduced the possibility to aggregate FDD and TDD carriers, which complicates the timing relations further in case of TDD on the primary component carrier. If the timing of the hybrid-ARQ acknowledgment would follow the primary component carrier only, there would be no timing relation for the secondary component carrier in some subframes (subframe 4 in Figure 12.7) and, consequently, it would not be possible to use those subframes for downlink data transmissions. Therefore, to handle this case, a reference configuration, compatible with the hybrid-ARQ timing of both the primary component carrier and all secondary component carriers is used to derive the hybrid-ARQ timing relations for the secondary component carriers. The primary component carrier uses the same timing as in the absence of CA. See Figure 12.7 for an example of aggregation of primary and secondary component carriers using configurations 3 and 1, respectively, in which case configuration 4 is used as a reference for the timing of any secondary component carrier. The reference configuration for other combinations of configurations can be found in Table 12.1.

For cross-carrier scheduling, the timing of not only the hybrid-ARQ acknowledgments but also the scheduling grants and scheduling assignments can become quite involved. In case of the same uplink–downlink configuration across all component carriers, the scheduling timing is similar to that of self-scheduling as there is no timing conflict across the component carriers. However, cross-carrier scheduling with different uplink–downlink allocations requires special attention as the downlink scheduling timing for each component carrier follows that of the primary component carrier. Together with the fact that a scheduling assignment cannot point to a downlink subframe in the future, this implies that some downlink subframes on some component carriers cannot be scheduled with cross-carrier scheduling, see Figure 12.8. This is consistent with the fact that, for a TDD device not capable of simultaneous transmission and reception, only downlink (uplink) subframe on a secondary component carrier that coincides with a downlink (uplink) subframe on the primary component carrier may be used. Note though that in absence of cross-carrier scheduling all subframes can be scheduled.

Uplink scheduling timing in presence of cross-carrier scheduling is rather complex. For some configurations the timing is given by the configuration of the scheduling carrier—the carrier on which the (E)PDCCH is transmitted—while for other configurations, the timing follows the configuration of the scheduled carrier, that is, the carrier upon which the PUSCH was transmitted. The same relations hold for PHICH as well, as it in essence is a single-bit uplink grant.

In case of cross-carrier scheduling, no reference configuration is needed for hybrid-ARQ acknowledgments in response to downlink transmissions as the scheduling timing for each component carrier follows the primary component. The "problematic" subframe, subframe 4 in Figure 12.7, can therefore never be scheduled and the hybrid-ARQ timing relations can be

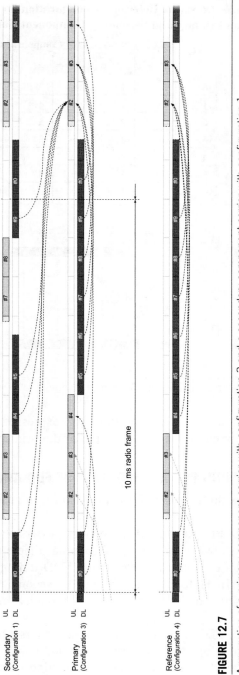

FIGURE 12.7

Aggregation of a primary component carrier with configuration 3 and a secondary component carrier with configuration 1.

Table 12.1 Uplink–Downlink Reference Configuration for Different Configurations on Primary and Secondary Component Carriers

		Secondary Component Carrier						
		0	**1**	**2**	**3**	**4**	**5**	**6**
Primary component carrier	**0**	0	1	2	3	4	5	6
	1	1	1	2	4	4	5	1
	2	2	2	2	5	5	5	2
	3	3	4	5	3	4	5	3
	4	4	4	5	4	4	5	4
	5	5	5	5	5	5	5	5
	6	6	1	2	3	4	5	6

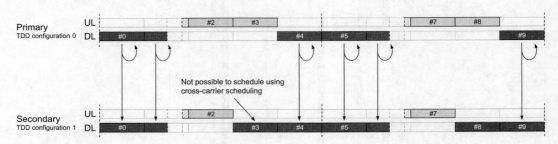

FIGURE 12.8

Example of cross-carrier scheduling with different TDD configurations.

derived from the primary component carrier alone (or, expressed differently, the reference timing configuration is identical to the configuration of the primary component carrier).

12.3.3 SCHEDULING TIMING FOR AGGREGATION OF FDD AND TDD CARRIERS

Many of the timing aspects discussed earlier apply also for the case of aggregation of FDD and TDD carriers.

In case of aggregation of FDD and TDD component carriers with the primary component carrier using FDD, there is always an uplink subframe available and consequently hybrid-ARQ acknowledgments can be transmitted using the FDD timing. Scheduling assignments and grants, including the PHICH, follow the timing of the secondary component carrier. This is obviously needed for self-scheduling but is, for simplicity, also used for cross-carrier scheduling.

In case of aggregation of FDD and TDD component carriers with the primary component carrier using TDD, the timing of downlink assignment and uplink grants is straight forward and the FDD timing is reused, that is, an uplink scheduling grant received in subframe n

implies uplink transmission in subframe $n + 4$. This holds for both self-scheduling and cross-carrier scheduling. Note that this implies that some subframes on the FDD component carriers cannot be scheduled from a TDD component carrier, similarly to the illustration in Figure 12.8.

However, the PHICH timing does not follow the FDD timing. Rather, the PHICH is transmitted six subframes after the uplink data being received. The reason for this, and not the normal FDD timing of $n + 4$, is to guarantee that there is a downlink subframe to transmit the PHICH in. Since $6 + 4 = 10$, the PHICH is transmitted one frame later than the uplink grant, which matches the periodicity of the uplink–downlink configuration. This is illustrated in Figure 12.9. The scheduling grant in subframe 0 triggers an uplink transmission in subframe 4 and a retransmission is requested using PHICH in the next subframe 0. In this example the FDD PHICH time would have worked as well (with a PHICH in subframe 8). However, for a grant in subframe 9, triggering an uplink transmission in subframe 3, the FDD PHICH time would not have worked as subframe 7 on the primary carrier is not a downlink subframe.

The timing of the hybrid-ARQ acknowledgment in the uplink in response to downlink data transmission is somewhat complex and is given by a table in the specifications, ensuring that the acknowledgment is transmitted in an uplink subframe on the primary component carrier. One example is given in Figure 12.10.

12.4 DRX AND COMPONENT-CARRIER DEACTIVATION

Discontinuous reception (DRX) was described in Section 9.7 as a means to reduce the device power consumption. This holds equally well for a device with CA and the same DRX mechanism is applied across all component carriers. Hence, if the device is in DRX it is not receiving on any component carrier, but when it wakes up, all (activated) component carriers will be woken up.

Although discontinuous reception greatly reduces the device power consumption, it is possible to go one step further in the case of carrier aggregation. From a power-consumption perspective, it is beneficial to receive on as few component carriers as possible. LTE therefore supports deactivation of downlink component carriers. A deactivated component carrier maintains the configuration provided by RRC but cannot be used for reception, neither PDCCH nor PDSCH. When the need arises, a downlink component carrier can be activated rapidly and used for reception within a few subframes. A typical use would be to configure several component carriers but deactivate all component carriers except the primary one. When a data burst starts, the network could activate several component carriers to maximize the downlink data rate. Once the data burst is delivered, the component carriers could be deactivated again to reduce device power consumption.

Activation and deactivation of downlink component carriers are done through MAC control elements. There is also a timer-based mechanism for deactivation such that a device may, after a configurable time with no activity on a certain component carrier, deactivate that

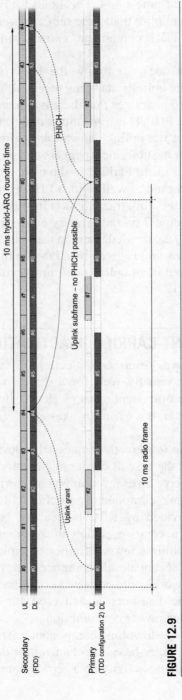

FIGURE 12.9

PHICH timing with the primary carrier using TDD.

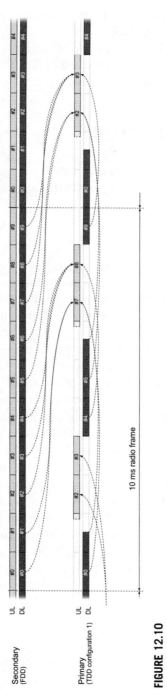

FIGURE 12.10

Example of hybrid-ARQ acknowledgment timing when aggregating FDD with TDD.

component carrier. The primary component carrier is always active as it must be possible for the network to communicate with the device.

In the uplink there is no explicit activation of uplink component carriers. However, whenever a downlink component carrier is activated or deactivated, the corresponding uplink component carrier is also activated or deactivated.

12.5 DOWNLINK CONTROL SIGNALING

Carrier aggregation uses the same set of L1/L2 control channels—PCFICH, (E)PDCCH, and PHICH—as in the nonaggregated case. However, there are some enhancements introduced, in particular to handle the larger number of hybrid-ARQ acknowledgments, as detailed in the following subsections.

12.5.1 PCFICH

The structure of each component carrier is the same as for the nonaggregated scenario, implying there is one PCFICH per component carrier. Independent signaling of the control-region size on the different component carriers is used, implying that the control region may be of different size on different component carriers. Hence, in principle the device needs to receive the PCFICH on each of the component carriers it is scheduled upon. Furthermore, as different component carriers may have different physical-layer cell identities, the location and scrambling may differ across component carriers.

If cross-carrier scheduling is used, that is, control signaling related to a certain PDSCH transmission is transmitted on a component carrier other than the PDSCH itself, the device needs to know the starting position for the data region on the carrier upon which the PDSCH is transmitted. Using the PCFICH on the component carrier carrying the PDSCH would be possible in principle, although it would increase the probability of incorrectly decoding the PDSCH since there are two PCFICH instances, one for the PDCCH decoding on one component carrier and one for PDSCH reception on the other component carrier. This would be problematic, especially since one use of cross-carrier scheduling is enhanced support of heterogeneous deployments (see Chapter 14), where some of the component carriers may be subject to strong interference. Therefore, for cross-carrier scheduled transmissions, the start of the data region is not obtained from the PCFICH on that component carrier, but is configured on a semi-static basis. The semi-statically configured value may differ from the value signaled on the PCFICH on the component carrier carrying the PDSCH transmission.

12.5.2 PHICH

The PHICH follows the same principles as for the nonaggregated case, see Chapter 6. This means that the PHICH resource to monitor is derived from the number of the first resource block upon which the corresponding uplink PUSCH transmission occurred in combination with the reference-signal phase rotation signaled as part of the uplink grant. As a general principle, LTE transmits the PHICH on the same component carrier that was used for the grant scheduling the corresponding uplink data transmission. Not only is this principle general in the sense that it can handle symmetric as well as asymmetric CA, it is also beneficial from a device power consumption perspective as the device only need monitor the component carriers it monitors for uplink scheduling grants (especially as the PDCCH may override the PHICH to support adaptive retransmissions, as discussed in Chapter 8).

For the case when no cross-carrier scheduling is used, that is, each uplink component carrier is scheduled on its corresponding downlink component carrier, different uplink component carriers will by definition have different PHICH resources. With cross-carrier scheduling, on the other hand, transmissions on multiple uplink component carriers may need to be acknowledged on a single downlink component carrier, as illustrated in Figure 12.11. Avoiding PHICH collisions in this case is up to the scheduler by ensuring that different reference-signal phase rotations or different resource-block starting positions are used for the different uplink component carriers. For semipersistent scheduling the reference-signal phase rotation is always set to zero, but since semipersistent scheduling is supported on the primary component carrier only, there is no risk of collisions between component carriers.

12.5.3 PDCCH AND EPDCCH

The (E)PDCCH carries downlink control information. For self-scheduling there are no major impacts to the (E)PDCCH processing as each component carrier in essence is independent, at least from a downlink control signaling perspective. However, if cross-carrier scheduling is configured by device-specific RRC signaling, the component carriers that a specific control message relates to must be indicated. This is the background to the carrier indication field mentioned in Chapter 6. Thus, most of the DCI formats come in two "flavors," with and

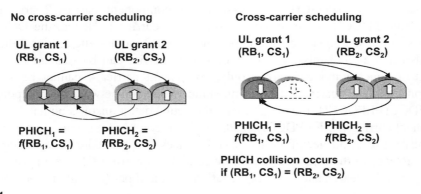

No cross-carrier scheduling

UL grant 1   UL grant 2
(RB_1, CS_1) (RB_2, CS_2)

$PHICH_1 =$ $PHICH_2 =$
$f(RB_1, CS_1)$ $f(RB_2, CS_2)$

Cross-carrier scheduling

UL grant 1   UL grant 2
(RB_1, CS_1) (RB_2, CS_2)

$PHICH_1 =$ $PHICH_2 =$
$f(RB_1, CS_1)$ $f(RB_2, CS_2)$

PHICH collision occurs
if $(RB_1, CS_1) = (RB_2, CS_2)$

FIGURE 12.11

PHICH association.

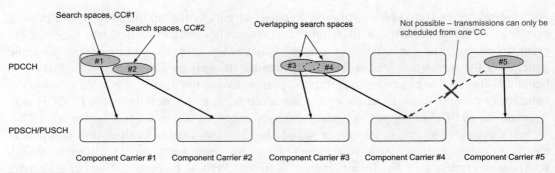

FIGURE 12.12

Example of scheduling of multiple component carriers.

without the carrier indication field, and which "flavor" the device is supposed to monitor is determined by enabling/disabling support for cross-carrier scheduling.

For signaling purposes, the component carriers are numbered. The primary component carrier is always given the number zero, while the different secondary component carriers are assigned a unique number each through device-specific RRC signaling. Hence, even if the device and the eNodeB have different understandings of the component-carrier numbering during a brief period of reconfiguration, at least transmissions on the primary component carrier can be scheduled.

Irrespective of whether cross-carrier scheduling is used or not, PDSCH/PUSCH on a component carrier can only be scheduled from *one* component carrier. Thus, for each PDSCH/PUSCH component carrier there is an associated component carrier, configured via device-specific RRC signaling, where the corresponding DCI can be transmitted. Figure 12.12 illustrates one example, where PDSCH/PUSCH transmissions on component carrier 1 are scheduled using PDCCHs/EPDCCHs transmitted on component carrier 1. In this case, as cross-carrier scheduling is not used, there is no carrier indicator in the corresponding DCI formats. PDSCH/PUSCH transmissions on component carrier 2 are cross-carrier scheduled from PDCCHs transmitted on component carrier 1. Hence, the DCI formats in the device-specific search space for component carrier 2 include the carrier indicator.

Note also that as transmissions on a component carrier can be scheduled by PDCCHs/ EPDCCHs on *one* component carrier only, component carrier 4 cannot be scheduled by PDCCHs on component carrier 5 as the semi-static association between the component carriers used for PDCCH/EPDCCH transmission and the actual data transmission has associated data on component carrier 4, with PDCCHs/EPDCCHs on component carrier 3 in this example.

With respect to blind decoding and search spaces, the procedures described in Section 6.4.5 apply to each of the activated[4] downlink component carriers. Hence, in principle there is one device-specific search space per aggregation level and per (activated) component carrier

---

[4]Individual component carriers can be activated/deactivated as discussed in Section 12.4.

upon which PDSCH can be received (or PUSCH transmitted), although there are some carrier aggregation-specific modifications. For devices configured to use carrier aggregation, this results in an increase in the number of blind decoding attempts compared to a device not using carrier aggregation, as scheduling assignments/grants for each of the component carriers need to be monitored. With the up to 32 component carriers supported in release 13, the number of blind decodes could be 1036 (or even higher if uplink spatial multiplexing is used). From a processing capability perspective this is not a major issue for a high-end device (remember, 32 carriers correspond to peak data rates up to 25 Gbit/s) but it can be challenging from a power-consumption perspective. Therefore, release 13 introduced the possibility to configure the number of decoding candidates and to restrict the number of DCI formats per secondary component carriers (based on the number of blind decoding attempts the devices indicates it supports).

The common search space is only defined for transmissions on the primary component carrier. As the main function of the common search space is to handle scheduling of system information intended for multiple devices, and such information must be receivable by all devices in the cell, scheduling in this case uses the common search space. For this reason, the carrier indication field is never present in DCI formats monitored in the common search space.

As mentioned earlier, there is one device-specific search space per aggregation level and component carrier used for scheduling the PDSCH/PUSCH. This is illustrated in Figure 12.12, where PDSCH/PUSCH transmissions on component carrier 1 are scheduled using PDCCHs transmitted on component carrier 1. No carrier indicator is assumed in the device-specific search space for component carrier 1 as cross-carrier scheduling is not used. For component carrier 2, on the other hand, a carrier indicator is assumed in the device-specific search space as component carrier 2 is cross-carrier scheduled from PDCCHs transmitted on component carrier 1.

Search spaces for different component carriers may overlap in some subframes. In Figure 12.12, this happens for the device-specific search spaces for component carriers 3 and 4. The device will handle the two search spaces independently, assuming (in this example) a carrier indicator for component carrier 4 but not for component carrier 3. If the device-specific and common search spaces relating to different component carriers happen to overlap for some aggregation level when cross-carrier scheduling is configured, the device only needs to monitor the common search space. The reason for this is to avoid ambiguities; if the component carriers have different bandwidths a DCI format in the common search space may have the same payload size as another DCI format in the device-specific search space relating to another component carrier.

12.6 **UPLINK CONTROL SIGNALING**

Carrier aggregation implies that a device needs to transmit acknowledgments relating to simultaneous reception of data on multiple DL-SCHs, that is, transmission of more than two hybrid-ARQ acknowledgments in the uplink must be supported. There is also a need to provide CSI feedback relating to more than one downlink carrier.

As a baseline, all the feedback is transmitted on the primary component carrier, motivated by the need to support asymmetric carrier aggregation with the number of downlink carriers supported by a device being unrelated to the number of uplink carriers. Carrier aggregation therefore calls for an increase in the payload capability of the PUCCH. This was the motivation for PUCCH formats 3, 4, and 5, described in Chapter 7.

With the introduction of aggregation of up to 32 carriers in release 13, the number of hybrid-ARQ acknowledgments can be fairly larger. To avoid overloading a single uplink component carrier with PUCCH transmission, a device can be configured with two carrier groups, see Figure 12.13. In each group one component carrier is used to handle PUCCH transmissions from the device, resulting in two PUCCHs for uplink control signaling from one device. The resulting structure is similar to the dual connectivity structure described in Chapter 16 and many details are common, for example, power headroom reports and power scaling. Cross-carrier scheduling between the two groups are not supported.

12.6.1 HYBRID-ARQ ACKNOWLEDGMENTS ON PUCCH

Carrier aggregation implies an increase in the number of hybrid-ARQ acknowledgments to convey to the eNodeB. PUCCH format 1 can be used to support more than two bits in the uplink by using resource selection where part of the information is conveyed by the PUCCH resource selected and part by the bits transmitted on the selected resource (Figure 12.14). The details on how to select the resources are rather complex and depend on the duplexing scheme (FDD or TDD) and whether cross-carrier scheduling is used or not, although the basic idea is the same.

As an example, assume FDD with no cross-carrier scheduling. Furthermore, assume four bits are to be transmitted in the uplink, that is, there are 16 possible combinations of positive and negative acknowledgments. For each of these 16 combinations, one PUCCH resource out of four possible resources is selected and upon this resource two bits are transmitted.

Two of the PUCCH candidate resources to select from are derived from the first (E)CCE using the same rule as in absence of carrier aggregation (assuming that the scheduling

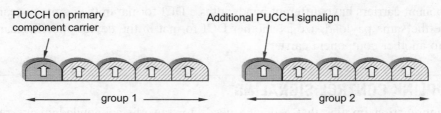

FIGURE 12.13

PUCCH on two uplink carriers in release 13.

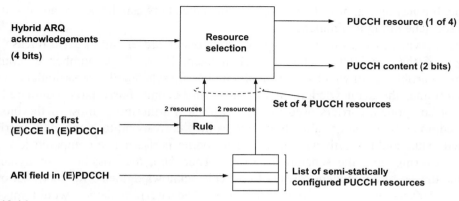

FIGURE 12.14

Example illustration of resource selection using PUCCH format 1 for carrier aggregation.

assignment is transmitted on, and relating to, the primary component carrier). The remaining two candidate resources are obtained by the *acknowledgment resource indicator* (ARI, see Section 6.4.6) on the (E)PDCCH pointing to a list of semi-statically configured PUCCH resources. For transmissions scheduled by the EPDCCH, the *acknowledgment resource offset* (ARO) is also included in the determination of the PUCCH resources. By the setting the ARO appropriately, the scheduler can ensure that multiple devices use noncolliding PUCCH resources. In presence of cross-carrier scheduling, all four resources are semi-statically configured and the ARI is not used.

For more than four bits, resource selection is less efficient and PUCCH formats 3, 4, or 5 are used. In PUCCH format 3, the set of multiple hybrid-ARQ acknowledgment bits (up to 10 or 20 bits for FDD or TDD, respectively, plus one bit reserved for scheduling requests) are jointly coded and transmitted as described in Chapter 7. PUCCH formats 4 and 5 support even larger number of bits. Not all devices support the newer PUCCH formats but for those that do support it, it can be used also for smaller numbers of acknowledgment bits.

The PUCCH resource to use is determined by the ARI. A device can be configured with four different resources for PUCCH format 3, 4, or 5 using RRC signaling. In the scheduling grant for a secondary carrier, the ARI informs the device which of the four resources to use. In this way, the scheduler can avoid PUCCH collisions between different devices by assigning them to different resources.

The set of hybrid-ARQ acknowledgments to transmit need to be coded into a set of bits and transmitted using PUCCH format 3, 4, or 5. Up to and including release 12, the ordering of the acknowledgments for the individual component carriers was determined by the semi-statically configured set of component carriers. Acknowledgment bits corresponding to configured but not scheduled component carriers are set to NAK. This means that there could be bits in a PUCCH message that relate to a component carrier not used in a particular subframe. It is a simple approach and avoids a mismatch between the device and eNodeB due to a missed scheduling assignment.

However, when increasing the number of component carriers up to 32 in release 13, the efficiency of the semi-static scheme is challenged. For a large number of configured component carriers out of which only a small number are scheduled to a particular device in a given subframe, the acknowledgment message can become fairly large compared to the number of component carriers used in the downlink. Furthermore, most of the bits in the acknowledgment message are already known as they correspond to component carriers not scheduled. This could negatively impact the decoding performance compared to a shorter message covering only the scheduled carriers. Therefore, the possibility to dynamically determine the set of component carriers to provide acknowledgments for such that only the scheduled carriers are included was introduced. Clearly, if a device would miss some scheduling assignments, the device and network would have different views on the number of acknowledgments to convey. The DAI, extended to cover also the carrier domain, is used to mitigate this disagreement.

12.6.2 CSI REPORTING ON PUCCH

CSI reports are typically needed for all the downlink component carriers. Since the baseline is to transmit all feedback on the primary component carrier, a mechanism handling multiple CSI reports is needed.

For periodic reporting, the basic principle is to configure the reporting cycles such that the CSI reports for the different component carriers are not transmitted simultaneously on PUCCH. Thus, CSI reports for different component carriers are transmitted in different subframes.

Handling of simultaneous channel-state information and acknowledgments varies a bit across the different releases. In release 10, simultaneous transmission of channel-state information and acknowledgments on PUCCH is not supported—neither for PUCCH format 2, nor for PUCCH format 3—and the CSI report is dropped. Release 11 and later releases, on the other hand, provide support for simultaneous transmission of acknowledgments and periodic CSI reports. The CSI bits are in this case concatenated with the acknowledgment bits and the scheduling request bit and transmitted using PUCCH format 3, 4, or 5. The PUCCH resource to use is determined by the ARI in the downlink control signaling, which selects the resource to use from a set of four resources preconfigured in the device by RRC signaling.

If the number of concatenated information bits is larger than 22 bits for PUCCH format 3 (the maximum number of information bits supported by PUCCH format 3), bundling of the acknowledgment bits are applied prior to concatenation in order to reduce the payload size. In case the payload still is larger than 22 bits, the CSI is dropped and only the acknowledgments (and potentially scheduling request) are transmitted. The process of bundling and dropping in case of a too large payload is illustrated in Figure 12.15. The same principle is used for PUCCH format 4 and 5 although with different limits on the number of bits. Furthermore, it is

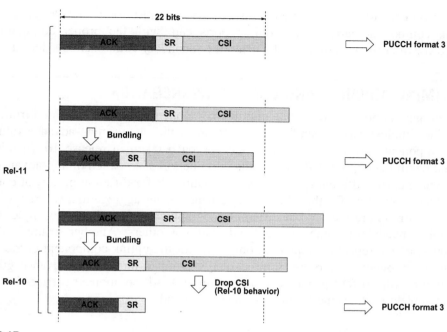

FIGURE 12.15

Multiplexing of acknowledgments, scheduling request, and CSI in release 11.

also possible to configure a device using PUCCH format 4/5 with *two* resources for these formats, one larger than the other. If the small resource results in a too high a code rate, the device uses the larger resource for PUCCH format 4/5.

12.6.3 CONTROL SIGNALING ON PUSCH

For carrier aggregation, control signaling is time multiplexed on one uplink component carrier only—that is, uplink control information on PUSCH cannot be split across multiple uplink component carriers. Apart from the aperiodic CSI reports, which are transmitted upon the component carrier that triggered the report, the primary component carrier is used for uplink control signaling if scheduled in the same subframe, otherwise one of the secondary component carriers is used.

For aperiodic CSI reporting, the CSI request field in the (E)PDCCH is extended to two bits (three bits in release 13), allowing for request of CSI reports for three (seven) combinations of downlink component carriers (the last bit combination represents no CSI request). Of these three (seven) alternatives, one is used to trigger a CSI report for the downlink component carrier associated with the uplink component carrier the scheduling grant relates to. The remaining alternatives point to configurable combinations of component carriers for which the CSI report should be generated. Thus, as an example, for a device capable of two downlink component carriers, aperiodic reports can, with the proper configuration, be requested for the primary component carrier, the secondary component carrier, or both.

12.7 TIMING ADVANCE AND CARRIER AGGREGATION

For carrier aggregation, there may be multiple component carriers transmitted from a single device. The simplest way of handling this is to apply the same timing-advance value for all uplink component carriers. This is also the approach taken in release 10. In release 11, additional flexibility is provided through the introduction of so-called timing-advance groups (TAGs) which allow different timing-advance commands for different groups of component carriers. One motivation for this could be interband carrier aggregation, where the different component carriers are received at different geographical locations, for example, by using remote radio heads for some of the bands but not others. Another example could be frequency-selective repeaters, repeating only some of the uplink component carriers.

Uplink component carriers are semi-statically grouped into timing-advance groups via RRC signaling (up to four groups can be configured). All component carriers in the same group are subject to the same timing-advance command.

MULTI-POINT COORDINATION AND TRANSMISSION

13

The principle of *spatial reuse* lies at the core of any cellular wireless-access system. Spatial reuse implies that the same communication resource, in the LTE case the same time—frequency resource, can be simultaneously used for communication at different, spatially-separated locations.

Inherently, transmissions carried out on the same time—frequency resource will cause interference to each other. To limit this interference, early cellular technologies relied on a static frequency separation between neighboring cells. As an example, in a *reuse-3* deployment the overall set of available frequency resources is divided into three groups. As illustrated in Figure 13.1, only frequencies of one of these groups are then used within a given cell, with frequencies of other groups being used in the most neighboring cells. Interference from transmissions in the most neighbor cells is then avoided and each communication link experiences a relatively high signal-to-interference ratio (SIR) regardless of the device position within the coverage area.

However, for modern wireless-access technologies such as LTE, which should be able to provide very high end-user data rates when the channel conditions so allow, this is not a good approach. Being hard limited to only a fraction of the overall available spectrum at a given transmission point would reduce the maximum achievable transmission bandwidth that can be used at the transmission point with a corresponding reduction in the maximum achievable data rates as a consequence.[1]

Even more important, in a wireless-access system dominated by highly dynamic packet-data traffic there is frequently no data available for transmission at a given transmission point. Having statically assigned a part of the overall available spectrum to that transmission point with no possibility to use the corresponding frequency resources to provide higher instantaneous transmission capacity at neighboring transmission points would imply an inefficient use of the available spectrum. Rather, in order to maximize system efficiency, as well as to enable end-user data rates as high as possible, a wireless-access technology should be deployed such that, fundamentally, all frequency resources are available for use at each transmission point.

---

[1]Instead of "cell" we here use the more general term "(network) transmission point." For a homogeneous deployment, which is the focus of this chapter, one can typically assume that each "transmission point" corresponds to a "cell." For the case of heterogeneous deployments, which is the topic of the next chapter, the distinction will in some cases be more important.

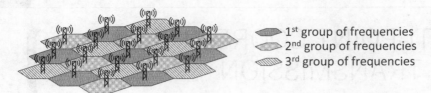

FIGURE 13.1

Reuse-3 deployment.

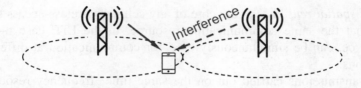

FIGURE 13.2

Downlink interference to device close to the border between two transmission points.

At the same time, for the specific case of transmission to a device close to the border between the coverage areas of two transmission points, see Figure 13.2, end-user quality and overall efficiency would be further improved if interference from the neighboring transmission point could be avoided.

Thus, even if all frequency resources should fundamentally be available for use at each transmission point, coordination across the transmission points can be beneficial. Such coordination could, for example, imply avoiding transmission, or transmitting with lower power or in a different direction (beam-forming), on a certain time–frequency resource in order to reduce the interference to devices served by other, neighboring transmission points if such a device would otherwise experience severe interference.

In certain cases one could even consider using both transmission points for transmission to the same device. This would not only avoid interference from the neighboring transmission point but would also boost the overall signal power available for transmission to the device.

The preceding discussion implicitly assumed downlink transmission with device reception being interfered by downlink transmissions from other network transmission points. However, the concept of coordination between network points as a means to better control the interference levels is applicable also to the uplink, although the interference situation in this case is somewhat different.

For the uplink, the interference level experienced by a certain link does not depend on where the transmitting device is located but rather on the location of the *interfering* devices, with interfering devices closer to the border between two, in this case, network *reception points* causing more interference to the neighboring reception point, see Figure 13.3. Still, the

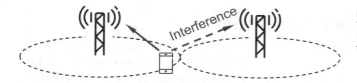

FIGURE 13.3

Uplink interference from device close to the border between two reception points.

fundamental goal of uplink coordination is the same as for the downlink, that is, to avoid the most severe interference situations.

One can envision two main deployment scenarios when considering coordination between network points:

- Coordination within a *homogeneous deployment*, for example between nodes in a macro deployment.
- Coordination within a *heterogeneous deployment*, such as between a macro node and under-laid lower-power nodes.

The focus of this chapter is on the first type of coordination—that is, coordination within homogeneous deployments. Heterogeneous deployments will create additional challenges and corresponding need for coordination between transmission points. This is further discussed in the next chapter as part of a more general discussion on heterogeneous deployments.

One factor that impacts the possibilities for coordination between network points is the available backhaul connectivity and especially its associated latency. Highly dynamic coordination requires low-latency connectivity between the points to be coordinated. One case when this is definitely available is when the points to be coordinated correspond to sectors of the same site ("intrasite coordination"). However very-low-latency connectivity may be available also in the case of geographically separated transmission points, especially if there are direct physical links (for example optical or wireless links) between the points. In other cases, only not-so-low latency internode connectivity, for example connectivity with latency in the order of several tens of milliseconds or more, may be available in which case one is limited to less dynamic coordination.

The 3GPP activities related to coordination between network points for LTE can be divided into two phases:

- Release 8 activities on *inter-cell interference coordination* (ICIC), primarily focusing on inter-eNB (X2) signaling to assist such coordination
- Release 10—13 activities on multi-point coordination/transmission targeting more dynamic coordination and focusing on new radio-interface features and device capabilities to enable/improve such coordination

Furthermore, as is discussed in the next chapter, additional schemes for interpoint coordination have been defined as part of enhancements for heterogeneous deployments.

13.1 INTER-CELL INTERFERENCE COORDINATION

The potential gains of coordination between transmission/reception points were extensively discussed in the early phase of LTE standardization with focus on coordination within homogeneous macro deployments. More specifically, the focus was on defining X2 signaling that could be used to enhance such coordination between cells corresponding to different eNodeB.[2] As the X2 interface is typically associated with not-so-low latency, the focus of the release 8 activities was on relatively slow coordination.

In the case of scheduling located at a higher-level node above the eNodeB, coordination between cells of different eNodeB would, at least conceptually, be straightforward as it could be carried out at the higher-level node. However, in the LTE radio-network architecture there is no higher-level node defined and scheduling is assumed to be carried out at the eNodeB. Thus, the best that can be done from an LTE specification point of view is to introduce messages that convey information about the local scheduling strategy/status between neighboring eNodeBs. An eNodeB can then use the information provided by neighboring eNodeBs as input to its own scheduling process. It is important to understand though that the LTE specifications do not specify how an eNodeB should react to this information. Rather, this is up to scheduler implementation.

To assist uplink interference coordination, two X2 messages were defined as part of LTE release 8, the *high-interference indicator* (HII) and the *overload indicator* (OI), see also Figure 13.4.

The *HII* provides information about the set of resource blocks within which an eNodeB has high sensitivity to interference. Although nothing is explicitly specified on how an eNodeB should react to the HII (or any other ICIC-related X2 signaling) received from a

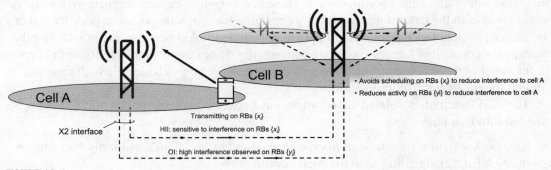

FIGURE 13.4

Illustration of uplink ICIC based on the HII and OI X2 signaling.

---

[2]As the focus was on X2, that is, inter-eNodeB signaling, the transmission/reception points relevant for the coordination would inherently correspond to different cells.

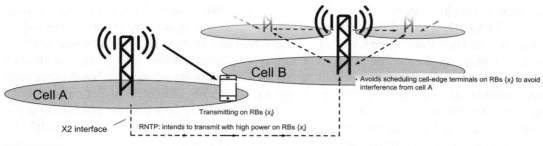

FIGURE 13.5

Illustration of downlink ICIC based on RNTP X2 signaling.

neighboring eNodeB, a reasonable action for the receiving eNodeB would be to try to avoid scheduling its own cell-edge devices on the same resource blocks, thereby reducing the uplink interference to cell-edge transmissions in its own cell as well as in the cell from which the HII was received. The HII can thus be seen as a *proactive* tool for ICIC, trying to prevent the occurrence of *too-low-SIR* situations.

In contrast to the HII, the OI is a *reactive* ICIC tool, essentially indicating, at three levels (low/medium/high), the uplink interference experienced by a cell on its different resource blocks. A neighboring eNodeB receiving the OI could then change its scheduling behavior to improve the interference situation for the eNodeB issuing the OI.

For the downlink, *the relative narrowband transmit power* (RNTP) was defined to support ICIC operation (see Figure 13.5). The RNTP is similar to the HII in the sense that it provides information, for each resource block, whether or not the relative transmit power of that resource block is to exceed a certain level. Similar to the HII, a neighboring cell can use the information provided by the received RNTP when scheduling its own devices, especially devices on the cell edge that are more likely to be interfered by the neighboring cell.

13.2 MULTI-POINT COORDINATION/TRANSMISSION

During the work on LTE release 10, the possibility for more dynamic coordination between network points was discussed under the term *Coordinated Multi Point* (CoMP) transmission/reception. Although initially discussed as part of the 3GPP activities on LTE release 10, the main features related to CoMP were introduced into the LTE specifications as part of release 11 and enhanced in later releases.

A main difference between the LTE release 8 ICIC activities described previously and the release 10/11 CoMP activities is that the latter focused on radio-interface features and device functionality to assist different coordination means. At the same time, there were no discussions on specific inter-eNodeB signaling to support CoMP. Rather, there was an assumption that low-latency backhaul was available for the coordination, in practice limiting release 11 CoMP features to either sectors of the same site or network points connected by direct low-latency links. There was also an implicit assumption that the different network points involved in the coordination were tightly synchronized and time aligned with each other. Extensions to relaxed backhaul scenarios with noncentralized baseband processing were introduced in release 12. These enhancements mainly consisted of defining new X2 messages for exchanging information about so-called *CoMP hypotheses*, essentially a potential resource allocation, and the associated gain/cost. In the same way as for ICIC, the eNodeBs can use this information for scheduling coordination.

The different approaches to CoMP considered for the LTE downlink can be divided into two main groups:

- Schemes where transmission is carried out from a specific transmission point but where the scheduling and link adaptation may be coordinated between transmission points. We refer to this as *multi-point coordination*.
- Schemes where transmission to a device may be carried out from different transmission points (*multi-point transmission*). The transmission can then either switch dynamically between the different transmission points or be carried out jointly from multiple points.

A similar distinction can be made for the uplink transmission direction in which case one would distinguish between (uplink) multi-point coordination, where the uplink scheduling is coordinated between different reception points, and *multi-point reception* where reception may be carried out at multiple points. It should be noted that, at least from a radio-interface point of view, uplink multi-point coordination/reception is very much a network implementation issue with very little impact on the device and very little visibility in the radio-interface specifications. The discussions in Sections 13.2.1 and 13.2.2 focus on coordination in the downlink transmission direction. Section 13.2.3 briefly discusses some aspects of uplink multi-point coordination/reception.

13.2.1 MULTI-POINT COORDINATION

As described earlier, multi-point coordination implies that transmission is carried out from a specific transmission point but functions such as link adaptation and/or scheduling are coordinated between multiple points.

13.2.1.1 Coordinated Link Adaptation

Link adaptation—the dynamic selection of data rate based on estimates/predictions of the instantaneous channel conditions to be experienced by a transmission—is one of the basic mechanisms for good system performance in LTE. Good link adaptation relies on the availability of good predictions of the interference level to be experienced by the transmission. However, in case of highly dynamic traffic conditions, the traffic activity of neighbor transmission points may vary rapidly. As a consequence the interference level may also vary rapidly and in an (apparently) unpredictable way.

Coordinated link adaptation is about using information related to transmission decisions of neighboring transmission points in the link-adaptation process, that is, in the decision with what data rate to transmit on a given resource. Note that this implies a multi-step process in the scheduling and link adaptation at transmission points:

1. For a given subframe, transmission points carry out transmission decisions. In the simplest case this may be decisions on whether or not to transmit data on a certain set of time–frequency resources, that is, a certain set of resource blocks within the subframe. In a more general case it may also include, for example, decisions on transmission power and/or beam-forming decisions for the given set of resources.
2. Information about the transmission decisions is shared between neighboring transmission points.
3. Transmission points use the information about transmission decisions of neighboring transmission points as input to the link-adaption decision for the transmission(s) to take place in the given subframe.

In LTE, link adaptation is carried out at the network side. However, as described in Chapter 9, the network typically bases the link-adaptation decisions on CSI reports provided by the devices. To enable coordinated link adaptation, that is, to allow for the network to take information about the transmission decisions made by neighboring transmissions into account in the rate selection, the device should provide multiple CSI reports corresponding to different *hypothesis* regarding the transmission decisions of neighboring transmission points. These CSI reports can then be used together with information about the actual transmission decisions of neighboring transmission points in the link adaptation.

In order for a device to be able to provide CSI reports corresponding to different hypothesis regarding the transmission decisions of neighboring transmission points, it should be configured with *multiple CSI processes*. As described in Chapter 9, each such process would correspond to a set of CSI-RS, one for each antenna port, and one CSI-IM resource for interference estimation. In order to support coordinated link adaptation, the set of CSI-RS should be the same for all processes and reflect the channel of the different antenna port(s) for the transmission point from which transmission is to be carried out. In contrast, the CSI-IM resources of the different CSI processes should be different and configured in such a way that they reflect the interference to be expected for different hypothesis regarding the transmission decisions of neighboring transmission points.

As an example, Figure 13.6 illustrates the case of coordinated link adaptation between two transmission points. The figure also illustrates three different CSI-RS resources on which there is either transmission, corresponding to ordinary (nonzero-power) CSI-RS, or no transmission, corresponding to zero-power CSI-RS, for the two transmission points.

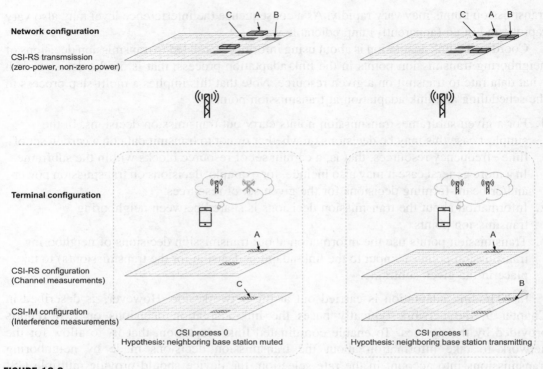

FIGURE 13.6

Example of using multiple CSI processes.

For a device associated with the left side transmission point, two CSI processes are configured:

- Process 0 with CSI-RS corresponding to resource A and CSI-IM corresponding to resource C (configured as zero-power CSI-RS at the neighboring transmission point). CSI-reporting by this CSI process will thus reflect the channel state under the hypothesis that there is no transmission from the neighboring transmission point.
- Process 1 with CSI-RS corresponding to resource A (same as for process 0) and CSI-IM corresponding to resource B (configured as nonzero-power CSI-RS at the neighboring transmission point). CSI reported by this process will thus reflect the channel state under the hypothesis that there is transmission from the neighboring transmission point.

The CSI reports delivered by the device for the different CSI processes would thus correspond to the different hypotheses regarding the transmission decision of the neighboring transmission point. Based on information regarding the expected transmission from the neighboring transmission point, the network can select the appropriate CSI report and use that in the link-adaptation decision.

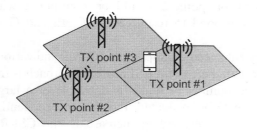

Resource	TX point #1	TX point #2	TX point #3
A	ZP CSI-RS	ZP CSI-RS	ZP CSI-RS
B	ZP CSI-RS	ZP CSI-RS	CSI-RS
C	ZP CSI-RS	CSI-RS	ZP CSI-RS
D	ZP CSI-RS	CSI-RS	CSI-RS
E	CSI-RS	ZP CSI-RS	ZP CSI-RS
F	CSI-RS	ZP CSI-RS	CSI-RS
G	CSI-RS	CSI-RS	ZP CSI-RS

FIGURE 13.7

CSI-RS/IM structure to support coordinated link adaption between three transmission points.

Coordinated link adaptation can also be carried out between more than two transmission points. As an example, consider a case where one would like to carry out coordinated link adaptation between three different transmission points (Figure 13.7). In this case there is a need for a total of seven CSI-RS resources, labeled A to G, configured as nonzero-power and zero-power CSI-RS at the different transmission points according to Figure 13.7.

A device associated with transmission point 1 should, in this case, be configured with four CSI processes, the CSI-IM resources of which would correspond to resources A to D in Figure 13.7. Measurements on these four CSI-IM resources would provide interference predictions that would correspond to different hypotheses regarding the transmission decisions of neighboring points. More specifically:

- Measurements on the CSI-IM resource corresponding to resource A would provide an interference prediction corresponding to the hypothesis that there is transmission from neither transmission point 2 nor transmission point 3 at the time of transmission.
- Measurements on the CSI-IM resource corresponding to resource B would provide an interference prediction corresponding to the hypothesis that there is transmission from transmission point 2 but not from transmission point 3.
- Measurements on the CSI-IM resource corresponding to resource C would provide an interference prediction corresponding to the hypothesis that there is transmission from transmission point 3 but not from transmission point 2.
- Finally, measurements on the CSI-IM resource corresponding to resource D would provide an interference prediction corresponding to the hypothesis that there is transmission from both transmission point 2 and transmission point 3.

Similarly, a device associated with transmission point 2 would be configured with CSI processes, the CSI-IM of which would correspond to resource A, B, E, and F in Figure 13.7. In this case, as an example, measurements on the CSI-IM resource corresponding to resource E would provide an interference prediction corresponding to the hypothesis that there is transmission from transmission point 1 but not from transmission point 3.

Likewise, a device associated with transmission point 3 would be configured with CSI processes, the CSI-IM of which would correspond to resources A, C, E, and G in Figure 13.7.

The support of multiple CSI processes greatly enhanced the support for CoMP and various beamforming schemes. However, one drawback with the current CSI process approach is the lack of scalability. If coordination across a large number of nodes is desirable or a large number of potential beam-forming candidates are to be evaluated, there is an exponential increase in the number of CSI processes with a corresponding increase in the CSI-RS overhead. Therefore, alternative way of reporting CSI for multiple transmission hypotheses may be needed in future LTE releases. Since the device can measure the signal strength from all points in the coordination set, one possibility could be to let the device compute and report the CSI under different hypotheses with the interference level computed from the signal strength measurements.

13.2.1.2 Coordinated Scheduling

Dynamic link adaption as described in the preceding paragraphs is about using information related to the transmission decisions made by neighboring transmission points in the link-adaptation decision, that is, in the selection of transmission rate to be used by a transmission point. Dynamic link adaptation is applicable and useful regardless of whether or not the actual transmission decisions are coordinated between the transmission points.

Coordinated scheduling is about coordinating the actual transmission decision(s) between transmission points. Thus, while coordinated link adaptation is about sharing of information between transmission points for better predictions of the interference levels, coordinated scheduling is about sharing of information and coordination between transmission points to reduce and control the actual interference levels.

In its most simple case, coordinated scheduling is about dynamically preventing transmission at certain time—frequency resource in order to reduce the interference to be experienced by a device served by a neighboring transmission point. In the LTE CoMP discussions, this has been referred to as *dynamic point blanking*. In the more general case it can also involve dynamically adjusting the transmit power (*coordinated power control*) or dynamically adjusting the transmission direction (*coordinated beam-forming*) for a specific set of resources. The enhancements part of FD-MIMO—see Chapter 10—can also be used for even better coordination gains.

In order to enable dynamic point blanking the network should be able to estimate/predict the impact to a device in terms of expected channel quality of transmissions from a neighboring transmission points and also be able to predict how much the channel quality would improve if transmissions from the neighboring transmission point would not take place. To enable this, multiple CSI processes configured in the same way as for coordinated link adaptation, discussed earlier, may be used. The different CSI processes provide different CSI reports reflecting different hypotheses regarding the transmission at the neighboring transmission points. By comparing these CSI reports, the network can estimate how much would be gained from blanking relevant time—frequency resources at a neighboring transmission point.

As an example, consider a device associated with transmission point 1 in the example scenario of Figure 13.7. If there were a large difference in the CQI, that is, the recommended data rate, of the CSI reports corresponding to resource B and D, this would be an indication that the device would be severely interfered by transmissions from transmission point 2 and it would be relevant to consider blanking of relevant time—frequency resources at that transmission point to improve the experienced channel quality and, as a consequence, the data rate that can be supported, for the device associated with transmission point 1.

On the other hand, if there were a very small difference in the CQI of the CSI reports corresponding to resource B and D, this would be an indication that the device is not severely interfered by transmissions from transmission point 2 and, at least in terms of channel quality for this specific device, it is not beneficial to apply blanking to transmission point 2.

13.2.2 MULTI-POINT TRANSMISSION

Multi-point coordination as described previously implies that transmissions carried out from neighboring transmission points are coordinated in terms of scheduling (if/when to transmit) and/or link adaption (with what rate to transmit). However, the transmission to a given device is still assumed to be carried out from one specific transmission point. In contrast, in case of *multi-point transmission*, the transmission to a given device can be carried out from different transmission points, either so that the point of transmission can change dynamically, referred to as *dynamic point selection*, or so that the transmission can be carried out jointly from multiple transmission points, referred to as *joint transmission*.

13.2.2.1 Dynamic Point Selection

As mentioned earlier, dynamic point selection implies transmission from a single transmission point but where the point of transmission can be changed dynamically as illustrated in Figure 13.8.

In the LTE context, dynamic point selection, and actually all CoMP schemes, are assumed to be based on the use of transmission mode 10. Thus, in case of dynamic point selection PDSCH transmission relies on DM-RS for channel estimation. As a consequence, the device does not need to be aware of the change of transmission point. What the device will see is simply a PDSCH transmission, the instantaneous channel of which may change abruptly as

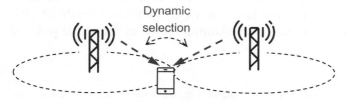

Dynamic
selection

FIGURE 13.8

Dynamic point selection between two transmission points.

the point of transmission is changed. In essence, from a device point of view, the situation would be identical to that of beam-forming based on non-codebook-based precoding.[3]

To assist in the dynamic selection of transmission point, a device should provide CSI reports corresponding to multiple transmission points. Similar to coordinated link adaptation and coordinated scheduling this may be achieved by configuring the device with multiple CSI processes.

As just described, in the case of coordinated link adaptation and coordinated scheduling the different CSI processes should correspond to the same transmission point, that is, the set of CSI-RS should be the same for the different processes. At the same time, the CSI-IM resources should be different for the different processes, allowing for the interference measurements and thus also the CSI reports to reflect different hypothesis regarding the transmission decisions of the neighboring transmission points.

In contrast, to support dynamic point selection the different CSI processes should provide CSI reports corresponding to different transmission points. Thus, the set of CSI-RS of the different processes should be different and correspond to CSI-RS transmitted by the different transmission points between which the dynamic point selection is carried out.

In addition to CSI reporting, the other main specification impact of dynamic point selection relates to PDSCH mapping and what a device can assume in terms of quasi-colocation relation between different reference signals.

In the normal case, PDSCH mapping to the resource blocks assigned for transmission avoids the resource elements used for CRS transmission within the serving cell of the device. However, in case of dynamic point selection the PDSCH transmission to a device may be carried out from a transmission point associated with a different cell than the serving cell. If this cell has a different CRS structure, in terms of number of CRS and/or CRS frequency shift, and the PDSCH mapping remained according to the CRS structure of the serving cell, CRS transmissions from the actual point of transmission would be severely interfered by the PDSCH transmission. Rather, in case of dynamic point selection, one would like the PDSCH mapping to dynamically match the CRS structure of the actual point of transmission.

A similar situation may arise for L1/L2 control signaling. The size of the control region, and thus the starting point for PDSCH transmission, of a certain cell may vary dynamically with information about the size of the control region provided to the device by means of the PCFICH (see Section 6.4). However, in case of dynamic point selection, if the size of the control region of the actual point of transmission differs from that of the serving cell and the PDSCH mapping remained according to the PCFICH of the serving cell, the L1/L2 control signaling of the actual point of transmission would run the risk of being severely interfered by the PDSCH transmission. Thus, similar to the case of CRS one would like the PDSCH mapping to dynamically match the size of the control region of the actual point of transmission.

---

[3]There would be a difference in what the device can assume in terms of quasi-colocation, as further discussed subsequently.

Furthermore, as described in Chapter 6, a device may be configured with a set of zero-power CSI-RS resources. From a device point of view the zero-power CSI-RS simply defines a set of resource elements to which PDSCH is not mapped, typically because these resources elements are used for other purposes, for example, as CSI-RS for other devices or as CSI-IM resources. However, if the PDSCH transmission is carried out from a different point one would typically like the PDSCH mapping to avoid a different set of resource elements as this transmission point would typically use different resource elements for CSI-RS and CSI-IM.

To handle these related problems in a unified way, transmission mode 10 allows for the dynamic reconfiguration of the PDSCH mapping by means of a *PDSCH-mapping-and-quasi-colocation indicator* provided as part of the downlink scheduling assignment, more specifically as part of DCI format 2D as described in Section 6.4.

A device can be provided with up to four different *PDSCH mapping and quasi-colocation configurations*.[4] Each such configuration specifies:

- a specific CRS configuration in terms of number of CRS and CRS frequency shift,
- a specific PDSCH starting point,
- a specific MBSFN subframe configuration,[5]
- a specific zero-power CSI-RS configuration.

The *PDSCH-mapping-and-quasi-colocation* indicator provided in the scheduling assignment then explicitly indicates which one of the up to four different configurations the device should assume for the PDSCH mapping for the corresponding subframe.

It should be noted that the PDSCH starting point indicator does not guarantee perfect match with the size of the control region of the actual point of transmission as the size of the control region can vary dynamically. The PDSCH starting point indication should be set to a sufficiently large value to guarantee that the PDSCH transmission does not overlap with the control region. As the size of the control region never exceeds three OFDM symbols (corresponding to control signaling in OFDM symbol 0, 1, and 2), the most straightforward way of achieving this would be to set the PDSCH starting point to three.[6] However, lower values can be used, allowing for a larger PDSCH payload, if one knows that, in a certain cell, the control region will always be limited to a lower value.

As the name suggests, the *PDSCH-mapping-and-quasi-colocation* configurations and the corresponding indicator in the scheduling assignment also provide information about what the device can assume in terms of quasi-colocation relation between antenna ports. As discussed in Chapter 6, for transmission mode 1 to 9, a device can assume that the antenna ports

---

[4]Note the different between the *PDSCH-mapping-and-quasi-colocation configuration* provided by means of higher layer signaling and the *PDSCH-mapping-and-quasi-colocation indicator* provided by the scheduling assignment.

[5]The MBSFN subframe configuration is related to the CRS configuration as it impacts the presence of CRS reference symbols in the data part of the subframe.

[6]In case of the smallest LTE cell bandwidths, the size of the control region could be up to four OFDM symbols.

corresponding to CRS of the serving cell, DM-RS, and the CSI-RS configured for the device are all jointly quasi-colocated. However, in case of dynamic point selection, a device may be configured with different sets of CSI-RS by means of different CSI processes. In practice these sets of CSI-RS correspond to different transmission points. Eventually, the PDSCH together with its corresponding DM-RS will be transmitted from one of these transmission points. The antenna ports used for the PDSCH transmission will then, in practice, be quasi-colocated with the set of CSI-RS corresponding to that specific transmission point. To provide this information to the device, which is not explicitly aware of from what transmission point the PDSCH transmission takes place, each *PDSCH mapping and quasi-colocation* configuration also indicates a specific set of CSI-RS for which the device can assume quasi-colocation with the DM-RS for the specific subframe.

13.2.2.2 Joint Transmission
While dynamic point selection implies transmission from a single transmission point but where the point of transmission can be changed dynamically, joint transmission implies the possibility for simultaneous transmission from multiple transmission points to the same device (Figure 13.9).

In case of joint transmission one can distinguish between two cases:

- Coherent joint transmission.
- Noncoherent joint transmission.

In case of coherent joint transmission it is assumed that the network has knowledge about the detailed channels to the device from the two or more points involved in the joint transmission and selects transmission weights accordingly, for example, to focus the energy at the position of the device. Thus, coherent joint transmission can be seen as a kind of beam-forming for which the antennas taking part in the beam-forming are not colocated but correspond to different transmission points.

There is currently no support in the LTE specifications for the device to report this kind of detailed channel knowledge for multiple transmission points and thus currently no explicit support for coherent joint transmission.

In contrast, for *noncoherent* joint transmission it is assumed that the network does not make use of any such detailed channel knowledge in the joint transmission. Thus, the only gain of noncoherent joint transmission is that the power of multiple transmission points is used for transmission to the same device, that is, in practice, a power gain. The benefit of this depends on whether or not the power of the second transmission point can be of better use for the transmission to other devices and also to what extent the extra transmission will cause

FIGURE 13.9

Joint transmission from two transmission points to the same device.

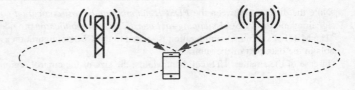

harmful interference to other transmissions. In practice, noncoherent joint transmission is only beneficial at low-load situations,

- where there is no other device available for which the second transmission point can be used,
- where the additional interference from the second transmission does not really cause any harm.

It should be noted that, in case of joint transmission, PDSCH may be jointly transmitted from points corresponding to two cells with, for example, different CRS configurations or different MBSFN configurations. In such a case one would like for the PDSCH mapping to match the configurations of both cells. However, currently each *PDSCH-mapping-and-quasi-colocation* configuration only corresponds to a single CRS configuration.

13.2.3 UPLINK MULTI-POINT COORDINATION/RECEPTION

The previous sections, Sections 13.2.1 and 13.2.2, focused on *downlink* multi-point coordination/transmission. However, as already mentioned, the same basic principles are also applicable for transmissions in the uplink direction (*uplink CoMP*). More specifically:

- Dynamic coordination of uplink transmissions in order to control uplink interference and achieve improved uplink system performance (uplink multi-point *coordination*).
- Reception of uplink transmissions at multiple points (uplink multi-point *reception* or uplink *joint reception*).

However, in contrast to downlink multi-point coordination/transmission, uplink multi-point coordination/reception has very little impact on the radio-interface specifications. Especially, any channel-state information needed for uplink scheduling coordination would be directly derived at the network (reception) side and would not require any specific device feedback.

Also, a device does not need to be aware at what point its uplink transmission is received as long as any downlink transmission corresponding to the uplink transmission, such as Hybrid ARQ feedback) is, for the device, transmitted in an expected way. In practice this means that even if the uplink is received at a point different from the transmission point associated with the serving cell, feedback such as Hybrid ARQ acknowledgments must still be transmitted from the transmission point of the serving cells. This would require the availability of low-latency connectivity between the reception and transmission points to ensure, for example, that the Hybrid-ARQ timing relations are retained. As already mentioned, the 3GPP release 10/11 CoMP discussions focused on the situation of low-latency connectivity between transmission/reception points.

Some aspects of the radio-interface design have taken the possibility for multi-point reception into account. For example, uplink multi-point reception was the main reason for introducing support for device-specific assignment of uplink reference-signal sequences as described in Section 7.2.

HETEROGENEOUS DEPLOYMENTS

14

The continuous increase in traffic within mobile-broadband systems and an equally continuous increase in terms of the data rates requested by end users will impact how cellular
networks are deployed in the future. In general, providing very high system capacity (traffic
volume per area unit) and very high per-user data rates will require a densification of the
radio-access network, that is, the deployment of additional network nodes (or transmission/
reception points). By increasing the number of nodes, the traffic per area unit can be increased
without requiring a corresponding increase in the traffic that needs to be supported per
network node. Also, by increasing the number of network nodes, the base-station-to-device
distances will, in general, be shorter, implying a link-budget improvement and a corresponding improvement in achievable data rates.

As illustrated at the top of Figure 14.1, uniform densification of the macro cell layer, that
is, reducing the coverage area of each cell and increasing the total number of macro-cell sites,[1]
is a path that has already been taken by many operators. As an example, in many major cities
the distance between macro-cell sites is less than a few hundred meters in many cases.

An alternative or complement to a uniform densification of the macro layer is to deploy
additional lower-power nodes, or "small cells", under the coverage area of the macro layer, as
illustrated at the bottom of Figure 14.1. In such a *heterogeneous deployment*, the low-power
nodes[2] provide very high traffic capacity and improved service experience (higher end-user
throughput) locally—for example, indoor and outdoor hot-spot positions—while the macro
layer provides full-area coverage. Thus, the layer with low-power nodes (pico nodes) can also
be referred to as providing *local-area access*, in contrast to the wide-area-covering macro
layer.

The idea of heterogeneous, or multilayer, deployments is in itself not new; "hierarchical
cell structures" have been used since mid-1990s. However, with extensive use of mobile
broadband, the interest in heterogeneous deployments as a means to increase capacity and
end-user data rates has increased significantly.

It is important to point out that the use of low-power nodes as a complement to a macro
network is a deployment strategy, not a technology component, and as such is already
possible in the first release of LTE. Nevertheless, LTE releases 10 and 11 provide additional

---

[1]Here, a macro node is defined as a high-power node with its antennas typically located above rooftop level.
[2]The term "pico node" is used to denote a low-power node, typically with the antennas located below rooftop level.

4G, LTE-Advanced Pro and The Road to 5G. http://dx.doi.org/10.1016/B978-0-12-804575-6.00014-5

Densification with additional macro nodes (homogeneous deployment)

Densification with *complementary* low-power nodes (heterogeneous deployment)

FIGURE 14.1

Homogeneous versus heterogeneous densification.

features improving the support for heterogeneous deployments, in particular in the area of handling interlayer interference.

14.1 INTERFERENCE SCENARIOS IN HETEROGENEOUS DEPLOYMENTS

One distinctive property of a heterogeneous deployment is the large difference in transmit power between the overlaid macro layer and the underlaid pico layer. Depending on the scenario, this may result in significantly more complex interference scenarios compared to a homogeneous network, more specifically interference between the layers. *Interlayer interference handling* is therefore a crucial aspect in most heterogeneous deployments.

If different frequency resources, in particular different frequency bands, are used for different layers, the interlayer interference can be avoided. Frequency separation is also the traditional way of handling interlayer interference in, for example, GSM where different carrier frequencies are used in the different layers. However, for a wideband radio-access technology such as LTE, using different carrier frequencies for different layers may lead to an undesirable spectrum fragmentation as discussed in the previous chapter. As an example, for an operator having access to 20 MHz of spectrum, a static frequency separation between two layers would imply that the total available spectrum had to be divided, with less than 20 MHz of spectrum being available in each layer. This could reduce the maximum achievable data rates within each layer. Also, assigning a substantial part of the overall available spectrum to a layer during periods of relatively low traffic may lead to inefficient spectrum utilization. Thus, with a wideband high-data-rate system such as LTE, it should preferably be possible to deploy a multilayered network with the same spectrum being available in all layers. This is in line with the motivation for single-frequency reuse in Chapter 13. Nevertheless, separate spectrum allocation for the two layers is a relevant scenario, especially if new spectrum becomes available at very high frequencies, less suitable for wide-area coverage. Furthermore, frequency-separated deployments imply that the duplex scheme can be chosen independently between the layers, for example, using FDD in the wide-area macro layer and TDD in the local-area pico layer.

Simultaneous use of the same spectrum in different layers implies interlayer interference. The characteristics of the interlayer interference depend on the transmission power in the respective layer, as well as in the node-association strategy used.

Traditionally, node association or cell association, that is, determining which network point the device should be connected to, is based on device measurements of the received power of some downlink signal—more specifically, the cell-specific reference signals in the case of LTE. Based on the device reporting those measurements to the network, the network decides whether a handover should take place or not. This is a simple and robust approach. In homogeneous deployments with all transmission points having the same power, downlink measurements reflect the uplink path loss and downlink-optimized network-point association is also reasonable from an uplink perspective. However, in a heterogeneous deployment, this approach can be challenged due to the large difference in transmission power between the layers. In principle, the best uplink reception point is not necessarily the best downlink transmission point, implying that uplink and downlink points ideally should be determined separately [83]. For example, downlink point selection could be based on received signal strength, while uplink point selection preferably is based on the lowest path loss. This is illustrated in Figure 14.2 where the "cell border" is

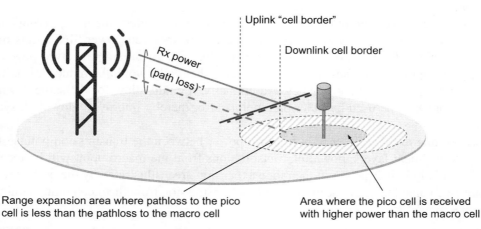

FIGURE 14.2

Illustration of high-interference area in a heterogeneous network deployment with range expansion.

different for uplink and downlink. However, since there are tight and time-critical dependencies between the uplink and downlink—for example, in the form of hybrid-ARQ acknowledgments transmitted in the downlink as a result of receiving uplink transmissions—the two links are in practice terminated in the same node.[3] Node association is therefore a compromise between the best choice from an uplink and downlink perspective.

From a single link perspective, associating the device with the transmission point with the highest received power implies that the device is often connected to a higher-power macro point even if the path loss to a pico point is significantly smaller. This will not be optimal from an uplink coverage and capacity point of view. It should also be noted that, even in terms of downlink system efficiency, it may not be optimal to select the transmission point with the highest received power in a heterogeneous network deployment. Although transmissions from the macro point are received with higher power than from a pico point, this is at least partly due to the higher transmit power of the macro point. In that case, transmission from the macro point is associated with a higher "cost" in terms of interference to other cells. Expressed alternatively, a transmission from the macro point will prohibit the use of the same physical resource in *any* of the underlaid pico points.

Alternatively, at the other extreme, node association could be based solely on estimates of the (uplink) path loss. In practice this can be achieved by applying an offset to the received-power measurements used in conventional cell association, an offset that would compensate for the difference in transmit power between different transmission points. Such an offset is supported by LTE already from the first release and possible to configure on a per-device basis. By using an offset in the node-association strategy, the area in which a pico point is selected is extended as illustrated in Figure 14.2. This is sometimes referred to as *range expansion*.

Selecting the network point to which the path loss is the smallest, that is, applying a large degree of range expansion, would maximize the uplink received power/SINR, thus maximizing the achievable uplink data rates. Alternatively, for a given target received power, the device transmit power, and thus the interference to other cells, would be reduced, leading to higher overall uplink system efficiency. Additionally, it could allow for the same downlink physical resource to be used by other pico points also, thereby improving downlink system efficiency.

However, due to the difference in transmit power between the transmission points of the different deployment layers, downlink transmissions from the macro point will be received with substantially higher power in the range expansion area (illustrated by the dashed region in Figure 14.2) than the actual desired downlink transmission from the pico point. Within this

---

[3]The downlink transmission point and the uplink reception point could be geographically separated if remote antennas are used, see further Section 14.5.

area, there is thus potential for severe downlink inter-cell interference from the macro point to devices receiving transmissions from a pico point. The interference has both a static load-independent component stemming from the cell-specific reference signals (CRS), synchronization signals (PSS, SSS) and system information (PBCH), and a dynamic load-dependent component stemming from data transmissions (PDSCH) and control signaling (PCFICH, PHICH, PDCCH, and EPDCCH).

The interference from PDSCH transmissions in the macro layer to lower-power PDSCH transmissions from a pico point can be relatively straightforwardly handled by scheduling coordination between the nodes according to the same principles as inter-cell interference coordination described in Section 13.1. As an example, an overlaid macro point could simply avoid high-power PDSCH transmission in resource blocks in which a device in the range expansion region of a pico point is to receive downlink data transmission. Such coordination can be more or less dynamic depending on to what extent and on what time scale the overlaid and underlaid nodes can be coordinated. The same coordination as for the PDSCH could also be applied to the EPDCCH. It should also be noted that, for a pico-network point located on the border between two macro cells, it may be necessary to coordinate scheduling between the pico point and both macro cells.

Less obvious is how to handle interference due to the macro-node transmissions that cannot use dynamic inter-cell interference coordination, such as the L1/L2 control signaling (PDCCH, PCFICH, and PHICH). Within a layer, for example, between two macro nodes, interference between such transmissions is not a critical issue as LTE, including its control channels, has been designed to allow for one-cell frequency reuse and a corresponding SIR down to and even below −5 dB. This inherent robustness allows for moderate range expansion, in the order of a couple of dB. In many scenarios this amount of range expansion is adequate and further increasing it would not improve performance, while in other scenarios a larger amount of range expansion may be useful. Using a large amount of range expansion can result in a signal-to-interference ratio that is too low for the control channels to operate correctly, calling for means to handle this interference situation. Note that the usefulness of range expansion is highly scenario dependent. One simplistic example where range expansion may not be useful is illustrated in the right part of in Figure 14.3 where the building walls provide isolation between the two cell layers.

In the following sections, four different approaches to heterogeneous deployments are described:

- Release 8 functionality, using features available in the first release of LTE to support a medium amount or range expansion. No inter-cell time synchronization or coordination is assumed.
- Frequency-domain partitioning, where an extensive amount of range expansion is supported through interference handling in the frequency domain, for example, by using carrier aggregation.

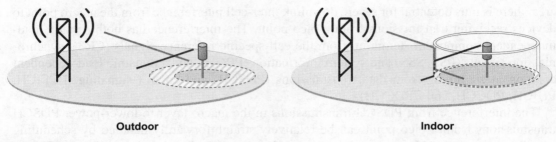

FIGURE 14.3

Illustration of range expansion in different scenarios.

- Time-domain partitioning, where an extensive amount of range expansion is supported through interference handling in the time domain.
- So called "Shared cell", using CoMP techniques from Chapter 13 to support a large amount of range expansion.

In the first three of these approaches, each transmission point defines a unique cell, that is, it has a unique cell identity and transmits all signals associated with a cell such as cell-specific reference signals and system information. The last approach differs in this respect as a transmission point does not necessarily define a unique cell. Instead, *multiple* geographically separated transmission points may belong to the same cell.

Finally, note that all schemes except the first assumes interlayer coordination and time-synchronization across (neighboring) transmission points.

14.2 HETEROGENEOUS DEPLOYMENTS USING REL-8 FUNCTIONALITY

Heterogeneous deployments are already possible from the first release of LTE using release 8 functionality. In this case, the transmission points define unique cells and point association, or cell selection, is typically based on the downlink received power as in the homogeneous case. Despite being simple—for example, there is no need for inter-cell time synchronization or inter-cell coordination—a fair amount of range expansion, up to several dBs, is easily achieved in this scheme by adjusting the cell selection offset. The amount of macro interference naturally limits the amount of range expansion possible, but the amount of range expansion possible is sufficient for many scenarios. Additional tools available in release 8 to obtain a fair amount of range expansion include PDCCH power boosting in the pico cell, fractional loading of the PDCCH in the overlaid macro cell to reduce interference, and adjusting the PDCCH operating point in terms of PDCCH error probability.

It is also worth pointing out that, in many scenarios, most of the gains are obtained by simply deploying the pico nodes with no or only a small amount of range expansion. However, in some specific scenarios, a larger amount of range expansion may be useful, calling for some of the schemes discussed later.

14.3 FREQUENCY-DOMAIN PARTITIONING

Frequency-domain partitioning attempts to reduce interference by using different parts of the frequency spectrum for different layers. The transmission points define unique cells and measurements of the received downlink power are used as the basis of point association or, equivalently, cell selection.

The simplest case is a static split, using different and nonoverlapping pieces of spectrum in the macro and pico layers as illustrated in Figure 14.4. Although simple, such a scheme suffers from not being able to dynamically reallocate resources between the layers to follow instantaneous traffic variations.

A more dynamic approach for handling the downlink interlayer interference in the range-expansion zone in case of a large amount of range expansion is to use carrier aggregation in combination with cross-carrier scheduling as outlined in Chapter 10. The basic idea is to split the overall spectrum into two parts through the use of two downlink carriers, f_1 and f_2, as illustrated in Figure 14.4 but without the loss of flexibility resulting from a static split.

In terms of data (PDSCH) transmission, both carriers are available in both layers and interference between the layers is handled by "conventional" inter-cell interference coordination (see Section 13.1). As already mentioned, such interference coordination can be more

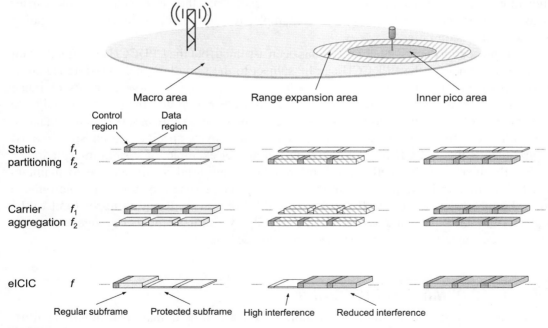

FIGURE 14.4

Frequency-domain and time-domain partitioning.

or less dynamic depending on the time scale on which the layers can be coordinated. Also, the possibility for carrier aggregation allows for both carriers, that is, the total available spectrum, to be assigned for transmission to a single device. Thus, at least for carrier-aggregation-capable devices, there is no spectrum fragmentation in terms of data (PDSCH) transmission. Legacy devices, on the other hand, will experience peak rates from a single carrier only. This may be an issue for an operator with a large fleet of legacy devices.

On the other hand, in terms of L1/L2 control signaling (PCFICH, PHICH, and PDCCH), there is at least partly a more semi-static frequency separation between the layers. More specifically, the macro layer should avoid high-power transmission within the control region on carrier f_1. Assuming a time-synchronized network, interference to the control region of the pico layer is reduced on this carrier and the pico cells can use the carrier for control signaling to devices in the range expansion region. Due to the possibility for cross-carrier scheduling, DL-SCH transmission can still be scheduled on both carriers, as well as an aggregation of these, subject to dynamic interlayer interference coordination even if the macro cell only transmits control signaling on carrier f_1. The same is true for a pico cell; even if the pico cell can only use carrier f_2 for transmission of scheduling assignments to devices in the range-expansion zone, DL-SCH transmissions can still be scheduled on both carriers.

It should be noted that, for devices in the inner part of a pico cell, carrier f_1 could also be used for L1/L2 control signaling. Similarly, macro cells could use also carrier f_2 for control signaling, assuming a reduced transmit power is used. Thus, the macro cell could use carrier f_2 for lower-power control signaling, for example, for devices close to the corresponding macro-transmission point.

In the preceding discussion, PDCCH has been assumed, but the EPDCCH can equally well be used. In principle, the EPDCCH can be subject to the same inter-cell interference coordination scheme as the PDSCH. This could be used to support large amounts of range expansion without carrier aggregation. In this case, the macro and pico layers simply use different sets of physical resource-block pairs, coordinated in a more or less dynamic manner. However, note that the EPDCCH is limited to the device-specific search spaces only. Consequently, the PDCCH is needed for scheduling, for example, of system information.

Finally, note that the (cell-specific reference) signals used by the device to maintain synchronization with the pico cell in the range expansion zone are subject to interference from the macro layer. How well the device can handle this will put an upper limit to the amount of range expansion possible. Hence, to fully exploit the benefits of range expansion, interference cancelling receivers are needed in the devices.

14.4 TIME-DOMAIN PARTITIONING

An alternative to frequency-domain partitioning is to use time-domain partitioning as illustrated at the bottom of Figure 14.4. In 3GPP, this is known as *(further) enhanced inter-cell interference coordination*, (F)eICIC. Work on eICIC started in release 10 and was finalized in release 11 under the name of FeICIC. Also in this case the transmission points correspond to separate cells, hence the FeICIC name used in 3GPP.

The basic idea with time-domain partitioning is to restrict the transmission power of the overlaid macro cell in some subframes. In these *reduced-power subframes* (or *almost blank subframes*), devices connected to the pico cell will experience less interference from the overlaid macro cell for both data and control. From a device perspective they serve as *protected subframes*. The pico cell can therefore schedule devices in the range expansion area using the protected subframes and devices in the inner part of the pico cell using all subframes. The macro cell, on the other hand, primarily schedules devices outside the protected subframes (see Figure 14.5 for an illustration). The gain from deploying the pico cells must be larger than the loss incurred by the macro cell reducing power in some subframes for the time-domain partitioning scheme to be attractive. Whether this holds or not is highly scenario dependent although using a reduced but nonzero transmission power in the macro cell for the protected subframes to limit the resource loss in the macro layer is often advantageous.

To support time-domain partitioning in a heterogeneous network, signaling of *protected-subframe patterns*, that is, information about the set of protected subframes, is supported between eNodeBs of different layers using the X2 interface. Note that the set of protected subframes could be different in different cells and more or less dynamic, once again depending on the time scale on which the deployment layers can be coordinated.

It should be noted that the macro cell must not necessarily avoid control-signaling transmission completely in the protected subframes. In particular, it could be beneficial to retain the possibility for a limited amount of control signaling related to uplink transmissions, for example, a limited amount of uplink scheduling grants and/or PHICH transmission, in order to not cause too much impact on the uplink scheduling. As long as the macro-cell control-signaling transmissions are limited and only occupy a small fraction of the overall control region, the interference to devices in the range expansion region of the pico cell could be kept at an acceptable level. However, the signaling of protected-subframe patterns is also

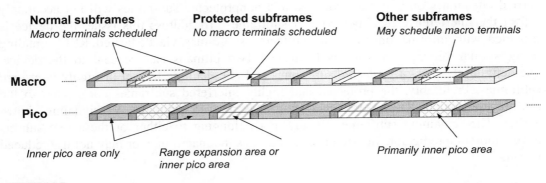

FIGURE 14.5

Protected subframes.

defined so that impact on the uplink scheduling is minimized even if no uplink scheduling grants and PHICH can be transmitted in protected subframes. This is achieved by having the protected-subframe patterns matched to the eight-subframe timing of the uplink hybrid-ARQ protocol. It should be noted that this implies that the pattern is not aligned to the 10 ms frame but only to a 40-ms four-frame structure for FDD. For TDD the periodicity also depends on the uplink—downlink configuration.

Up to four different patterns can be exchanged between eNodeBs: two patterns for scheduling and CSI measurement purposes, one pattern for RRM measurements in the serving cell, and one pattern for RRM measurements in neighboring cells. The purpose of having two patterns for CSI measurements is to handle load variations between macro- and pico layers without frequent reconfigurations of the devices as is further discussed later (see Figure 14.6 for an illustration). In case of a high load in the range expansion area, which is the case at time t_1 in the figure, it may be beneficial to have a relatively large number of protected subframes to allow these devices to be served by the pico cell. At a later time instant, t_2 in the figure, the majority of the devices have moved from the range expansion area into the macro area, calling for a reduction of the number of protected subframes. Thus, by varying the size of the set of currently protected subframes, the configurations can be adjusted to match changes in the scenario. At the same time, there must be a set of subframes that are always protected in order for the pico cell to be able to contact a device in the range expansion area as the connection to a device otherwise might be lost. Therefore, exchange of two subframe patterns are supported on X2, one which is intended to be used for the *currently* protected subframes and one for the *always* protected subframes. The intention is to use the former for scheduling coordination across the cells, allowing for relatively frequent updates, while the latter is updated infrequently and used as the basis for configuring protected subframes in the devices as described later. Subframes belonging to neither of the two sets earlier can be thought of as "never protected" subframes.

Clearly, the interference experienced by devices connected to the pico cell may vary significantly between protected and nonprotected subframes. Thus, CSI measurements carried out jointly on both the protected and nonprotected subframes will not accurately reflect the interference of either type of subframes. To address this issue, the device is provided with information about the protected subframes via dedicated RRC signaling using similar bitmaps as described earlier.[4] Two bitmaps can be sent to the device, one defining the sets of *protected* subframes and one defining the set of *highly interfered* subframes. Preferably, the protected and highly interfered subframes correspond to the always-protected and never-protected subframes derived from the X2 signaling shown earlier. The remaining subframes, if any, not belonging to either of these two subsets have a more unpredictable interference situation as the macro may or may not use reduced power.

---

[4]Note that transmission mode 10 with multiple CSI processes can be used as an alternative to the two bitmaps.

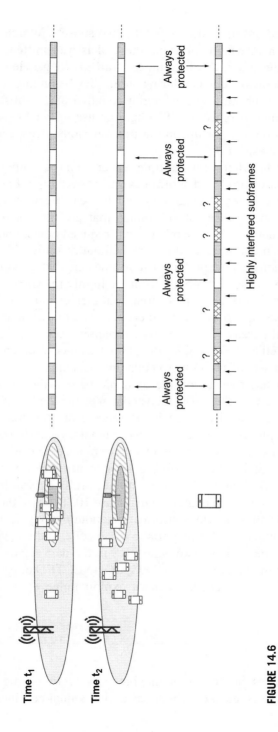

FIGURE 14.6

Exchange of subframe patterns between macro and pico nodes and the corresponding configuration in the device.

CSI reporting is carried out individually for the two subsets. Which subset a certain CSI report reflects depends on in which subframe the CSI is transmitted; the CSI reflects the interference situation in the subset to which the subframe belongs. Hence, the device should only average interference measurements during subframes belonging to the same subset. A CSI report transmitted in a subframe not belonging to either of the subsets is undefined from an interference measurement perspective. Through the use of two subsets, the network can predict the radio-channel quality for upcoming transmissions, irrespective of whether they occur in protected subframes or not.

Having two subsets is beneficial for multiple reasons. One example is the situation outlined above, where the set of protected subframes varies over time. Frequent updates of the configurations in all the affected devices may not be feasible with reasonable overhead. Instead, measuring CSI on a subset of subframes that are always protected is typically preferred as it allows the network to dynamically use reduced power and schedule devices in the range expansion zone in additional subframes without reconfiguring all the devices. CSI reports reflecting the situation in the protected subframes are then used for link adaptation in the range expansion zone, while CSI reports from the highly interfered subframes are useful when scheduling devices in the inner pico area. Another example is when a pico cell is located at the border between, and subject to interference from, two macro cells. If the macro cells have differently configured and only partly overlapping sets of protected subframes, the pico-cell scheduling as well as the configuration of the CSI measurement sets need to take the structure of the protected sets of both macro cells into account.

The discussion so far has focused on the dynamic part of the interference, that is, the interference part that varies with the traffic load and can be handled by ICIC and time-domain partitioning. However, there is also static interference from the macro cell in the range expansion area. For example, cell-specific reference signals, synchronization signals, and the PBCH still need to be transmitted. To support extensive range expansion, despite the presence of these signals and channels, the interfering signals need to be cancelled. Therefore, cancellation of CRS, PSS/SSS, and PBCH is required to fully exploit the features described previously—functionality that is not mandated in release 10. To assist the device in cancelling the interference, RRC signaling provides information about the physical-layer identity of the neighboring cells, the number of antenna ports in those cells, and the MBSFN configuration (MBSFN configuration is needed as there is no CRS in the data region in those subframes).

Time-domain partitioning is supported in RRC_CONNECTED only. Idle devices will still be able to connect to the pico cell, but cannot exploit range expansion until they enter RRC_CONNECTED.

14.5 SHARED CELL

In the partitioning schemes described in the previous sections, the transmission points correspond to individual cells, each of which has an individual cell identity that is different

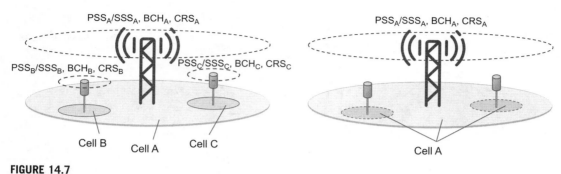

FIGURE 14.7

Independent cells (left) versus shared cell (right).

from neighboring cells in either of the network layers. Consequently, each pico node transmits unique system information, synchronization signals, and cell-specific reference signals. This is illustrated on the left in Figure 14.7.

Alternatively, CoMP techniques (see Chapter 13) can be used to realize heterogeneous deployments. To understand this approach, remember the distinction between a *cell* and *transmission point*. A cell has a unique physical-layer cell identity from which the position of the cell-specific reference signals is derived. By acquiring the cell identity, a device can determine the CRS structure of the cell and receive system information necessary to access the network. A transmission point, on the other hand, is in this context simply one or more colocated antennas from which a device can receive data transmissions.

By exploiting the DM-RS introduced in release 10, the PDSCH does not have to be transmitted from the same point as the cell-specific reference signals. Data can instead be transmitted from one of the pico transmission points when beneficial and the time–frequency resources can be reused between spatially separated pico transmission points. Since the pico-transmission points neither transmit (unique) cell-specific reference signals, nor system information, they do not define cells but are part of the overlaid macro cell. This CoMP approach to heterogeneous deployments is therefore commonly referred to as *shared-cell ID*, illustrated on the right in Figure 14.7.

In Figure 14.8, data is transmitted to device number two from the rightmost transmission point. Since the associated DM-RS is transmitted from the same transmission point as the data, the point used for data transmission does not need to be known by the device. Spatial reuse gains, that is, reusing the time–frequency resources used for the data transmission across multiple pico nodes within the same macro cell, are hence obtained similarly to the resource-partitioning schemes described in the previous sections.

The control information required in release 10 is based on CRS and the control information therefore needs to be transmitted from (at least) the macro site as is the case for the first device in Figure 14.8. Thus, in many cases, data and the associated control signaling originates from *different* transmission points. This is in theory transparent to the device; it

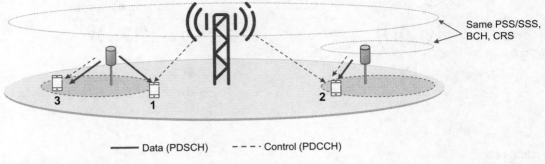

Same PSS/SSS, BCH, CRS

———— Data (PDSCH) – – – – Control (PDCCH)

FIGURE 14.8

Heterogeneous deployment using shared cell.

only needs to know which reference signal to use for which piece of information but not from which transmission point the information originate. The quasi-colocation mechanism introduced in release 11 (see Chapter 5) is preferably used to ensure that the device only exploits the relevant reference signals.

In Figure 14.8, multiple ways of transmitting the control information are illustrated. The first case, involving device 1 with control signaling originating from the macro site only, has already been described. Alternatively, identical CRS and control channels can be transmitted from the macro *and* the pico node as shown for device 2. To the device this will appear as one composite node as the *same* signal is transmitted from both nodes. The first case is beneficial from a network power-consumption perspective as the pico-transmission point is active only when there is data to transmit. The second case, on the other hand, provides an improvement in the signal-to-noise ratio for the control signaling via over-the-air combining of the macro and pico transmissions. Furthermore, as an LTE device estimates the uplink path loss for power control purposes from the received CRS signal strength (see Chapter 7 for a discussion on uplink power control), the second case can sometimes result in a more accurate uplink power control.

In both these cases there is no spatial-reuse gain for the control signaling as the macro site is involved in all these transmissions and time—frequency resources cannot be reused across pico nodes. This can be addressed by using the DM-RS-based EPDCCH for control signaling and transmitting it from the same node as used for the PDSCH as is the case of the third device in Figure 14.8.

Devices not supporting DM-RS-based transmission can still operate in the shared-cell scheme. Data transmissions to these devices are CRS-based and thus handled in the same manner as the PDCCH control signaling described previously. Although no spatial reuse gain will be possible for these devices, they will benefit from the pico nodes through an improved signal-to-noise ratio.

Channel-state feedback used for scheduling decisions is preferably based on the CSI-RS. Different pico nodes, as well as the macro node, can be configured to use different and noninterfering CSI-RS configurations to allow devices to estimate the channel conditions to the transmission points corresponding to the different nodes. For devices not supporting CSI-RS, the channel-state feedback is based on the CRS. In these cases, the eNodeB may need to scale the received reports to account for the difference in the set of transmission points used for CRS and PDSCH.

Deploying a shared-cell scheme can be done by connecting one or multiple remote radio units (RRUs) as well as the macro site to the same main unit using optical fibers. One reason for this is the tight coupling between the macro and pico nodes with control and data originating from different transmission points, calling for low-latency connections.

Centralization of the processing provides benefits also in terms of uplink performance and, in many cases, this alone can motivate the usage of RRUs with centralized processing. Any combination of transmission points, not necessarily those used for downlink transmission to a device, can be used for receiving the transmissions from that device. By combining the signals from the different antennas in a constructive manner at the central processing node—in essence, uplink CoMP (see Chapter 13)—a significant improvement in the uplink data rates can be achieved. In essence, uplink reception and downlink transmission have been *decoupled* and it is possible to do "uplink range expansion" without causing a downlink interference problem as in separate-cell-ID deployments. The uplink gains can be achieved also for release 8 devices.

Heterogeneous deployments using shared cells can also provide additional mobility robustness compared to deployments with separate cells. This can be an important aspect, especially when moving from a pico node to the macro node. In a separate-cell deployment, a handover procedure is required to change the serving cell. If, during the time it takes to perform the handover procedure, the device has moved too far into the macro area, it may drop the downlink connection from the pico node before the handover is complete, leading to a radio-link failure. In a shared-cell deployment, on the other hand, the transmission point to use for downlink transmission can be rapidly changed *without a handover procedure*. Thus, the probability of dropping connections is reduced.

14.6 CLOSED SUBSCRIBER GROUPS

The discussion in the previous sections assumes that the devices are allowed to connect to the low-power pico node. This is known as *open access* and typically the low-power nodes are operator-deployed in such a scenario. Another scenario, giving rise to a similar interference problem, is user-deployed home base stations. The term *closed subscriber groups* (CSG) is commonly used to refer to cases when access to such low-power base stations is limited to a small set of devices, for example the family living in the house where the home base station is located. CSG results in additional interference scenarios. For example, a device located close to but not permitted to connect to the home base station will be subject to strong interference and may not be able to access the macro cell. In essence, the presence of a home base station may cause a coverage hole in the operator's macro network, a problem that is particularly worrisome as the home base stations typically are user deployed and their locations are not controlled by the operator. Similarly, reception at the home base station may be severely impacted by uplink transmissions from the device connected to the macro cell. In principle, these problems can to some extent be solved by the same means as mentioned earlier—by relying on interference coordination between the scheduling in the home-eNodeB layer and an overlaid macro. However, note that the interference avoidance must be two-way, that is, one must not only avoid interference from the macro cell to home-eNodeB devices in the high-interference outer region of the home-eNodeB coverage area, but also home-eNodeB interference to devices close to the home-eNodeB but not being part of the home-eNodeB CSG. Furthermore, most interference coordination scheme assumes the presence of an X2 interface between the macro and home eNodeB, which may not always hold. Therefore, if closed subscriber groups are supported, it is preferable to use a separate carrier for the CSG cells to maintain the overall performance of the radio-access network. Interference handling between CSG cells, which typically lack backhaul-based coordination schemes, could rely on distributed algorithms for power control and/or resource partitioning between the cells.

SMALL-CELL ENHANCEMENTS AND DYNAMIC TDD

15

Low-power nodes, or small cells, have attracted a lot of interest, the reason being the need for a dense deployment in order to provide very high data rates and high capacity. LTE is already from its first release capable of providing high performance in a wide range of scenarios, including both wide-area and local-area access. However, with most users being stationary or slowly moving, there is an increasing focus on high data rates in (quasi-)stationary situations. A number of features have been introduced to the LTE standard aiming at improving the support for low-power nodes in release 12. In this chapter, two of these features, namely small-cell on/off and dynamic TDD, are described. Before describing these features, note that there are also other features, not described in this chapter but related to small-cell improvements. Chapter 14 discusses how low-power nodes can be deployed together with macro nodes, resulting in a so-called heterogeneous deployment. Dual connectivity, described in Chapter 16, was developed in 3GPP under the umbrella of small-cell enhancements, as was the extension to 256QAM described in Chapter 6. Interworking with 802.11-based WLAN and license-assisted access, see Chapter 17, are other examples of features mainly relevant in a deployment with low-power nodes.

15.1 SMALL-CELL ON/OFF

In LTE, cells are continuously transmitting cell-specific reference signals and broadcasting system information, regardless of the traffic activity in the cell. One reason for this is to enable idle-mode devices to detect the presence of a cell; if there were no transmissions from a cell there is nothing for the device to measure upon and the cell would therefore not be detected. Furthermore, in a large macro-cell deployment there is a relatively high likelihood of at least one device being active in a cell, motivating continuous transmission of reference signals. LTE was designed with universal frequency reuse in mind and can handle the inter-cell interference from these transmissions.

However, in a dense deployment with a large number of relatively small cells, the likelihood of not all cells serving device at the same time can sometimes be relatively high. The downlink interference scenario experienced by a device may also be more severe with devices experiencing very low signal-to-interference ratios due to interference from neighboring, potentially empty, cells, especially if there is a large amount of line-of-sight propagation. In such dense, small-cell scenarios, selectively turning off cells can provide significant gains in reduced interference and reduced power consumption. The faster a cell is turned on or off, the more efficiently it can follow the traffic dynamics and the larger the gains are.

In principle, turning off a cell is straightforward and can be handled using existing network management mechanisms. However, even if there are no devices in a particular cell, turning off that cell has an impact on idle mode devices. In order not to impact these devices, other cells must provide basic coverage in the area otherwise handled by the cell turned off. Furthermore, the idle mode procedures in LTE are not designed under the assumption of cells being turned on or off frequently and it may take quite some time before a device discovers a cell that has been turned on. Transition from a dormant state to a fully active state for a cell takes many hundreds of milliseconds which is too slow to track any dynamic traffic variations and would have a noticeable performance impact.

With this as a background, mechanisms for significantly more rapid on/off operation of small cells[1] in a dense deployment, including on/off on a subframe level, were extensively discussed during the development of release 12. Based on these discussions, it was decided to base the release 12 small-cell on/off mechanism on the activation/deactivation mechanism in the carrier aggregation framework (see Chapter 12). This means that on/off is restricted to secondary cells in active mode only—that is, the primary carrier is always on. Restricting the on/off operation to secondary carriers greatly simplifies the overall design as idle mode compatibility is not impacted while the secondary carriers can be activated/deactivated on a fast basis.

When a secondary component carrier is turned off, a device should in principle not expect any transmissions on that carrier. This means that the device should not expect any synchronization signals, cell-specific reference signals, CSI reference signals, or system information from a deactivated cell. Although a carrier being completely silent would lead to the best energy savings and the lowest interference, it would also imply that the device cannot maintain synchronization to that carrier or perform any measurements, for example, mobility-related measurements. Without any active-mode mobility handling there is a significant risk that a device may have left the coverage area of a secondary cell without the network being aware of it. Therefore, to address these aspects, a new form of reference signal, the *discovery reference signal*, was introduced in release 12. The discovery signal is transmitted with a low-duty cycle and used by the device to perform mobility measurements and to maintain synchronization. The discovery signal, although designed as part of the small-cell on/off work, is useful also without switching off cells. For example, it can be used to assist shared-cell operation, described in Chapter 14, or full-dimension MIMO, discussed in Chapter 10. An example of small-cell on/off is shown in Figure 15.1.

---

[1]Although the feature is referred to as "small cell on/off" and a small-cell deployment was assumed in the discussions, the feature is not restricted to small cells from a specification perspective.

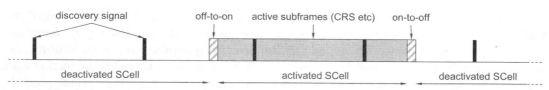

FIGURE 15.1

Small-cell on/off.

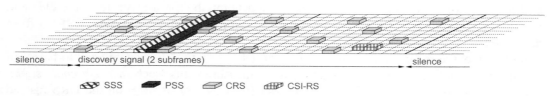

FIGURE 15.2

Example of discovery signal transmission for FDD.

15.1.1 DISCOVERY SIGNAL AND ASSOCIATED MEASUREMENTS

The *discovery reference signal* (DRS), although described as a new signal, actually consists of a combination already existing signals, namely

- synchronization signal (PSS and SSS) to assist in obtaining the cell identity and coarse frequency and time synchronization;
- cell-specific reference signals (CRS) to assist in obtaining fine frequency and time synchronization;
- CSI reference signals (optional) useful in determining the transmission point identity within the cell.

In the specifications, the discovery signal is defined from a device point of view. More specifically, a so-called *discovery signal occasion* is one to five subframes (two to five for TDD) where the device may assume the above-mentioned signals being present, starting with the synchronization signals in the first subframe.[2] This is illustrated in Figure 15.2. The periodicity of the DRS occasions can be set to 40, 80, or 160 ms.

A discovery signal occasion always starts in subframe 0 or 5. This follows directly from the definition of synchronization signals where the secondary synchronization signal is transmitted in subframe 0 and 5 in both FDD and TDD. The CSI-RS, which is an optional part of the DRS, can be transmitted on antenna port 15 in any of the subframes in the discovery

---

[2]For TDD the secondary synchronization signal is in the first subframe and the primary synchronization signal in the second subframe as a consequence of the synchronization signal design. This is also the reason behind the shortest possible discovery signal duration of two subframes in TDD.

signal occasion, subject to any restrictions in each of the subframes. The purpose of the CSI-RS is to be able to identify individual transmission points belonging to the same physical-layer cell identity. This can be used for selectively turning on certain transmission points in the cell in response to a device measurement report as discussed in conjunction with shared-cell operation in Chapter 14. As shown in Figure 15.2 there are several possible locations for a CSI-RS in a subframe. In FDD, there are up to 96 different positions, assuming a five subframes long discovery signal occasion starting in subframe 5,[3] thus allowing a large number of transmission points to be identified. The different CSI-RS locations can also be used to create orthogonality between transmission points by, on a given resource element, transmit nonzero-power CSI-RS from one transmission point and zero-power CSI-RS from the others.

Radio-resource management can be based on DRS—that is, the device need to base cell identification and radio-resource management measurements such as *reference signal received power* (RSRP) and *reference signal received quality* (RSRQ) on the DRS instead of the PSS/SSS/CRS mentioned in Chapter 11. The device is configured whether to use DRS-based measurements or not through RRC signaling. To assist these measurements, the device is provided with a *DRS measurements timing configuration* (DMTC) which indicates a 6 ms window within which a DRS may occur. If the CSI-RS are part of the DRS, the network also provides the device with information on, for each physical-layer cell identity, which CSI-RS configurations to measure upon and the virtual cell identity used for generating the CSI-RS sequences (the sequences for PSS/SSS and CRS are tied to the physical-layer cell identity). Within the DMTC, the device will search for discovery signals. For each of the discovery signals found fulfilling the triggering condition the device will report the RSRP and/or RSRQ together with information for cell and transmission point identification.

15.2 DYNAMIC TDD AND EIMTA

In LTE, seven different uplink—downlink configurations are possible as described in Chapter 5 where each subframe is either an uplink subframe or a downlink subframe (the special subframe can, to a large extent, be seen as a downlink subframe in this context). This configuration is in practice static, which is a reasonable assumption in larger macro cells. However, with an increased interest in local-area deployments, TDD is expected to become more important compared to the situation for wide-area deployments to date. One reason is that unpaired spectrum allocations are more common at higher-frequency bands less suitable for wide-area coverage but usable for local-area coverage. Furthermore, some of the problematic interference scenarios in wide-area TDD networks are less pronounced in local-area deployments with lower-transmission power and below-rooftop antenna installations. To better handle the high traffic dynamics in a local-area scenario, where the number of devices

---

[3]In this case there are 16 different configurations in subframe 5 and 20 different configurations in each of the following subframes; $16 + 4 \cdot 20 = 96$.

transmitting to/receiving from a local-area access node can be very small, dynamic TDD is beneficial. In dynamic TDD, the network can dynamically use resources for either uplink or downlink transmissions to match the instantaneous traffic situation, which leads to an improvement of the end-user performance compared to the conventional static split of resources between uplink and downlink. The more isolated a cell is, the better the traffic dynamics can be exploited. To harvest these benefits, LTE release 12 includes support for dynamic TDD, or *enhanced interference mitigation and traffic adaptation* (eIMTA) as is the official name for this feature in 3GPP.

One simple approach to dynamic TDD would be to, from a device perspective, treat every subframe as a downlink subframe, including the monitoring for control signaling, unless there is an uplink transmission explicitly scheduled. However, for various reasons 3GPP has chosen a somewhat different approach where the uplink–downlink allocation is signaled at the beginning of each frame (or set of frames) to enable dynamically varying uplink–downlink usage.

15.2.1 BASIC PRINCIPLES OF EIMTA

With the introduction of eIMTA, the uplink–downlink configuration is not static but can vary on a frame-by-frame basis. This is handled by the network broadcasting the *current uplink–downlink configuration* to use for each frame (or set of frames as discussed later).

The broadcasting allows the uplink–downlink configuration to change and meet different requirements on uplink and downlink traffic. However, there is also a need to handle uplink feedback such as hybrid-ARQ acknowledgments in response to downlink traffic, as well as uplink-related downlink control signaling. Having some subframes that are guaranteed to be downlink or uplink, irrespective of the dynamic reconfiguration, is therefore beneficial. For example, hybrid-ARQ feedback resulting from downlink transmissions is preferably transmitted in subframes guaranteed to be uplink to avoid error cases. Random-access transmissions also need a subframe guaranteed to be in the uplink direction. Therefore, eIMTA is making use of three different types of uplink–downlink configurations:

- the *uplink reference configuration*,
- the *downlink reference configuration*,
- the *current uplink-downlink configuration*.

The first two of these quantities are semi-statically configured and, among other things, determine the timing for the hybrid-ARQ signaling while the last one determines the usage of subframes in the current frame and can be dynamically changed on a frame-by-frame basis.

The *uplink reference configuration* is obtained from SIB1. It is also the configuration used by non-eIMTA-capable devices, simply known as the uplink–downlink configuration in earlier release. Preferably, to allow for the maximum flexibility for eIMTA-capable devices, this configuration is uplink heavy, hence its name. Downlink subframes in this reference configuration are guaranteed to be downlink subframes despite any dynamic reconfiguration

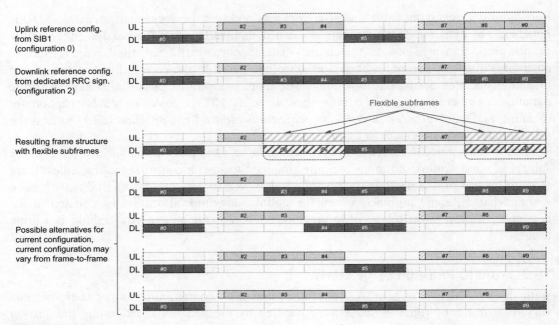

FIGURE 15.3

Example of flexible subframes.

and therefore useful for downlink transmission of, for example, the PHICH. In Figure 15.3, configuration 0 has been used as an example.

The *downlink reference configuration* is obtained from dedicated RRC signaling, specific to eIMTA-capable devices. As suggested by the name a downlink-heavy configuration is a good choice for maximum flexibility. A key property of this reference configuration is that uplink subframes are guaranteed to be uplink subframes despite any dynamic reconfiguration and therefore useful for hybrid-ARQ feedback as is discussed later. In Figure 15.3, configuration 2 has been used as an example for the downlink reference configuration. From the two reference configurations an eIMTA-capable device can compute the so-called *flexible subframes* as the difference between the two reference configurations. A flexible subframe can be used in either transmission direction as is described in the following.

The *current uplink—downlink configuration* determines which subframes that are uplink and which that are downlink in the current frame.[4] It must be chosen among the seven possible uplink—downlink allocations described in Chapter 5 and be within the limits set by the flexible subframes obtained from the reference configurations. This is the configuration that is broadcasted regularly and can be dynamically changed in order to follow traffic

---

[4]In case the uplink reference configuration and the current subframe configuration indicating a certain subframe as special and downlink subframe, respectively, the subframe is a downlink subframe.

variations. In Figure 15.3, the four different possibilities for the current uplink–downlink configuration in the example are illustrated.

The current uplink–downlink allocation is broadcasted using DCI format 1C on the PDCCH to all eIMTA-enabled devices. A special identity, the eIMTA-RNTI, is used on the control channel to indicate the current configuration. Multiple three-bit fields are used in DCI format 1C, each field indicating one of the seven uplink–downlink configurations in Chapter 5 for each of the component carriers the device is configured with, subject to any restrictions arising from the reference configurations.

From the perspective of dynamically adapting to varying traffic conditions, it is beneficial to broadcast the current configuration as often as possible—that is for each frame. From a signaling overhead perspective, on the other hand, less-frequent signaling results in a lower overhead. Therefore, the signaling periodicity for the current configuration can be set to once per 10, 20, 40, or 80 ms. It is also possible to configure in which of the subframes the device will monitor for the DCI format 1C carrying the current subframe allocation.

Upon detecting the DCI format 1C using the eIMTA-RNTI, the device will set the current uplink–downlink configuration accordingly. However, a device may occasionally not succeed in receiving the current uplink–downlink allocation and thus may not know which subframes that are uplink and which that are downlink. Therefore, a device not detecting the current uplink–downlink configuration will assume the current allocation being equal to the uplink reference configuration for the coming frame(s). In other words, the device behaves in the same way as a non-eIMTA-enabled device in this situation.

Figure 15.4 illustrates an example of eIMTA operation with the same reference configurations as in Figure 15.3. The current uplink–downlink configuration is broadcasted to all

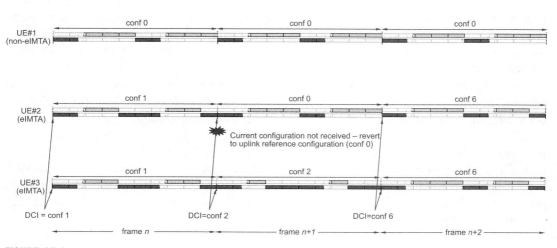

FIGURE 15.4

Example of eIMTA operation.

eIMTA-enabled devices one per frame. In frames n and $n+2$ both devices received the current configuration correctly and applied the corresponding uplink—downlink configuration. In frame $n+1$, however, the first device did not receive the current configuration. Hence, as a fallback, it applied the uplink reference configuration for that frame.

15.2.2 SCHEDULING AND HYBRID-ARQ RETRANSMISSIONS

The basic principle described earlier, to dynamically configure the uplink—downlink allocation for each frame, works nicely from a downlink scheduling perspective. The eNodeB can, using any downlink subframe in the current configuration, schedule downlink data. In response to the downlink data transmission, hybrid-ARQ acknowledgments need to be transmitted from the device in an uplink subframes. The timing of those acknowledgments is given by the downlink reference configuration in a similar way as for carrier aggregation with different uplink—downlink configurations—that is, they are transmitted in subframes guaranteed to be uplink subframes. Restricting hybrid-ARQ feedback to guaranteed uplink subframes instead of using "any" uplink subframe is beneficial as it maintains a fixed timing also in case the device has not properly received the configuration for the current frame. Improper reception of the current configuration is also the reason why bundling of hybrid-ARQ acknowledgments is not supported in eIMTA. Furthermore, the downlink reference configuration is used to derive the number of hybrid-ARQ processes.

Uplink scheduling is somewhat more complicated as the uplink grant received in one downlink subframe controls uplink transmission in a later subframe, the timing of which is given by the specifications. To be able to reuse the already developed timing relations—that is, which downlink subframe to use to schedule uplink transmission in a certain subframe—uplink grants are restricted to be transmitted in the guaranteed downlink subframes only. Similarly, the PHICH, which essentially is an uplink retransmission grant, follows the same principle. Obviously, if the scheduler has scheduled an uplink (re) transmission in a certain (flexible) subframe, that subframe cannot be used for downlink transmission.

From the discussion on the downlink and uplink reference configurations in the previous section, a relevant question is why these configurations, at least the downlink reference configuration which is of no relevance for legacy devices, are configurable. One alternative solution would be to hard code the downlink reference configuration to configuration 5, the most downlink-heavy configuration. The answer lies in the hybrid-ARQ latency and the fact that a downlink-heavy configuration results in fewer subframes available for hybrid-ARQ feedback and consequently a larger delay for hybrid-ARQ feedback. With configurable reference configurations, it is possible to balance the hybrid-ARQ latency against the number flexibility in subframe allocation.

15.2.3 RRM MEASUREMENTS AND CSI REPORTING

In a dynamic TDD network, the transmission direction of subframes is not necessarily aligned across multiple cells. Consequently, the interference scenario may be substantially different between subframes guaranteed to be downlink and subframes that are flexibly assigned to downlink. This will impact not only measurements for radio-resource management, for example, handover decisions, but also rate control.

Handover decisions should be consistent and not be influenced by short-term traffic variations. Therefore, measurements such as RSRP and RSRQ used for mobility handling are made upon guaranteed downlink subframes and not impacted by changes in the current uplink—downlink configuration. The mobility and handover behavior of an eIMTA-enabled device and a (legacy) device following the semi-statically signaled configuration (the uplink reference) is therefore identical.

Rate control, on the other hand, should reflect the instantaneous channel conditions at the device. Since the interference behavior can be quite different between guaranteed and flexible downlink subframes, interference is measured separately for the two sets of subframes and CSI reports provided separately for each of the two sets.

15.2.4 UPLINK POWER CONTROL

Uplink interference can, similarly to the downlink case, be radically different in subframes guaranteed to be uplink and subframes dynamically assigned for uplink transmissions. Different transmission power setting can therefore be useful, allowing the uplink transmission power to be set to a higher value to counteract interference from simultaneous downlink transmissions in neighboring cells in case of uplink subframes flexibly assigned compared to guaranteed uplink subframes.

In eIMTA, this is handled through independent power control loops: one for dynamically assigned uplink subframes and one for guaranteed uplink subframes. For each of the two sets of subframes, power control is handled as described in Chapter 7 with the parameters separately configured for each of the two subframe sets.

15.2.5 INTER-CELL INTERFERENCE COORDINATION

Dynamic TDD allows the uplink—downlink configuration to change dynamically on a per-cell basis. Although one of the reasons for dynamic TDD is to follow rapid changes in traffic behavior in a cell, it may not always be feasible to dynamically adapt the uplink—downlink configuration without coordination with neighboring cells. In an isolated cell, completely independent adaptation is possible, while in a large macro network a more or less static allocation as LTE was originally designed for may be the only possibility. However, there is a large range of scenarios between these two extremes were a dynamic TDD is possible with some degree of inter-cell coordination.

Between cells belonging to the same eNodeB this is purely as implementation issue in the scheduling algorithm. However, when the cells to coordinate between belong to different eNodeBs, coordination across the X2 interface is required. To assist interference coordination across cells in eIMTA, a new X2 message, the *intended uplink—downlink configuration* is introduced, and the *overload indicator* part of the release 8 inter-cell interference coordination framework (ICIC, see Chapter 13) is extended.

The intended uplink–downlink configuration is an X2 message where one cell can indicate the uplink–downlink configuration it intends to use for the coming period to neighboring cells. The scheduler in the cell receiving this massage can take this into account when determining the configuration to use in that cell. For example, if the neighboring cell indicates a flexible subframe will be used for uplink transmission, the cell receiving this message may try to avoid assigning the same flexible subframe for downlink transmissions.

The *overload indicator* in ICIC, indicates the uplink interference experienced by a cell on its different resource blocks. For TDD, the overload indicator refers to the uplink reference configuration—that is, the uplink–downlink configuration used by non-eIMTA devices. With the introduction of eIMTA, an extended overload indicator is added. The extended overload indicator is identical to the release 8 overload indicator with the addition of information about which uplink subframes it relates to, thus allowing interference information related to the current uplink–downlink configuration.

DUAL CONNECTIVITY

16

For a device to communicate at least one connection between the device and the network is required. As a baseline, the device is connected to one cell handling all the uplink as well as downlink transmissions. All data flows, user data as well as RRC signaling, is handled by this cell. This is a simple and robust approach, suitable for a wide range of deployments and the basis for LTE.

However, allowing the device to connect to the network at multiple cells can be beneficial in some scenarios (Figure 16.1):

- User-plane aggregation, where the device is transmitting and receiving data to/from multiple sites, in order to increase the overall data rate. Note that there are *different* streams transmitted/received to/from the different nodes unlike, for example, uplink CoMP where the *same* stream is received at multiple antenna sites.
- Control-plane/user-plane separation, where control plane communication is handled by one node and user plane by another. This can, for example, be used to maintain a robust control-plane connection to the macro cell while offloading the user-plane data to the pico cell.

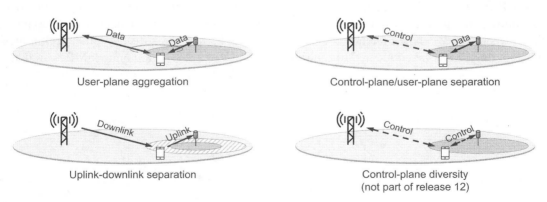

User-plane aggregation

Control-plane/user-plane separation

Uplink-downlink separation

Control-plane diversity
(not part of release 12)

FIGURE 16.1

Usage scenarios for dual connectivity.

4G, LTE-Advanced Pro and The Road to 5G. http://dx.doi.org/10.1016/B978-0-12-804575-6.00016-9

- Uplink—downlink separation where downlink and uplink are handled by separate sites. In heterogeneous deployments, the best downlink is not necessarily the best uplink which motivates handling the two separately as discussed in Chapter 14.
- Control-plane diversity where RRC commands are sent from two nodes. One example of when this can be beneficial is mobility in a heterogeneous deployment. A device connected to a pico cell but moving into the macro cell will receive a handover command from the network. To reduce the likelihood of not receiving this command due to rapid degradation of the pico-link quality, the handover command can be transmitted from two nodes. Although this scenario was discussed as part of release 12, it is not supported by the final specifications due to time constraints in the development of release 12.

Note that control plane in this context refers to the control signaling transmitted from higher layers, for example, RRC signaling, and not to the L1/L2 control signaling. The device is connected to two cells and each of these cells handles their own L1/L2 control signaling.

Some of these scenarios can be supported partially already in LTE release 8. For example, uplink—downlink separation can be achieved by using remote radio units connected to a central baseband unit as discussed in Chapter 14. Carrier aggregation can also be used for support of these scenarios, for example, by transmitting different carriers from different antenna points connected to a common baseband unit. However, support for these scenarios is very rudimentary, if at all existing, and requires centralized baseband processing with correspondingly high backhaul requirements in terms of low latency and high capacity.

To address these shortcomings, *dual connectivity* was introduced in release 12 and further refined in release 13. Dual connectivity implies a device is simultaneously connected to two eNodeBs, the *master eNodeB* and the *secondary eNodeB*, each with its own scheduler and interconnected using the X2 interface, see Figure 16.2. Note that this is as seen from a device perspective or, expressed differently, a master eNodeB for one device could act as a secondary eNodeB for another device. Since both the master and secondary eNodeBs handle their own scheduling and has their own timing relations, the normal X2 interface can be used to connect the two sites with relaxed latency requirements. This is in contrast to two other techniques where a device can be connected to multiple cells, carrier aggregation and CoMP, which are

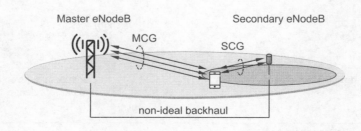

FIGURE 16.2

Dual connectivity.

more demanding on the interaction between the involved cells and typically requires low-latency, high-capacity backhaul and have stringent requirements on synchronization.

Carrier aggregation can be used in each of the two eNodeBs—that is, a device may be connected to multiple cells in each of the two nodes. The master eNodeB is responsible for scheduling transmissions in the *master cell group* (MCG) while the secondary eNodeB handles the *secondary cell group* (SCG). Within each cell group there is one primary component carrier[1] and, if carrier aggregation is used, one or more secondary component carriers. Similarly to carrier aggregation, dual connectivity is supported only when the device is active—that is, in the RRC_CONNECTED state, and not in idle mode.

It is assumed that the device is capable of simultaneous transmission (and reception) toward both the eNodeBs involved in dual connectivity. This is more or less a consequence of the fundamental assumption that dual connectivity should not require any dynamic coordination between the two eNodeBs. Furthermore, it is assumed that the two cell groups use different, nonoverlapping frequency bands—that is, only interband dual connectivity is supported. Although the dual-connectivity framework as such would allow same-frequency deployment, the resulting interference scenarios would require additional mechanisms not part of release 13. Finally, although the specifications are eNodeB-type agnostic, a heterogeneous scenario with the master cell group handled by a macro base station and the secondary cell group handled by a pico base station has often been assumed in the development of the overall solution. This is however not required. The first not the device connects to take the role of the master eNodeB and the secondary eNodeB is added subsequently; in case the device connected to the pico node first the pico node takes the role of the master eNodeB and the macro node may be added as a secondary eNodeB.

The dual-connectivity framework has turned out to be applicable also for inter-RAT aggregation. One example hereof is aggregation of LTE and WLAN, specified in release 13, where a primary cell group using LTE is aggregated with one or more WLAN carriers on a secondary basis. It is also expected that an evolution of the dual connectivity framework will play an important role for the tight interworking between LTE and a new 5G radio-access technology as discussed in Chapter 23.

16.1 ARCHITECTURE

The overall architecture described in Chapter 4 is used also for dual connectivity—that is, an X2 interface between the eNodeBs and an S1-c interface between the master eNodeB and the core network, see Figure 16.3. The master eNodeB always has a direct S1-u interface to S-GW, while the secondary eNodeB may or may not have a direct S1-u interface to S-GW depending on the architecture. Note that the figure illustrates the interfaces relevant for

---

[1]In the specifications, the primary component carrier is denoted PCell in the MCG and PSCell in the SCG, and the secondary component carriers are denoted SCell in both the cell groups.

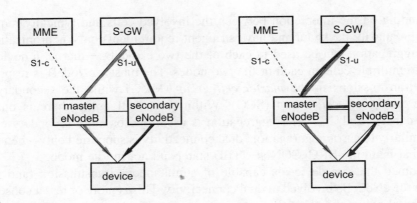

FIGURE 16.3

Architecture.

handling a certain device only. The secondary eNodeB is usually connected to the core network with S1-u and S1-c as it may act as a master eNodeB for another device.

For the user plane part, there are two possibilities of routing data to a device via the secondary eNodeB. In the first alternative, data to be transmitted via the secondary eNodeB is sent on the S1 interface to the secondary eNodeB directly (left part of Figure 16.3); in the second alternative the data to be transmitted by the secondary eNodeB is routed via the master eNodeB (right part of the Figure 16.3).

The first alternative is less demanding on the backhaul but results in dual connectivity being visible above the radio-access network which can be a drawback from a mobility perspective, for example, in scenarios with frequent changes of the secondary cell group. User-plane aggregation can be achieved with mechanisms such as multipath TCP [60]; that is, aggregation of the two data flows is done in the TCP transport layer above the core network and not in the radio-access network.

The second alternative, while more demanding on the backhaul as data between the secondary eNodeB and the device is routed via the master eNodeB, can provide higher performance as the aggregation is done closer to the radio interface and usage of multipath TCP is not necessary. Mobility resulting in a change of the secondary eNodeB is invisible to the core network which can be beneficial as handover between pico cells may be common in a heterogeneous deployment.

The protocol architecture for the user plane is shown in Figure 16.4 and is identical to the architecture for the single eNodeB case with the addition of support for split bearers. There are three different types of radio bearers: bearers transmitted from the master eNodeB, bearers transmitted from the secondary eNodeB, and split bearers transmitted across both eNodeBs. As seen in the figure, there is a single PDCP protocol entity for the split bearer, located in the master eNodeB, in which the data are split up and parts of the downlink data are forwarded over the X2 interface to the secondary eNodeB. There is a flow control mechanism defined on the X2 interface to control the amount of data forwarded to the secondary eNodeB.

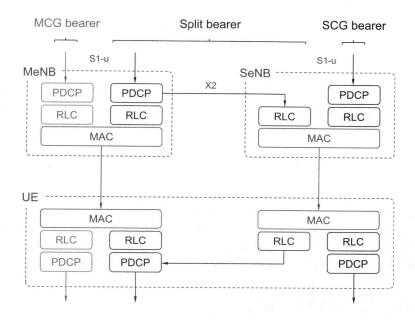

FIGURE 16.4

User-plane architecture.

In release 12, split bearers are supported in the downlink only and semi-static configuration assigns handling of uplink traffic of a split bearer to either the master eNodeB or the secondary eNodeB. This restriction is removed in release 13 where split bearers are supported also in the uplink.

On the device side, there are two MAC entities, one for each of the master and secondary cell groups. For split bearers, the received data packets may be received in the incorrect order, for example, if the secondary eNodeB had to perform retransmissions while this was not needed for the master eNodeB. Therefore, for RLC-AM the PDCP entity for a split bearer performs reordering in a similar way as the RLC handles reordering after the hybrid ARQ mechanism, see Chapter 8.

For the control plane, only the master eNodeB is connected to the MME[2] via the S1 interface, see Figure 16.5. This means that the secondary eNodeB is not visible to the MME, which is beneficial from a signaling perspective toward the core network as handovers are expected to be more frequent between secondary eNodeBs handling pico cells than for master eNodeBs handling macro cells in a heterogeneous deployment.

Each of the master and secondary eNodeBs independently handles scheduling and controls all the resources in the cells belonging to the respective eNodeB. Radio-resource control

---

[2]As discussed previously, this is from the perspective of a single device. The secondary eNodeB is typically connected to the MME as well as it may be a master eNodeB for another device.

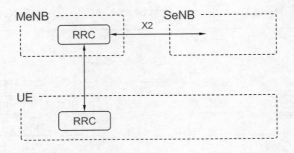

FIGURE 16.5

Control-plane architecture.

(RRC) is handled by the master eNodeB, transmitting RRC messages to the device. To handle radio resources in cells belonging to the secondary eNodeB, inter-eNodeB RRC messages have been defined. The master eNodeB does not modify the contents of the messages from the secondary eNodeB but simply encapsulates them into the RRC messages sent to the device.

16.2 PHYSICAL-LAYER IMPACT

Dual connectivity is mainly affecting the architecture while the impact on the physical-layer structure is modest. The two physical layers, one for each of the cell groups, operate independently of each other with their own data transmission, associated control signaling, random access, and so on. Each of the cell groups also has its own MAC layer which schedules physical-layer transmission independently. However, there are some aspects, in particular timing and power handling, where there is a mutual dependence between the physical layers in the two cell groups.

16.2.1 TIMING

LTE is designed to work both with and without inter-cell synchronization[3] and dual connectivity is not different in this respect. Within a cell group, the same requirements as for any other carrier aggregation scenario are applied—that is, all component carriers received are within a 33 μs window. Between cell groups, on the other hand, there is no requirement on synchronization, resulting in two cases:

- synchronous dual connectivity, where subframe boundaries of the two cell groups are aligned at the device within 33 μs;
- asynchronous dual connectivity, where subframe boundaries of the two cell groups have an arbitrary[4] timing relation at the device.

---

[3]In carrier aggregation all the involved cells are time synchronized.
[4]Actually, the specifications mention a 500-μs limit; time differences larger than this are handled by an offset in the subframe numbering.

In principle, there is no fundamental difference between these two cases but the actual implementation may differ. For example, if the carriers from the master cell group and secondary cell group are sufficiently close in frequency, a single RF chain may be capable of handling both if they are synchronized but not otherwise. Setting of the uplink transmission power is one example, which in many implementations can occur at the start of a subframe only. Such an implementation would be capable of synchronous dual connectivity where the subframe boundaries coincide but not asynchronous dual connectivity. To allow for different implementations, a device reports for each band combination whether it is capable of asynchronous operation or not. An asynchronous device can operate in either synchronous or asynchronous dual connectivity while a synchronous device can operate in synchronous dual connectivity only.

From a network perspective, it is up to the implementation to account for differences in SFN and subframe offset; for example, in conjunction with discontinuous reception in the device. Release 13 adds the possibility for the device to measure and report the difference between the two cell groups to assist network operation.

16.2.2 POWER CONTROL

Transmission power is the main area where there is an inter-cell-group dependency in the physical layer. Although the power setting is individual for each cell group, regulations specify the maximum transmission power *per device* and hence create a dependency between the cell groups when it comes to power sharing. Thus, when the device reaches its maximum transmission power there is a need to scale the power of the individual channels in the different cell groups. This may sound straightforward but the fact that the cell groups can be unsynchronized complicates the picture. For a given cell group, changes in the transmission power should occur at the subframe boundaries only as the receiver may assume that the transmission power is constant across a subframe.

The synchronous case, illustrated to the left in Figure 16.6, has all the subframe boundaries aligned across the cell groups. When setting the transmission power for subframe m in the master cell group the activity in the overlapping subframe in the secondary cell group is known and scaling the transmission power for the different channels is straightforward. Furthermore, changes in the transmission power occur at the subframe boundaries only.

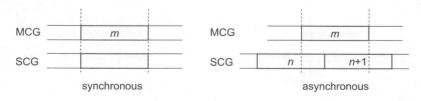

FIGURE 16.6

Synchronous and asynchronous operation.

The asynchronous case, illustrated to the right in Figure 16.6, is more complicated. As an example, consider the master cell group (the situation is similar for the secondary cell group). The available transmission power in subframe m of the master cell group may now depend on *two* subframes in the secondary cell group, the ongoing subframe n and the future subframe $n+1$.

Since the power setting for the master cell group can be done at the corresponding subframe boundaries only, some margin for what might happen in the secondary cell group is necessary.

With the situation described earlier in mind, two methods for sharing the transmission power across cell groups are defined. They mainly differ in whether the scaling in case of a power limitation is done across all cells in all cell groups or separately per cell group. The power-control mode to use is configured by RRC signaling.

Dual-connectivity power-control mode 1 scales the power across cell groups as illustrated to the left of Figure 16.7. In case of a power limitation, the transmission power is scaled across all cells, regardless of the group they belong to, in the same way as in carrier aggregation. The only exception is that uplink control information in the master cell group is prioritized over uplink control information in the secondary cell group in case the same UCI type is used in both cell groups. In essence, this power-control mode does not differentiate between the cell groups and treats all cells in the same way. Power-control mode 1 is possible in synchronous operation only as the transmission power can be changed at subframe boundaries only. In an asynchronous scenario, the power of the master cell group would need to change as a result of power allocation done at the beginning of the subframes in an secondary cell group and vice versa, something which is not possible.

Dual-connectivity power-control mode 2 scales the power across carriers within each cell group but not between cell groups as illustrated to the right of Figure 16.7. The minimum guaranteed power available per cell group, expressed as a fraction of the maximum power, is configured through RRC signaling. In case of power limitation, each cell group is given at least its minimum guaranteed power. The remaining power is then first given to the cell group associated with the earlier transmission. In Figure 16.6 this means that, at the beginning of

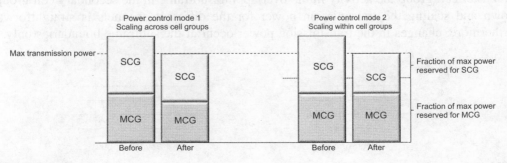

FIGURE 16.7

Power-control behavior for mode 1 and 2 when reaching the maximum transmission power.

subframe m, the secondary cell group can use the amount of the remaining power required to maintain the power constant during subframe n. Any power left after this is given to the master cell group in subframe m. Similarly, at the beginning of subframe $n+1$ in the secondary cell group, the master cell group uses the amount of power from the remaining power required to maintain the transmission power during subframe m. Since asynchronous operation implies that the subframe boundaries are not time aligned, transmission power for one cell group may need change at the subframe boundary for that cell group while it should be kept unchanged for the other cell group, power-control mode two is the only mode supported for asynchronous operation.

16.3 SCHEDULING IN DUAL CONNECTIVITY

Scheduling in dual connectivity is done independently in each eNodeB in the same way as described in earlier chapters—that is, the master cell group and the secondary cell group can independently transmit a scheduling request when needed. Similarly, discontinuous reception, DRX, is configured separately for each of the cell groups. Thus, as the two cell groups are scheduled independently, there is no need for tight coordination between the two eNodeBs although various forms of inter-cell coordination mechanism can be implemented if desired—neither is there a need to specify how to split downlink data across the two cell groups for a split bearer.

However, there are some aspects of scheduling, in particular handling of split bearers in the uplink in release 13, that are impacted by the introduction of dual connectivity. To determine how data for an uplink split bearer should be split across cell groups a threshold can be configured in the device. If the amount of data in the PDCP buffers for the split bearer is larger than the threshold, data is transmitted across both cell groups, otherwise it is transmitted only on one of the cell groups similarly to release 12.

Device reports used to support scheduling are also impacted by the introduction of dual connectivity. More specifically, the amount of data awaiting transmission and the available power, has to be split across the two cell groups with the corresponding impact on the reporting mechanisms.

Buffer status is reported per cell group. This is a natural choice as an eNodeB is only interested in the cell group it is scheduling. Knowing the buffer status for the master cell group does not help the eNodeB scheduling the secondary cell group and vice versa. For a split bearer, release 12 relies on semi-static configuration to determine in which cell group the split bearer should be included and reporting buffer status separately for the two cell groups works fine also for split bearers. In release 13, if the amount of data exceeds the previously mentioned threshold, the full amount of data is reported for the split bearer for both cell groups and it is up to the implementation to coordinate the scheduling decisions if needed.

Power headroom reporting is somewhat more complicated than buffer status reporting since the power is a resource shared across both cell groups as described earlier. Power

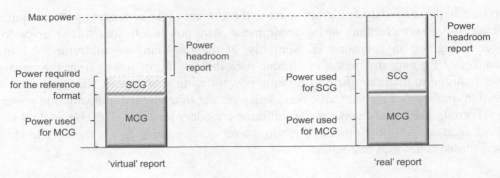

FIGURE 16.8

Power headroom report for master cell group (MCG) [the report for secondary cell group (SCG) is similar].

headroom reports for one cell group therefore need to take the activity in the other cell group into account. There are two possibilities for this with higher-layer configuration determining which one to use:

- the power used by the other cell group is given by a configured reference format (virtual report);
- the actual power used by the other cell group is used if it is transmitting (real report).

This is illustrated for the master cell group in Figure 16.8 (the same principle applies for the secondary cell group). The power headroom is the maximum transmission power of the device from which the actual master cell group transmission power and, depending on the reporting type, the actual power used (or virtual power assumed to be used) for the secondary cell group is subtracted.

UNLICENSED SPECTRUM AND LICENSE-ASSISTED ACCESS

17

Spectrum is fundamental for wireless communication, and there is a never-ending quest for more spectrum to meet the ever-increasing demands of increased capacity and higher data rates. Increasing the amount of spectrum available to LTE is thus highly important. LTE and previous cellular systems were designed for licensed spectrum where the operator has an exclusive license for a certain frequency range. Licensed spectrum offers many benefits since the operator can plan the network and control the interference. It is thus instrumental to providing quality-of-service (QoS) guarantees and wide-area coverage. However, the amount of licensed spectrum an operator has access to may not be sufficient, and there is typically a cost associated with obtaining a spectrum license.

Unlicensed spectrum, on the other hand, is open for anyone to use at no cost, subject to a set of rules, for example, on maximum transmission power. Since anyone can use the spectrum, the interference situation is typically much more unpredictable than for licensed spectrum. Consequently, QoS and availability cannot be guaranteed. Furthermore, the maximum transmission power is modest, making it less suitable for wide-area coverage. Wi-Fi and Bluetooth are two examples of communication systems exploiting unlicensed spectrum.

From the discussion earlier, it is seen that these two spectrum types have different benefits and drawbacks. An attractive option is to combine the two such that licensed spectrum is used to provide wide-area coverage and QoS guarantees with unlicensed spectrum as a local-area complement to increase user data rates and overall capacity without compromising on coverage, availability, and reliability.

Using unlicensed spectrum to complement LTE in licensed spectrum is in itself not new. Many operators are already using Wi-Fi to boost capacity in local areas. In such mixed deployments, the selection of the radio access to use, LTE or Wi-Fi, is currently handled autonomously by the device. This has some drawbacks as the device may select Wi-Fi even if staying on LTE would provide a better user experience. One example of such a situation is when the Wi-Fi network is heavily loaded while the LTE network enjoys a light load. To address such situations, the LTE specifications have been extended in release 12 with means for the network to assist the device in the selection procedure [82]. Basically, the network configures a signal-strength threshold controlling when the device should select LTE or Wi-Fi.

4G, LTE-Advanced Pro and The Road to 5G. http://dx.doi.org/10.1016/B978-0-12-804575-6.00017-0

Furthermore, release 13 also supports LTE–WLAN aggregation where LTE and WLAN are aggregated at the PDCP level using a framework very similar to dual connectivity.

The primary reason for Wi-Fi integration is to cater to operators with existing Wi-Fi deployments. However, a tighter integration between licensed and unlicensed spectrum can provide significant benefits. For example, operating one LTE network covering both spectrum types is simpler than operating two different technologies, one for licensed and the other for unlicensed spectrum. Mobility is another benefit. LTE was designed with mobility in mind from the start while Wi-Fi performs best when the users are more or less stationary. QoS handling and the possibility for increased spectral efficiency in a scheduled system are additional examples of the benefits of using LTE for both licensed and unlicensed spectrum.

With these points in mind, 3GPP has specified *license-assisted access* (LAA) as part of release 13. The basis for LAA is carrier aggregation with some component carriers using licensed spectrum and some unlicensed spectrum, see Figure 17.1. Mobility, critical control signaling and services demanding high QoS rely on carriers in licensed spectrum, while (parts of) less demanding traffic can be handled by the carriers in unlicensed spectrum. This is the reasoning behind the name "license-assisted access" where licensed spectrum is used to assist access to unlicensed spectrum.

LAA targets operator-deployed low-power nodes in the 5 GHz band in, for example, dense urban areas, indoor shopping malls, offices, and similar scenarios. It is not intended as a replacement for user-deployed Wi-Fi nodes at home as it requires access to licensed spectrum, nor is it intended for wide-area coverage as the allowed transmission power in unlicensed bands is fairly low. Initially, in release 13, LAA will support downlink traffic with release 14 extending it to also handle uplink traffic.

One important characteristic of LAA is the fair sharing of unlicensed spectrum with other operators and other systems, in particular Wi-Fi. There are several mechanisms that enable this. First, *dynamic frequency selection* (DFS), where the LAA node searches and finds a part

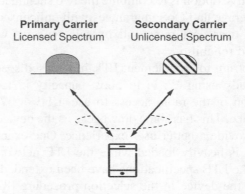

Primary Carrier
Licensed Spectrum

Secondary Carrier
Unlicensed Spectrum

FIGURE 17.1

License-assisted access.

of the unlicensed spectrum with low load, is used, thereby avoiding other systems if possible. Release 13 also supports a *listen-before-talk* (LBT) mechanism,[1] where the transmitter ensures that there are no ongoing transmissions on the carrier frequency prior to transmitting. With these mechanisms, fair coexistence between LAA and Wi-Fi is possible, and LAA can in fact be a better neighbor to Wi-Fi than another Wi-Fi network [56].

17.1 SPECTRUM FOR LAA

Unlicensed spectrum exists in multiple frequency bands. In principle, any unlicensed band can be used for LAA although the 5 GHz band is the main target. One reason is the availability of fairly large amounts of bandwidth in the 5 GHz band and a reasonable load compared to the 2.4 GHz band.

The 5 GHz band is available in most parts of the world, see Figure 17.2, although there are some differences between the different regions. In the following, a brief overview of the regulatory requirements in different parts of the world is given. For a more detailed overview, see [57] and the references therein.

The lower part of the band, 5150−5350 MHz, is typically intended for indoor usage with a maximum transmission power of 23 dBm in most regions. In total 200 MHz is available, divided in two parts of 100 MHz each with regulations calling for DFS and *transmit power control* (TPC) in the 5250−5350 MHz range.

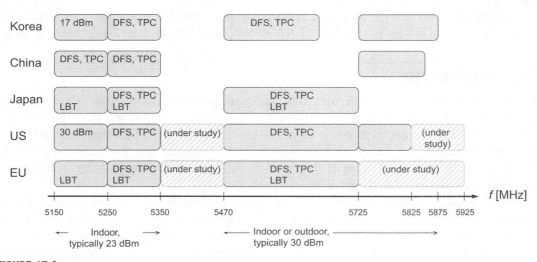

FIGURE 17.2

Overview of unlicensed frequency bands in different regions.

---

[1]Sometimes the term *clear-channel assessment* (CCA) is used instead of LBT although herein CCA is used for assessing whether the channel is available not including any backoff procedure.

DFS means that the transmitter continuously must assess whether the spectrum is used for other purposes. If such usage is detected, the transmitter must vacate the frequency within a specific time (e.g., 10 s) and not use it again until at least a certain time (e.g., 30 min) has passed. The purpose is to protect other systems, primarily radars, which have higher priority for the usage of unlicensed spectrum. TPC means that a transmitter should be able to reduce its transmission power below the maximum allowed power with the intention to reduce the overall interference level when needed.

In the part of the band above 5470 MHz, transmission powers up to 30 dBm and outdoor usage is allowed in many regions. The amount of spectrum differs across regions but up to 255 MHz can be available. DFS and TPC are mandatory.

In several regions, particularly Europe and Japan, LBT is mandatory. LBT is a mechanism where the transmitter listens to any activity on the channel prior to each transmission in order not to transmit if the channel is occupied. It is thus a much more dynamic coexistence mechanism than DFS. Other regions, for example, the United States, do not have any LBT requirements.

There may also be requirements on the minimum transmission bandwidth, how long a single transmitter may use the channel, and the fraction of time a transmitter must leave the channel idle. For example, a channel occupancy time of at most 6 ms[2] is allowed within Europe while the corresponding number for Japan is 4 ms. The channel occupancy time sets a limit on how long a transmission burst can be. Europe also has two sets of rules for unlicensed spectrum usage described in [58], one for frame-based equipment and one for load-based equipment. The two sets of rules were specified to be applicable to the now-defunct Hiperlan/2 standard and Wi-Fi, respectively. LAA is designed with the load-based rules in mind.

17.2 WI-FI BASICS

Wi-Fi is a well-known system operating in unlicensed spectrum. Although unlicensed spectrum is not allocated to Wi-Fi but free to use by anyone following the regulatory requirements, Wi-Fi is, and is likely to continue to be, a common radio-access technology in unlicensed spectrum. A brief overview of the Wi-Fi behavior and the underlying 802.11 standard with focus on 802.11ac is therefore given in the following in order to better understand some of the LAA design choices.

Wi-Fi divides the available spectrum into several 20 MHz frequency channels. Transmissions use one, or in case of channel bonding, multiple such frequency channels. Coordinating those transmissions across nodes can be done either in a centralized or in a distributed manner. In practice, centralized coordination is seldom used in Wi-Fi and distributed coordination is by far the most common. The following description therefore focuses on *enhanced distributed channel access* (EDCA), an enhancement providing QoS

---

[2]Under certain conditions, 8 ms or 10 ms is allowed.

enhancements to the *distributed coordination function* (DCF) part of the original 802.11 specifications.

Nodes using EDCA use LBT which includes a backoff procedure prior to transmitting. First, the transmitter listens and waits until the frequency channel is available during a period of time known as *arbitration inter-frame space* (AIFS). A frequency channel is declared available if the power level is lower than −62 dBm and no Wi-Fi preamble with a power level of −82 dBm or higher is detected, otherwise it is unavailable.

Once the frequency channel has been declared available during (at least) AIFS, the transmitter starts the backoff procedure. A backoff timer is initialized with a random number, representing the duration in multiples of the 9 μs slot time the channel must be available before a transmission can take place. The backoff timer is decreased by one for each 9 μs slot the channel is sensed idle, whereas whenever the channel is sensed busy the backoff timer is put on hold until the channel has been idle for a period of AIFS.

Once the backoff timer has expired, the node has acquired a *transmission opportunity* (TXOP). Multiple packets can be transmitted back to back during the TXOP without any need for LBT between them as long as the maximum TXOP duration is not violated. If the TXOP duration is set to zero, only a single packet is allowed, and a new backoff procedure has to be used for each packet.

Upon reception of a packet (or a set of contiguous packets[3]) the receiver responds with an acknowledgment message. The acknowledgment is transmitted a *short inter-frame space* (SIFS) duration of 16 μs after the reception of the packet. Since the SIFS is shorter than the AIFS, no other Wi-Fi user can grab the channel during this period. If no acknowledgment is received either the data or the acknowledgment itself was lost and a retransmission is performed. After completing the TXOP and prior to transmitting the next packet(s) from the transmission buffer, regardless of whether it is a retransmission or a new packet, a random backoff is performed by using the same procedure as described earlier. The reason for the backoff procedure is to avoid collisions between multiple transmitters. Without the random backoff, two nodes waiting for the channel to become available would start transmitting at the same time, resulting in a collision and most likely both transmissions being corrupted. With the random backoff, the likelihood of multiple transmitters simultaneously trying to access the channel is greatly reduced.

The random value used to initialize the backoff timer must be within the *contention window* and is drawn from a uniform distribution with exponentially increasing size for each retransmission attempt. For the nth retransmission attempt, the backoff time is drawn from the uniform distribution $[0, \min(2^{n-1}CW_{\min}, CW_{\max})]$. The larger the contention window, the larger the average backoff value, and the lower the likelihood of collisions.

---

[3]In the original 802.11 standard the receiver responds with an acknowledgment after each packet, while the 802.11e amendment, used in 802.11ac, introduced block acknowledgments enabling a single acknowledgment message to cover multiple packets. This is often used in combination with TXOP > 0.

Table 17.1 Default Parameters for the Different Access Classes (for an access point)

Priority Class	CW_{min}	CW_{max}	AIFS	TXOP
Voice	3	7	25 µs	3.008 ms
Video	7	15	25 µs	1.504 ms
Best effort	15	63	43 µs	0
Background	15	1023	79 µs	0
Legacy DCF	15	1023	34 µs	0

The original 802.11 standard relied on the distributed coordination function and did not support handling of different traffic priorities; all traffic was treated with the same priority. This was addressed by the introduction of EDCA where one of the major enhancements was handling of different traffic priorities. This is done using priority-class-dependent values for CW_{min} and CW_{max} as shown in Table 17.1 for the four priority classes.[4] High-priority traffic uses a smaller contention window to get faster access to the channel while low-priority traffic uses a larger contention window, increasing the likelihood of high-priority data being transmitted before low-priority data. Likewise, different duration of AIFS are used for the different priority classes, resulting in high-priority traffic sensing the channel for a shorter period of time and grabbing the channel quicker than low-priority traffic. For comparison, the last row in Table 17.1 shows the corresponding values when using the legacy DCF functionality instead of EDCA.

LBT and the associated backoff procedures are illustrated in Figure 17.3 for three different users. The first user gets access to the channel relatively quick as there are no other users actively transmitting. Once the backoff timer has expired, the packet is transmitted. The third user, for which a packet arrived while the first user is transmitting, found the channel to be occupied. The backoff timer is held and not decremented until the channel is available again. However, in the meantime the second user grabbed the channel and again the third user's backoff timer is put on hold, deferring the transmission further. Once the backoff timer for the third user expires the data are transmitted.

One benefit of EDCA lies in its distributed nature—any device can communicate with any other device without the need for a centralized coordination node. However, the use of LBT with backoff timers implies a certain overhead. At higher loads such a distributed protocol is less efficient and higher efficiency would be possible with a centralized scheduling function. This would also be well in line with the common scenario of multiple devices communicating with a central access point.

---

[4]The numbers given are for an access point, clients use slightly different values (larger AIFS for voice and video and larger CW_{max} for best effort). See 802.11e for a description on how the parameters relate. For 802.11ac, the parameters are given as AIFS = SIFS + $n\sigma$ where n is an access-class dependent parameter and $\sigma = 9$ µs is the slot time.

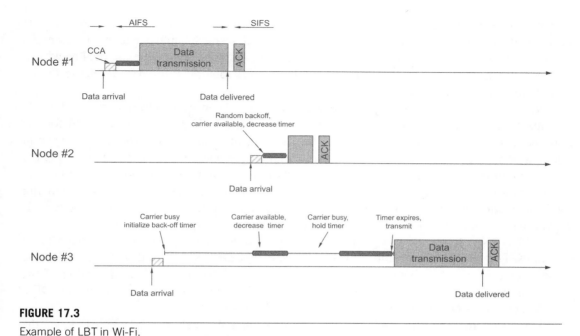

FIGURE 17.3

Example of LBT in Wi-Fi.

Another aspect of the LBT mechanism in Wi-Fi is that it, in essence, leads to a time reuse larger than one across multiple access points, each providing coverage in a certain area. A transmission from one access point is typically detected at the neighboring access points, all of which will find the frequency channel occupied and therefore defer their transmissions. This is less efficient from a capacity perspective as it is known that reuse one results in higher capacity.

The overview in the preceding paragraphs is brief as the intention is to provide some background to some of the LAA design choices. For a more detailed description of Wi-Fi and aspects such as the hidden node problem occurring in distributed schemes, the reader is referred to [59].

17.3 TECHNOLOGY COMPONENTS FOR LAA

LAA is, as already mentioned, based on the carrier aggregation framework in LTE. The primary component carrier, and optionally, one or more secondary component carriers operate in licensed spectrum, aggregated with one or more secondary component carriers operating in unlicensed spectrum. With the carrier aggregation enhancements in release 13 enabling aggregation of up to 32 carriers, up to 620 MHz of unlicensed spectrum can be exploited by a single device in combination with 20 MHz of licensed spectrum. In practice, the use of carrier aggregation implies that licensed and unlicensed spectrum is handled by the same node, possibly with remote radio heads connected over low-latency backhaul. To handle separate nodes for licensed and unlicensed spectrum, interconnected with nonideal backhaul, LAA could be based on the dual-connectivity framework, but this is not part of release 13 and is left for potential introduction in future releases.

No changes are required for the component carriers in licensed spectrum as this is the type of spectrum for which LTE was designed. Many of the LTE design choices hold for the component carriers in unlicensed spectrum as well, although some aspects are different compared to licensed spectrum, primarily due to multiple operators and/or systems using the same spectrum. DFS, TPC, LBT, discontinuous transmission, and radio-resource management are some of the larger areas affected by the introduction of LAA, although some smaller enhancements were made in other areas as well. A quick overview of these components is given with a more detailed description in the following sections.

DFS is used to vacate the frequency channel upon detecting interference from radar systems. This is a requirement in some frequency bands. DFS is also used when activating the node, for example, at power up, in order to find an unused or lightly used portion of the spectrum for future transmissions. No specification enhancements are needed to support DFS; implementation-specific algorithms in the eNodeB are sufficient.

TPC is required in some bands and regions, requiring the transmitter to be able to lower the power by 3 or 6 dB relative to the maximum output power. This is purely an implementation aspect and is not visible in the specifications.

LBT ensures that the carrier is free to use prior to transmission. It is a vital feature that allows fair sharing of the spectrum between LAA and other technologies such as Wi-Fi. In some regions, in particular Europe and Japan, it is a mandatory feature. The introduction of LBT to LTE impacts the specifications and is a completely new feature in LTE which has been extensively discussed in 3GPP.

Downlink discontinuous transmission is required not only to comply with regulations, since some regions limit the maximum transmission duration, but also to nicely coexist with other users of the unlicensed spectrum. Only when the channel is declared available should transmissions take place. In particular, continuous transmission of cell-specific reference signals, as originally required by LTE, is not possible, impacting not only data demodulation but also RRM functionality, and requiring additions to the specifications. To some extent, discontinuous transmission can be seen as small-cell on/off operating on a subframe basis. The introduction of discontinuous transmission will impact time and frequency synchronization, *automatic gain control* (AGC) setting, and CSI measurements as these functionalities typically rely on certain reference signals always being present.

The fact that multiple operators in the same area may use the same carrier frequency, unlike the licensed case where the carrier frequency is unique to an operator, also needs to be considered. In addition, multiple operators may also end up using the same physical-layer cell identity. The structure of many signals and channels are linked to the physical-layer cell identity (e.g., CSI-RS, downlink scrambling, and CRS sequences). This would lead to the situation that a device connected to operator A may successfully receive signals and channels originating from operator B. However, the likelihood of this is very small [57] and the eNodeB implementation can avoid this by trying to detect the other physical-layer cell identities being used on the targeted carrier and by selecting an unused cell identity.

17.3.1 DYNAMIC FREQUENCY SELECTION

The purpose of DFS is to determine the carrier frequencies for the secondary carriers in order to find an available or at least lightly loaded carrier frequency. If this is successful, there are no coexistence problems with other systems using the unlicensed spectrum. Since around 25 frequency channels, each 20 MHz wide, are part of the 5 GHz band, and the output power is fairly low, there is a reasonably high likelihood to find unused or lightly loaded frequency channels.

DFS is performed at power-up of an LAA cell. In addition to power-up, DFS can also be performed on an event-triggered basis. For example, the base station can periodically measure the interference or power level when not transmitting in order to detect whether the carrier frequency is used for other purposes and if a more suitable carrier frequency is available. If this is the case, the base station can reconfigure the secondary component carriers to a different frequency range (essentially an inter-frequency handover).

DFS is, as already mentioned, a regulatory requirement for some frequency bands in many regions. One example motivating DFS being mandated is radar systems, which often have priority over other usage of the spectrum. If the LAA base station detects radar usage, it must stop using this carrier frequency within a certain time (typically 10 s). The carrier frequency is not to be used again until at least 30 min has passed.

The details of DFS are up to the implementation of the base station, and there is no need to mandate any particular solution in the specifications. RSSI measurement from the device can be used by the implementation-specific DFS algorithm.

17.3.2 LISTEN BEFORE TALK

In LTE, all transmissions are scheduled and the scheduler is in complete control of when transmissions occur on a carrier in licensed spectrum. Scheduling is also used for LTE transmissions in unlicensed bands, but an inherent consequence of unlicensed spectrum is that there can be multiple transmitters (potentially belonging to different operators) using the same spectrum without any coordination between them. Although channel selection aims to find one or more frequency channels with no or very light usage, and in many cases succeeds in this, simultaneous usage of the same frequency channel by multiple systems cannot be precluded. LBT refers to the mechanism used by LAA to check the availability of a channel prior to using it. In LAA, the transmitter listens to potential transmission activity on the channel prior to transmitting.

Regulatory requirements vary across regions with some regions, for example, Japan and Europe, mandating the use of LBT in unlicensed bands while some other regions being more relaxed. In regions where there are no regulatory requirements on LBT, one could in principle deploy LTE release 12 and use small-cell on/off of the secondary carriers in unlicensed spectrum to realize a fractional loading mechanism for Wi-Fi coexistence. When the secondary carriers are active, Wi-Fi would detect the channel as busy and not transmit. Similarly, when the secondary carriers are inactive Wi-Fi can use the spectrum. The length of the on and off periods could vary according to the load but may be in the order of 100 ms. Although such a mechanism works and can be tuned for fair spectrum sharing with Wi-Fi, it would not be suitable for global deployment as some regions require LBT. This is the reason why LBT is a vital part of LAA since LAA is targeted to provide a single global solution framework. Note that the intention with LBT is to ensure fair sharing and coexistence with *other* networks, for example, Wi-Fi, and not to coordinate transmissions *within* an LAA cell. Coordinating transmissions between multiple devices in the same LAA cell is done through scheduling in the same way as in licensed spectrum.

Downlink LBT in LAA is based on the same underlying principles as Wi-Fi. Note that LBT is a much more dynamic operation than channel selection because it is performed prior to each transmission burst. It can therefore follow variations in the channel usage on a very fast time scale, basically in milliseconds.

A transmission burst is a contiguous transmission spanning one or more subframes on a given component carrier from a given eNodeB with no transmission immediately before or after. Before transmitting a burst, the eNodeB assesses whether the frequency channel is available or not by performing the LBT procedure with a random backoff as illustrated in Figure 17.4. Whenever the eNodeB has found the channel to be idle after previously being busy, it executes a defer period. The defer period starts with a delay of 16 μs, after which the eNodeB measures the energy for a certain number of 9 μs slots which depends on the priority class as shown in Table 17.2. The total length of the defer period, whose purpose is to avoid collisions with potential Wi-Fi acknowledgments transmitted 16 μs after data reception, is at least equal to 25 μs.[5] For each observation of the channel in a slot during the defer period, the frequency channel is declared available if the received energy is less than a threshold, which among other things depends on the regulatory requirements, the maximum transmission power, and whether the LAA is the only technology using the frequency channel.

After executing the defer period, the eNodeB performs a full random backoff procedure similarly to what was discussed in Section 17.2. A backoff timer is initialized with a random number, drawn from a uniform distribution [0, *CW*], representing the duration in multiples of 9 μs the channel must be available for before transmission can take place. The availability of the channel within a 9 μs time slot is subject to the same rules as in the defer period described

---

[5]The time 25 μs equals the sum of the 16 μs SIFS and the 9 μs slot duration.

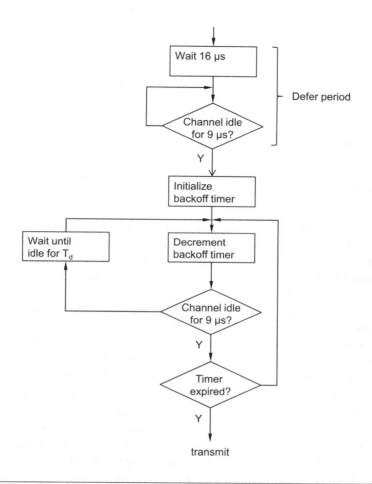

FIGURE 17.4

The LBT procedure for LAA.

<table>
<tr><td colspan="5">**Table 17.2 Contention-Windows Sizes for Different Priority Classes**</td></tr>
<tr><td></td><td>**Priority Class**</td><td>**Defer Period**</td><td>**Possible *CW* Values {*CW*_{min}, ..., *CW*_{max}}**</td><td>**Max Burst Length**[a]</td></tr>
</table>

	Priority Class	Defer Period	Possible CW Values $\{CW_{min}, ..., CW_{max}\}$	Max Burst Length[a]
1	Signaling, voice, real-time gaming	25 µs	{3,7}	2 ms
2	Streaming, interactive gaming	25 µs	{7,15}	3 ms
3	Best-effort data	43 µs	{15,31,63}	10 ms or 8 ms
4	Background traffic	79 µs	{15,31,63,127,255,511,1023}	10 ms or 8 ms

[a]*Regulatory requirements may limit the burst length to smaller values than in the table. If no other technology is sharing the frequency channel, 10 ms is used, otherwise 8 ms.*

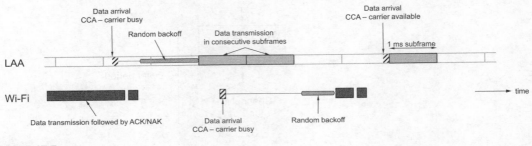

FIGURE 17.5

Example of coexistence between Wi-Fi and LAA.

earlier. Once the timer has expired, the random backoff is completed and the burst is transmitted. When the timer expires, the node may postpone transmission and transmit at a later time if the channel is found to be free for a period of 25 μs. If the channel is found to be busy during that period of 25 μs, the node executes another defer period and then performs a full random backoff as described earlier.

The size of the contention window is adjusted based on hybrid-ARQ acknowledgments received from the device with the contention window CW (approximately) doubling up to a limit CW_{max} if a negative hybrid-ARQ is received.[6] If the hybrid-ARQ acknowledgment is positive, the contention window is reset to its minimum value, $CW = CW_{min}$.

There are four different priority classes defined, each with individual contention windows and with different maximum and minimum values of the contention window as depicted in Table 17.2. The intention with different priority classes is to use a shorter backoff for high-priority data, thereby taking priority over low-priority transmissions. The high-priority classes also have a shorter maximum burst length to not block the channel for too long.

The purpose of LBT is, as already discussed, to avoid transmitting when the channel is already in use. One reason is to handle coexistence with other radio-access technologies using the unlicensed spectrum, for example, Wi-Fi as illustrated in Figure 17.5. However, as already discussed in conjunction with the Wi-Fi overview in Section 17.2, LBT may detect transmissions also in neighboring cells, in essence resulting in a time-reuse scheme as can be observed in dense Wi-Fi networks. To maintain the benefits of reuse-one operation, for which LTE was originally designed and which in many scenarios leads to higher capacity, LBT should preferably be blind to LAA transmissions in neighboring cells of the same network but still monitor activity from other radio-access technologies and other operators. One possibility to achieve this is to time synchronize neighboring cells and select a common start time for the LBT procedure such that all cells that are part of the same network perform LBT at the

---

[6]The description is slightly simplified; the specifications describe in detail how to handle different cases of bundled acknowledgments, see [26] for details.

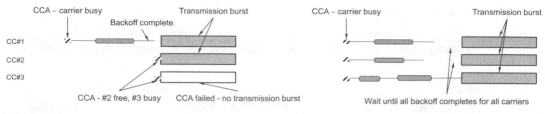

FIGURE 17.6

LBT for multiple carriers, single backoff counter (left), individual backoff counters (right).

same time. The pseudo-random generators should also be aligned across nodes to ensure the same backoff duration in neighboring cells.

The LBT procedure above has been described for the case of one carrier using unlicensed spectrum. If there are multiple carriers using unlicensed spectrum, backoff in conjunction with LBT can be handled in two ways illustrated in Figure 17.6:

- A single backoff value, valid for one of the component carriers in unlicensed spectrum. When the counter has expired, transmissions may take place on all component carriers subject to passing a clear-channel assessment (CCA) of duration 25 μs on the other carriers prior to transmission.
- Multiple backoff values, one per carrier. Transmissions take place once all backoff counters have reached zero. Note that the backoff completes at different time instants for the different carriers. However, "early" carriers cannot start to transmit as this would make listening on the other carriers impossible and the eNodeB therefore has to wait for the longest backoff counter.

Uplink transmissions in unlicensed spectrum is not part of release 13 but is planned for release 14. In principle, if the uplink transmission follows immediately after a downlink transmission, there is no need for LBT in the uplink as long as the maximum channel occupancy is not exceeded. A 25 μs CCA check prior to transmission would be sufficient, assuming the maximum channel occupancy is not exceeded. When needed, LBT in the uplink can be handled in a similar way as for the downlink. This would imply that a device sometimes may need to ignore an uplink grant received if the device finds the frequency channel to be busy when uplink transmission is to take place.

17.3.3 FRAME STRUCTURE AND BURST TRANSMISSION

When the channel is found to be available, the eNodeB can initiate a transmission burst the format of which is described later. However, before going into the details, a description of the LAA frame structure as such is beneficial.

The unlicensed 5 GHz band is an unpaired band and hence TDD is the relevant duplex scheme, making frame structure type 1 unsuitable for LAA. However, since LBT is used and

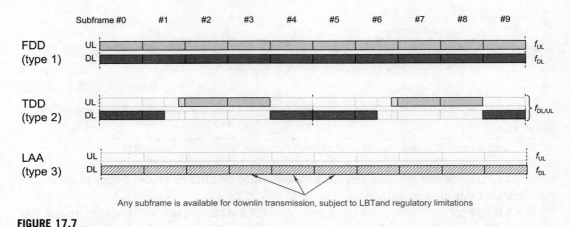

Any subframe is available for downlin transmission, subject to LBTand regulatory limitations

FIGURE 17.7

Frame structure type 3.

transmission burst can start in any subframe, frame structure type 2 with its fixed[7] split into uplink and downlink is not suitable either, especially since release 13 is supporting downlink only. Hence, a frame structure allowing downlink transmissions to start in any subframe, subject to LBT, and not enforcing a fixed allocation of uplink and downlink subframes is needed. For this reason, frame structure type 3, illustrated in Figure 17.7 was introduced in release 13. From most perspectives, frame structure type 3 has the same mapping of signals and channels as frame structure type 1 and is treated as such in various procedures: for example, carrier aggregation procedures for transmitting hybrid-ARQ acknowledgments in the uplink.

As described in the previous section, transmissions may start immediately after the LBT procedure is finished, alternatively at a later point in time as long as the channel is found available immediately before the transmission. Being able to start data transmissions only at one instant every 1 ms clearly has limitations in terms of LAA operation in highly loaded channels as another transmitter may grab the channel in the meantime. One possibility to avoid this would be for an eNodeB implementation to start transmission of an arbitrary "reservation signal" until the subframe starts to ensure the channel is available once the data transmission starts. To reduce the impact from the limitation of data transmissions starting at subframe boundaries only, the possibility to start data transmission also at the slot boundary inside a subframe, a partially filled subframe, is supported for LAA, see Figure 17.8. If the possibility to start downlink transmission at the slot boundary is enabled by RRC signaling, any transmissions starting in the second slot uses the same mapping for the second slot as otherwise used in the first slot of a normal subframe—that is, there is a control region at the

---

[7]With the introduction of eIMTA some flexibility in the uplink—downlink allocation is possible although only on a frame basis.

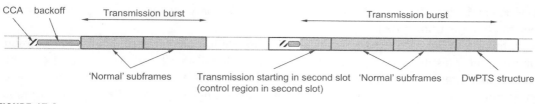

FIGURE 17.8

Illustration of transmission bursts partially filled subframes.

beginning of the second slot in this case. Neither synchronization signals, nor CSI-IM resources are assumed to be part of such a partially filled subframe. A device configured to support PDSCH transmissions starting in the second slot obviously need to monitor PDCCH/EPDCCH in the second slot in addition to the first slot.

The length of a transmission burst depends on the amount of data to transmit and regulatory requirements. As a consequence, PDSCH transmission may need to end prior to the end of a subframe. The last subframe in a burst can therefore use one of the DwPTS structures, in addition to occupying a full subframe. Unlike TDD operation in licensed spectrum, where the device knows in advance where the DwPTS is located from the semi-static uplink—downlink configuration, this is not the case in LAA as the burst length is a dynamic quantity. Instead, the eNodeB signals to the LAA terminal the occurrence of an ending partial subframe in that subframe and in the prior subframe. The signaling uses DCI format 1C and a reserved RNTI value, the CC-RNTI.

Data demodulation can be based on DM-RS or CRS, although the device can only assume these signals are present when a burst is transmitted. Transmission modes 1, 2, 3, 4, 8, 9, and 10 are supported. Only normal cyclic prefix can be used for LAA which is reasonable as the transmit power limitations for unlicensed spectrum implies relatively small cells with modest time dispersion.

17.3.4 REFERENCE SIGNALS AND DISCONTINUOUS TRANSMISSION

Discontinuous transmission—that is, the possibility for completely blank subframes—is a fundamental requirement and a consequence of LBT. Clearly, if the LBT mechanism declares the channel to be occupied, the base station (or device) should not transmit. Consequently, periodically transmitted signals need special attention. In the downlink this impacts the CRS, CSI-RS, and discovery reference signal (DRS) design.

Cell-specific reference signals cannot occur in every subframe as is the baseline for LTE. Hence, functionality such as AGC setting, time—frequency synchronization, RRM measurements, channel-state feedback, and demodulation of data and control channels need to rely on signals that are part of a transmission burst only. AGC setting and time—frequency synchronization can be handled by the CRS both in subframes carrying PDSCH as well as the discovery signal. The DM-RS may also be additionally used. For data demodulation both

DM-RS and CRS can be used, although the device can only assume these signals are present when a burst is transmitted.

Similarly to the CRS, CSI-RS may not be possible to transmit periodically. Furthermore, intermittent transmissions as a result of LBT as well as other non-LAA usage of the unlicensed spectrum typically lead to a rapidly fluctuating interference environment. Interference measurements when the serving cell is not transmitting may not reflect the interference characteristics when the device is receiving data. Hence, interference measurements for CSI purposes should only occur when the serving cell is transmitting. Transmission occasions for CSI-RS are configured in the same way as for a licensed carrier—that is, according to a certain repetitive pattern—but only those CSI-RS overlapping with a transmission burst are transmitted. Since the device knows when CSI-RS are transmitted in a burst to that device, it can use these CSI-RS as a basis for its CSI reports in the same way as for a licensed carrier. The CSI reports are then transmitted on PUSCH on a licensed carrier.

The DRS was introduced in release 12 to support small-cell on/off where secondary component carriers can be turned off except for periodic transmission of DRS as described in Chapter 15. The device is configured with a time window within which it expects the DRS to be transmitted. DRS are used also in LAA and are the basis for radio-resource management including cell identification. However, as a consequence of transmitting on unlicensed spectrum, DRS transmission must be preceded by a CCA.

If the DRS are transmitted together with the PDSCH—that is, inside a transmission burst—it is automatically subject to the LBT mechanism performed prior to the burst and the discovery signal can be multiplexed[8] with ongoing data transmission in the burst.

On the other hand, if the discovery signal is not part of a transmission burst, the DRS must be preceded by a CCA spanning 25 μs. The DRS is transmitted only if the channel is detected as being available. Since the exact time when the DRS is transmitted depends on the CCA, the discovery signal may move around in time and the device needs to detect whether a DRS is transmitted or not before basing measurements on it, something that is not necessary in licensed spectrum.

The DRS structure is the same as in previous releases, but the duration of a discovery signal is limited to 12 OFDM symbols. Since the discovery signal is potentially moving around in time as a result of the CCA, the PSS/SSS being part of a discovery signal may occur outside subframe 0 and 5. Consequently, frame timing cannot be obtained from the PSS/SSS on the secondary carrier—only subframe timing can be obtained—however, this is not a problem as the frame timing can be obtained from the licensed carrier. Finally, to simplify the DRS detection in the device, the CRS/CSI-RS/PSS/SSS sequences do not vary with the subframe number but are kept unchanged across subframes 0—4 and 5—9, respectively.

The release 12 DRS structure is not contiguous in time. A consequence of this is that another node in principle can find the channel being available during the unused OFDM

---

[8]Simultaneous transmission of a discovery signal and PDSCH is possible in subframes 0 and 5 only.

symbols between two CRS symbols and start transmitting. One possibility for an implementation to avoid this is by transmitting a "dummy signal" when needed to make the DRS transmission contiguous in time.

Discontinuous transmission for a future extension to the uplink is straightforward as all transmissions are scheduled. Periodic transmissions of uplink sounding reference signals can easily be avoided by relying on aperiodic SRS only.

17.3.5 SCHEDULING, HYBRID-ARQ, AND RETRANSMISSIONS

Scheduling requires no major changes but is handled the same way as carrier aggregation in general. Both self-scheduling and cross-carrier scheduling can be used. Self-scheduling can be beneficial for downlink transmissions to better spread the control signaling load across all the carriers; if the channel is available for control signaling it will also be available for downlink data transmission and a single LBT is sufficient. Only cross-carrier scheduling from the primary carrier, which operates on licensed spectrum, is supported; scheduling an unlicensed carrier from another unlicensed carrier is neither supported, nor useful. Furthermore, in case of cross-carrier scheduling, the PDSCH always starts in the first subframe—that is, partially filled subframes are not supported.

Hybrid-ARQ requires no changes for the downlink; the existing asynchronous hybrid-ARQ scheme can be used. Hybrid-ARQ retransmissions are subject to LBT in the same way as an original transmission. Depending on the transmission duration, retransmissions may follow in the same burst as the original transmission. A device transmits hybrid-ARQ acknowledgments on the uplink primary carrier using the same mechanism as in carrier aggregation in general.

For uplink transmission, which is not part of release 13 but to be considered in later releases, cross-carrier scheduling from licensed spectrum can be beneficial as self-scheduling may require two successful LBTs, one for the downlink control signaling and the other for the actual uplink data transmission. Furthermore, asynchronous hybrid-ARQ operation is required in the uplink, calling for enhancements to the uplink hybrid-ARQ protocol. A synchronous protocol implies a fixed timing for the retransmission and the channel may not be available at a fixed retransmission instant. Asynchronous operation requires the hybrid-ARQ process to be signaled. Consequently, retransmissions must be scheduled with PDCCH or EPDCCH and not PHICH in this case.

17.3.6 RADIO BEARER MAPPING AND QOS CONTROL

QoS handling in LTE is, as discussed in Chapter 4, handled by different radio bearers. Multiple radio bearers are multiplexed and, in case of carrier aggregation, the multiplexed data stream is distributed across the component carriers. Carrier aggregation is thus invisible to the RLC and PDCP layers and a certain radio bearer may be transmitted on an arbitrary subset of component carriers. In case all component carriers are transmitted in licensed spectrum, which was the assumption when carrier aggregation was developed, this is not an issue as all component carriers have essentially the same conditions.

For LAA, where some component carriers are transmitted in unlicensed spectrum, the situation is quite different. Depending on the interference situation, other usage of the unlicensed spectrum, and the outcome of LBT, the component carriers in unlicensed spectrum may have substantially different conditions than the ones in licensed spectrum. At some points, the unlicensed component carriers may have very low availability or be subject to a long latency due to LBT. Hence, for LAA there is a need to control on which component carriers a certain radio bearer is mapped.

In the downlink this is an implementation issue. The scheduler in the eNodeB can control which data from the different radio bearers is mapped to the different component carriers, thus controlling which data is transmitted on licensed spectrum and which is transmitted on unlicensed spectrum.

In the uplink, which is not part of release 13, the situation is more complex. The scheduler cannot control which component carrier a certain piece of data is transmitted upon. Critical data may therefore end up on less reliable unlicensed spectrum, possibly resulting in guaranteed bitrate requirements not being fulfilled. Furthermore, LBT mechanisms may impact when transmission occur on a certain component carrier, thereby affecting any latency requirements. One possible approach could be to schedule data on unlicensed spectrum only when the buffer status reports indicate that no critical data is awaiting transmission, but this could result in inefficient utilization of unlicensed spectrum. Enhancements to control upon which component carriers a certain radio bearer is mapped can therefore be beneficial. However, this is not part of release 13 but may be considered in future releases.

17.4 ENHANCEMENTS BEYOND RELEASE 13

Uplink transmission in unlicensed spectrum was studied to some extent in release 13, but in order to complete release 13 in time it was decided to focus on downlink transmissions. In release 14, support for uplink transmissions on unlicensed spectrum will be added. Some of the aspects to consider for uplink transmissions have been discussed in the previous sections.

Another enhancement for future releases could be to support LAA also in the dual-connectivity framework. Carrier aggregation, which is the basis for LAA in release 13, in practice requires licensed and unlicensed spectrum to be handled in the same node or across an ideal backhaul as a consequence of the tight timing relations between the component carriers. Dual connectivity with more relaxed timing relations between the two nodes would allow greater deployment flexibility in this respect. For example, an existing macro site in licensed spectrum could be complemented by small nodes handling unlicensed spectrum only even without ideal backhaul between the macro site and the small nodes. One could even envision multiple operators with separate macro networks sharing a common set of nodes in unlicensed spectrum—for example, an indoor deployment in an office building.

Stand-alone operation, where LTE is extended to operate in unlicensed spectrum without support for licensed spectrum, is another possible enhancement. This would broaden the applicability of LTE also to entities not owning any licensed spectrum (e.g., small business, venues, and landlords). Examples of the technical enhancements required include mobility, random access, and system information distribution. Most of these enhancements are also needed for extending LAA to use the dual-connectivity framework as a complement to the carrier-aggregation-based design.

RELAYING

18

The possibility of a device communicating with the network, and the data rate that can be used, depends on several factors, including the path loss between the device and the base station. The link performance of LTE is already quite close to the Shannon limit and from a pure link-budget perspective, the highest data rates supported by LTE require a relatively high signal-to-noise ratio (SNR). Unless the link budget can be improved—for example, with different types of beam-forming solutions—a denser infrastructure is required to reduce the device-to-base-station distance and thereby improve the link budget.

A denser infrastructure is mainly a deployment aspect, but in later releases of LTE, various tools enhancing the support for low-power base stations were included. One of these tools is *relaying*, which can be used to reduce the distance between the device and the infrastructure, resulting in an improved link budget and an increased possibility for high data rates. In principle this reduction in device-to-infrastructure distance could be achieved by deploying traditional base stations with a wired connection to the rest of the network. However, relays with a shorter deployment time can often be an attractive alternative, as there is no need to deploy a specific backhaul.

A wide range of relay types can be envisioned, some of which could already be deployed in release 8.

Amplify-and-forward relays, commonly referred to as *repeaters*, simply amplify and forward the received analog signals and are, on some markets, relatively common as a tool for handling coverage holes. Traditionally, once installed, repeaters continuously forward the received signal regardless of whether there is a device in their coverage area or not, although more advanced repeaters can be considered as well. Repeaters are transparent to both the device and the base station and can therefore be introduced in existing networks. The fact that the basic principle of a repeater is to amplify whatever it receives, including noise and interference as well as the useful signal, implies that repeaters are mainly useful in high-SNR environments. Expressed differently, the SNR at the output of the repeater can never be higher than at the input.

Decode-and-forward relays decode and re-encode the received signal prior to forwarding it to the served users. The decode-and-re-encode process results in this class of relays not amplifying noise and interference, as is the case with repeaters. They are therefore also useful in low-SNR environments. Furthermore, independent rate adaptation and scheduling for the

4G, LTE-Advanced Pro and The Road to 5G. http://dx.doi.org/10.1016/B978-0-12-804575-6.00018-2

403

base station—relay and relay—device links is possible. However, the decode-and-re-encode operation implies a larger delay than for an amplify-and-forward repeater, longer than the LTE subframe duration of 1 ms. As for repeaters, many different options exist depending on supported features (support of more than two hops, support for mesh structures, and so on) and, depending on the details of those features, a decode-and-forward relay may or may not be transparent to the device.

18.1 RELAYS IN LTE

LTE release 10 introduced support for a decode-and-forward relaying scheme (repeaters require no additional standardization support other than RF requirements and are already available in release 8). A basic requirement in the development of LTE relaying solutions was that the relay should be transparent to the device—that is, the device should not be aware of whether it is connected to a relay or to a conventional base station. This ensures that release 8/9 devices can also be served by relays, despite relays being introduced in release 10. Therefore, so-called *self-backhauling* was taken as the basis for the LTE relaying solution. In essence, from a logical perspective, a relay is an eNodeB wirelessly connected to the rest of the radio-access network by using the LTE radio interface. It is important to note that, even though the relay from a device perspective is identical to an eNodeB, the physical implementation may differ significantly from a traditional base station, for example, in terms of output power.

In conjunction with relaying, the terms *backhaul link* and *access link* are often used to refer to the base station—relay connection and the relay—device connection, respectively. The cell to which the relay is connected using the backhaul link is known as the *donor cell* and the donor cell may, in addition to one or several relays, also serve devices not connected via a relay. This is illustrated in Figure 18.1.

Since the relay communicates both with the donor cell and devices served by the relay, interference between the access and backhaul links must be avoided. Otherwise, since the power difference between access-link transmissions and backhaul-link reception at the relay can easily be more than 100 dB, the possibility of receiving the backhaul link may be completely ruined. Similarly, transmissions on the backhaul link may cause significant

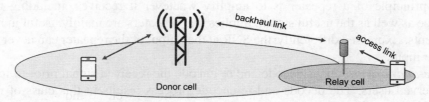

FIGURE 18.1

Access and backhaul links.

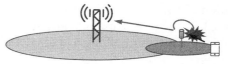

Access-to-backhaul interference   Backhaul-to-access interference

FIGURE 18.2

Interference between access and backhaul links.

interference to the reception of the access link. These two cases are illustrated in Figure 18.2. Therefore, isolation between the access and backhaul links is required—isolation that can be obtained in one or several of the frequency, time, and/or spatial domains.

Depending on the spectrum used for access and backhaul links, relaying can be classified into *outband* and *inband* types.

Outband relaying implies that the backhaul operates in a spectrum separate from that of the access link, using the same radio interface as the access link. Provided that the frequency separation between the backhaul and access links is sufficiently large, interference between the backhaul and access links can be avoided and the necessary isolation is obtained in the frequency domain. Consequently, no enhancements to the release 8 radio interface are needed to operate an outband relay. There are no restrictions on the activity on the access and backhaul links, and the relay can in principle operate with full duplex.

Inband relaying implies that the backhaul and access links operate in the same spectrum. Depending on the deployment and operation of the relay, this may, as the access and backhaul link share the same spectrum, require additional mechanisms to avoid interference between the access and backhaul links. Unless this interference can be handled by proper antenna arrangements, for example, with the relay deployed in a tunnel with the backhaul antenna placed outside the tunnel, a mechanism to separate activity on the access and backhaul links in the time domain is required. Such a mechanism was introduced as part of release 10 and is described in more detail below. Since the backhaul and access links are separated in the time domain, there is a dependency on the transmission activity and the two links cannot operate simultaneously.

The RF requirements for decode-and-forward relays were introduced in release 11. Because of the similarities with operation of the access and backhaul links with base stations and UEs, respectively, the requirements are to a large extent very similar to the corresponding ones for base stations and UEs. This is discussed in more detail in Chapter 22.

18.2 OVERALL ARCHITECTURE

From an architectural perspective, a relay can, on a high level, be thought of as having a "base-station side" and a "device side". Toward devices, it behaves as a conventional eNodeB using the access link, and a device is not aware of whether it is communicating with a relay or a "traditional" base station. Relays are therefore transparent for the devices and devices from the first LTE release, release 8, can also benefit from relays. This is important from an operator's perspective, as it allows a gradual introduction of relays without affecting the existing device fleet.

Toward the donor cell, a relay initially operates as a device, using the LTE radio interface to connect to the donor cell. Once connection is established and the relay is configured, the relay uses a subset of the "device side" functionality for communication on the backhaul link. In this phase, the relay-specific enhancements described in this chapter may be used for the backhaul.

In release 10, the focus is on two-hop relaying and scenarios with a relay connected to the network via another relay are not considered. Furthermore, relays are stationary—that is, handover of a relay from one donor cell to another donor cell is not supported. The case for using mobile relays is not yet clear, and therefore it was decided in release 10 not to undertake the relatively large task of adapting existing core-network procedures to handle cells that are moving over time, something that could have been a consequence of a mobile relay.

The overall LTE relaying architecture is illustrated in Figure 18.3. One key aspect of the architecture is that the donor eNodeB acts as a proxy between the core network and the relay. From a relay perspective, it appears as if the relay is connected directly to the core network as the donor eNodeB appears as an MME for the S1 interface and an eNodeB for X2 toward the relay. From a core-network perspective, on the other hand, the relay cells appear as if they belong to the donor eNodeB. It is the task of the proxy in the donor eNodeB to connect these two views. The use of a proxy is motivated by the desire to minimize the impact to the core network from the introduction of relays, as well as to allow for features such as tight coordination of radio-resource management between the donor eNodeB and the relay.

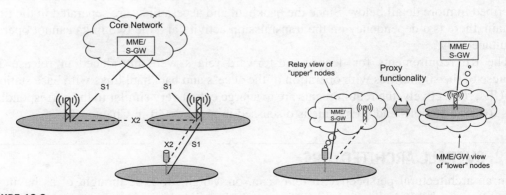

FIGURE 18.3

LTE relaying architecture.

18.3 BACKHAUL DESIGN FOR INBAND RELAYING

In the case of inband relaying, the backhaul and access links operate in the same spectrum. As discussed in the previous section, a mechanism to separate activity on the access and backhaul links in the time domain is required unless sufficient isolation between the two links can be achieved in other ways, for example, through appropriate antenna arrangements. Such a mechanism should ensure that the relay is not transmitting on the access link at the same time as it is receiving on the backhaul link (and vice versa).

One way to handle this is to "blank" some subframes on the access link to provide the relay with the possibility to communicate with the donor eNodeB on the backhaul link. In the uplink, the scheduler in the relay can in principle schedule such that there is no access-link activity in certain subframes. These subframes can then be used for uplink transmissions on the backhaul link as the relay does not need to receive anything on the access link in these subframes. However, blanking subframes on the access downlink is not possible. Although a release 10 device in principle could have been designed to cope with blank subframes, devices from earlier releases expect at least cell-specific reference signals (CRS) to be present in all downlink subframes. Hence, to preserve the possibility of also serving releases 8/9 devices, which was an important requirement during standardization of release 10, the design of the backhaul link must be based on the assumption that the access link can operate with release 8 functionality only.

Fortunately, from the first release LTE included the possibility of configuring MBSFN subframes (see Chapter 5). In an MBSFN subframe, devices expect CRS and (possibly) L1/L2 control signaling to be transmitted only in the first one or two OFDM symbols, while the remaining part of the subframe can be empty. By configuring some of the access-link subframes as MBSFN subframes, the relay can stop transmitting in the latter part of these subframes and receive transmissions from the donor cell. As seen in Figure 18.4, the gap during which the relay can receive transmissions from the donor cell is shorter than the full subframe duration. In particular, as the first OFDM symbols in the subframe are unavailable for reception of donor-cell transmissions, L1/L2 control signaling from the donor to the relay cannot be transmitted using the regular PDCCH. Instead, a relay-specific control channel, the R-PDCCH, is introduced in release 10.

Not only are transmission gaps in the access downlink required in order to receive transmissions from the donor cell, but also reception gaps in the access link are needed in

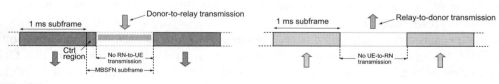

FIGURE 18.4

Multiplexing between access and backhaul links.

order to transmit on the backhaul from the relay to the donor cell. As already mentioned, such gaps can be created through proper scheduling of uplink transmissions.

The detailed specifications of the physical-layer enhancements introduced in release 10 to support the backhaul can be found in [37].

18.3.1 ACCESS-LINK HYBRID-ARQ OPERATION

The access-link gaps discussed earlier, MBSFN subframes in the downlink and scheduling gaps in the uplink, used in order to be able to receive and transmit, respectively, on the backhaul link, affect the hybrid-ARQ operation. Note that hybrid-ARQ is used on both the access and backhaul links. Since compatibility with release 8 was a fundamental requirement in the development of the LTE relaying solution, there are no changes to access-link hybrid-ARQ operation.

For uplink transmissions on PUSCH, hybrid-ARQ acknowledgments are transmitted on PHICH. Since the PHICH can be transmitted by the relay even in MBSFN subframes, the operation is identical to that in earlier releases of the LTE standard. However, although the hybrid-ARQ acknowledgment can be received, the subframe where the retransmission should take place (8 ms after the initial transmission for FDD, configuration dependent for TDD) may be used by the backhaul link and not be available for the access link. In that case the corresponding uplink hybrid-ARQ process needs to be suspended by transmitting a positive acknowledgment on the PHICH, irrespective of the outcome of the decoding. By using PDCCH, a retransmission can instead be requested in a later subframe available for the same hybrid-ARQ process, as described in Chapter 8. The hybrid-ARQ round-trip time will be larger in those cases (e.g., 16 ms instead of 8 ms for FDD).

Downlink transmissions on PDSCH trigger hybrid-ARQ acknowledgments to be sent on PUCCH and, for proper operation, the relay should be able to receive those acknowledgemnts. The possibility to receive PUCCH on the access link depends on the backhaul operation, more specifically on the allocation of subframes for backhaul communication.

In FDD, backhaul subframes are configured such that an uplink subframe occurs 4 ms after a downlink subframe. This is chosen to match the access-link hybrid-ARQ timing relations, where an uplink subframe follows 4 ms after a downlink subframe. As the relay cannot transmit on the access link simultaneously with the backhaul link, there is no access-link transmission in subframe n and, consequently, no hybrid-ARQ transmission in subframe $n + 4$. Hence, the inability to receive access-link hybrid-ARQ acknowledgments in some subframes is of no concern as the corresponding downlink subframes cannot be used for access-link transmission anyway. Downlink retransmissions are not an issue as they are asynchronous and can be scheduled in any suitable downlink subframe on the access link.

In TDD, the relay node may not be able to receive hybrid-ARQ feedback on PUCCH in uplink subframes used for transmission on the backhaul link. One possibility is to restrict the downlink scheduler such that no devices transmit PUCCH in uplink subframes the relay cannot receive. However, such a restriction may be too limiting. Alternatively, the relay can schedule without restrictions in the downlink and ignore the hybrid-ARQ acknowledgment. Retransmissions can then either be handled blindly—that is, the relay has to make an educated "guess" on whether a retransmission is required based on, for example, CSI feedback or RLC retransmissions are used to handle missing packets. Another possibility is to configure repetition of the hybrid-ARQ acknowledgments such that at least some of the repeated acknowledgments are receivable by the relay.

18.3.2 BACKHAUL-LINK HYBRID-ARQ OPERATION

For the backhaul link, the underlying principle in the design is to maintain the same timing relations as in release 8 for scheduling grants and hybrid-ARQ acknowledgments. As the donor cell may schedule both relays and devices, such a principle simplifies the scheduling implementation, as scheduling and retransmission decisions for devices and relays are taken at the same point in time. It also simplifies the overall structure, as release 8 solutions can be reused for the relay backhaul design.

For FDD, the subframes configured for downlink backhaul transmission therefore follow a period of 8 ms in order to match the hybrid-ARQ round-trip time to the extent possible. This also ensures that the PUCCH can be received in the access link, as discussed in the previous section. However, as the possible configurations of MBSFN subframes have an inherent 10 ms structure while the hybrid-ARQ timing follows an 8 ms periodicity, there is an inherent mismatch between the two. Hence, as illustrated in Figure 18.5, some backhaul subframes may be spaced 16 ms apart, as subframes 0, 4, 5, and 9 cannot be configured as MBSFN subframes (see Chapter 5). Uplink backhaul subframes follow 4 ms after a downlink backhaul subframe, following the principle discussed in the previous paragraph.

For TDD, there is an inherent 10 ms component in the hybrid-ARQ timing relations, which matches the 10 ms MBSFN structure and makes it possible to keep a regular spacing of the backhaul transmission attempts. Subframes 0, 1, 5, and 6 cannot be configured as MBSFN subframes. Hence, TDD configuration 0, where subframes 0 and 5 are the only downlink subframes, cannot be used in a relay cell since this configuration does not support any MBSFN subframes. For configuration 5, there is only a single uplink subframe and in order to support both the backhaul and access links at least two uplink subframes are needed.

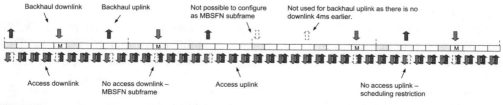

FIGURE 18.5

Example of backhaul configuration for FDD.

Therefore, of the seven TDD configurations supported in LTE, only configurations 1, 2, 3, 4, and 6 are supported in relay cells. For each TDD configuration, one or several backhaul configurations are supported, as shown in Table 18.1.

The underlying timing principles for hybrid-ARQ acknowledgments and uplink scheduling grants are, as mentioned earlier, to keep the same principles as for the access link. However, backhaul transmissions may occur in backhaul subframes only. Therefore, for TDD, the acknowledgment of a transport block on the backhaul link in subframe n is transmitted in subframe $n + k$, where $k \geq 4$ and is selected such that $n + k$ is an uplink *backhaul* subframe if the acknowledgment is to be transmitted from the relay and a downlink *backhaul* subframe if the acknowledgment is transmitted from the eNodeB.

The numbering of uplink hybrid-ARQ processes on the backhaul link is similar to the TDD numbering on the access link, where uplink hybrid-ARQ process numbers are assigned sequentially to the available backhaul occasions as shown in Figure 18.6, taking into account the same processing times as for the access link (see Chapter 8). This is in contrast to the FDD access link, where the uplink hybrid-ARQ process number can be directly derived from the subframe number. The reason for adopting a somewhat different strategy is to minimize the maximum hybrid-ARQ round-trip time. Due to the fact that a pure 8 ms periodicity does not always match the MBSFN allocation, the actual uplink round-trip time, unlike the FDD access link, is not constant but, similar to the TDD access link, is dependent on the subframe number.

18.3.3 BACKHAUL DOWNLINK CONTROL SIGNALING

The gap during which the relay can receive transmissions from the donor cell is, as seen in Figure 18.4, shorter than the full subframe duration. In particular, as the first OFDM symbols in the subframe are unavailable for reception of transmissions from the donor cell, L1/L2 control signaling from the donor to the relay cannot be transmitted using the regular PDCCH.[1] Instead, a relay-specific control channel, the R-PDCCH, was introduced in release 10.

The R-PDCCH carries downlink scheduling assignments and uplink scheduling grants, using the same DCI formats as for the PDCCH. However, there is no support for power control commands using DCI formats 3/3A. The main function of DCI formats 3/3A is to support semi-persistent scheduling, a feature mainly targeting overhead reduction for low-rate services and not supported for the backhaul link.

In the time domain, the R-PDCCH is, as already mentioned, received in the "MBSFN region" of the subframe, while in the frequency domain, transmission of the R-PDCCH occurs in a set of semi-statically allocated resource blocks. From a latency perspective it is

---

[1]In principle, the PDCCH could be received if the subframe structures of the access and backhaul links are offset by two to three OFDM symbols, but with the drawback that relay and donor cells would not be time aligned, which is beneficial, for example, in heterogeneous deployments.

Table 18.1 Supported Backhaul Configurations for TDD

Backhaul Subframe Configuration	Uplink–Downlink Configuration in Relay Cell	Backhaul DL:UL Ratio	0	1	2	3	4	5	6	7	8	9
0	1	1:1					D				U	
1		2:1				U	D					D
2							D				U	D
3	2	2:2				U	D				D	D
4		1:1			U	U	D					D
5		2:1				U					U	D
6					U	D				U	D	D
7		2:1			U	D	D			U		
8	3					D				U	D	D
9		3:1			U	D	D			D		
10						U				U	D	D
11		2:1				U				D	D	D
12		3:1				U				D	D	D
13	4	1:1				U					D	D
14		2:1				U				D		D
15						U					D	D
16		3:1				U	D			D	D	D
17		4:1				U				D	D	D

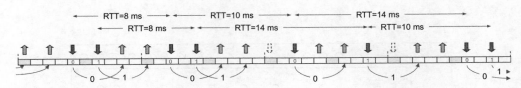

FIGURE 18.6

Example of hybrid-ARQ process numbering for FDD.

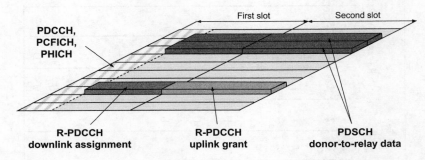

FIGURE 18.7

Example of R-PDCCH transmission.

beneficial to locate transmissions of downlink scheduling assignments as early as possible in the subframe. As discussed in Chapter 6, this was the main motivation for dividing normal subframes into a control region and a data region. In principle, a similar approach could be taken for the R-PDCCH, namely dividing the set of resource blocks used for R-PDCCH transmission into a control part and a data part. However, because it is not possible to exploit fractions of a subframe for transmission of PDSCH to devices connected directly to the donor cell, transmission of a single R-PDCCH could block usage of a relatively large number of resource blocks. From an overhead and scheduling flexibility perspective, a structure where the frequency span of the R-PDCCH is minimized (while still providing sufficient diversity) and resources are allocated mainly in the time dimension is preferable. In the release 10 design of the R-PDCCH, these seemingly contradicting requirements have been addressed through a structure where downlink assignments are located in the first slot and uplink grants, which are less time critical, in the second slot of a subframe (see Figure 18.7). This structure allows the time-critical downlink assignments to be decoded early. To handle the case when there is no uplink grant to transmit to the relay, the R-PDCCH resources in the second slot may be used for PDSCH transmission *to the same relay*.

Coding, scrambling, and modulation for the R-PDCCH follows the same principles as for the PDCCH (see Chapter 6), with the same set of aggregation levels supported (one, two, four,

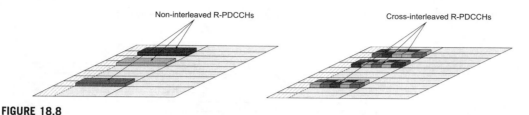

FIGURE 18.8

R-PDCCH mapping types, no cross-interleaving (left) and cross-interleaving (right).

and eight CCEs). However, the mapping of the R-PDCCH to time—frequency resources is different. Two different mapping methods, illustrated in Figure 18.8, are supported:

- without cross-interleaving,
- with cross-interleaving.

Without cross-interleaving, one R-PDCCH is mapped to one set of virtual resource blocks, where the number of resource blocks (one, two, four, or eight) depends on the aggregation level. No other R-PDCCHs are transmitted using the same set of resource blocks. If the resource blocks are located sufficiently apart in the frequency domain, frequency diversity can be obtained, at least for the higher aggregation levels. Non-interleaved mapping is, for example, useful for beam-forming of the backhaul transmissions or when applying frequency-selective scheduling to the R-PDCCH. Either CRS or demodulation reference signals (DM-RS) can be used for demodulation.

Cross-interleaved mapping is similar to the strategy used for the PDCCH and reuses most of the PDCCH processing structures except for the mapping to resource elements. A set of R-PDCCHs is multiplexed together, interleaved, and mapped to a set of resource blocks allocated for R-PDCCH transmission. As transmissions to multiple relays may share the same set of resource blocks, CRS are the only possibility for demodulation. The motivation for this mapping method is to obtain frequency diversity also for the lowest aggregation level. However, it also comes at the cost of blocking additional resource blocks from PDSCH transmission as, even at low aggregation levels, several resource blocks in the frequency domain are used for the R-PDCCH.

For both mapping cases, cross-interleaved as well as non-cross-interleaved, a set of candidate R-PDCCHs is monitored by the relay node. The set of resource blocks upon which the relay monitors for R-PDCCH transmission is configurable by the donor cell by signaling a set of virtual resource blocks using resource allocation type 0, 1, or 2 (see Chapter 6 for a discussion on resource allocation types). The sets may or may not overlap across multiple relay nodes. In the subframes used for backhaul reception, the relay attempts to receive and decode each of the R-PDCCHs candidates as illustrated in Figure 18.9 and, if valid downlink control information is found, applies this information to downlink reception or uplink transmission. This approach is in essence similar to the blind decoding procedure used in the devices, although there are some differences. First, there are no common search

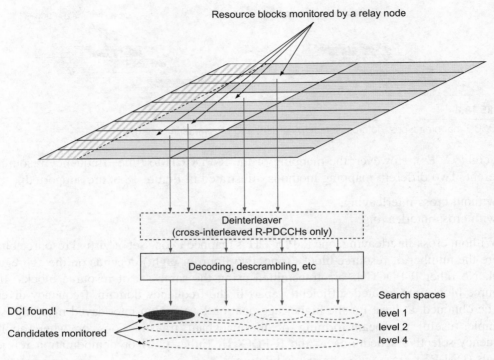

FIGURE 18.9

Principle illustration of R-PDCCH monitoring.

spaces for the relays as there is no need to receive broadcast information. Any information necessary for relay operation is transmitted using dedicated signaling. Secondly, the search spaces for the non-interleaved mapping are not time varying as in devices, but remain static in time.

The number of blind decoding attempts is the same as for a device—that is, six, six, two, and two attempts for aggregation levels one, two, four, and eight, respectively. However, note that an R-PDCCH can be transmitted in either the first or second slot. Hence, the total number of decoding attempts performed by a relay is 64.[2]

No PHICH channel is defined for the backhaul. The main reason for the PHICH in release 8 was efficient support of nonadaptive retransmissions for delay-sensitive low-rate applications such as voice-over IP. The backhaul from a relay, on the other hand, typically uses a higher data rate as multiple devices are served by the relay. Hence, as control signaling overhead is less of an issue, the PHICH was omitted from the backhaul in order to simplify the overall design. Retransmissions are still supported through the use of the R-PDCCH.

---

[2]Two slots and two DCI formats per transmission mode results in $2 \cdot 2 \cdot (6 + 6 + 2 + 2) = 64$.

18.3.4 REFERENCE SIGNALS FOR THE BACKHAUL LINK

Backhaul reception at the relay can use CRS or DM-RS, described in Chapter 6. Different reference-signal types can be used for R-PDCCH and PDSCH, but if the R-PDCCH is received using DM-RS, then DM-RS should be used for PDSCH as well. This is a reasonable restriction as DM-RS for R-PDCCH is motivated by beam-forming. If beam-forming is used for the R-PDCCH, there is no incentive not to use beam-forming also for the PDSCH. The opposite scenario, CRS for R-PDCCH and DM-RS for the PDSCH, does make sense though. One example is interleaved mapping of the control signaling, where multiple R-PDCCHs are multiplexed and individual beam-forming cannot be used, together with beam-forming of the PDSCH. The different combinations of reference signals supported for the backhaul link are summarized in Table 18.2.

Note also that in the case of (global) time alignment between the donor and relay cell, the last OFDM symbol cannot be received by the relay as it is needed for reception—transmission switching. Hence, the DM-RS on the last OFDM symbols in the subframe cannot be received. For transmission ranks up to 4 this is not a problem, as the necessary reference signals are also available earlier in the subframe. However, for spatial multiplexing with five or more layers, the first set of reference signals in the subframe is used for the lower layers while the second set of reference signals, located at the end of the subframe and that cannot be received, is used for the higher layers. This implies that reference signals for rank 5 and higher cannot be received by the relay, and backhaul transmissions are therefore restricted to at most four-layer spatial multiplexing, irrespective of the timing relation used.

18.3.5 BACKHAUL—ACCESS LINK TIMING

To ensure that the relay is able to receive transmissions from the donor cell, some form of timing relation between the downlink transmissions in the donor and relay cells must be defined, including any guard time needed to allow the relay to switch between access-link transmission to backhaul-link reception and vice versa.

Table 18.2 Combinations of Reference Signals and R-PDCCH Mapping Schemes

Reference Signal Type used for Demodulation of		R-PDCCH Mapping Scheme
R-PDCCH	**PDSCH**	
CRS	CRS	Cross-interleaved or non-cross-interleaved
CRS	DM-RS	Cross-interleaved or non-cross-interleaved
DM-RS	DM-RS	Non-cross-interleaved

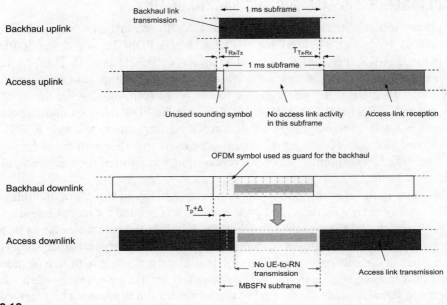

FIGURE 18.10

Backhaul timing relations in case the relay cell timing is derived from the backhaul timing.

A natural choice for the timing of the access link is to synchronize it to the frame timing of the backhaul link as observed by the relay. From this backhaul downlink timing reference, the timing of the access-link transmission is derived as shown at the bottom of Figure 18.10. The backhaul uplink timing is subject to the normal timing advance controlled by the donor cell, ensuring that the backhaul-uplink transmissions are time aligned with other uplink transmissions received by the donor base station.

In the backhaul downlink, the first OFDM symbol in the data region is left unused to provide the guard time for relay switching, and a small time offset is used to distribute the guard between Tx—Rx and Rx—Tx switching at the relay. This case is shown at the bottom of Figure 18.11. Locating the guard symbol at the beginning of the data region instead of at the end is beneficial as the guard symbol is needed at the relay side only and can therefore still be used for transmission of PDCCHs to devices in the donor cell. In principle, the guard time comes "for free" from a donor cell perspective and the freedom in shifting the relay node frame timing relative to the donor cell timing is used to move the "free" guard period to where it is needed.

The backhaul uplink is subject to the normal timing advance controlled by the donor cell, ensuring that the backhaul uplink transmissions are time aligned with other uplink transmissions received by the donor base station. Similarly to the guard time needed to switch from access-link transmission to backhaul-link reception, which influenced the downlink

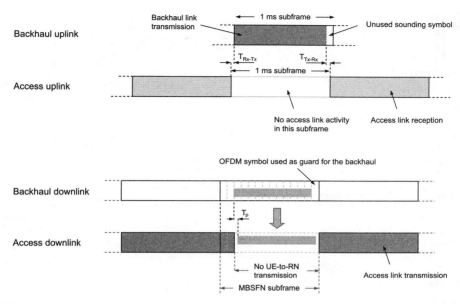

FIGURE 18.11

Backhaul timing relations in access link transmission in the relay and donor cells are time synchronized.

timing relation between the access and backhaul links, there may also be the need for a guard time in the uplink direction to switch from access-link reception to backhaul-link transmission. However, unlike the downlink case, how to handle this is not standardized but left for implementation, noting that functionality already present in release 8 is sufficient for providing the necessary guard time.

In principle, if the relay could switch from access-link reception to backhaul-link transmission within the cyclic prefix, no provisions for additional switching time would be necessary. However, the switching time is implementation dependent and typically larger than the cyclic prefix. For larger switching times, one possibility is to use the shortened transmission format on the access link, originally intended for sounding, as shown in the top part of Figure 18.10. By configuring all the devices in the relay cell to reserve the last OFDM symbol of the preceding subframe for sounding-reference signals but not to transmit any sounding-reference signals, a guard period of one OFDM symbol is created. This guard time can then be divided into Rx−Tx and Tx−Rx switching times through a time offset between the frame timing of the backhaul and access links.

For some deployments, it is desirable to align the access-link transmission timing of the relay with the transmission timing of the donor cell—that is, to use a global timing reference for all the cells. One example hereof is TDD. In such deployments, the necessary guard times are obtained in a slightly different manner compared to the case of using reception timing of

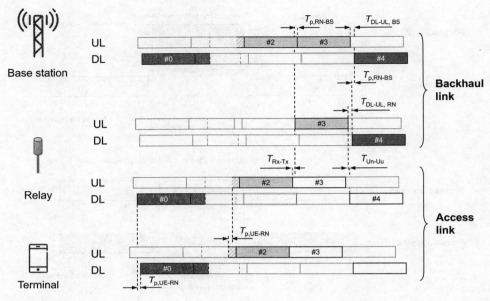

FIGURE 18.12

Example of uplink timing relation for TDD.

the backhaul downlink. In this case it is not possible to obtain the necessary guard time by shifting the subframe timing at the relay. Hence, the guard time for switching from backhaul-link reception to access-link transmission will also be visible at the donor cell, as the last OFDM symbol in the resource blocks used for the backhaul transmission cannot be used for other transmissions in the relay cell. If the time for Tx–Rx switching is longer than the donor-cell-to-relay-node propagation delay, then the first OFDM symbol has to be left unused as well. This case is shown in the bottom part of Figure 18.11.

In the backhaul uplink, the guard time necessary, similar to the previous timing case, is obtained through configuration of (unused) sounding instances. However, unlike the previous case, sounding is configured in the *backhaul* link, as shown at the top of Figure 18.11. Note that this implies that sounding cannot be used for the backhaul link as the OFDM symbol intended as a sounding-reference symbol is used as guard time.

In the case of TDD operation, guard time for the access–backhaul switch can, in addition to the methods discussed earlier, be obtained from the guard period required for TDD operation itself. This is shown in Figure 18.12 and is a matter of using the appropriate settings of timing advance and timing offsets.

Backhaul downlink transmissions consist of data transmitted on the PDSCH and L1/L2 control signaling transmitted on the R-PDCCH, as already discussed. Both these types of transmission must follow one of the timing scenarios discussed earlier. In order to allow for different implementations and deployments, the LTE specifications provide not only the possibility to configure which of the two access–backhaul downlink timing relations to use, but also flexibility in terms of the time span of the channels transmitted on the backhaul link.

PDSCH transmissions intended for a relay can be semi-statically configured to start on the second, third, or fourth OFDM symbol to cater for different control region sizes in the donor cell and relay cells. The PDSCH transmission ends at the last or second last OFDM symbol, depending on which of the two timing cases mentioned earlier is used.

R-PDCCH transmissions intended for a relay always start at the fourth OFDM symbol. A fixed starting position was chosen to simplify the overall structure. Since the amount of resource blocks occupied by an R-PDCCH is relatively small compared to the PDSCH, the overhead reduction possible with a configurable starting position is small and does not justify the additional specification and testing complexity.

MULTIMEDIA BROADCAST/ MULTICAST SERVICES

19

In the past, cellular systems have mostly focused on transmission of data intended for a single user and not on broadcast/multicast services. Broadcast networks, exemplified by the radio and TV broadcasting networks, have on the other hand focused on covering very large areas with the same content and have offered no or limited possibilities for transmission of data intended for a single user. *Multimedia broadcast multicast services* (MBMS) support multicast/broadcast services in a cellular system, thus combining the provision of multicast/ broadcast and unicast services within a single network. The provision of broadcast/multicast services in a mobile-communication system implies that the same information is *simultaneously* provided to multiple terminals, sometimes dispersed over a large area corresponding to a large number of cells, as shown in Figure 19.1. In many cases, it is better to broadcast the information across the area rather than using individual transmissions to each of the users.

MBMS, introduced in LTE release 9, supports transmission of the same content to multiple users located in a specific area, known as the *MBMS service area*, and possibly comprising multiple cells. Two mechanisms for MBMS delivery are available in LTE, *single-cell point to multipoint* (SC-PTM) and *multicast-broadcast single frequency network* (MBSFN).

SC-PTM, introduced in release 13, is in essence very similar to unicast and intended as a complement to MBSFN for services of interest in a single cell only. All transmissions are dynamically scheduled but instead of targeting a single device, the same transmission

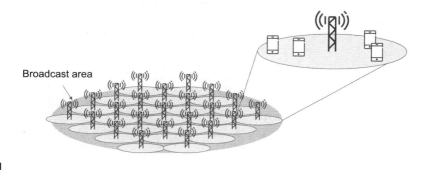

FIGURE 19.1

Broadcast scenario.

4G, LTE-Advanced Pro and The Road to 5G. http://dx.doi.org/10.1016/B978-0-12-804575-6.00019-4

is received by multiple devices simultaneously. SC-PTM is described in more detail in Section 19.4.

MBSFN, introduced in release 9, targets the case when the MBMS service is of interest over a larger area. In such a case, the resources (downlink transmit power) needed to provide a certain broadcast data rate can be considerably reduced if terminals at the cell edge can utilize the received power from broadcast transmissions from multiple cells when detecting/decoding the broadcast data. One way to achieve this and further improve the provision of broadcast/multicast services in a multi-cell network is to ensure that the broadcast transmissions from different cells *are truly identical* and *transmitted mutually time aligned*. In this case, the transmissions received from multiple cells will, as seen from the terminal, appear as a single transmission subject to severe multi-path propagation, see Figure 19.2. As long as the cyclic prefix is sufficiently large, the OFDM receiver can easily handle the equivalent time dispersion "for free" without knowing which cells are involved in the transmission. The transmission of identical time-aligned signals from multiple cells, especially in the case of provision of broadcast/multicast services, is sometimes referred to as single-frequency network (SFN) or, in LTE terminology, MBSFN operation [9]. MBSFN transmission provides several benefits:

- Increased received signal strength, especially at the border between cells involved in the MBSFN transmission, as the device can utilize the signal energy received from multiple cells.
- Reduced interference level, once again especially at the border between cells involved in the MBSFN transmission, as the signals received from neighboring cells will not appear as interference but as useful signals.
- Additional diversity against fading on the radio channel as the information is received from several, geographically separated locations, typically making the overall aggregated channel appear highly time-dispersive or, equivalently, highly frequency selective.

Altogether, this allows for significant improvements in the multicast/broadcast reception quality, especially at the border between cells involved in the MBSFN transmission, and, as a consequence, significant improvements in the achievable multicast/broadcast data rates.

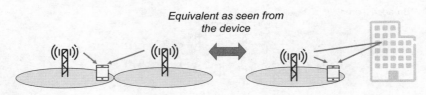

FIGURE 19.2

Equivalence between simulcast transmission and multi-path propagation.

When providing multicast/broadcast services for mobile devices there are several aspects to take into account, of which two deserve special attention and will be elaborated upon later: good coverage and low device power consumption.

The coverage, or more accurately the data rate possible to provide, is basically determined by the link quality of the worst-case user, as no user-specific adaptation of transmission parameters can be used in a multicast/broadcast system providing the same information to multiple users.

Providing for power-efficient reception in the device in essence implies that the structure of the overall transmission should be such that data for a service-of-interest is provided in short high-data-rate bursts rather than longer low-data-rate bursts. This allows the device to occasionally wake up to receive data with long periods of DRX in between. In LTE, this is catered for by time-multiplexing unicast and broadcast transmissions, as well as by the scheduling of different MBMS services, as discussed later in this chapter.

The rest of the chapter focuses on MBSFN transmission, including the overall architecture and scheduling in an MBSFN network, with SC-PTM briefly covered towards the end of the chapter.

19.1 ARCHITECTURE

MBMS services can be delivered using MBSFN or SC-PTM. In the following, the MBSFN mechanisms are described with the release 13 addition of SC-PTM discussed in Section 19.4.

An *MBSFN area* is a specific area where one or several cells transmit the same content. For example, in Figure 19.3, cells 8 and 9 both belong to MBSFN area C. Not only can an MBSFN area consist of multiple cells, a single cell can also be part of multiple, up to eight, MBSFN areas, as shown in Figure 19.3 where cells 4 and 5 are part of both MBSFN areas A and B. Note that, from an MBSFN reception point of view, the individual cells are invisible, although the device needs to be aware of the different cells for other purposes, such as reading

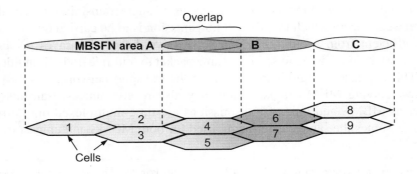

FIGURE 19.3

Example of MBSFN areas.

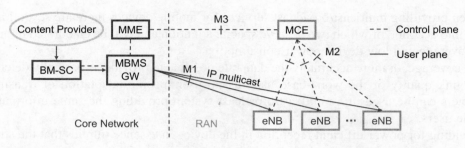

FIGURE 19.4

LTE MBMS architecture.

system information and notification indicators, as discussed in the following. The MBSFN areas are static and do not vary over time.

The usage of MBSFN transmission requires not only time synchronization among the cells participating in an MBSFN area, but also usage of the same set of radio resources in each of the cells for a particular service. This coordination is the responsibility of the *multi-cell/multicast coordination entity* (MCE), which is a logical node in the radio-access network handling allocation of radio resources and transmission parameters (time—frequency resources and transport format) across the cells in the MBSFN area. The MCE is also responsible for selecting between SC-PTM and MBSFN for MBMS services. As shown in Figure 19.4, the MCE[1] can control multiple eNodeBs, each handling one or more cells.

The *broadcast multicast service center* (BM-SC), located in the core network, is responsible for authorization and authentication of content providers, charging, and the overall configuration of the data flow through the core network. The MBMS gateway (MBMS-GW) is a logical node handling multicast of IP packets from the BM-SC to all eNodeBs involved in transmission in the MBSFN area. It also handles session control signaling via the MME.

From the BM-SC, the MBMS data is forwarded using IP multicast, a method of sending an IP packet to multiple receiving network nodes in a single transmission, via the MBMS gateway to the cells from which the MBMS transmission is to be carried out. Hence, MBMS is not only efficient from a radio-interface perspective, but it also saves resources in the transport network by not having to send the same packet to multiple nodes individually unless necessary. This can lead to significant savings in the transport network.

A device receiving MBMS transmission may also receive unicast transmission on the same carrier as MBMS and unicast transmissions are time-multiplexed onto different subframes. This assumes the same carrier being used for both MBMS and unicast transmission,

---

[1]There is an alternative architecture supported where MCE functionality is included in every eNodeB. However, as there is no communication between MCEs in different eNodeBs, the MBSFN area would in this case be limited to the set of cells controlled by a single eNodeB.

which may limit the deployment flexibility in case an operator uses multiple carriers (multiple frequency bands) in an MBSFN area. In release 11, enhancements were introduced to improve operation in such deployments. Briefly, the device informs the network about its MBMS interest and capabilities. The network can take this information into account and ensure that the device is able to receive the relevant MBMS service, for example, by handover of the device to the carrier providing the MBMS transmission. A carrier-aggregation capable device may receive unicast transmissions on one component carrier and MBMS on another component carrier.

19.2 MBSFN CHANNEL STRUCTURE AND PHYSICAL-LAYER PROCESSING

The basis for MBSFN transmission is the *multicast channel* (MCH), a transport-channel-type supporting MBSFN transmission. Two types of logical channels can be multiplexed and mapped to the MCH:

- *multicast traffic channel* (MTCH);
- *multicast control channel* (MCCH).

The MTCH is the logical channel type used to carry MBMS data corresponding to a certain MBMS service. If the number of services to be provided in an MBSFN area is large, multiple MTCHs can be configured. As no acknowledgments are transmitted by the devices, no RLC retransmissions can be used and consequently the RLC unacknowledged mode is used.

The MCCH is the logical channel type used to carry control information necessary for reception of a certain MBMS service, including the subframe allocation and modulation-and-coding scheme for each MCH. There is one MCCH per MBSFN area. Similarly to the MTCH, the RLC uses unacknowledged mode.

One or several MTCHs and, if applicable,[2] one MCCH are multiplexed at the MAC layer to form an MCH transport channel. As described in Chapter 4, the MAC header contains information about the logical-channel multiplexing, in this specific case the MTCH/MCCH multiplexing, such that the device can de-multiplex the information upon reception. The MCH is transmitted using MBSFN in one MBSFN area.

The transport-channel processing for MCH is, in most respects, the same as that for DL-SCH as described in Chapter 6, with some exceptions:

- In the case of MBSFN transmission, the same data is to be transmitted with the same transport format using the same physical resource from multiple cells typically belonging to different eNodeBs. Thus, the MCH transport format and resource allocation cannot be dynamically adjusted by the eNodeB. As described in the preceding

---

[2]One MCCH per MBSFN area is needed, but it does not have to occur in every MCH TTI, nor on all MCHs in the MBSFN area.

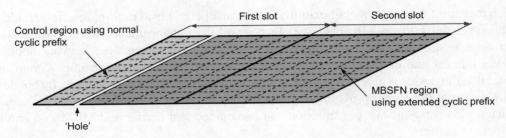

FIGURE 19.5

Resource-block structure for MBSFN subframes, assuming normal cyclic prefix for the control region.

paragraphs, the transport format is instead determined by the MCE and signaled to the devices as part of the information sent on the MCCH.

- As the MCH transmission is simultaneously targeting multiple devices and therefore no feedback is used, hybrid ARQ is not applicable in the case of MCH transmission.
- As already mentioned, multi-antenna transmission (transmit diversity and spatial multiplexing) does not apply to MCH transmission.

Furthermore, as also mentioned in Chapter 6, the PMCH scrambling should be *MBSFN-area specific*—that is, identical for all cells involved in the MBSFN transmission. There are also a number of smaller differences. For example, 256QAM is only supported when the PMCH carries the MTCH and not the MCCH,[3] spatial multiplexing is not supported for the PMCH, nor is any transmit-diversity scheme specified.

The MCH is mapped to the PMCH physical channel and transmitted in MBSFN subframes, illustrated in Figure 19.5. As discussed in Chapter 5, an MBSFN subframe consists of two parts: a *control region*, used for transmission of regular unicast L1/L2 control signaling; and an *MBSFN region*, used for transmission of the MCH.[4] Unicast control signaling may be needed in an MBSFN subframe, for example, to schedule uplink transmissions in a later subframe, but is also used for MBMS-related signaling, as discussed later in this chapter.

As discussed in at the beginning of this chapter, in the case of MBSFN-based multicast/broadcast transmission, the cyclic prefix should not only cover the main part of the actual channel time dispersion but also the timing difference between the transmissions received from the cells involved in the MBSFN transmission. Therefore, MCH transmissions, which can take place in the MBSFN region only, use an extended cyclic prefix. If a normal cyclic prefix is used for normal subframes, and therefore also in the control region of MBSFN subframes, there will be a small "hole" between the two parts of the MBSFN subframe, as

---

[3]This is a consequence of 256QAM being introduced at a later stage and not supported by all devices while the MCCH burst be received by all devices using MBMS services.

[4]As discussed in Chapter 5, MBSFN subframes can be used for multiple purposes and not all of them have to be used for MCH transmission.

illustrated in Figure 19.5. The reason is to keep the start timing of the MBSFN region fixed, irrespective of the cyclic prefix used for the control region.

As already mentioned, the MCH is transmitted by means of MBSFN from the set of cells that are part of the corresponding MBSFN area. Thus, as seen from the device point of view, the radio channel that the MCH has propagated over is the aggregation of the channels of each cell within the MBSFN area. For channel estimation for coherent demodulation of the MCH, the device can thus not rely on the normal cell-specific reference signals transmitted from each cell. Rather, in order to enable coherent demodulation for MCH, special MBSFN reference symbols are inserted within the MBSFN part of the MBSFN subframe, as illustrated in Figure 19.6. These reference symbols are transmitted by means of MBSFN over the set of cells that constitute the MBSFN area—that is, they are transmitted at the same time—frequency position and with the same reference-symbol values from each cell. Channel estimation using these reference symbols will thus correctly reflect the overall aggregated channel corresponding to the MCH transmissions of all cells that are part of the MBSFN area.

MBSFN transmission in combination with specific MBSFN reference signals can be seen as transmission using a specific antenna port, referred to as *antenna port 4*.

A device can assume that all MBSFN transmissions within a given subframe correspond to the same MBSFN area. Hence, a device can interpolate over all MBSFN reference symbols within a given MBSFN subframe when estimating the aggregated MBSFN channel. In contrast, MCH transmissions in different subframes may, as already discussed, correspond to different MBSFN areas. Consequently, a device cannot necessarily interpolate the channel estimates across multiple subframes.

As can be seen in Figure 19.6, the frequency-domain density of MBSFN reference symbols is higher than the corresponding density of cell-specific reference symbols. This is needed as the aggregated channel of all cells involved in the MBSFN transmission will be equivalent to a highly time-dispersive or, equivalently, highly frequency-selective channel. Consequently, a higher frequency-domain reference-symbol density is needed.

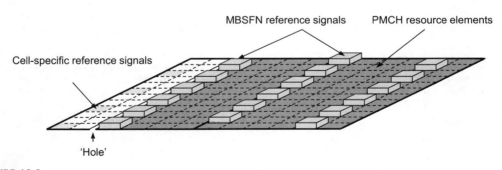

FIGURE 19.6

Reference-signal structure for PMCH reception.

There is only a single MBSFN reference signal in MBSFN subframes. Thus, multi-antenna transmission such as transmit diversity and spatial multiplexing is not supported for MCH transmission. The main argument for not supporting any standardized transmit-diversity scheme for MCH transmission is that the high frequency selectivity of the aggregated MBSFN channel in itself provides substantial (frequency) diversity. Transmit-diversity schemes transparent to the device and hence not requiring any specific support in the specifications can still be applied if beneficial.

19.3 SCHEDULING OF MBSFN SERVICES

Good coverage throughout the MBSFN area is, as already explained, one important aspect of providing broadcast services. Another important aspect, as mentioned in the introduction, is to provide for energy-efficient reception. In essence, for a given service, this translates into transmission of short high-rate bursts in between which the device can enter a DRX state to reduce power consumption. LTE therefore makes extensive use of time-multiplexing of MBMS services and the associated signaling, as well as provides a mechanism to inform the device *when* in time a certain MBMS service is transmitted. Fundamental to the description of this mechanism are the *common subframe allocation* (CSA) period and the *MCH scheduling period* (MSP).

All MCHs that are part of the same MBSFN area occupy a pattern of MBSFN subframes known as the CSA. The CSA is periodic, as illustrated in Figure 19.7. The subframes used for transmission of the MCH must be configured as MBSFN subframes, but the opposite does not

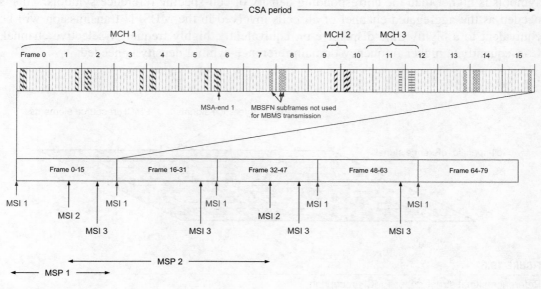

FIGURE 19.7

Example of scheduling of MBMS services.

hold—MBSFN subframes can be configured for other purposes as well, for example, to support the backhaul link in the case of relaying, as described in Chapter 18. Furthermore, the allocation of MBSFN subframes for MCH transmission should be identical across the MBSFN area as otherwise there will not be any MBSFN gain. This is the responsibility of the MCE.

Transmission of a specific MCH follows the *MCH subframe allocation* (MSA). The MSA is periodic and at the beginning of each MCH scheduling period, an MAC control element is used to transmit the *MCH scheduling information* (MSI). The MSI indicates which subframes are used for a certain MTCH in the upcoming scheduling period. Not all possible subframes need to be used; if a smaller number than allocated to an MCH is required by the MTCH(s), the MSI indicates the last MCH subframe to be used for this particular MTCH (*MSA end* in Figure 19.7), while the remaining subframes are not used for MBMS transmission. The different MCHs are transmitted in consecutive order within a CSA period—that is, all subframes used by MCH n in a CSA are transmitted before the subframes used for MCH $n+1$ in the same CSA period.

The fact that the transport format is signaled as part of the MCCH implies that the MCH transport format may differ between MCHs but must remain constant across subframes used for the same MCH. The only exception is subframes used for the MCCH and MSI, where the MCCH-specific transport format, signaled as part of the system information, is used instead.

In the example in Figure 19.7, the scheduling period for the first MCH is 16 frames, corresponding to one CSA period, and the scheduling information for this MCH is therefore transmitted once every 16 frames. The scheduling period for the second MCH, on the other hand, is 32 frames, corresponding to two CSA periods, and the scheduling information is transmitted once every 32 frames. The MCH scheduling period can range from 80 ms to 10.24 s.

To summarize, for each MBSFN area, the MCCH provides information about the CSA pattern, the CSA period, and, for each MCH in the MBSFN area, the transport format and the scheduling period. This information is necessary for the device to properly receive the different MCHs. However, the MCCH is a logical channel and is itself mapped to the MCH, which would result in a chicken-and-egg problem—the information necessary for receiving the MCH is transmitted on the MCH. Hence, in TTIs when the MCCH (or MSI) is multiplexed into the MCH, the MCCH-specific transport format is used for the MCH. The MCCH-specific transport format is provided as part of the system information (SIB13; see Chapter 11 for a discussion about system information). The system information also provides information about the scheduling and modifications periods of the MCCH (but not about CSA period, CSA pattern, and MSP, because those quantities are obtained from the MCCH itself). Reception of a specific MBMS service can thus be described by the following steps:

- Receive SIB13 to obtain knowledge on how to receive the MCCH for this particular MBSFN area.
- Receive the MCCH to obtain knowledge about the CSA period, CSA pattern, and MSP for the service of interest.
- Receive the MSI at the beginning of each MSP. This provides the device with information on which subframes the service of interest can be found in.

After the second step, the device has acquired the CSA period, CSA pattern, and MSP. These parameters typically remain fixed for a relatively long time. The device therefore only needs to receive the MSI and the subframes in which the MTCH carrying the service of interest are located, as described in the third bulleted item in the preceding list. This greatly helps to reduce the power consumption in the device as it can sleep in most of the subframes.

Occasionally there may be a need to update the information provided on the MCCH, for example when starting a new service. Requiring the device to repeatedly receive the MCCH comes at a cost in terms of device power consumption. Therefore, a fixed schedule for MCCH transmission is used in combination with a change-notification mechanism, as described in the following paragraph.

The MCCH information is transmitted repeatedly with a fixed repetition period and changes to the MCCH information can only occur at specific time instants. When (part of) the MCCH information is changed, which can only be done at the beginning of a new modification period, as shown in Figure 19.8, the network notifies the devices about the upcoming MCCH information change in the preceding MCCH modification period. The notification mechanism uses the PDCCH for this purpose. An eight-bit bitmap, where each bit represents a certain MBSFN area, is transmitted on the PDCCH in an MBSFN subframe using DCI format 1C and a reserved identifier, the M-RNTI. The notification bitmap is only transmitted when there are any changes in the services provided (in release 10, notification is also used to indicate a counting request in an MBSFN area) and follows the modification period, as described earlier.

The purpose of the concept of notification indicators and modification periods is to maximize the amount of time the device may sleep to save battery power. In the absence of any changes to the MCCH information, a device currently not receiving MBMS may enter DRX and only wake up when the notification indicator is transmitted. As a PDCCH in an MBSFN subframe spans at most two OFDM symbols, the duration during which the device needs to wake up to check for notifications is very short, translating to a high degree of power saving. Repeatedly transmitting the MCCH is useful to support mobility; a device entering a

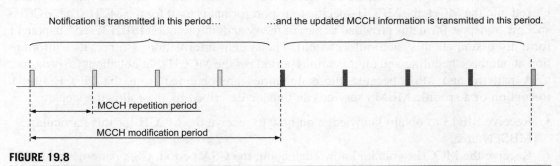

FIGURE 19.8

MCCH transmission schedule.

new area or a device missing the first transmission does not have to wait until the start of a new modification period to receive the MCCH information.

19.4 SINGLE-CELL POINT TO MULTIPOINT TRANSMISSION

Single-cell point to multipoint, SC-PTM, was introduced in release 13 in order to support MBMS services of interest in a single cell (or a small area consisting of a few cells) and complementing the MBSFN transmission scheme described in the preceding section. The same overall architecture is reused with the MCE being responsible for selecting between MBSFN and SC-PTC transmission.

The SC-MTCH, mapped to the DL-SCH, carriers the MBMS information to be transmitted using SC-PTM. The same mechanism as unicast transmission is used—that is, dynamic scheduling of the DL-SCH. This allows efficient support of very bursty traffic compared to the slower scheduling used for MBSFN transmission. Since the DL-SCH transmissions target multiple devices, hybrid-ARQ feedback and CSI reports are not used. Consequently, transmission modes 1—3, which do not rely on feedback, are supported. A new group RNTI, the G-RNTI, is used for dynamic scheduling with a different G-RNTI for each MBMS service.

Control information such as the G-RNTIs used for different MBMS services are provided on the SC-MCCH logical channel type, mapped to the DL-SCH and dynamically scheduled using the SC-RNTI. SIB20 contains information assisting the device to receive the SC-MCCH.

LTE FOR MASSIVE MTC APPLICATIONS

20

20.1 INTRODUCTION

Applications such as mobile telephony, mobile broadband, and media delivery are fundamentally about information being communicated to and/or from human beings. However, wireless communication is increasingly also being used to provide end-to-end connectivity between nonhumans—that is, different types of "things" or "machines." In 3GPP this is referred to as *machine-type communications* (MTC). The term *internet of things* (IoT) is also often used in this context.

Although MTC applications span a very wide range of different applications, on a high level one is often talking about two main categories, *massive-MTC* applications and *critical-MTC* applications.

Massive-MTC applications are applications typically associated with a very large, or *massive*, number of connected devices, such as different types of sensors, actuators, and similar devices. Massive-MTC devices typically have to be of very low cost and have very low average energy consumption enabling very long battery life. At the same time, the amount of data generated by each device is typically small and the data-rate and latency requirements are often relatively relaxed.

For some massive-MTC applications it is important that one can provide connectivity also in locations such as deep within the basement of buildings and in very rural or even deserted areas with very sparse network deployments. A radio-access technology supporting such massive-MTC applications must therefore be able to operate properly with very high path loss between base stations and devices.

Critical-MTC applications, on the other hand, are applications typically associated with requirements on extremely high reliability and extremely high availability within the area where the application is to be supported. Examples of critical-MTC applications include traffic safety, control of critical infra-structure, and wireless connectivity for industrial processes. Many of these applications also have requirements on very low and predictable latency. At the same time, very low device cost and very low device energy consumption is typically less important for these kinds of applications.

4G, LTE-Advanced Pro and The Road to 5G. http://dx.doi.org/10.1016/B978-0-12-804575-6.00020-0

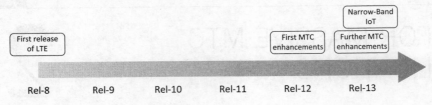

FIGURE 20.1

Steps in the LTE evolution targeting enhanced support for massive-MTC applications.

Many MTC applications can be well supported already with the first releases of LTE. However, recently there have been updates as part of the evolution of LTE specifically targeting enhanced support for massive-MTC applications.

As outlined in Figure 20.1, the first step in this evolution was taken in 3GPP release 12. This was then followed by additional steps in release 13. The different updates to the LTE specifications addressing massive-MTC applications in releases 12 and 13 are discussed in more detail in Section 20.2 and 20.3, respectively.

In addition to this more direct evolution of LTE toward enhanced support for massive-MTC applications, there is an ongoing parallel 3GPP activity referred to as *narrow-band IoT* (NB-IoT). NB-IoT started as a 3GPP technology track separated from the evolution of LTE. However, recently the NB-IoT technology has been aligned with LTE and it can now be seen as part of the overall LTE evolution. NB-IoT is further discussed in Section 20.4

20.2 MTC ENHANCEMENTS FOR LTE RELEASE 12

As mentioned earlier, the first step to enhance the support for MTC applications, specifically targeting lower device cost and reduced device energy consumption, was taken as part of LTE release 12. This included a new UE category with reduced data-rate capability, modified half-duplex operation, and possibility for devices with only one receive antenna. It also included a new *power-saving mode* targeting reduced device energy consumption.

20.2.1 DATA-RATE CAPABILITY AND UE CATEGORY 0

As described in Chapter 3, LTE defines different UE categories, where each category is associated with a maximum supported data rate.

To enable lower-cost devices for MTC applications, LTE release 12 introduced a new, lower-rate UE category. The new UE category was labeled *Category 0* (zero) in line with the typical association of higher-numbered UE categories with more extensive capabilities.[1]

---

[1]Strictly speaking, there are, from release 12, different categories for downlink and uplink.

The data-rate limitations of different UE categories are in the LTE specifications expressed as upper limits for the size of a transport block. For UE category 0, this limit was set to 1000 bits for both uplink and downlink. In combination with a TTI of 1 ms, this leads to a maximum supported data rate of 1 Mbit/s for category 0 devices assuming full-duplex FDD operation. It can be noted that the 1 Mbit/s data-rate limitation corresponds to the transport-channel data rate, including overhead added on higher layers (MAC and above). The maximum data rate from an application point of view is thus somewhat lower.

It should also be pointed out that that the 1000-bit limit for the transport-block size is only valid for user-data transmission. In order to be able to support category 0 devices without more extensive updates on the network side, these devices must still be able to receive system information, paging messages, and random-access responses with a transport-block size of 2216 bits which is the maximum size of SIB1.

Similar to category 1, category 0 does not include support for spatial multiplexing. Furthermore, uplink modulation is limited to QPSK and 16QAM.

It should be pointed out that category 0 devices still have to support the full carrier bandwidth, that is, up to 20 MHz. As is seen in Section 20.3, this requirement was relaxed as part of the further MTC enhancements in LTE release 13.

20.2.2 **TYPE-B HALF-DUPLEX OPERATION**

Already from its first release, the LTE specifications have allowed for terminals only capable of half-duplex FDD operation. As described in Chapter 5, half-duplex operation implies that there is no simultaneous transmission and reception on the device side. This allows for relaxed duplex-filter requirements, enabling lower-cost devices.

Due to timing advance, the transmission of an uplink subframe will start before the end of the previous downlink subframe. As described in Chapter 5, when uplink transmission follows directly upon downlink reception, a device only capable of half-duplex operation is therefore "allowed" to skip reception of the last OFDM symbol(s) of the downlink subframe.

To allow for further complexity/cost reduction, LTE release 12 introduced a new half-duplex mode, *half-duplex type B*, specifically targeting category 0 devices. Half-duplex type B allows for a much larger idle time between downlink reception and uplink transmission by specifying that a device is not expected to receive the last downlink subframe before an uplink subframe nor the first downlink subframe *after* an uplink subframe, see Figure 20.2.

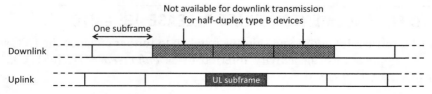

FIGURE 20.2

Scheduling restrictions for half-duplex type B.

By providing such a large idle time when switching between transmission and reception, half-duplex type B allows for more reuse of RF functionality—for example, in terms of oscillators—between the transmitter and the receiver, thereby further reducing the complexity of category 0 devices.

20.2.3 POSSIBILITY FOR DEVICES WITH A SINGLE RECEIVE ANTENNA

The LTE technical specifications do not explicitly mandate any specific receiver implementation such as the number of receive antennas to be used at the device side. However, already since the first release, the LTE performance requirements have been such that, implicitly, two-antenna reception has been mandatory for all UE categories.

To enable further complexity reduction for devices targeting massive-MTC applications, this requirement was relaxed for UE category 0. Rather, the performance requirements for category 0 devices are such that they can be fulfilled with only a single receive antenna at the device side.

20.2.4 POWER-SAVING MODE

In addition to the above-mentioned steps to enable lower-cost devices, LTE release 12 also included a new *power-saving mode* (PSM) to enable reduced energy consumption and corresponding extended battery life for massive-MTC devices.

Entering PSM is from a device point of view similar to powering off. The difference is that the device remains registered in the network and does not need to reattach or reestablish PDN (packet data network) connections.

As a consequence, a device that has entered PSM cannot be reached by the network and reestablishing connectivity must be initiated by the device. PSM is therefore mainly relevant for "monitoring" devices for which the need for data transfer is typically triggered by events on the device side.

To enable network-originating data transfer, devices using PSM must regularly reconnect to the network and stay awake for a brief period to allow the network to make paging attempts to the UE. Such reconnection should be infrequent in order to retain the energy-saving benefits of PSM, implying that PSM is primarily beneficial for infrequent network-originating data traffic that is not latency critical.

20.3 MTC ENHANCEMENTS FOR LTE RELEASE 13: eMTC

As part of LTE release 13, additional steps were taken to further enhance the LTE support for massive-MTC applications, an activity referred to as *enhanced MTC* (eMTC). The main targets for this activity were to

- Enable further device cost reduction beyond what was achieved with release 12 UE Category 0.

- Extend the coverage for low-rate massive-MTC devices with a target to enable operation with at least 15 dB higher coupling loss, compared to pre-release 13 devices.

On a high level, the main new features of eMTC were the possibility for

- more narrow RF bandwidth on the device side, enabling further reduction in device complexity and corresponding cost;
- extensive repetition for both downlink and uplink, enabling extended coverage for low-rate services;
- extended DRX.

Devices that are limited to narrow RF bandwidth are, in the LTE specification referred to as *bandwidth-reduced low-complexity* (BL) UEs. However, we will use the term eMTC rather than BL in the following.

It should also be pointed out that the limitation to narrow RF bandwidth and the support for coverage extension by means of repetition are actually independent properties in the sense that the coverage extension may be used also for devices not limited to narrow RF bandwidth.

In Sections 20.3.1 and 20.3.2 some higher-level aspects of narrowband operation and coverage extension by means of repetition are discussed. The following sections (Sections 20.3.3−20.3.7) then describe how these features impact, in more detail, different aspects of the LTE radio interface. Finally, Section 20.3.8 gives an overview of extended DRX.

Similar to release 12 category 0 devices, a limit on the transport-block size to 1000 bits, no support for spatial multiplexing, half-duplex type B operation, and the possibility for single-antenna reception at the device side, is valid also for release 13 eMTC. For eMTC devices, modulation is limited to QPSK and 16QAM for both uplink and downlink.

It should be pointed out that, for eMTC devices, the 1000-bit limit on the transport-block size is valid for *all* transmissions, including system information, paging message, and random-access responses. Due to the narrowband characteristics of eMTC devices there was anyway a need to redesign the mechanisms for delivering system information and therefore no need to retain the release 12 requirement that devices must be able to receive the maximum SIB1 transport-block size of 2216 bits.

It should also be pointed out that, due to an increase in HARQ round-trip time, see Section 20.3.3.4, the maximum sustainable downlink data rate for full-duplex eMTC devices is limited to 800 kbps.

20.3.1 NARROW−BAND OPERATION

From its first release, the LTE specifications have required that all devices should support all LTE carrier bandwidths, that is, up to 20 MHz. The reason for this requirement was to simplify specification and network operation by ensuring that all devices accessing the network were able to transmit and receive over the full network carrier bandwidth.

Taking into account the overall complexity of a device, including high-resolution screens and application processors, the RF complexity associated with having to support all carrier bandwidths is not critical for typical mobile-broadband devices, However, the situation is

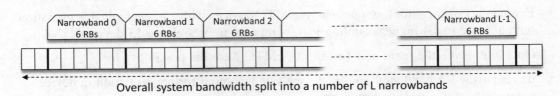

FIGURE 20.3

Overall carrier bandwidth split into *L* narrowbands of size six resource blocks.

different for low-cost MTC devices for which the radio part contributes a much larger fraction of the overall device complexity and associated cost. Support for a wide bandwidth also impacts the energy consumption, and thus the battery life, of the device. For these reason it was decided that release-13 eMTC devices were only required to support transmission and reception over an instantaneous bandwidth corresponding to the minimum LTE carrier bandwidth, that is, about 1.4 MHz.

To allow for such narrowband eMTC devices to access a more wideband LTE carrier, the concept of *narrowbands* was introduced in LTE release 13. As outlined in Figure 20.3, the overall wideband carrier is split into a number of narrowbands, each consisting of six resource blocks in the frequency domain. As the overall carrier bandwidth measured in number of resource blocks may not always be a multiple of six, there may be one or two resource blocks not part of any narrowband at each edge of the carrier. Also, the center resource block of the carrier will not be part of any narrowbands in case of an odd number of resource blocks within the overall carrier bandwidth.

At a given time instant, an eMTC device can only transmit over a bandwidth corresponding to a single narrowband, that is, six consecutive resource blocks. Similarly, at a given time instant, an eMTC device can only receive over a bandwidth corresponding to a single narrowband. Consequently, physical channels inherently spanning more than one narrowband, that is, more than six consecutive resource blocks, cannot be received by an eMTC device. This is, for example, the case for the PDCCH, PCFICH, and PHICH. The functions of these channels thus have to be provided by other means for eMTC devices.

Although an eMTC device can only transmit/receive a single narrowband, the device should be able to switch narrowband between subframes. Assuming that the device has an RF front end with a bandwidth corresponding to the bandwidth of a single narrowband, switching between different narrowbands will require retuning of the RF front end, something that may take as long time as 150 μs or up to two OFDM symbols. How such a retuning time is made available depends on if the retuning is to be done between subframes used for downlink reception or between subframes used for uplink transmission. In short one can say that

- receiver retuning between downlink subframes is assumed to take place during the control region at the start of the subframe;
- time for transmitter retuning between uplink subframes is made available by not transmitting the last symbol(s) just before retuning and/or the first symbols(s) just after retuning, see also Section 20.3.4.

20.3.2 COVERAGE ENHANCEMENTS BY MEANS OF REPETITION

The second aim of the 3GPP eMTC activities was to enable significantly extended coverage for low-rate MTC applications. The explicit target in the design of eMTC was to enable operation with at least 15 dB higher coupling loss compared to pre-release 13 devices.

It is important to understand that the aim of eMTC was not to extend coverage for a given data rate. Rather, coverage is extended by reducing the data rate. The key task was then to ensure that

- the lower data rates could be provided with sufficient efficiency;
- there is sufficient coverage for the different control channels and signals needed to establish and retain connectivity.

The following should be noted:

- For RF-complexity reasons, there is an assumption that eMTC devices may have a maximum output power limited to 20 dBm—that is, 3 dB lower than the typical maximum output power of LTE devices. To allow for at least 15 dB higher coupling loss, the uplink link budget for eMTC devices thus needs to be improved by at least 18 dB.
- Use of single-antenna reception implies a loss in downlink link performance, a loss which also has to be compensated for by the downlink coverage enhancements.
- The different LTE channels and signals are not fully balanced in the sense that they do not have exactly the same coverage. This means that coverage does not necessarily need to be improved the same amount for all signals and channels to reach an overall network coverage gain of 15 dB.
- Certain signals and channels, such as the PSS/SSS and the PBCH, are transmitted over only a fraction of the carrier bandwidth and are normally assumed to share the overall base-station power with other, parallel transmissions. The coverage of such signals/channels can partly be extended by *power boosting*—that is, by assigning them a larger fraction of the overall base-station power at the expense of other, parallel transmissions.

The main tool to extend the coverage of eMTC devices is the use of *multi-subframe repetition*. This means that a single transport block is transmitted over multiple, in some cases a very large number of, subframes (N_{rep} subframes), thereby providing higher transmit energy per information bit for a given transmit power.

The LTE specifications distinguish between two *modes* of coverage enhancements for eMTC

- *Coverage enhancement mode A (CE mode A)* targeting relatively modest coverage enhancement;
- *CE mode B* targeting more extensive coverage enhancement.

The two modes differ, for example, in the number of repetitions supported, where CE mode B supports much more extensive repetition compared to CE mode A.

To some extent one can see the aim of CE mode A to compensate for the lower (−3 dB) eMTC device output power and the degraded receiver performance due to single-antenna

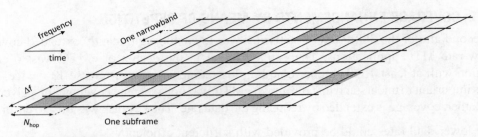

FIGURE 20.4 Frequency hopping for eMTC.

The figure assumes frequency hopping between four different narrowbands (possible only for downlink) with a frequency-hopping block length $N_{hop} = 2$.

reception of eMTC devices. From that point of view, the aim of CE mode A is, at least partly, to ensure the same coverage for low-complexity eMTC devices as for non-eMTC devices. CE mode B then provides the full coverage extension up to at least the targeted 15 dB higher maximum coupling loss.

Each eMTC device is individually configured by the network to operate in either CE mode A or CE mode B.

Repetitions will, by default, take place in consecutive subframes.[2] However, the network can explicitly configure certain subframes as not being available for repetition ("invalid subframes") by means of a bitmap provided as part of the eMTC-specific system information. In case of FDD, the set of invalid subframes are independently configured for downlink and uplink.

If the set of invalid subframes is not explicitly configured by the network, the device can assume that all uplink subframes are valid subframes. Furthermore, the device can assume that all downlink subframes are valid subframes except subframes configured as MBSFN subframes.

In case of an invalid subframe, repetition is postponed until the next valid subframe. Thus, the presence of invalid subframes does not reduce the number of repetitions but simply prolong the time over which the overall set of repetitions is carried out.

The concept of valid and invalid subframes is also applicable for the first transmission in a sequence of repetitions. On the other hand, in case of no repetitions ($N_{rep} = 1$), the transmission is carried out even if the corresponding subframe is configured as an invalid subframe.

In case of repetition over multiple subframes, *inter-subframe frequency hopping* can optionally be applied on both uplink and downlink. As illustrated in Figure 20.4, frequency hopping is carried out between different narrowbands and in blocks of N_{hop} subframes, where N_{hop} can be configured from $N_{hop} = 1$ (frequency hopping between every subframe) to $N_{hop} = 8$ (frequency hopping between every eight subframe) for CE mode A and from

---

[2]In case of TDD, consecutive downlink and uplink subframes for downlink and uplink transmission, respectively.

$N_{hop} = 2$ to $N_{hop} = 16$ for CE mode B.[3] The reason to allow for frequency hopping on a multi-subframe basis ($N_{hop} > 1$), rather than to hop every subframe, is to allow for inter-subframe channel estimation.

In order to align the frequency hopping between different transmissions and ensure that all devices "hop" at the same time instant, the hopping instants are not UE specific. Rather, the hopping instants are common for all frequency-hopping eMTC devices within the cell and depend on the absolute subframe number counted from the first subframe in the frame with frame number (SFN) equal to 0. In other words, if one labels the first subframe of the frame with SFN equal to 0 as subframe number 0, hopping takes place between subframe $k \cdot N_{hop} - 1$ and subframe $k \cdot N_{hop}$ for all frequency-hopping transmissions regardless of the start of the transmission. This also means that the subframe counter in case of frequency hopping takes into account also invalid subframes.

The frequency-hopping block length N_{hop} and the hopping offset Δf are cell-specific parameters that are separately configured for downlink and uplink and also separately configured for CE mode A and CE mode B. In case of uplink frequency hopping, hopping is carried out between just two different narrowbands. For the downlink, frequency hopping can be configured to be carried out over two narrowbands or four equally spaced narrowbands.

20.3.3 DOWNLINK TRANSMISSION: PDSCH AND MPDCCH

As described in Chapter 6, conventional (non-eMTC) downlink data transmission is carried out on the PDSCH with the associated control signaling (downlink control information, DCI) provided by means of the PDCCH. The PDCCH is transmitted in the *control region* of the subframe, consisting of the up to three first OFDM symbols of each subframe.[4] The size of the control region, or equivalently, the start of the *data region* in which PDSCH is transmitted, is dynamically signaled on the PCFICH.

As also described in Chapter 6, DCI can alternatively be provided by means of the EPDCCH which is confined to a limited set of resource blocks.

Both the PDCCH and PCFICH are inherently wideband and can thus not be received by a narrowband eMTC device. DCI for eMTC devices is therefore provided by means of a new physical control channel, the *MPDCCH*. The MPDCCH can be seen as an EPDCCH extended to support narrowband operation and coverage extension by means of repetition. This also includes the introduction of new eMTC-specific DCI formats for downlink scheduling assignments and uplink scheduling grants.

Even if the PDCCH is not used for eMTC devices, there will typically still be PDCCH transmissions targeting other (non-eMTC) devices within the cell. Thus, there will still be a control region within each subframe, and even though an eMTC device does not need to read the different control channels transmitted within the control region, it still needs to know the

---

[3]This is true for FDD. For TDD the possible values for N_{rep} are {1,5,10,20} for CE mode A and {5,10,20,40} for CE mode B.
[4]Or up to four OFDM symbols in case of 1.4 MHz carrier bandwidth.

starting point for the data region, that is, the starting point for PDSCH/MPDCCH trans-missions. This information is provided to eMTC devices as part of the eMTC-specific system information.[5] In contrast to non-eMTC devices, for eMTC devices the starting point for PDSCH/MPDCCH transmissions is thus semi-static and assumed to change only on a very slow basis.

Note that the semi-static configuration of the PDSCH/MPDCCH starting point for eMTC devices does not prevent dynamic variations of the size of the control/data regions for other devices. The only constraint is that, in subframes with downlink eMTC transmissions (PDSCH or MPDCCH), *these* transmissions must start at the semi-statically configured po-sition. As a consequence, in subframes with downlink eMTC transmissions, the control re-gion of the subframe cannot extend *beyond* this position. However, it may very well be shorter, with a corresponding earlier starting position for PDSCH/EPDCCH transmissions to other, non-eMTC devices scheduled in the same subframe. Furthermore, in subframes with no eMTC downlink transmissions, the size of the control region can be fully flexible.

20.3.3.1 Downlink Transmission Modes
As described in Chapter 6, there are ten different downlink transmission modes defined for LTE. Of these ten modes, transmission modes 1, 2, 6, and 9 are applicable for eMTC devices. As eMTC devices are not assumed to support multi-layer transmission, transmission mode 9 is limited to single-layer precoding in case of eMTC.

In Section 20.3.2 it was described how frequency hopping is carried out in blocks of N_{hop} subframes. This was done in order to enable inter-subframe channel estimation. For the same reason, an eMTC device can assume that the transmission mode 9 precoder does not change over the same block of N_{hop} subframes. Note that the assumption that the precoder is un-changed for a block of N_{hop} subframes is valid even if frequency hopping is disabled.

20.3.3.2 PDSCH/MPDCCH Repetition
Repetition can be applied to the PDSCH with a very wide range of different repetition numbers, ranging from a single transmission ($N_{rep} = 1$) to 2048 repetitions ($N_{rep} = 2048$). The number of repetitions to use for a certain PDSCH transmission is a combination of semi-static configuration and dynamic selection on a per-transmission basis.

For each of the two coverage-enhancement modes, the network configures a set of possible number of repetitions on cell level, where each set consists of four and eight different values for CE mode A and CE mode B, respectively. The actual number of repetitions for a specific PDSCH transmission is then dynamically selected by the network from the four/eight values available in the corresponding configured set. The device is informed about the selected number of PDSCH repetitions in the scheduling assignment (two/three bits of signaling indicating one of four/eight values for CE mode A and CE mode B, respectively).

---

[5]The starting of the data region in subframes in which the corresponding system information is transmitted is predefined to the fifth OFDM symbol for carrier bandwidths equal to 1.4 MHz and the fourth OFDM symbols for carrier bandwidths larger than 1.4 MHz.

	CE Mode A	CE Mode B
Table 20.1 Set of PDSCH Repetition Numbers for CE Mode A and CE Mode B		
Set 1	{1,2,4,8}	{4,8,16,32,64,128,256,512}
Set 2	{1,4,8,16}	{1,4,8,16,32,64,128,192}
Set 3	{1,4,16,32}	{4,8,16,32,64,128,192,256}
Set 4	—	{4,16,32,64,128,192,256,384}
Set 5	—	{4,16,64,128,192,256,384,512}
Set 6	—	{8,32,128,192,256,384,512,768}
Set 7	—	{4,8,16,64,128,256,512,1024}
Set 8	—	{4,16,64,256,512,768,1024,1536}
Set 9	—	{4,16,64,128,256,512,1024,2048}

The table is also valid for PUSCH repetition, see Section 20.3.4.

Table 20.1 shows the different sets of repetition numbers that can be configured on cell-level for CE mode A and CE mode B, respectively. Note that the maximum number of repetitions for CE mode A is 32 while the corresponding number of CE mode B is 2048. For CE mode A, the minimum number of transmissions is always one (no repetitions).

Repetition can also be applied to the MPDCCH. Similar to PDSCH transmissions, the number of repetitions to use for a certain MPDCCH transmission is a combination of semi-static configuration and dynamic selection on a per-transmission basis.

On cell level, the network configures the maximum number of MPDCCH repetitions (R_{max}) from the set {1,2,4,8,16,32,64,128,256} and broadcast it as part of the eMTC-specific system information. The network then dynamically selects the actual number of repetitions to use for a specific MPDCCH transmission from the set {$R_{max}, R_{max}/2, R_{max}/4, R_{max}/8$}.[6]

Information about the number of repetitions used for a specific MPDCCH transmission is included in the scheduling assignment as a 2-bit parameter indicating one of four different values. In other words, information about the number of MPDCCH repetitions is carried within the MPDCCH itself. An MPDCCH transmission carried out with a certain number of repetitions R can only start at, at most, every R subframe. Consequently, a device that has correctly decoded an MPDCCH transmission with a DCI indicating a certain number of repetitions can, without ambiguity, determine the first subframe, and consequently also the last subframe, in which the MPDCCH was transmitted. As we will see later, in case of DCI carrying downlink scheduling assignments, the device needs to know the last subframe in which the MPDCCH was transmitted in order to be able to determine the starting subframe for the corresponding scheduled PDSCH transmission. Similarly, in case of DCI carrying uplink scheduling grants, the device needs to know the last subframe in which the MPDCCH

---

[6]If R_{max} is less than eight, the number of repetitions can only be within a subset of the configured values.

was transmitted in order to be able to determine the starting subframe for the corresponding scheduled PUSCH transmission.

20.3.3.3 PDSCH Scheduling

In the normal case, downlink scheduling assignment is *intra-subframe*. This means that a scheduling assignment provided on PDCCH or EPDCCH in a certain subframe relates to a PDSCH transmission *in the same subframe*.

In contrast, scheduling assignments for eMTC devices are *inter-subframe*. More specifically, also taking into account the possibility for repetition on both MPDCCH and PDSCH, a scheduling assignment on MPDCCH ending in subframe n relates to a corresponding PDSCH transmission starting in subframe $n + 2$, see Figure 20.5. The delay between the end of the scheduling assignment and the start of the corresponding PDSCH transmission allows for decoding of the scheduling assignment without having to buffer the received signal, thereby enabling lower device complexity.

In order to determine the subframe in which the PDSCH transmission starts, the device must know the last subframe of the corresponding MPDCCH transmission. As mentioned earlier, this is possible due to the inclusion of the number of MPDCCH repetitions in the DCI, in combination with the restriction that an MPDCCH transmission carried out with a certain number of repetitions R can only start at, at most, every R subframe.

In general, DCI for eMTC devices is provided by means of a set of new DCI formats carried on EPDCCH. Among these, *DCI format 6-1A* and *DCI format 6-1B* are used for scheduling assignments for PDSCH transmissions. The content of these DCI formats are shown in Table 20.2 where the values in the right column indicates the number of bits used for the specific parameter.

DCI format 6-1A is used for scheduling assignments for devices operating in CE Mode A. It contains similar information as the DCI formats used for normal (non-eMTC) scheduling assignments, but extended with information related to PDSCH and MPDCCH repetition as well as a frequency-hopping flag to dynamically enable/disable PDSCH frequency hopping (see later). DCI format 6-1B is a more compact format used for scheduling assignments for devices operating in CE Mode B.

The resource assignment included in DCI format 6-1A and 6-1B is split into two parts. A *narrowband indicator* specifies the narrowband in which the downlink PDSCH is located. A

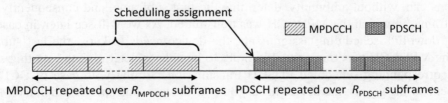

FIGURE 20.5

Timing relation between scheduling assignment on MPDCCH and corresponding PDSCH transmission.

Table 20.2 DCI Formats 6-1A and 6-1B

Field	6-1A	6-1B
Frequency-hopping flag	1	—
Resource information (resource assignment)		
Narrowband indicator	Var[a]	Var[a]
Resource-block indicator	5	1
Number of PDSCH repetitions	2	3
HARQ process number	3/4[b]	1
MCS	4	4
RV	2	—
New data indicator	1	1
PMI confirmation	1	—
Precoding information	1/2[c]	—
DM-RS scrambling/antenna ports	2	—
Downlink assignment index	2	—
PUCCH power control	2	—
SRS request	0/1	—
Ack/Nack offset	2	2
Number of MPDCCH repetitions	2	2
Flag for 6-0/6-1 differentiation	1	1

[a]*Number of bits depends on the system bandwidth.*
[b]*3 bits for FDD; 4 bits for TDD.*
[c]*Number of bits depends on number of antenna ports.*

resource-block indicator then specifies the exact set of resource blocks used for the PDSCH transmission given the narrowband specified by the narrowband indicator.

For DCI format 6-1A, that is, for CE mode A, the resource-block indicator consists of 5 bits, thus being able to point at any set of consecutive resource blocks within the six resource blocks of the narrowband.[7] Note that this is essentially the same as downlink resource-allocation type 2 used for normal (non-eMTC) scheduling assignments, see Chapter 6, where the bandwidth of the carrier has been replaced by the "bandwidth" of a single narrowband, that is, six resource blocks.

For DCI format 6-1B, used for CE mode B, the resource-block indicator consists of a single bit indicating two sets of consecutive resource blocks:

- all six resource blocks within the narrowband;
- resource block 0—3, that is, four consecutive resource blocks.

In case of frequency hopping, the resource assignment provides the narrowband of the first transmission. Narrowbands for subsequent repetitions are then given by cell-specific offsets

---

[7]There are a total of $1 + 2 + 3 + 4 + 5 + 6 = 21$ such combinations.

relative to the first transmission. The set of resource blocks within the narrowband is the same for each hop. For the downlink, frequency hopping can take place between either two or four narrowbands.

The hopping period, the hopping offset, and the number of hopping frequencies are configured on a cell level.

Even if frequency hopping for PDSCH is configured, for CE mode A frequency hopping can be dynamically disabled on a per-transmission basis by means of a *frequency-hopping flag* in the scheduling assignment, see Table 20.2. It is not possible to dynamically disable frequency hopping for CE mode B and, consequently, there is no frequency-hopping flag in DCI format 6-1B.

The "6-0/6-1 differentiation" flag indicates if the DCI is of format 6-1A/6-1B (downlink scheduling assignment) or 6-0A/6-0B (uplink scheduling grant, see Section 20.3.4.1). A device configured in CE mode A only has to receive DCI formats 6-0A and 6-1A. Likewise, a device configured in CE mode B only has to receive DCI formats 6-0B and 6-1B. To reduce the number of blind decoding when detecting MPDCCH, DCI formats 6-0A and 6-1A are of the same size; likewise, DCI formats 6-0B and 6-1B are of same size. Once an MPDCCH of the expected size has been decoded, the device can, from the 6-0/6-1 differentiation flag, determine if the DCI corresponds to a downlink scheduling assignment ("6-1" format) or an uplink scheduling grant ("6-0" format).

20.3.3.4 Downlink Hybrid ARQ

Downlink HARQ for eMTC is asynchronous and adaptive as in pre-release-13. This means that the network can make a retransmission at any time and using a different frequency resource compared to the prior transmission.

The two-subframe shift between scheduling assignment on MPDCCH and the corresponding PDSCH transmission, as outlined in Figure 20.5, implies a two-subframe increase in the HARQ round-trip time. Thus, even without any repetitions the eMTC downlink HARQ round-trip will be ten subframes, compared to the eight-subframe round-trip time for normal (non-eMTC) transmission. As eMTC devices still only need to support eight HARQ processes, the maximum duty cycle for transmission to a specific device is 80% for full-duplex devices. This will reduce the maximum sustainable data rate to 800 kbit/s, compared to 1 Mbit/s for UE category 0 devices.

20.3.4 UPLINK TRANSMISSION: PUSCH AND PUCCH

As described in Chapter 7, conventional (non-eMTC) uplink data transmission is carried out on the PUSCH physical channel while uplink control signaling (uplink control information, UCI) is provided by means of the PUCCH physical channel.

PUSCH and PUCCH are used also for eMTC in essentially the same way as for non-eMTC devices, with the following extensions/modifications:

- Possibility for repetition for extended coverage for both PUSCH and PUCCH.
- Modified hybrid ARQ with adaptive and asynchronous retransmissions, see Section 20.3.4.2.

Repetitions for PUSCH follows a similar approach as for downlink PDSCH, see Section 20.3.3.2. On a cell level the network configures a set of possible repetition numbers where each set consists of four and eight different repetition numbers for CE mode A and CE mode B, respectively. The actual number of repetitions to be used for the uplink PUSCH transmission, from the four/eight values of the configured set, is then dynamically selected by the network and provided to the device as part of the scheduling grant.

In case of retuning between subframes used for uplink transmission, the time for retuning is made available by not transmitting two symbols of the uplink transmissions before and/or after the retuning. Exactly what symbols are used for retuning depends on what is being transmitted in the subframes before and after the retuning. In general, symbols are, if possible, taken from subframes with PUSCH transmission, rather than PUCCH transmission. Thus, if PUSCH is transmitted in the last subframe before retuning and/or in the first subframe after retuning, two symbols of the PUSCH subframe are used for retuning.[8] In case of PUCCH transmission in both the last subframe before retuning and the first subframe after retuning, one OFDM symbol of each subframe is used for retuning.

20.3.4.1 PUSCH Scheduling

Similar to downlink scheduling assignments, uplink scheduling grants for eMTC devices are carried on the MPDCCH using two new DCI formats, see Table 20.3.

DCI format 6-0A is used for scheduling grants for devices operating in CE Mode A. It contains similar information as DCI format 0 (see Chapter 6) but extended with information related to PUSCH and MPDCCH repetition:

- A frequency flag allowing for dynamic enabling/disabling of PUSCH frequency hopping.
- A HARQ process number needed for asynchronous uplink HARQ, see Section 20.3.4.2.
- The number of repetitions for the scheduled PUSCH.
- The number of repetitions for the MPDCCH transmission.

DCI format 6-0B is a more compact DCI format used for scheduling grants for devices operating in CE Mode B.

Similar to downlink scheduling assignments, the resource assignment included in DCI formats 6-0A and 6-0B is split into two parts. A *narrowband indicator* specifies the narrowband in which the uplink resource to be used for the PUSCH transmission is located. A *resource-block indicator* then specifies the exact set of resource blocks to use for the PUSCH transmission given the narrowband specified by the narrowband indicator.

For DCI format 6-0A, used for CE mode A, the resource-block indicator consists of 5 bits, thus being able to point at any combination of consecutive resource blocks within the six

---

[8]If PUSCH is transmitted in both the last subframe before the retuning and in the first subframe before the retuning, one symbol from each subframe is used for retuning.

Table 20.3 DCI Format 6-0A and 6-0B

Field	6-0A	6-0B
Frequency-hopping flag	1	—
Resource information (resource assignment)		
Narrowband indicator	Var[a]	Var[a]
Resource-block indicator	5	3
Number of PUSCH repetitions	2	3
HARQ process number	3	1
MCS	4	4
RV	2	—
New data indicator	1	1
CSI request	1	—
SRS request	0/1	—
Downlink assignment index/uplink index	2	—
PUSCH power control	2	—
Number of MPDCCH repetitions	2	2
Flag for 6-0/6-1 differentiation	1	1

[a]*Number of bits depends on the system bandwidth.*

resource blocks of the narrowband. Note that this is essentially the same as uplink resource-allocation type 0 used for normal (non-eMTC) uplink grants, see Chapter 6, where the bandwidth of the carrier has been replaced by the "bandwidth" of a single narrowband, that is, six resource blocks.

For DCI format 6-0B, used for CE mode B, the resource-block indicator consists of 3 bits indicating eight sets of consecutive resource block. Thus, there are some restrictions in what resource-block combinations that can be assigned. This is similar to downlink scheduling assignments for CE mode B (DCI format 6-1B), although the restrictions are less for DCI format 6-0B (eight different combinations) compared to DCI format 6-1B (only two combinations).

The timing for uplink scheduling grants is the same as for other (non-eMTC) PUSCH. In other words, a scheduling grant *ending* in downlink subframe n is valid for an uplink PUSCH transmission starting in uplink subframe $n + 4$. Note that, due to repetition, the eMTC scheduling grant may be transmitted over several downlink subframes. Similarly, the scheduled PUSCH transmission may extend over several uplink subframes.

Similar to the downlink, in case of frequency hopping the resource grant provides the narrowband of the first transmission. Narrowbands for subsequent repetitions are then given by cell-specific offsets relative to the first transmission. For the uplink, frequency hopping, if configured, always takes place between only two narrowbands.

Also similar to the downlink, in case of CE mode A, frequency hopping can be dynamically disabled by means of a frequency-hopping flag in the scheduling grant.

20.3.4.2 Uplink Hybrid ARQ

As described in Chapter 8, the baseline LTE uplink hybrid ARQ is *synchronous* and *non-adaptive*. More specifically,

- a single-bit hybrid-ARQ acknowledgment transmitted on the downlink PHICH physical channel is provided at a specific time instant relative to the uplink PUSCH transmission to be acknowledged;
- depending on the detected hybrid-ARQ acknowledgment, there is a well-defined retransmission carried out at a specific relative time instant, more exactly eight subframes after the prior transmission;
- the retransmission is carried out on the same frequency resource as the original transmission.

As the PHICH is a wideband transmission spanning the entire carrier bandwidth, it cannot be received by an narrowband eMTC device. Thus, the pre-release 13 uplink HARQ procedure cannot be applied to eMTC devices.

As also described in Chapter 8, there is, in general, a possibility to override the PHICH by explicitly scheduling a retransmission using a scheduling grant on PDCCH or EPDCCH. This allows for *adaptive* retransmissions—that is, the retransmission can be scheduled on a different frequency resource, compared to the prior transmission. However, the retransmission will still occur at a specific time instant, that is, retransmissions are still synchronous.

For eMTC devices uplink retransmissions are *always* explicitly scheduled, that is, eMTC device are not assumed to receive HARQ acknowledgments on the PHICH. As there is no PHICH to which the explicit retransmission requests have to be time aligned, eMTC uplink retransmissions are also *asynchronous*, that is, there is no strict relation between the timing of the initial transmissions and the timing of the requested retransmission.

As shown in Table 6.7 in Chapter 6, DCI format 0 and 4, providing scheduling grants for uplink non-eMTC transmissions, include a new data indicator (NDI) to support the explicit scheduling of uplink retransmissions. Due to the asynchronous HARQ, the scheduling grant for eMTC devices (DCI format 6-0A and 6-0B, see earlier) also includes a HARQ process indicator. For CE mode A, there are a total of eight HARQ processes, (corresponding to a 3-bit HARQ-process indicator) while CE mode B is limited to two HARQ processes (1-bit HARQ-process indicator).

20.3.4.3 PUCCH

As described in Chapter 7, UCI, including CSI reports, scheduling requests and hybrid-ARQ acknowledgments, are carried on the PUCCH physical channel. Each PUCCH transmission covers one subframe and is frequency-wise located at the edge of the carrier with frequency hopping on slot level. There are several different PUCCH formats associated with different payload sizes and, in practice, addressing different types of UCI.

UCI is needed also for eMTC. However, due to the limited set of supported transmission modes, as well as other limitations, not all PUCCH formats need to be supported for eMTC.

More specifically, only PUCCH formats 1, 1A, and 2 are supported in case of FDD while, for TDD, also PUCCH format 2A is supported.

The structure of each PUCCH transmission is also somewhat different, compared to the conventional (non-eMTC) PUCCH transmission.

Similar to other physical channels, repetition can be applied to PUCCH transmission, where the number of repetitions is configured by the network:

- For CE mode A, the number of repetitions can be 1 (no repetitions), 2, 4, and 8.
- For CE mode B, the number of repetitions can be 4, 8, 16, and 32.

In contrast to PDSCH, MPDCCH, and PUSCH, it is thus not possible to dynamically vary the number of repetitions for PUCCH.

PUCCH frequency hopping in case of eMTC PUCCH is carried out in blocks of N_{hop} subframes, rather than on slot level which is used for PUCCH for non-eMTC devices. This means that, in case the number of repetitions is configured to be smaller or equal to the frequency hopping-block length N_{hop}, there may be no PUCCH frequency hopping for eMTC. Especially, in case of no repetitions ($N_{rep} = 1$), there is no frequency hopping for eMTC devices.

20.3.4.4 Uplink Power Control

Uplink power control (PUSCH and PUCCH) for eMTC devices differs depending on if the device is configured for coverage extension mode A or coverage extension mode B.

In case of coverage extension mode A, uplink power control is essentially the same as for non-eMTC devices as described in Chapter 7 where, in case of eMTC devices, power control commands are provided within DCI format 6-0A (uplink scheduling grants) and 6-1A (downlink scheduling assignments). eMTC devices can also receive the power-control-specific DCI formats 3 and 3A. In case of repetition, the transmit power is unchanged for the set of subframes over which the repetition is carried out.

In case of coverage extension mode B, which is assumed to be used in the most severe propagation conditions, the transmit power is always set to the maximum per-carrier transmission power for both PUCCH and PUSCH transmissions. For this reason there are also no power-control commands included in DCI formats 6-0B and 6-1B.

20.3.5 SYNCHRONIZATION SIGNALS AND BCH

The LTE synchronization signals (PSS/SSS) and the PBCH are confined within the 72 center subcarriers of a carrier. Thus they can be received also by eMTC devices having an RF front end with a bandwidth limited to 1.4 MHz, despite that they are not necessarily confined within a single narrowband.

The synchronization signals are used unchanged for eMTC devices. As these signals do not vary in time, extended coverage can be achieved by having devices accumulating the received signal for longer time when searching for PSS/SSS. This may result in longer search

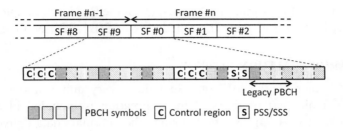

FIGURE 20.6

PBCH mapping for eMTC (FDD).

time, both at initial access and for mobility. However, as latency is not considered a critical parameter for massive-MTC applications, and the devices can be assumed to be stationary or of low mobility at least in case of the massive-MTC applications for which a large amount of coverage extension is important, this has been considered acceptable.

As described in Chapter 11, the coded BCH transport block is mapped to the first subframe of four consecutive frames. In each of these subframes, the PBCH is transmitted within the first four OFDM symbols of the second slot. Each BCH transport block is thus transmitted over a total of 16 OFDM symbols (four subframes and four OFDM symbols in each subframe).

To extend coverage, the PBCH is repeated a factor 5, that is, each BCH transport block is transmitted over a total of 80 OFDM symbols. Exactly how this is done differs somewhat between FDD and TDD.

For FDD, the four OFDM symbols of subframe 0 are repeated in five additional symbols in subframe 0 and eleven symbols in subframe 9 of the preceding frame, see Figure 20.6. Each symbol is thus repeated four times. For TDD, PBCH is likewise transmitted in subframes 0 and 5.[9]

If the five times repetition does not provide sufficient coverage, additional power boosting can be used to further extend the coverage of the PBCH.

The mapping of the repeated PBCH symbols is done such that a modulation symbol is repeated on the same subcarrier as the original symbol. This allows for the PBCH to be used for frequency tracking by the device, without having to decode the PBCH.

Note that the PBCH structure described above is not a new physical channel. Rather, the pre-release-13 PBCH is extended in a backward compatible way to support extended coverage. Legacy devices can still detect the PBCH and thus acquire the master information block (MIB), by just receiving the four pre-release 13 OFDM symbols in the second slot.

This also means that the MIB transmitted on the BCH still contains all the legacy information expected by pre-release 13 devices. Additional eMTC-specific information on the MIB

---

[9]This is true for the case of normal cyclic prefix. In case of an extended cyclic prefix, each symbol is repeated only three times and in somewhat different symbols, compared to normal cyclic prefix.

is provided by using 5 of the original 10 "spare bits" of the BCH, see Chapter 11. These 5 bits are used to convey information about the scheduling of SIB1-BR, see the following section.

20.3.6 SYSTEM-INFORMATION BLOCKS

As discussed in Chapter 11, the MIB only includes a very minor amount of system information, while the main part of the system information is included in different system-information blocks (SIBs) transmitted using the normal data-delivery mechanisms (DL-SCH marked with SI-RNTI). The scheduling of SIB1 is fixed in the specification while information about the time-domain scheduling of the remaining SIBs is included in SIB1.

As the legacy SIB1 may have a bandwidth exceeding six resource blocks and may consist of up to 2216 bits, it cannot be received by eMTC devices which are limited to a 1.4 MHz bandwidth and a maximum transport-block size of 1000 bits. Thus, for eMTC, a new SIB1, referred to as SIB1 *bandwidth-reduced* (SIB1-BR) was introduced.

SIB1-BR is transmitted over six resource blocks (one narrowband) and repeated a number of times per 80 ms interval. Although SIB1-BR is formally transmitted on PDSCH, repetition is done in a different way compared to other PDSCH transmissions. As described in earlier sections, in normal case PDSCH repetitions are carried out over contiguous valid subframes. However, for SIB1-BR the repetitions are equally spaced in time during the 80 ms period:

- For a repetition factor of 4, SIB1-BR is transmitted in one subframe every second frame.
- For a repetition factor of 8, SIB1-BR is transmitted in one subframe every frame.
- For a repetition factor of 16, SIB1-BR is transmitted in two subframes every frame.

The exact set of frames/subframes in which SIB1-BR is transmitted is provided in Table 20.4.

Information about the SIB1-BR repetition factor (4, 8, or 16) and transport block size (six different sizes) are included in the MIB, using 5 of the original 10 spare bits.

SIB1-BR then includes scheduling information for the remaining SIBs relevant for eMTC devices.

Table 20.4 Set of Subframes in Which SIB1-BR is Transmitted

Repetition Factor	PCID	FDD		TDD	
		SFN	Subframe	SFN	Subframe
4	Even	Even	4	odd	5
	Odd	Odd	4	odd	0
8	Even	Any	4	any	5
	Odd	Any	9	any	0
16	Even	Any	4 and 9	any	0 and 5
	Odd	Any	0 and 9	any	0 and 5

20.3.7 RANDOM ACCESS

As described in Chapter 11, the LTE random-access procedure consists of four steps:

Step 1: Uplink preamble transmission.
Step 2: Downlink random-access response providing timing-advance command and uplink resource for step 3.
Step 3: Uplink transmission of mobile-terminal identity.
Step 4: Downlink transmission of contention-resolution message.

As also described in Chapter 11, steps 2—4 use the same physical-layer functionality as normal uplink and downlink data transmission. They can thus directly rely on the repetition for PDSCH, MPDCCH, PUCCH, and PUSCH for CE described earlier.

To provide CE for the entire random-access procedure, repetition can also be applied to the preamble transmission.

As described in Chapter 11, the random-access (preamble) resource consists of a frequency block with a bandwidth corresponding to six resource blocks occurring in a set of subframes. In each cell there is one PRACH configuration defining

- the preamble format;
- the exact frequency resource used for PRACH transmissions;
- the exact set of subframes in which PRACH can be transmitted.

For the preamble transmission, the device selects a preamble from the set of available preambles and transmits it with the specified power. If a random-access response (step 2) is detected within the configured time window, the random-access procedure continues with steps 3 and 4. If no random-access response is detected, the procedure will be repeated, possibly with an increased transmission power.

For eMTC devices, up to four different random-access *CE levels* can be defined, each associated with its own PRACH configuration and corresponding PRACH resource.[10] Especially, the different CE levels can be associated with different frequency resources. A device selects the CE level based on its estimated path loss. In this way, random-access attempts from devices that are in very different coverage situations within a cell can be kept separate and not interfere with each other.

Each CE level is also associated with a specific repetition number indicating the number of repetitions to be used for the preamble transmission. The device selects a corresponding CE level, carries out a sequence of consecutive preamble transmissions according to the indicated number of repetitions and waits for a random-access response in the same way as for non-eMTC devices. If a random-access response in detected within the configured window, the random-access procedure continues with steps 3 and 4. If no random-access response is detected, the procedure is repeated a specified number of times. If no random-access response

---

[10]Note that the up to four CE levels are different from the two CE modes discussed earlier.

is detected within the specified number of attempts, the device moves the next higher CE level and repeats the procedure.

As discussed earlier, each active eMTC device is configured with a CE mode, limiting, for example, how many repetitions are carried out and determining, for example, what DCI formats are valid.

During random access, a device is not yet in RRC_CONNECTED state and has thus not been configured with a specific CE mode. Still, the different messages transmitted during steps 2−4 of random-access procedure are carried using the normal CE mechanisms for PDSCH, MPDCCH, PUCCH, and PUSCH assuming a certain CE mode. More specifically, for steps 2−4 of the random-access procedure, the device should assume.

- CE mode A if the latest PRACH (preamble) transmission used the resources associated with CE level 0 or 1.
- CE mode B if the latest PRACH transmission used the resources associated with CE level 2 or 3.

20.3.8 EXTENDED DRX

To reduce energy consumption and extend battery life, a device can enter DRX as described in Chapter 9. In DRX, a device monitors downlink control signaling in one subframe per *DRX cycle* and can be asleep during the remaining time of the cycle. A device can be in DRX in both connected and idle state, where DRX in idle state corresponds to a device being asleep and only waking up to check for paging messages, see Chapter 13.

From the first release of LTE, the DRX cycle has been limited 256 frames or 2.56 s. This has been sufficient for conventional mobile broadband devices for which longer time for access is typically anyway not acceptable. However, for some massive MTC applications for which very long battery life is a requirements and which, at the same time, can accept very long latency when accessing the network, the possibility for longer DRX/sleep cycles would be desirable.

As described in Section 20.1, LTE release 12 introduced PSM for reduced energy consumption and corresponding extended battery life. PSM is an excellent mechanism for applications characterized by device-triggered communication. However, it is not suitable for applications where the network needs to initiate communication as a device that has entered PSM cannot be paged by the network.

To further reduce device energy consumption for applications characterized by network-initiated traffic that is not latency critical, *extended DRX* was introduced as part of LTE release 13.

In extended DRX, the DRX cycle can be extended to 1024 frames corresponding to 10.24 s for devices in connected state and up 262144 frames corresponding to 2621.44 s or close to 44 min for devices in idle state.

To handle such very long DRX cycles that exceed the range of the SFN, a new 10-bit hyper-SFN (HSFN) has been defined. In contrast to the SFN, which is provided on the MIB, the HSFN is provided within SIB1.

20.4 NARROW-BAND INTERNET OF THINGS
20.4.1 BACKGROUND

In parallel to the release 12/13 activities discussed earlier to enhance the LTE support for massive-MTC applications, a separate activity related to low-cost MTC devices was initiated within the 3GPP GERAN group—that is, the 3GPP group responsible for the GSM technical specifications. Even today, GSM is by far the most dominating cellular technology for truly low-cost MTC applications, and the aim of the GERAN activities was to develop a technology that could eventually replace GSM for these applications. For reasons explained in the following, this activity was referred to as NB-IoT.

A key requirement for NB-IoT was that it should be truly narrowband, with an RF bandwidth in the order of 200 kHz or less, in order to be able to replace GSM carriers with NB-IoT carriers on a carrier-by-carrier basis. It should also be possible to deploy an NB-IoT carrier within the guard bands of LTE carriers, see Figure 20.7.

Several different technologies were considered as part of the NB-IoT work in GERAN, most of which were very different from LTE. However, late in the process it was decided to move the NB-IoT activities from GERAN to 3GPP RAN—that is, the 3GPP group responsible for the LTE technical specifications. At the same time an additional NB-IoT requirement was introduced, namely that, in addition to being able to be deployed in LTE guard bands, an NB-IoT carrier should also be able to efficiently coexist *within* an LTE carrier, see Figure 20.8.

While deployment in LTE guard bands essentially just requires a sufficiently narrowband carrier, the later requirement puts much stronger constraints on the NB-IoT physical-layer design. As a consequence, it was concluded that at least the NB-IoT downlink should have a physical-layer structure aligned with LTE, that is, OFDM with a subcarrier spacing of

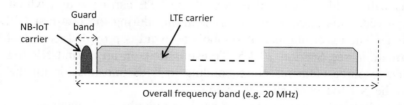

FIGURE 20.7

NB-IoT deployed in LTE guard band.

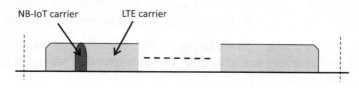

FIGURE 20.8

NB-IoT deployed within an LTE carrier.

15 kHz. This turned the NB-IoT activity from being a completely separate technology track to becoming much more integrated with the main stream evolution of LTE.

The rest of this section provides an overview of the current technical status of NB-IoT. It should be pointed out though that the work on NB-IoT is at the time of this writing still ongoing in 3GPP and details may still be changed/updated.

20.4.2 NB-IOT DEPLOYMENT MODES

As discussed earlier, there are three different deployment modes for NB-IoT:

- Deployment in spectrum of its own, for example, in spectrum refarmed from GSM. This is referred to as *stand-alone deployment*.
- Deployment within the guard band of an LTE carrier, referred to as *guard-band deployment*.
- Deployment within an LTE carrier, referred to as *inband deployment*.

It should be pointed out that, even in case of inband deployment, NB-IoT should be seen as a carrier of its own, separate from the LTE carrier.

20.4.3 DOWNLINK DATA TRANSMISSION

On the downlink, NB-IoT is based on OFDM transmission fully aligned with LTE, that is, with a subcarrier spacing of 15 kHz and the same basic time-domain structure as LTE.

Each NB-IoT carrier consists of 12 subcarriers. In other words, each NB-IoT carrier corresponds to a single LTE resource block in the frequency domain.

In case of stand-alone and guard-band deployment the entire resource block is available for NB-IoT transmissions.[11] On the other hand, in case of inband deployment NB-IoT transmissions will avoid the control region of the LTE carrier within which the NB-IoT carrier is deployed. This is done by not transmitting during the few first OFDM symbols of the subframe. The exact number of symbols to avoid is provided as part of the NB-IoT system information, see Section 20.4.5. Transmissions on an inband NB-IoT carrier will also avoid the resource elements corresponding to CRS transmission on the LTE carrier within which the NB-IoT carrier is deployed.

NB-IoT downlink data transmission is based on two physical channels:

- The *narrowband PDCCH* (NPDCCH) carrying scheduling information, that is, scheduling assignments/grants for downlink and uplink transport-channel transmission.
- The *narrowband PDSCH* (NPDSCH) carrying the actual downlink transport-channel data.

Within a resource block, either NDPCCH or NDPSCH is transmitted. In addition, NB-IoT reference signals (NRS) corresponding to up to two antenna ports are included in the last two OFDM symbols of each slot, see Figure 20.9.

---

[11]Not true for subframes carrying the NPBCH physical channel, see Section 20.4.5.

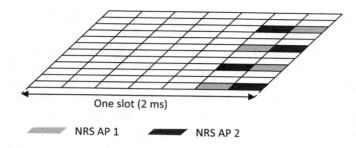

One slot (2 ms)

NRS AP 1   NRS AP 2

FIGURE 20.9

NRS structure.

Up to two NDPCCH can be frequency multiplexed within a subframe. Downlink scheduling assignment for NB-IoT is inter-subframe, that is, a scheduling assignment on NPDCCH ending in subframe n relates to a corresponding NPDSCH transmission starting in subframe $n + \varDelta$. This is similar to eMTC, see Section 20.3.3.3. However, while for eMTC the time offset between the end of the NPDCCH transmission and the start of the corresponding NPDSCH transmission is fixed to two subframes, in case of NB-IoT the time offset $\varDelta$ can vary dynamically with information about the time offset provided as part of the scheduling assignment. The possibility for different time offset allows for the transmission of scheduling assignments to two different devices in the same subframe even though their corresponding NPDSCH transmissions have to occur in different subframes.[12]

Channel coding for downlink data (the DL-SCH transport channel) is based on the same tail-biting convolutional code as used for downlink control signaling in LTE, see Chapter 6. The main reason for using tail-biting convolutional coding, rather than Turbo coding as is used for DL-SCH in LTE, is to reduce the channel-decoding complexity at the device side. NB-IoT downlink modulation is limited to QPSK.

In terms of multi-antenna transmission, NB-IoT supports transmission from one antenna port or two antenna ports. In case of two antenna ports transmission is based on space-frequency block codes (SFBC) as is also used for LTE transmission mode 2, see Chapter 6.

20.4.4 UPLINK TRANSMISSION

In contrast to the downlink there are two different modes with different numerologies for the NB-IoT uplink:

- one mode based on a 15 kHz subcarrier spacing;
- one mode based on 3.75 kHz subcarrier spacing.

The numerology of the 15-kHz mode is fully aligned with LTE. However, in contrast to LTE, uplink transmissions from a device can be carried out over only a subset of the

---

[12]There can only be one NPDSCH transmission per subframe within an NB-IoT carrier.

subcarriers of a resource block. More specifically, uplink transmission can be carried out over one, three, six, or twelve subcarriers, where transmission over twelve subcarriers corresponds to the full NB-IoT carrier bandwidth. The reason for allowing for transmission over only a fraction of the total NB-IoT carrier bandwidth is that, in extreme coverage situations, a device may not be able to transmit with a data rate that justifies the use of such large bandwidths. By transmitting over only a fraction of the subcarriers, that is, over only a fraction of the NB-IoT carrier bandwidth, multiple devices can be frequency multiplexed within one uplink carrier, thereby allowing for more efficient resource utilization.

In case of transmission over 12 subcarriers, the minimum scheduling granularity in the time domain is 1 ms (one subframe or two slots). For smaller assigned bandwidths, the scheduling granularity is increased to 2 ms (two subframes), 4 ms (four subframes), and 8 ms (eight subframes) for transmission over six, three, and one subcarrier, respectively.

In case of 3.75 kHz subcarrier spacing, there can be up to 48 subcarriers within the NB-IoT uplink bandwidth. However, each uplink transmission only consists of *a single sub-carrier*. As a consequence, the 3.75-kHz uplink mode only supports very low uplink data rates.

The time-domain structure of the 3.75-kHz uplink mode is outlined in Figure 20.10. As can be seen, the time-domain structure is not a direct factor-of-four scaling of the LTE numerology. Especially, even though the cyclic prefix for the 3.75-kHz mode is longer than the cyclic prefix of the 15-kHz mode, it is not four times longer. As a consequence, there is an idle time at the end of each subframe. In case of inband deployment, this idle time could, for example, be used to reduce interference to SRS transmissions on the uplink LTE carrier.

For the 3.75-kHz uplink mode the time-domain scheduling granularity is 16 slots or 32 ms.

Scheduling grants for uplink transmissions are provided on the NDPCCH. Similar to the downlink, the time offset between the NDPCCH transmission and the start of the corresponding uplink transmission can vary dynamically with the exact offset provided as part of scheduling grant.

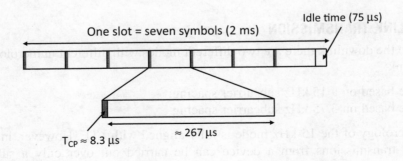

FIGURE 20.10

Time-domain structure of 3.75-kHz uplink mode.

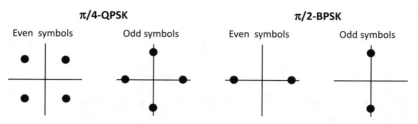

FIGURE 20.11

Constellations for π/4-QPSK and π/2-BPSK modulation.

Channel coding for UL-SCH is, for NB-IoT, based on the same Turbo coding as for LTE. Although decoding of Turbo codes is relatively complex, thus the use of tail-biting convolutional coding for the NB-IoT downlink, encoding for Turbo codes is of low complexity.

In case of transmission over multiple subcarriers, uplink modulation is based on conventional QPSK. However, in case of single-subcarrier transmission (both 15 kHz and 3.75 kHz), modulation is based on π/4-QPSK or π/2-BPSK. As shown in Figure 20.11, π/4-QPSK is the same as QPSK but where the entire constellation is shifted an angle of π/4 for odd-numbered symbols. Likewise, π/2-BPSK is the same as BPSK with the constellation shifted an angle of π/2 for odd-numbered symbols. The error performance of π/4-QPSK and π/2-BPSK is identical to that of QPSK and BPSK. However, the phase shift between consecutive symbols leads to a further reduced cubic metric, thereby enabling higher power-amplifier efficiency.

After modulation, DFT-precoding is applied in the same way as for the LTE uplink. Note that in case of transmission using a single subcarrier the DFT precoding has no effect.

20.4.5 NB-IOT SYSTEM INFORMATION

Similar to LTE, NB-IoT system information consists of two parts:

- A MIB transmitted on a special physical channel (NPBCH).
- SIBs transmitted in, essentially, the same way as any other downlink data. Information about the scheduling of SIB1 is provided in the MIB while the scheduling of the remaining SIBs is provided on SIB1.

As illustrated in Figure 20.12, the NPBCH is transmitted in subframe 0 in every frame with each transport block transmitted over a total of 64 subframes (640 ms). The TTI of the NPBCH (strictly speaking, the TTI of the BCH transport channel transmitted on the NPBCH) is thus 640 ms.

The transmission of NPBCH always "assumes" inband deployment in the sense that the NPBCH transmissions:

- avoid the first three symbols of the subframe;
- avoid the possible location of LTE CRS.

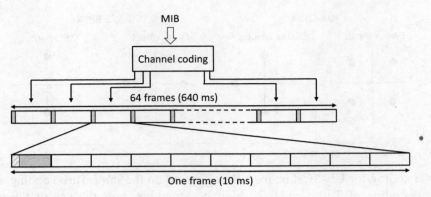

FIGURE 20.12

NB-IoT system information.

This makes it possible for a device to detect and decode the corresponding system information, without knowing if the NB-IoT carrier is deployed inband or not. Information about the deployment mode is then provided as part of the system information on SIB1.

DEVICE-TO-DEVICE CONNECTIVITY

21

21.1 OVERVIEW

Support for direct *device-to-device* (D2D) connectivity was first introduced in release 12 of the 3GPP LTE specifications. As the name suggests, direct D2D connectivity implies a *direct* radio link between devices.

For obvious reasons, D2D connectivity is only possible between devices in relatively close proximity of each other. Services based on D2D connectivity are therefore sometimes also referred to as *proximity services*, or *"ProSe."*

One reason for introducing support for D2D connectivity in the LTE specifications was an explicitly expressed interest to use the LTE radio-access technology for public-safety-related communication services. For the public-safety use case it is seen as important and in some cases even a requirement to support at least a limited degree of local connectivity between devices even when there is no infrastructure available. Thus, support for direct D2D connectivity was seen as a critical component to ensure LTE fulfillment of all the requirements of the public-safety use case. However, support for D2D connectivity may also enable new types of commercial services, thus expanding the usability of the LTE radio-access technology in general.

LTE distinguishes between two types of D2D connectivity:

- *D2D communication*, implying exchange of user data directly between devices. At this stage, including also LTE release 13, D2D communication is only targeting the public-safety use case. Details of D2D communication are provided in Section 21.2.
- *D2D discovery*, implying the possibility for a device to transmit signals that enable its presence to be directly detected by other devices in its neighborhood.[1] In contrast to D2D communication, D2D discovery has already from the beginning targeted a wider range of use cases, including commercial services. Details of D2D discovery are provided in Section 21.3.

For a direct D2D radio link, the notion of downlink and uplink transmission directions is obviously not applicable. Instead, 3GPP has introduced the term *sidelink* to characterize the

---

[1] As described in Section 21.3, LTE discovery is not really about discovering devices as such but rather about the discovering *services* announced by devices.

4G, LTE-Advanced Pro and The Road to 5G. http://dx.doi.org/10.1016/B978-0-12-804575-6.00021-2

direct D2D link. In order to align with 3GPP terminology we will from now on use the term sidelink rather than D2D.

21.1.1 SIDELINK TRANSMISSION

LTE sidelink connectivity should be possible in normal cellular (LTE) spectrum, including both paired (FDD) and unpaired (TDD) spectrum. Consequently, good co-existence between sidelink transmissions and normal cellular (downlink/uplink) transmissions in the same spectrum has been a key requirement in the design of LTE sidelink connectivity.

However, sidelink connectivity may also take place in spectrum not used by commercial cellular networks. An example of this is the public-safety use case for which specific spectrum has been assigned in several countries/regions.

In case of paired spectrum, sidelink connectivity takes place in the uplink part of the spectrum. Consequently, devices supporting sidelink connectivity for a certain FDD band need to be able to receive also in the uplink band.

There are several reasons why sidelink connectivity takes place in the uplink band in case of FDD spectrum:

- Regulatory rules are typically concerned with what and how devices transmit but do not restrict what and how devices receive. From a regulatory point of view, sidelink connectivity in uplink spectrum is therefore more straightforward compared to sidelink connectivity in downlink spectrum, as the later would imply device transmission in spectrum assumed to be used for network transmission.
- From a device-implementation point of view, it is less complex to include additional receiver functionality (support for reception in an uplink band) compared to the additional transmitter functionality needed in case sidelink connectivity would take place in downlink bands.

In a similar way, in case of TDD spectrum sidelink connectivity is assumed to take place in uplink subframes. It should be noted though that, while the 3GPP specifications define if a specific (FDD) frequency band is for downlink or uplink, the TDD downlink/uplink configuration is defined by the network, in principle on cell level. As a consequence, at least in principle different cells may have different downlink/uplink configurations, something which needs to be taken into account, for example, in case of sidelink connectivity between devices in different cells.

It should also be understood that sidelink connectivity is fundamentally unidirectional in the sense that all current LTE sidelink transmissions are, essentially, broadcast transmissions with, for example, no associated control signaling in the opposite direction. There may of course be sidelink transmissions from a device A received by a device B and, simultaneously, sidelink transmissions from device B received by device A. But these are then, radio-wise, completely independent transmissions.

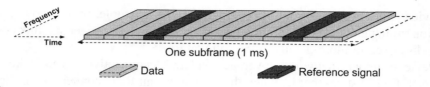

FIGURE 21.1

Subframe structure for sidelink transmission.

In addition to using uplink spectrum, sidelink connectivity also reuses the basic uplink transmission structure, more specifically the basic structure of PUSCH transmission. Thus, all sidelink transmissions, with the exception of sidelink synchronization signals (SLSS), see Section 21.4, are based on DFT-spread OFDM with a subframe structure as illustrated in Figure 21.1. Note that, in case of sidelink transmission, the last OFDM symbol of the subframe is not transmitted. This is done in order to create the guard time needed when switching between sidelink transmission and sidelink reception, as well as between sidelink transmission/reception and regular uplink transmission.

21.1.2 IN-COVERAGE VS. OUT-OF-COVERAGE SIDELINK CONNECTIVITY

As illustrated in Figure 21.2, devices involved in sidelink connectivity may be under network *coverage* ("in-coverage" scenario). However, sidelink connectivity is also possible for devices outside of network coverage ("out-of-coverage" scenario). There could also be situations when some devices involved in sidelink connectivity are under network coverage and some devices are outside network coverage. For the in-coverage scenario, the device receiving a sidelink transmission may be within the same cell as the transmitting device (intra-cell) or in different cells (inter-cell).

For release 12, only sidelink communication was supported out of coverage while sidelink discovery was only possible under network coverage. However, support for out-of-coverage sidelink discovery for the public-safety use case has recently been introduced as part of 3GPP release 13.

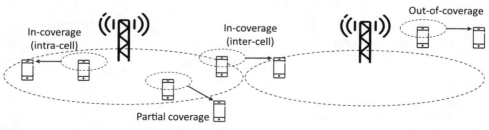

FIGURE 21.2

Different coverage scenarios for sidelink connectivity.

If a device is in coverage or out of coverage will, for example, impact how it acquire its transmission timing and the configuration parameters needed for proper sidelink connectivity as will be discussed in more details in the following.

For devices under network coverage, sidelink connectivity can take place in RRC_CONNECTED state, that is, when the device has an RRC connection to the network. However, sidelink connectivity can also take place in RRC_IDLE state, in which case the device does not have a dedicated connection to the network. It should be noted that being in RRC_IDLE state is not the same thing as being out of coverage. A device in RRC_IDLE state may still be under network coverage and will then have access to, for example, the network system information even if there is no RRC connection established.

21.1.3 SIDELINK SYNCHRONIZATION

Before devices can establish sidelink connectivity, they should be reasonably well synchronized to each other and to the overlaid network if present.

One reason for this is to ensure that sidelink transmissions will take place within intended time—frequency resources, thereby reducing the risk for uncontrolled interference to other sidelink and non-sidelink (cellular) transmissions in the same band.

As indicated in Figure 21.3, a device under network coverage should use the ordinary cell synchronization signals (PSS/SSS, see Chapter 11) of the serving cell (RRC_CONNECTED state) or the cell the device is camping on (RRC_IDLE state) as timing reference for its sidelink transmissions.

However, to allow for network control of transmission timing to extend beyond the area of direct network coverage LTE sidelink connectivity also includes the possibility for devices to transmit special *sidelink synchronization signals* (SLSSs). A device under network coverage may transmit SLSS in line with the transmission timing acquired from the network. This signal can then be received and used as timing reference for sidelink transmissions by near-by out-of-coverage devices. These devices can then, in turn, transmit their own SLSS that can be

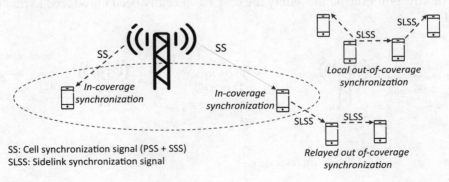

SS: Cell synchronization signal (PSS + SSS)
SLSS: Sidelink synchronization signal

FIGURE 21.3

Transmission-timing acquisition for sidelink transmissions.

detected and used as timing references by other out-of-coverage devices. In this way, the area over which devices are synchronized to and derive their transmission timing from the overlaid network can be further expanded beyond the area of direct network coverage.

A device not within network coverage and not detecting any sufficiently strong SLSS will autonomously transmit SLSS which can then be detected and forwarded by other out-of-coverage devices. In this way, local synchronization between out-of-coverage devices can be achieved even without the presence of an overlaid network.

In addition to its function as a timing reference for sidelink transmissions for out-of-coverage devices, a SLSS can also serve as a timing reference for sidelink *reception*.

To ease the reception of sidelink transmissions, a receiving device should preferably have good knowledge of the timing of the signal to be received. For sidelink connectivity between devices using the same transmission-timing reference, for example, in case of in-coverage devices having the same serving cell, a receiving device can use its own transmission timing also for reception.

To enable sidelink connectivity between devices that do not rely on the same reference for transmission timing, for example, sidelink connectivity including devices in different non-time-aligned cells, a device may transmit SLSS in parallel to its other sidelink transmissions. These synchronization signals can then be used as reference for *reception* timing by receiving devices.

An example of this is illustrated in Figure 21.4. In this case, device A uses the synchronization signal of its serving cell (SS_A) as timing reference for its sidelink transmissions. Similarly, device B uses SS_B as timing reference for its sidelink transmissions. However, as timing reference for the reception of sidelink transmissions from device A, device B will use the synchronization signal $SLSS_A$ transmitted by device A and derived from SS_A. Likewise, device A will use $SLSS_B$ as timing reference for the reception of sidelink transmissions from device B.

Further details of sidelink synchronization, including details of the structure of SLSS, are provided in Section 21.4.

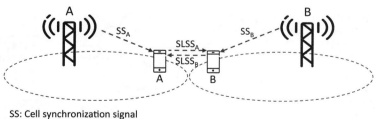

SS: Cell synchronization signal
SLSS: Sidelink synchronization signal

FIGURE 21.4

Use of sidelink synchronization signals (SLSS) as timing references for sidelink reception.

21.1.4 CONFIGURATION FOR SIDELINK CONNECTIVITY

Before a device can take part in sidelink connectivity, it has to be properly configured. Such configuration includes, for example, parameters defining the set of resources (subframes and resource blocks) that are available for different types of sidelink transmission.

Sidelink-related configuration parameters are partly provided as part of the cell system information, see Chapter 11. More specifically, two new SIBs have been introduced for sidelink-related configuration parameters:

- SIB18 for configuration parameters related to sidelink communication.
- SIB19 for configuration parameters related to sidelink discovery.

In addition to this *common configuration* provided via the cell system information, devices in RRC_CONNECTED state that are to engage in sidelink connectivity will also be individually configured by means of dedicated RRC signaling.

Configuration by means of system information or dedicated RRC signaling is obviously not possible for devices that are not under network coverage. Such devices instead have to rely on *pre-configured* sidelink-related configuration parameters. This pre-configuration essentially serves the same purpose as the common configuration provided as part of the sidelink-related system information.

An out-of-coverage device may, for example, have been provided with the pre-configured parameters at an earlier stage when it was under network coverage. Other possibilities include providing the pre-configuration on the SIM card or hard-coded into the device. Note that out-of-coverage operation is currently only targeting the public-safety use case. Out-of-coverage operation is thus typically associated with special devices/subscriptions.

21.1.5 ARCHITECTURE FOR SIDELINK

Figure 21.5 illustrates the network architecture related to sidelink connectivity. To support sidelink connectivity a new *ProSe Function* has been introduced in the core network together with a number of new network interfaces. Among these interfaces, PC5 corresponds to the

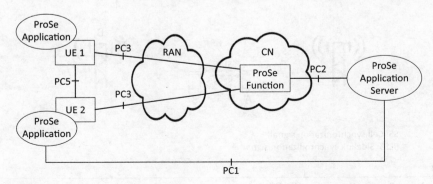

FIGURE 21.5

Architecture for sidelink (ProSe) connectivity.

direct link between devices while PC3 is the interface between sidelink-capable devices and the ProSe Function.

The ProSe Function is responsible for all sidelink functionality within the core network. It, for example, provides devices with the parameters needed to establish sidelink connectivity (discovery or communication). The ProSe Function also provides the mapping between discovery message codes and the actual discovery message, see Section 21.3.

21.1.6 SIDELINK CHANNEL STRUCTURE

Figure 21.6 illustrates the channel structure related to sidelink connectivity, including logical channels, transport channels and physical channels/signals.

The *sidelink traffic channel* (STCH) is the logical channel carrying user data for sidelink communication. It is mapped to the *sidelink shared channel* (SL-SCH) transport channel which, in turn, is mapped to the *physical sidelink shared channel* (PSSCH). In parallel to the PSSCH, there is the *physical sidelink control channel* (PSCCH) carrying *sidelink control information* (SCI) which enables a receiving device to properly detect and decode the PSSCH.

The *sidelink discovery channel* (SL-DCH) is the transport channel used for discovery announcements. On the physical layer, it is mapped to the *physical sidelink discovery channel* (PSDCH). Note that there is no logical channel related to sidelink discovery, that is, the discovery message is inserted directly into the SL-DCH transport block on the MAC layer. There are thus no RLC and PDCP layers for sidelink discovery.

Finally, sidelink synchronization is based on two signals/channels:

- The already mentioned SLSS which is associated with a specific *sidelink identity* (SLI).
- The *sidelink broadcast control channel* (S-BCCH) with corresponding transport channel (the *sidelink broadcast channel*, SL-BCH) and physical channel (the *physical sidelink broadcast channel*, PSBCH). This channel is used to convey some very basic sidelink-related "system information," referred to as the sidelink master information block (SL-MIB), between devices, see further Section 21.4.2.

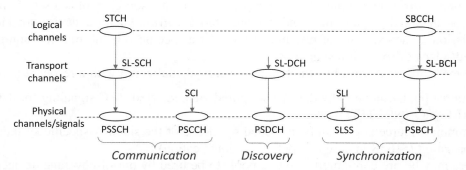

FIGURE 21.6

Sidelink channel structure.

21.2 SIDELINK COMMUNICATION

As already mentioned, sidelink communication implies the exchange of user data directly between close-by devices. In release 12, sidelink communication is limited to *group communication*. In practice this implies that

- the sidelink transmission is broadcast by the device with no assumptions regarding the link properties related to a certain receiving device,
- the sidelink transmission may be received and decoded by any sidelink-communication-capable device in the proximity of the transmitting device, and
- a *group identity* included in the control part of the sidelink transmission allows for a receiving device to determine if it is among the intended receivers of the data.

It should be noted that nothing prevents the group of devices involved in the sidelink communication to consist of only two devices, implying that there is, in practice, only a single intended receiver.

As previously mentioned, sidelink communication is based on two physical channels:

- The *physical sidelink shared channel* (PSSCH) carries the actual transport-channel (SL-SCH) data.
- The *physical sidelink control channel* (PSCCH) carries control information that enables receiving devices to properly detect and decode the PSSCH.

The PSCCH thus serves a similar purpose as the PDCCH/EPDCCH (Chapter 6) which, among other things, carries control information that enables receiving devices to properly detect and decode a corresponding PDSCH carrying downlink transport-channel data.

21.2.1 RESOURCE POOLS AND ASSIGNMENT/SELECTION OF TRANSMISSION RESOURCES

For sidelink communication (as well as for sidelink discovery, see Section 21.3), the concept of *resource pools* has been introduced. Simply speaking, a resource pool is a set of physical resources, in practice subframes and resource blocks, *available* to a device for sidelink transmissions. The exact set of resources to use for a specific sidelink transmission is then assigned/selected from the resource pool.

There are different ways by which a device can be configured with a resource pool:

- Resource pools can be individually configured via dedicated RRC signaling for devices in RRC_CONNECTED mode.
- Common resource pools can be provided by means of the sidelink-specific system information (SIB18 in case of sidelink communication).
- There may be pre-configured resource pools to be used by out-of-coverage devices.

For sidelink communication each resource pool consists of:

- a *PSCCH subframe pool* defining a set of subframes available for PSCCH transmission;
- a *PSCCH resource-block pool* defining a set of resource blocks available for PSCCH transmission within the PSCCH subframe pool;
- a *PSSCH subframe pool* defining a set of subframes available for PSSCH transmission;
- a *PSSCH resource-block pool* defining a set of resource blocks available for PSSCH transmission within the PSSCH subframe pool.

There are two types or *modes* of sidelink communication. The two modes differ in terms of how a device is assigned or selects the exact set of resources to use for the sidelink transmission from a configured resource pool. This includes resources for PSCCH transmission as well as for the actual data (transport-channel) transmission using PSSCH:

- In case of sidelink communication mode 1, a device is explicitly assigned, by means of a *scheduling grant* received from the network, a specific set of PSCCH/PSSCH resources.
- In case of sidelink communication mode 2, a device by itself selects the set of PSCCH/PSSCH resources.

As it relies on explicit scheduling grants provided by the network, mode 1 sidelink communication is only possible for in-coverage devices in RRC_CONNECTED state. In contrast, mode 2 sidelink communication is possible in coverage as well as out of coverage and in both RRC_IDLE and RRC_CONNECTED state.

21.2.2 PHYSICAL SIDELINK CONTROL CHANNEL PERIODS

In the time domain, sidelink communication is based on so-called *PSCCH periods*. Each SFN period, see Chapter 11, consisting of 1024 frames or 10240 subframes, is divided into equal-lengths PSCCH periods.

If the set of resources to be used for transmission is explicitly assigned by the network, that is, in case of sidelink communication mode 1, the assignment is carried out on a PSCCH-period basis. Similarly, if a device, by itself, selects the transmission resources (sidelink communication mode 2), the selection is done on a PSCCH-period basis.

In case of FDD, the length of the PSCCH period can be configured to 40, 80, 160, or 320 subframes. In case of TDD, the set of possible lengths for the PSCCH period depends on the downlink/uplink configuration.

21.2.3 SIDELINK CONTROL INFORMATION/PHYSICAL SIDELINK CONTROL CHANNEL TRANSMISSION

PSCCH is transmitted once every PSCCH period. As already mentioned, the PSCCH carries control information, referred to as *sidelink control information* (SCI), which enables a receiving device to properly detect and decode the data transmission on PSSCH. The SCI

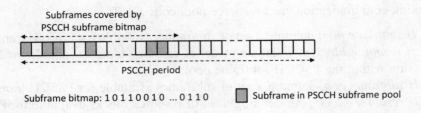

Subframe bitmap: 1 0 1 1 0 0 1 0 ... 0 1 1 0   ▨ Subframe in PSCCH subframe pool

FIGURE 21.7

PSCCH subframe pool within a PSCCH period.[2]

includes, for example, information about the time—frequency resources (subframes and resource blocks) used for the PSSCH transmission. The content of the SCI are described in more detail in Section 21.2.5 after a discussion on the structure of PSSCH transmission.

Channel coding and modulation for SCI is done in essentially the same way as for DCI (Chapter 6) and consists of the following steps:

- 16-bit CRC calculation;
- rate 1/3 tail-biting convolutional coding;
- rate matching to match to the number of coded bits to the size of the PSCCH resource;
- bit-level scrambling with a predefined seed;
- QPSK modulation.

The modulated symbols are then DFT precoded before being mapped to the physical resources (subframes and resource blocks) assigned/selected for the PSCCH transmission.

The *PSCCH subframe pool*, that is, the set of subframes *available* for PSCCH transmission within each PSSCH period is given by a subframe bitmap provided as part of the sidelink configuration, see Figure 21.7.[2] In case of sidelink connectivity in FDD spectrum, the bitmap is of length 40. For TDD, the length of the bitmap depends on the downlink/uplink configuration.

The *PSCCH resource-block pool*, that is, the set of resource blocks *available* for PSCCH transmission within the subframe pool, consists of two equal-size sets of frequency-wise consecutive resource blocks, see Figure 21.8. The resource-block pool can thus be fully described by

- the index S_1 of the first resource block in the "lower" set of resource blocks;
- the index S_2 of the last resource block in the "upper" set of resource blocks;
- the number M of resource blocks in each of the two sets.

As illustrated in Figure 21.9, a PSCCH transmission is carried out over two subframes and within one resource block pair[3] in each subframe. Exactly what subframes and resource

---

[2]Note that the figure assumes sidelink communication in paired/FDD spectrum. In case of unpaired/TDD spectrum, the bitmap only covers the uplink subframes as defined by the current DL/UL configuration.

[3]Remember that one subframe consists of two resource blocks.

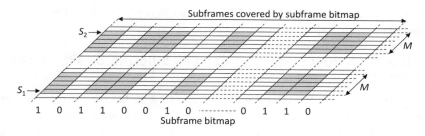

FIGURE 21.8

Structure of PSCCH resource pool.

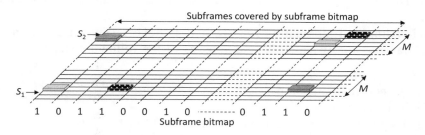

FIGURE 21.9

PSCCH transmission.

blocks, within the configured resource pool, to use for a certain PSCCH transmission is jointly given by a parameter n_{PSCCH}. n_{PSCCH} is either provided in the scheduling grant delivered by the network (for sidelink communication mode 1) or autonomously selected by the transmitting device (sidelink communication mode 2).

The mapping from n_{PSCCH} to actual set of PSCCH resources is such that if the transmission in the first subframe takes place in the lower set of resource blocks, the transmission in the second subframe will take place in the upper set of resource blocks and vise versa. The mapping is also such that if two different values of n_{PSCCH} imply mapping to the same first subframe, the second transmission will take place in different subframes or vise versa. Thus, PSCCH transmissions corresponding to different values of n_{PSCCH} will, time-wise, only collide in one of the two subframes. This has two benefits:

- Due to near-far effects, simultaneous sidelink transmissions from multiple devices in the same subframe may severely interfere with each other even if they are frequency-wise separated, that is, taking place in different resource blocks. The PSCCH mapping ensures that such collisions will only occur in one of the two subframes as long as the devices have been assigned or selected different values for n_{PSCCH}.
- A device cannot transmit and receive PSCCH in the same subframe. The PSCCH mapping ensures that a device could still transmit and receive PSCCH *in the same*

PSCCH period, assuming that different values of n_{PSCCH} are used for the two transmissions.

21.2.4 SIDELINK SHARED CHANNEL/PHYSICAL SIDELINK SHARED CHANNEL TRANSMISSION

Actual transport channel (SL-SCH) data is transmitted in form of transport blocks on the PSSCH physical channel. Each transport block is transmitted over four consecutive subframes within the PSSCH subframe pool. Transmission of *M* transport blocks within a PSCCH period thus requires 4*M* subframes. Note that a single SCI on PSCCH carries control information related to PSSCH transmission for the entire PSCCH period. This is in contrast to DCI which normally only defines the PDSCH transmission within the same subframe.

Channel-coding and modulation for SL-SCH is done in the same way as for uplink (UL-SCH) transmission (Chapter 7) and consists of the following steps:

- CRC insertion;
- code-block segmentation and per-code-block CRC insertion;
- rate 1/3 Turbo coding;
- rate matching (based on physical-layer hybrid-ARQ functionality);
- bit-level scrambling;
- data modulation (QPSK/16QAM).

The rate matching matches the set of coded bits to the size of the physical resource assigned/selected for the transmission of the transport block, taking the modulation scheme into account. There is no Hybrid ARQ for SL-SCH. However, rate matching and mapping of a coded transport block to the four subframes is done in the same way as the selection of redundancy versions for Hybrid-ARQ retransmissions.

The bit-level scrambling depends on the group identity, that is, the identity of the group that the sidelink transmission targets.

PSSCH data modulation is limited to QPSK and 16QAM. The network may impose a specific modulation scheme to use for PSSCH transmission as part of the sidelink configuration. If the network does not impose a specific modulation scheme, the transmitting device autonomously selects the modulation scheme. Information about the assigned/selected modulation scheme is then provided to receiving devices as part of the SCI.

After channel coding and modulation, DFT precoding is applied followed by mapping to the physical resource assigned/selected for the PSSCH transmission.

In case of sidelink communication mode 1, the PSSCH subframe pool, that is, the set of subframe *available* for PSCCH transmission, consists of all uplink subframes after the last subframe of the PSCCH subframe pool, see Figure 21.10.

The exact set of subframes to use for PSSCH transmission in a PSCCH period is given by a *time repetition pattern index* (TRPI) provided as part of the scheduling grant. As illustrated in Figure 21.11, the TRPI points to a specific *time repetition pattern* (TRP) within a *TRP table*

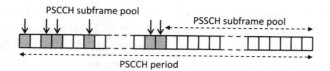

FIGURE 21.10

Subframes available for PSSCH (data) transmission within a PSCCH period (the "PSSCH subframe pool") for sidelink communication mode 1.

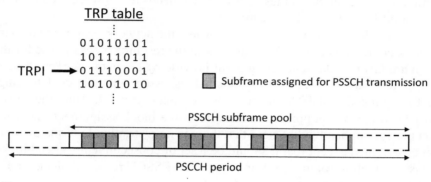

FIGURE 21.11

Assignment of subframes to PSSCH (sidelink communication mode 1).

explicitly defined within the LTE specifications.[4] Periodic extension of the indicated TRP then gives the uplink subframes assigned for the PSSCH transmission.

The TRPI is then included in the SCI in order to inform receiving devices about the set of subframes in which the PSSCH is transmitted.

In case of sidelink communication mode 2 the PSSCH subframe pool, that is, the set of (uplink) subframes available for PSSCH transmission, consists of a subset of the mode 1 subframe pool. More specifically, a periodic extension of a bitmap defined in the sidelink configuration indicates what subframes are included in the PSSCH subframe pool. In this way the network can ensure that certain subframes will not be used for PSSCH transmissions.

The device then autonomously decides on the exact set of subframes to use for the PSSCH transmission by randomly selecting a TRP from the TRP table. Similar to sidelink communication mode 1, the receiving device is informed about the selected TRP by including the corresponding TRPI in the SCI.

In addition to limiting the set of subframes that are part of the PSSCH subframe pool, in case of sidelink communication mode 2 there are also limitations in the TRP selection.

---

[4]For FDD the TRP table consists of 106 entries, with each TRP consisting of eight bits. For TDD the size of the TRP table, as well as the length of the TRP, depends on the downlink/uplink configuration.

In general, the TRP table consists of TRPs with different number of ones, corresponding to different fractions of subframes assigned for the PSSCH transmission. This includes, for example, the all-one TRP corresponding to assigning all subframes of the PSSCH pool for PSSCH transmission from a specific device. However, in case of sidelink communication mode 2 the TRP selection is restricted to TRPs with a limited number of ones, thus limiting the PSSCH transmission duty cycle. For example, in case of FDD, the TRP selection is limited to TRPs with a maximum of four ones, corresponding to a 50% duty cycle for the PSSCH transmission.

In addition to the set of subframes, a device also needs to know the exact set of resource blocks to be used for the PSSCH transmission.

In case of sidelink communication mode 1, where the network assigns the resources to use for the sidelink communication, information about the resource blocks to use for the PSCCH transmission are given in the scheduling grant provide by the network. The structure of that resource, and the way by which it is signaled, is essentially identical to single-cluster allocation for uplink (PUSCH) transmissions, see Section 6.4.7. Thus the resource grant includes a 1-bit frequency-hopping flag and a resource-block assignment, the size of which depends on the system bandwidth. Note that there is no restriction in terms of what resource blocks can be assigned except that it should be a set of consecutive resource blocks. In other words, in case of sidelink communication mode 1, the PSSCH resource-block pool consists of all resource blocks within the carrier bandwidth.

In case of sidelink communication mode 2, there are restrictions in terms of what resource blocks are available for PSSCH transmission. This PSSCH resource block pool has the same structure as the PSCCH resource-block pool, that is, it consists of two sets of frequency-wise consecutive resource blocks defined by three parameters S_1, S_2, and M, compare Figure 21.8. Note that the parameters defining the PSSCH resource-block pool are configured separately from those defining the PSCCH resource-block pool. A device configured to operate in sidelink communication mode 2 will then autonomously select a set of consecutive resource blocks from the PSSCH resource-block pool.

Information about the assigned/selected set of resource blocks is provided to receiving devices as part of the SCI.

21.2.5 SIDELINK CONTROL INFORMATION CONTENT

As discussed in the preceding section, the SCI carries information needed by a receiving device to properly detect and decode the PSSCH and extract the SL-SCH data. This includes information about the exact set of resources (subframes and resource blocks) in which the PSSCH is transmitted:

- The TRPI, indicating the set of subframes used for the PSSCH transmission.
- *A frequency hopping flag* indicating whether or not frequency hopping is used for the PSSCH transmission.

- *A resource-block and hopping-resource allocation* indicating what resource blocks, within the subframes indicated by the TRPI, are used for the PSSCH transmission.

The last parameter is essentially identical to the corresponding parameters in the uplink scheduling grant in DCI format 0.
In addition, the SCI includes

- a five-bits indicator of the modulation and coding scheme (MCS) used for the PSSCH transmission;
- an eight-bit *group destination ID*, indicating the group for which the sidelink communication is intended;
- an eleven-bit timing-advance indicator.

21.2.6 SCHEDULING GRANTS AND DCI FORMAT 5

As described above, devices within network coverage can be configured to only initiate sidelink communication when having been provided with an explicit scheduling grant by the network (sidelink communication mode 1). This is similar to devices being allowed to transmit on the uplink only when having an explicit scheduling grant for uplink transmission, see Chapter 7. As described in Chapter 6, such scheduling grants are provided via PDCCH/ePDCCH using DCI format 0 or DCI format 4. In a similar way, sidelink scheduling grants are provided via the PDCCH/ePDCCH using a new *DCI format 5*.
DCI format 5 includes the following information:

- The parameter n_{PSCCH} indicating the physical resource (subframes and resource blocks) on which PSCCH is to be transmitted.
- The TRPI indicating what subframes within the PSSCH subframe pool to use for the PSSCH transmission.
- *A frequency hopping flag* indicating whether or not frequency hopping should be applied for the PSSCH transmission.
- *A resource-block and hopping-resource allocation* indicating what resource blocks, within the subframes indicated by the TRPI, should be used for the PSSCH transmission.

The last parameter is, essentially, identical to the corresponding parameters of the scheduling grant for conventional uplink (PUSCH) transmission, see Chapter 7.
In addition, DCI format 5 includes a 1-bit transmit power control (TPC) command that applies to both PSCCH and PSSCH.
As outlined in Figure 21.12, the sidelink scheduling grant is valid for the next PSCCH period starting at least four subframes after the arrival of the scheduling grant. Note that this provides the same amount of time from the arrival of the scheduling grant to the actual scheduled transmission, as for normal uplink (PUSCH) transmission.
The transmission of scheduling grants for sidelink communication is supported by buffer-status reports (BSRs) provided to the network by devices involved in sidelink

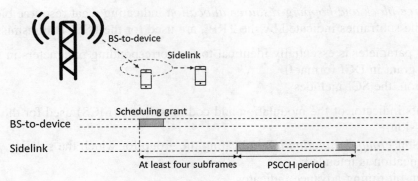

FIGURE 21.12

Timing relation for uplink scheduling.

communication. Similar to uplink buffer status reports, see Chapter 9, the sidelink BSRs are conveyed as MAC control elements and indicate the amount of data available for transmission at the device.

21.2.7 RECEPTION RESOURCE POOLS

In the preceding section the concept of a resource pool, defining the set of resources (subframes and set of resource blocks) that are available for transmissions related to sidelink communication (PSCCH and PSSCH) has been described.

In addition to this *transmission resource pool*, a device that is to take part in sidelink communication is also configured with one or several *reception resource pools* related to sidelink communication.

A reception resource pool describes the set of resources (subframes and resource blocks) in which a device can expect to *receive* sidelink-communication-related transmissions. Especially, the PSCCH part of the reception resource pool describes the set of resources in which the device should search for PSCCH transmissions. Furthermore, the PSSCH part of the resource is needed for the receiver to be able to properly interpret the resource information in the SCI.

The reason why a device may be configured with multiple reception pools is that it may receive sidelink communication from multiple devices and these devices may be configured with different transmission pools. This may be the case regardless of whether the devices are within the same cell or within different cells.[5] In principle, one can say that a device should be configured with a reception pool that is the union of the transmission resource pools of the devices with which it is to communicate. In practice, this is realized by configuring the device

---

[5]Devices in RRC_IDLE state within the same cell will use the same transmission resource pool provided by SIB 18. However, devices within the same cell in RRC_CONNECTED state may be individually configured with different transmission pools. Devices in RRC_IDLE state in different cells may also be configured with different transmission pools.

with multiple reception pools that jointly covers the transmission pools of the relevant devices.

Reception pools for sidelink communication are provided as part of the sidelink-related system information (SIB18) for in-coverage devices and as part of the pre-configuration for out-of-coverage devices.

21.3 SIDELINK DISCOVERY

Sidelink discovery is about devices repeatedly broadcasting short fixed-size messages that can be directly detected ("discovered") by other nearby devices. These messages could, for example, be announcements of "services," such as a restaurant announcing a special menu to by-passers, or requests for information such as asking for near-by people with specific competence.

It is important to understand that the actual message to be conveyed is not explicitly included in the broadcast message. Instead, the broadcast message consists of a user identity and a message code. Mapping from message codes to actual message is then provided by the network ProSe Function, see Section 21.1.5.

As described in Section 21.1.6, discovery messages are transmitted in form of transport blocks on the SL-DCH transport channel which, in turn is mapped to the PSDCH (physical sidelink discovery channel) physical channel. Thus, in contrast to sidelink communication there is no PSCCH/PSSCH structure with associated control information carried on a separate physical channel for discovery transmissions.

Comparing with sidelink communication, the transmission of discovery messages (SL-DCH transport blocks) on PSDCH is in many respects more similar to the transmission of control information (SCI) on the PSCCH, than the SL-SCH transmission on PSSCH:

- Similar to PSCCH, the PSDCH payload (the SL-DSCH transport block) is of fixed size (232 bits).
- Similar to PSCCH, a receiving device "searches" for PSDCH in a reception pool. In contrast, for PSSCH reception the receiver is informed about the exact resource by means of the SCI.

As will be seen in the following section, the resource-pool structure for PSDCH transmission is also, in many ways, similar to the resource-pool structure for PSCCH transmission.

21.3.1 RESOURCE POOLS AND SELECTION/ASSIGNMENT OF TRANSMISSION RESOURCES

In the time domain, discovery is based on equal-sized *discovery periods* similar to the PSCCH periods used for sidelink communication, see Section 21.2.2.

Similar to sidelink communication, for sidelink discovery a device is configured with one or several *resource pools* defining the resources available for the discovery (PSDCH) transmission. In case of discovery, each resource pool consists of

- a *PSDCH subframe pool* defining a set of subframes available for discovery transmission;
- a *PSDCH resource-block pool* defining a set of resource blocks available for discovery transmission within the subframe pool.

The PSDCH subframe pool is given by a subframe bitmap, similar to the PSCCH subframe pool (Section 21.2.3). However, while the bitmap directly gives the PSCCH subframe pool, the discovery subframe pool is given by a *periodic repetition* of the subframe bitmap.

The discovery resource-block pool consists of two sets of frequency-wise consecutive resource blocks defined by three parameters S_1, S_2, and M, that is, the same structure as the PSCCH resource-block pool (Section 21.2.3 and Figure 21.8).

Similar to sidelink communication there are two types or modes of sidelink discovery that differ in terms of how a device is assigned/selects the exact set of resources to use for the discovery (PSDCH) transmission[6]:

- In case of discovery type 1, a device by itself selects the set of physical resources to use for the discovery transmission from a configured resource pool.
- In case of discovery type 2B, a device is explicitly assigned, by means of RRC signaling, the set of resources to use for the discovery transmission from a configured resource pool.[7]

While discovery type 1 can be used by devices in both RRC_IDLE and RRC_CON-NECTED state, discovery type 2B is only possible in RRC_CONNECTED state.

It should be noted that, for discovery type-2B, the assignment of discovery resources is done by means of RRC signaling. The assignment is then valid until explicitly changed. This is in contrast to sidelink communication mode 1, for which the transmission resources are *dynamically* assigned by means of scheduling grants (DCI format 5) on PDCCH/EPDCCH and with the assignment only valid for the PSCCH period in which it is provided. [8] This also means that while, for sidelink communication, the device is first configured with a resource pool (by means of RRC signaling) and then dynamically assigned the specific resource to use for the sidelink transmission by means of DCI on PDCCH/EPDCCH, for sidelink discovery the configuration of resource pool and the assignment of the exact resource to use for a discovery transmission, is done jointly as part of the sidelink configuration.

In case of discovery type 1, that is, when the device selects the exact set of resources to use for discovery transmission, each device may be configured with multiple resource pools

---

[6]The specification somewhat arbitrarily uses the term *mode* for sidelink communication and the term *type* for discovery. To align with the specification we will do the same here.

[7]At an early stage of the 3GPP work on sidelink connectivity, there was discovery type 2, with a special case referred to as type 2B. In the end, only the special case remained.

[8]Note that the numbering of the discovery types is reversed compared to the case of sidelink communication. In case of discovery, "type 1" refers to the type/mode where the device selects the resources while, in case of sidelink communication, the corresponding mode is referred to as mode 2.

where each pool is associated with a certain *RSRP range*, where RSPR (Reference Signal Receiver Power) is essentially a measure of the path loss to a certain cell. The device selects the resource pool from which to select the discovery resources based on the measured RSRP for the current cell. This allows for the separation of devices into non-overlapping resource pools depending on the path loss to the current cell and thus, indirectly, depending on the distance to other cells.

21.3.2 DISCOVERY TRANSMISSION

As already mentioned, a discovery message, that is, the SL-DSCH transport block, is of a fixed size of 232 bits.

Channel-coding and modulation for SL-DCH is done in the same way as for uplink (UL-SCH) transmission (Chapter 7) and consists of the following steps:

- CRC insertion;
- code-block segmentation and per-code-block CRC insertion;
- rate 1/3 Turbo coding;
- rate matching;
- bit-level scrambling with a predefined seed (510);
- data modulation (QPSK only).

DFT precoding is then applied before the mapping to the time—frequency resource selected/assigned for the PSDCH transmission.

Each SL-DCH transport block, is transmitted over $N_{RT} + 1$ consecutive subframes within the discovery subframe pool, where the *"number of retransmissions"* N_{RT} is part of the discovery configuration provided by the network.

Within each subframe, two frequency-wise consecutive resource blocks of the resource-block pool are used for the discovery transmission, with the resource blocks changing for each subframe.

21.3.3 RECEPTION RESOURCE POOLS

Similar to sidelink communication, there is also for discovery a set of common *reception* resource pools provided to devices as part of the sidelink-related system information, that is, SIB19 in case of discovery. Similar to sidelink communication, devices that are to receive sidelink discovery messages search for PSDCH transmissions within the configured set of resource pools.

21.4 SIDELINK SYNCHRONIZATION

As already mentioned in Section 21.1.3, the aim of sidelink synchronization is to provide timing references for sidelink transmission as well as for sidelink reception.

In general, an in-coverage device should use the synchronization signal (PSS + SSS) of the serving cell (for devices in RRC_CONNECTED state) or the cell the device is camping on (for devices in RRC_IDLE state) as timing reference for its sidelink transmissions.

Out-of-coverage devices may acquire transmission timing from special SLSS transmitted by other devices. Those devices could, themselves, be in coverage, implying that their transmission timing, including the timing of their SLSS transmissions, is derived directly from the network. However, they may also be out of coverage, implying that their transmission timing has either been derived from SLSS transmissions from yet other devices or been selected autonomously.

Selecting an SLSS as timing reference for sidelink transmission, is, in the LTE specifications, referred to as selecting a *synchronization reference* UE *or SyncRef UE*.

It should be understood that, despite the term "SyncRef UE," what a device selects as timing reference is not a device as such but *a received SLSS*. This may seem like semantics but is an important distinction. Within a cluster of out-of-coverage devices, several devices may transmit the same SLSS. A device using that SLSS as timing reference will thus not synchronize to an SLSS transmission of a specific device but to the aggregated SLSS corresponding to multiple devices.

SLSS may also be used as timing reference for sidelink reception by both out-of-coverage devices and in-coverage devices. Such a reception timing reference is needed when a device is to receive a sidelink transmission originating from a device with a different transmission-timing reference, for example, an in-coverage device having a different serving cell. Each reception pool is associated with a certain *synchronization configuration*, in practice with a specific SLSS. When receiving sidelink transmissions according to a certain reception pool, a device should use the corresponding SLSS as timing reference for the reception.

21.4.1 SIDELINK IDENTITY AND STRUCTURE OF THE SIDELINK SYNCHRONIZATION SIGNAL

Similar to cell synchronization signals corresponding to different cell identities, an SLSS is associated with a *sidelink identity* (SLI). There are 336 different SLIs divided into two groups with 168 SLIs in each group:

- The first group, consisting of SLI number 0 to SLI number 167, is used by devices that are either in coverage or out of coverage but have a SyncRef UE corresponding to a device that is in coverage. We refer to this as the *in-coverage group*.
- The second group, consisting of SLI number 168 to SLI number 335, is used by the remaining devices, that is, out-of-coverage devices that have a SyncRef UE corresponding to a device that is also out of coverage or out-of-coverage devices that have no SyncRef UE at all. We will refer to this as the *out-of-coverage group*.

One can also group the 336 SLIs into *SLI pairs*, each consisting of one SLI from the in-coverage group and the corresponding SLI from the out-of-coverage group. Comparing with

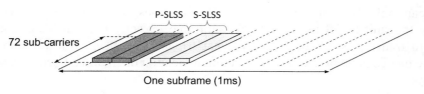

P-SLSS S-SLSS

72 sub-carriers

One subframe (1ms)

FIGURE 21.13

Structure of sidelink synchronization signal.

the cell synchronization signals, the 168 different SLI pairs consisting of two SLIs each can be seen as corresponding to the 168 different cell-identity groups consisting of three cell identities each (see Chapter 11).

Similar to a cell synchronization signal, an SLSS actually consists of two components—a *primary sidelink synchronization signal* (P-SLSS) and a *secondary sidelink synchronization signal* (S-SLSS).

As outlined in Figure 21.13, the P-SLSS consists of two OFDM symbols transmitted within the second and third[9] symbol of a subframe while the S-SLSS consists of two OFDM symbols transmitted within the fifth and sixth symbol. Similar to the cell synchronization signal, each SLSS covers the 72 center subcarriers of the carrier.[10]

The two P-SLSS symbols are identical and are generated in the same way as the PSS. As described in Chapter 11 there are three different PSS derived from three different Zadoff-Chu (ZC) sequences, where each PSS corresponds to one specific cell identity from each of the 168 cell-identity groups. In the same way, there are two different P-SLSS derived from two different ZC-sequences (different from the ZC-sequences of the PSS). The two different P-SLSS correspond to SLIs in the in-coverage group and out-of-coverage group, respectively.

The two S-SLSS symbols are also identical and generated in the same way as the SSS, see Chapter 11. There are 168 different S-SLSS, where each S-SLSS corresponds to one of the 168 different SLI pairs.

SLSS can only be transmitted in certain *SLSS subframes* corresponding to every 40th subframe. The exact set of SLSS subframes is given by a *subframe offset* that locates the SLSS subframes relative to the first subframe of the frame with SFN = 0. For in-coverage devices the subframe offset is provided as part of the sidelink-related system information (SIB 18 and SIB 19 for devices involved in sidelink communication and sidelink discovery, respectively). For out-of-coverage devices there are to two offsets, corresponding to two different sets of SLSS subframes, provided as part of the pre-configuration. The reason for providing two offsets is to allow for out-of-coverage devices to transmit and receive SLSS in the same 40 ms period, see further Section 21.4.4.

---

[9]The first and second symbol in case of extended cyclic prefix.
[10]In contrast to the downlink carrier, on the sidelink carrier there is no non-transmitted DC carrier, compare Figure 11.3.

SLSS can only be transmitted in SLSS subframes. However, a device does not necessarily transmit SLSS in every SLSS subframe. Exactly in what SLSS subframes a device is to transmit SLSS depends on what triggers the SLSS transmission and also whether the device is involved in sidelink communication or sidelink discovery, see further Section 21.4.4.

21.4.2 THE SIDELINK BROADCAST CHANNEL AND SIDELINK MASTER INFORMATION BLOCK

A device that serves as a possible synchronization source, that is, a device that transmits SLSS may also transmit the sidelink broadcast channel (SL-BCH) mapped to the PSBCH. The SL-BCH carries some very basic information, contained within the *sidelink master information block* (SL-MIB), needed for out-of-coverage devices to establish sidelink connectivity. More specifically, the SL-MIB carries the following information:

- Information about carrier bandwidth assumed by the device transmitting the SL-MIB.
- Information about the TDD configuration assumed by the device transmitting the SL-MIB.
- Information about the frame number (SFN) and subframe number of the frame/subframe in which the SL-BCH is transmitted. This allows for devices to synchronize to each other also on frame/subframe level.
- An *in-coverage indicator*, indicating whether or not the device transmitting the SL-BCH is within network coverage. As described further in Section 21.4.3, the in-coverage indicator is used by out-of-coverage devices when selecting SyncRef UE.

After convolutional coding and modulation (QPSK), the PSBCH is transmitted in the same subframe and the same resource blocks as used for the SLSS transmission.

An out-of-coverage device, after acquiring an SLSS, decodes the corresponding SL-BCH and acquires the SL-MIB. Based on, among other things, the in-coverage indicator on the SL-MIB, the device decides if the acquired SLSS is to be used as synchronization reference, that is, as SyncRef UE. In that case, the device uses the remaining SL-MIB information (carrier bandwidth, TDD configuration, and SFN/subframe number) as assumptions for subsequent sidelink transmissions.

21.4.3 SYNCREF UE SELECTION

There are well-specified rules for how an out-of-coverage device should select a SyncRef UE, that is, select a timing reference for sidelink transmissions, see also Figure 21.14:

- If no sufficiently strong cell can be found, implying that the device is out of coverage, the device should first search for SLSS corresponding to devices that in themselves are under network coverage. If a sufficiently strong such SLSS can be found the device should use that SLSS as SyncRef UE.

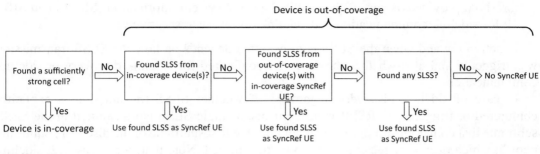

FIGURE 21.14

SyncRef UE selection.

- If no such SLSS can be found the device should instead search for SLSS corresponding to devices that are out of coverage but in themselves have a SyncRef UE corresponding to in-coverage devices. If a sufficiently strong such SLSS can be found the device should use that SLSS as SyncRef UE.
- If no such SLSS can be found the device should search for any SLSS. If a sufficiently strong SLSS can be found the device should use that SLSS as SyncRef UE.
- If no SLSS can be found, the device will autonomously decide its transmission timing, that is, the device will have no SyncRef UE.

Note that the this procedure assumes that a device can determine

- that a found SLSS corresponds to devices that are under network coverage and
- that a found SLSS corresponds to devices that are out of coverage but in themselves have an in-coverage SyncRef UE.

As will be seen in the following section, this is possible by means of a combination of the in-coverage indicator on the sidelink broadcast channel (SL-BCH) and the rule by which an out-of-coverage device selects its Sidelink Identity (SLI) depending on the SLI of the selected SyncRef UE.

21.4.4 TRANSMISSION OF SIDELINK SYNCHRONIZATION SIGNALS

21.4.4.1 In-coverage Devices

Transmission of SLSS by in-coverage devices can be triggered in different ways as follows:

- A device in RRC_CONNECTED state can be explicitly configured by the network to transmit SLSS.
- If not explicitly configured to transmit SLSS, transmission of SLSS can be triggered by the measured received signal strength (RSRP) of the current cell being below a certain

threshold provided as part of the sidelink-specific system information (SIB 18 and SIB 19 for sidelink communication and sidelink discovery, respectively).

Exactly how and when the SLSS is transmitted depends on how the SLSS transmission was triggered and also whether the device is configured for sidelink discovery or sidelink communication.

In case of sidelink discovery, regardless of whether SLSS transmission is explicitly configured or triggered by RSPR measurements, a single SLSS is transmitted in the SLSS subframe that comes closest in time and not after the first subframe of the discovery subframe pool in which discovery messages are to be transmitted. Note that, for release 12, sidelink discovery will only take place in coverage. Transmission of SLSS by devices involved in sidelink discovery is therefore only done to provide timing references for sidelink reception.[11]

In case of sidelink communication, if SLSS transmission is explicitly configured the device will transmit SLSS in every SLSS subframe regardless of if actual sidelink communication is carried out in a PSCCH period or not. On the other hand, if SLSS transmission is triggered by RSRP measurements the device will only transmit SLSS in the SLSS subframes contained within the PSCCH period(s) in which actual sidelink communication will be carried out.

Exactly what SLSS to transmit or, more specifically, the SLI, is provided as part of the sidelink-related system information.

21.4.4.2 Out-of-coverage Devices

An out-of-coverage device should transmit SLSS if it has no selected SyncRef UE or if the RSRP of the selected SyncRef UE is below a certain threshold, where the threshold is provided as part of the pre-configured information.

In general, a device that has a selected SyncRef UE should either use the same SLI as the SyncRef UE or the corresponding paired SLI (the SLI from the out-of-coverage group if the SLI of the SyncRef UE is from the in-coverage group or vice versa).

Furthermore, in general the SLSS should be transmitted assuming one of the two different SLSS subframe offset that are provided as part of the pre-configuration. More specifically, the device should select the offset so that the transmitted SLSS does not collide with the reception of the SLSS of the received SyncRef UE.

The rules for selecting the SLI and for setting the in-coverage indicator of the SL-MIB are as follows, see also Table 21.1:

- An out-of-coverage device with a SyncRef UE for which the in-coverage indicator is set to TRUE should, regardless of the SyncRef UE SLI,
  - use the same SLI as the SyncRef UE,
  - set the SL-MIB in-coverage indicator to FALSE.

---

[11]This is changed in release 13, see Section 21.5.

Table 21.1 Rules for Selection SLI and Setting In-Coverage Indicator

		SyncRef UE In-coverage Indicator	
		TRUE	**FALSE**
SyncRef UE SLI	From in-coverage group	Set in-coverage indicator to FALSE Set SLI to $SLI_{SyncRef\ UE}$	Set in-coverage indicator to FALSE Set SLI to $SLI_{SyncRef\ UE} + 168$
	From out-of-coverage group		Set in-coverage indicator to FALSE Set SLI to $SLI_{SyncRef\ UE}$

- An out-of-coverage device with a SyncRef UE which has an SLI from the in-coverage group and for which the in-coverage indicator is set to FALSE should
  - use the corresponding SLI from the out-of-coverage group,
  - set the SL-MIB in-coverage indicator to FALSE.
- An out-of-coverage device with a selected SyncRef UE which has an SLI from the out-of-coverage group and for which the in-coverage indicator is set to FALSE should
  - use the same SLI as the SyncRef,
  - set the SL-MIB in-coverage indicator to FALSE.

Thus, from the SyncRef UE in-coverage indicator and SLI, a device can conclude on the in-coverage/out-of-coverage status of a candidate SyncRef UE, see also Figure 21.15:

- If the in-coverage indicator is "TRUE," the candidate SyncRef UE is in-coverage regardless of the SLI of the SyncRef UE.
- If the in-coverage indicator is "FALSE" and the SLI is from the in-coverage group, the candidate SyncRef UE is out of coverage but, in itself, has a SyncRef UE that is in coverage.
- If the in-coverage indicator is "FALSE" and the SLI is from the out-of-coverage group, the candidate SyncRef UE is out of coverage and, in itself, also has a SyncRef UE that is out-of-cover-coverage.

As discussed in Section 21.4.3, this information is needed for a device to make a proper selection of SyncRef UE.

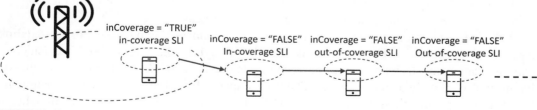

FIGURE 21.15

Sidelink synchronization.

21.5 DEVICE-TO-DEVICE EXTENSIONS IN LTE RELEASE 13

The focus of the previous sections has been on the initial LTE sidelink functionality in release 12. The sidelink functionality is further extended in release 13 with the main features being:

- The possibility for sidelink discovery also for devices that are not within network coverage.
- The extension of network coverage by means of Layer-3-based relaying via an intermediate device.

21.5.1 OUT-OF-COVERAGE DISCOVERY

Release-13 out-of-coverage discovery only targets the public-safety use case. It relies on the same mechanisms as the release-12 in-coverage discovery although limited to discovery type 1, that is, the mode when the device autonomously selects the resources to use for the discovery transmission from, in this case, a *pre-configured* resource pool.

Out-of-coverage discovery also impacts transmission of SLSS, that is, the sidelink synchronization channel, and the sidelink broadcast channel (SL-BCH) carrying the sidelink MIB.

For release 12, the only task of SLSS transmission in the context of discovery is to provide a timing reference for discovery reception, for example, for reception of discovery transmissions by devices in neighbor cells. As a consequence, in the context of sidelink discovery SLSS is only transmitted in direct combination with actual discovery transmissions.

This is in contrast to sidelink communication where SLSS is also used to provide a timing reference for sidelink transmissions for out-of-coverage devices. Thus, devices configured for sidelink communication can be configured to transmit SLSS also when not having any actual data to transmit.

For release 13, the same is possible also for devices configured for sidelink discovery. Furthermore, devices configured for sidelink discovery may, in release 13, also transmit the SL-BCH providing the SL-MIB information to out-of-coverage devices.

21.5.2 LAYER-3 RELAYING

The introduction of Layer-3 relaying has very little impact on the actual radio-access specifications as it very much relies on the functionality of the sidelink communication and sidelink discovery mechanisms introduced already in release 12.

A device capable of serving as a Layer-3 relay announces this by means of the sidelink-discovery mechanism. The D2D link of the relaying then relies on the sidelink-communication mechanisms while the relaying-device-to-network communication relies on the normal LTE cellular mechanisms. The only impact of the Layer-3 relaying functionality to the radio specifications is in terms of RRC functionality, for example, functionality to configure the relaying device.

SPECTRUM AND RF CHARACTERISTICS

22

Spectrum flexibility is, as mentioned in Chapter 3, a key feature of LTE radio access and is set out in the LTE design targets [6]. It consists of several components, including deployment in different-sized spectrum allocations and deployment in diverse frequency ranges, both in paired and unpaired frequency bands. There are a number of frequency bands identified for mobile use and specifically for IMT today. These are presented in detail in Chapter 2. The use of OFDM in LTE gives flexibility both in terms of the size of the spectrum allocation needed and in the instantaneous transmission bandwidth used. The OFDM physical layer also enables frequency-domain scheduling, as briefly discussed in Chapter 3. Beyond the physical-layer implications described in Chapters 6 and 7, these properties also impact the RF implementation in terms of filters, amplifiers, and all other RF components that are used to transmit and receive the signal. This means that the RF requirements for the receiver and transmitter will have to be expressed with flexibility in mind.

22.1 FLEXIBLE SPECTRUM USE

Most of the frequency bands identified above for deployment of LTE are existing IMT-2000 bands and some bands also have legacy systems deployed, including WCDMA/HSPA and GSM. Bands are also in some regions defined in a "technology neutral" manner, which means that coexistence between different technologies is a necessity.

The fundamental LTE requirement to operate in different frequency bands [40] does not, in itself, impose any specific requirements on the radio-interface design. There are, however, implications for the RF requirements and how those are defined, in order to support the following:

- *Coexistence between operators in the same geographic area in the band.* Coexistence may be required with other operators that deploy LTE or other IMT-2000 technologies, such as UMTS/HSPA or GSM/EDGE. There may also be non-IMT-2000 technologies deployed. Such coexistence requirements are to a large extent developed within 3GPP, but there may also be regional requirements defined by regulatory bodies in some frequency bands.
- *Colocation of base station equipment between operators.* There are in many cases limitations to where base-station (BS) equipment can be deployed. Often, sites must be

shared between operators or an operator will deploy multiple technologies in one site. This puts additional requirements on both BS receivers and transmitters.

- *Coexistence with services in adjacent frequency bands and across country borders.* The use of the RF spectrum is regulated through complex international agreements, involving many interests. There will therefore be requirements for coordination between operators in different countries and for coexistence with services in adjacent frequency bands. Most of these are defined in different regulatory bodies. Sometimes the regulators request that 3GPP includes such coexistence limits in the 3GPP specifications.

- *Coexistence between operators of TDD systems* in the same band is provided by inter-operator synchronization, in order to avoid interference between downlink and uplink transmissions of different operators. This means that all operators need to have the same downlink/uplink configurations and frame synchronization, not in itself an RF requirement, but it is implicitly assumed in the 3GPP specifications. RF requirements for unsynchronized systems become much stricter.

- *Release-independent frequency-band principles.* Frequency bands are defined regionally and new bands are added continuously. This means that every new release of 3GPP specifications will have new bands added. Through the "release independence" principle, it is possible to design terminals based on an early release of 3GPP specifications that support a frequency band added in a later release.

- *Aggregation of spectrum allocations.* Operators of LTE systems have quite diverse spectrum allocations, which in many cases do not consist of a block that easily fits exactly one LTE carrier. The allocation may even be noncontiguous, consisting of multiple blocks spread out in a band. Many operators also have allocations in multiple bands to use for LTE deployment. For these scenarios, the LTE specifications support *carrier aggregation* (CA), where multiple carriers in contiguous or noncontiguous blocks within a band, or in multiple bands, can be combined to create larger transmission bandwidths.

22.2 FLEXIBLE CHANNEL BANDWIDTH OPERATION

The frequency allocations for LTE (see Chapter 2) are up to 2×75 MHz, but the spectrum available for a single operator may be from 2×20 MHz down to 2×5 MHz for FDD and down to 1×5 MHz for TDD. Furthermore, the migration to LTE in frequency bands currently used for other radio-access technologies (RATs) must often take place gradually to ensure that a sufficient amount of spectrum remains to support the existing users. Thus, the amount of spectrum that can initially be migrated to LTE can be relatively small, but may then gradually increase, as shown in Figure 22.1. The variation of possible spectrum scenarios implies a requirement for spectrum flexibility for LTE in terms of the transmission bandwidths supported.

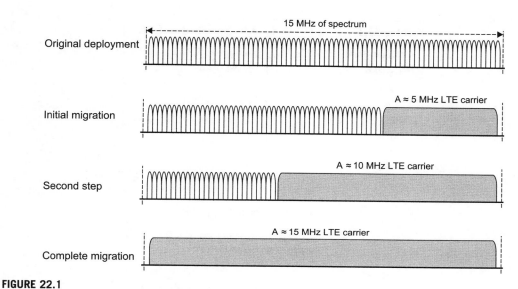

FIGURE 22.1

Example of how LTE can be migrated step by step into a spectrum allocation with an original GSM deployment.

The spectrum flexibility requirement points out the need for LTE to be scalable in the frequency domain. This flexibility requirement is stated in [6] as a list of LTE spectrum allocations from 1.25 to 20 MHz. Note that the final channel bandwidths selected differ slightly from this initial assumption.

As shown in Chapter 5, the frequency-domain structure of LTE is based on resource blocks consisting of 12 subcarriers with a total bandwidth of 12×15 kHz $= 180$ kHz. The basic radio-access specification including the physical-layer and protocol specifications enables *transmission bandwidth configurations* from 6 up to 110 resource blocks on one LTE RF carrier. This allows for channel bandwidths ranging from 1.4 MHz up to beyond 20 MHz in steps of 180 kHz and is fundamental to providing the required spectrum flexibility.

In order to limit implementation complexity, only a limited set of bandwidths are defined in the RF specifications. Based on the frequency bands available for LTE deployment today and in the future, as described in the preceding paragraphs, and considering the known migration and deployment scenarios in those bands, a limited set of six channel bandwidths is specified. The RF requirements for the BS and terminal are defined only for those six channel bandwidths. The channel bandwidths range from 1.4 to 20 MHz, as shown in Table 22.1. The lower bandwidths, 1.4 and 3 MHz, are chosen specifically to ease migration to LTE in spectrum where CDMA2000 is operated, and also to facilitate migration of GSM and TD-SCDMA to LTE. The specified bandwidths target relevant scenarios in different frequency bands. For this reason, the set of bandwidths available for a specific band is not necessarily the same as in other bands. At a later stage, if new frequency bands are made available that have

Table 22.1 Channel Bandwidths Specified in LTE	
Channel Bandwidth ($BW_{channel}$)	**Number of Resource Blocks (N_{RB})**
1.4 MHz	6
3 MHz	15
5 MHz	25
10 MHz	50
15 MHz	75
20 MHz	100

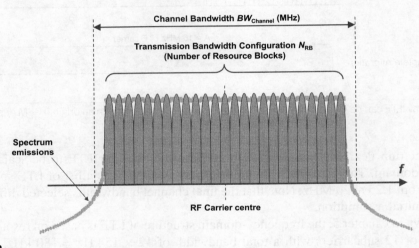

FIGURE 22.2

The channel bandwidth for one RF carrier and the corresponding transmission bandwidth configuration.

other spectrum scenarios requiring additional channel bandwidths, the corresponding RF parameters and requirements could be added in the RF specifications, without actually having to update the physical-layer specifications. The process of adding new channel bandwidths would in this way be similar to adding new frequency bands.

Figure 22.2 illustrates in principle the relationship between the channel bandwidth and the number of resource blocks N_{RB} for one RF carrier. Note that for all channel bandwidths except 1.4 MHz, the resource blocks in the transmission bandwidth configuration fill up 90% of the channel bandwidth. The spectrum emissions shown in Figure 22.2 are for a pure OFDM signal, while the actual transmitted emissions will also depend on the transmitter RF chain and other components. The emissions outside the channel bandwidth are called *unwanted emissions* and the requirements for those are discussed later in this chapter.

22.3 **CARRIER AGGREGATION FOR LTE**

The possibility from 3GPP Release 10 to aggregate two or more component carriers in order to support wider transmission bandwidths has several implications for the RF characteristics. The impacts for the BS and terminal RF characteristics are also quite different. Release 10 had some restrictions on carrier aggregation in the RF specification, compared to what has been specified for physical layer and signaling, while in later releases there is support for carrier aggregation within and between a much larger number of bands and also between more than two bands.

There is, from an RF point of view, a substantial difference between the two types of carrier aggregation defined for LTE (see also Chapter 12 for more details):

- *Intra-band contiguous carrier aggregation* implies that two or more carriers within the same operating band are aggregated (see also the first two examples in Figure 12.1). Since aggregated carriers from an RF perspective have similar RF properties as a corresponding wider carrier being transmitted and received, there are many implications for the RF requirements. This is especially true for the terminal. For the BS, it corresponds in practice to a multi-carrier configuration (nonaggregated) already supported in earlier releases, which also means that the impact is less than that for the terminal.
- *Intra-band noncontiguous carrier aggregation* implies that there is a gap between the aggregated carriers, making the set of carriers noncontiguous. For the UE, this is declared to be the case for any set of aggregated carriers where the spacing between carriers is larger than the nominal spacing. For the BS, carriers are considered to be noncontiguous if there is a need to define special coexistence requirements in the "gap" between aggregated subblocks.
- *Inter-band carrier aggregation* implies that carriers in different operating bands are aggregated (see also the last example in Figure 12.1). Many RF properties within a band can, to a large extent, remain the same as for a single carrier case. There is, however, impact for the terminal, due to the possibility for intermodulation and cross-modulation within the terminal when multiple transmitter and/or receiver chains are operated simultaneously. For the BS it has very little impact, since in practice it corresponds to a BS supporting multiple bands. There is however additional BS impact, if the inter-band carrier aggregation is deployed with a multi-band Base Station, see Section 22.12.

Intra-band contiguous carrier aggregation is in Release 13 for up to three component carriers aggregated within a band in 12 different bands. There is also support for intra-band noncontiguous carrier aggregation in nine bands. Inter-band carrier aggregation is specified for up to four bands, including both paired bands for FDD and unpaired bands for TDD, and also between paired and unpaired bands. Because of the varying impact on the RF properties for the UE, each band combination has to be specified separately.

Close to 150 different band combinations with two, three, or four bands are defined in Release 13 of the 3GPP specifications. The band or set of bands over which carriers are

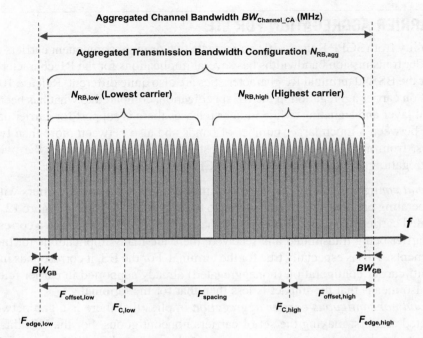

FIGURE 22.3

Definitions for intra-band carrier aggregation RF parameters, example with two aggregated carriers.

aggregated is defined as a UE capability (the term UE, User Equipment, is used in 3GPP specifications instead of terminal). For all band combinations, downlink operation is defined for the terminal. There are only a few bands with uplink operation defined for the terminal. The reason is that transmission in multiple bands from a terminal has large implications in terms of potential intermodulation products created, which creates restrictions in terms of how the UE can operate. This is solved through an allowed reduction of the terminal output power called MPR (*maximum power reduction*), in order to mitigate the intermodulation products. The allowed MPR depends on the number of resource blocks that are transmitted in each of the aggregated component carriers, the modulation format, and the terminals capability in terms of maximum number of resource blocks it can transmit per band (see also the CA bandwidth classes defined in the following).

For intra-band carrier aggregation, the definitions of $BW_{channel}$ and N_{RB} shown in Figure 22.2 still apply for each component carrier, while new definitions are needed for the aggregated channel bandwidth ($BW_{Channel\_CA}$) and the aggregated transmission bandwidth configuration ($N_{RB,agg}$) shown in Figure 22.3. In connection with this, a new capability is defined for the terminal called *CA bandwidth class*. There are six classes in Release 13, where each class corresponds to a range for $N_{RB,agg}$ and a maximum number of component carriers,

Table 22.2 UE Carrier Aggregation Bandwidth Classes (Release 13)

Channel Aggregation Bandwidth Classes	Aggregated Transmission BW Configuration	Number of Component Carriers
A	≤100	1
B	≤100	2
C	101 to 200	2
D	201 to 300	3
E, F	Under study (301 to 500 and 701 to 800)	Under study

as shown in Table 22.2. The classes corresponding to aggregation of more than two component carriers or consisting of more than 300 RBs are under study for later releases.

The terminal capability *E-UTRA CA configuration* [38] is defined as a combination of the operating band (or bands) where the terminal can operate with carrier aggregation, and a bandwidth class. For example, the terminal capability to operate with inter-band carrier aggregation in bands 1 and 5 in bandwidth class A is called CA_1A_5A. For each E-UTRA CA configuration, one or more *bandwidth combination sets* are defined, setting the channel bandwidths that can be used in each band, and what the maximum aggregated bandwidth is. A terminal can declare capability to support multiple bandwidth combination sets.

A fundamental parameter for intra-band carrier aggregation is the channel spacing. A tighter channel spacing than the nominal spacing for any two single carriers could potentially lead to an increase in spectral efficiency, since there would be a smaller unused "gap" between carriers. On the other hand, there is also a requirement for the possibility to support legacy single-carrier terminals of earlier releases. An additional complication is that the component carriers should be on the same 15 kHz subcarrier raster in order to allow reception of multiple adjacent component carriers using a single FFT instead of an FFT per subcarrier.[1] As discussed in Section 5.6, this property, together with the fact that the frequency numbering scheme is on a 100 kHz raster, results in the spacing between two component carriers having to be a multiple of 300 kHz, which is the least common denominator of 15 and 100 kHz.

For the specification, RF requirements are based on a nominal channel spacing that is derived from the channel bandwidth of the two adjacent carriers $BW_{Channel(1)}$ and $BW_{Channel(2)}$ as follows[2]:

$$F_{Spacing,Nominal} = \left\lfloor \frac{BW_{Channel(1)} + BW_{Channel(2)} - 0.1 \left| BW_{Channel(1)} - BW_{Channel(2)} \right|}{2 \cdot 0.3} \right\rfloor 0.3 \quad (22.1)$$

---

[1] In case of independent frequency errors between component carriers, multiple FFTs and frequency-tracking functionality may be needed anyway.

[2] ⌊....⌋ denotes the "floor" operator, which rounds the number down.

In order to allow for a tighter packing of component carriers, the value of $F_{Spacing}$ can be adjusted to any multiple of 300 kHz that is smaller than the nominal spacing, as long as the subcarriers do not overlap.

RF requirements for LTE are normally defined relative to the channel bandwidth edges. For intra-band carrier aggregation, this is generalized so that requirements are defined relative to the edges of the aggregated channel bandwidth, identified in Figure 22.3 as $F_{edge,low}$ and $F_{edge,high}$. In this way many RF requirements can be reused, but with new reference points in the frequency domain. The aggregated channel bandwidth for both terminal and BS is defined as:

$$BW_{Channel\_CA} = F_{edge,high} - F_{edge,low} \tag{22.2}$$

The location of the edges is defined relative to the carriers at the edges through a new parameter F_{offset} (see Figure 22.3) using the following relation to the carrier center positions F_C of the lowest and highest carriers:

$$F_{edge,low} = F_{C,low} - F_{offset,low} \tag{22.3}$$

$$F_{edge,high} = F_{C,high} + F_{offset,high} \tag{22.4}$$

The value of F_{offset} for the edge carriers and the corresponding location of the edges are, however, not defined in the same way for terminal and BS.

For the BS, there are legacy scenarios where the BS receives and transmits adjacent independent carriers, supporting legacy terminals of earlier releases using single carriers. This scenario will also have to be supported for a configuration of aggregated carriers. In addition, for backward compatibility reasons, a fundamental parameter such as channel bandwidth and the corresponding reference points (the channel edge) for all RF requirements will have to remain the same. The implication is that the channel edges shown in Figure 22.2 for each component carrier will also remain as reference points when the carriers are aggregated. This results in the following BS definition of F_{offset}, for carrier aggregation, which is "inherited" from the single carrier scenario:

$$F_{offset} = \frac{BW_{channel}}{2} \quad \text{(for base station)} \tag{22.5}$$

Unlike the BS, the terminal is not restricted by legacy operation, but rather from the nonlinear properties of the PA and the resulting unwanted emissions mask. At both edges of the aggregated channel bandwidth, a guard band BW_{GB} will be needed, in order for the emissions to reach a level where the out-of-band (OOB) emissions limits in terms of an emission mask are applied. Whether a single wide carrier or multiple aggregated carriers of the same or different sizes are transmitted, the guard band needed will have to be the same at both edges, since the emission mask roll-off is the same. A problem with the backward-compatible BS definition is that the resulting guard BW_{GB} is proportional to the channel BW and would therefore be *different* if carriers of different channel BW are aggregated.

For this reason, a different definition is used for the terminal, based on a "symmetrical" guard band. For the edge carriers (low and high), F_{offset} is half of the transmission bandwidth configuration, plus a symmetrical guard band BW_{GB}:

$$F_{offset} = \frac{0.18 \text{ MHz} \cdot N_{RB}}{2} + BW_{GB} \quad \text{(for terminal uplink)} \tag{22.6}$$

where 0.18 MHz is the bandwidth of one resource block and BW_{GB} is proportional to the channel BW of the largest component carrier. For the CA bandwidth classes defined where the edge carriers have the same channel bandwidth, F_{offset} will in principle be the same for terminals and BSs and $BW_{Channel\_CA}$ will be the same.

It may look like an anomaly that the definitions may potentially lead to slightly different aggregated channel BW for the terminal and the BS, but this is in fact not a problem. Terminal and BS requirements are defined separately and do not have to cover the same frequency ranges. The aggregated channel BW for both terminal and BS do, however, have to be within an operator's license block in the operating band.

Once the frequency reference point is set, the actual RF requirements are to a large extent the same as for a single carrier configuration. Which requirements are affected is explained for each requirement in the discussion later in this chapter.

22.4 OPERATION IN NONCONTIGUOUS SPECTRUM

Some spectrum allocations used for LTE deployments consist of fragmented parts of spectrum for different reasons. The spectrum may be recycled 2G spectrum, where the original licensed spectrum was "interleaved" between operators. This was quite common for original GSM deployments, for implementation reasons (the original combiner filters used were not easily tuned when spectrum allocations were expanded). In some regions, operators have also purchased spectrum licenses on auctions and have for different reasons ended up with multiple allocations in the same band that are not adjacent.

For deployment of noncontiguous spectrum allocations there are a few implications:

- If the full spectrum allocation in a band is to be operated with a single BS, the BS has to be capable of operation in noncontiguous spectrum.
- If a larger transmission bandwidth is to be used than what is available in each of the spectrum fragments, both the terminal and the BS have to be capable of *intra-band noncontiguous carrier aggregation* in that band.

Note that the capability for the BS to operate in noncontiguous spectrum is not directly coupled to carrier aggregation as such. From an RF point of view, what will be required by the BSs is to receive and transmit carriers over an RF bandwidth that is split in two (or more) separate subblocks, with a subblock gap in-between as shown in Figure 22.4. The spectrum in the subblock gap can be deployed by any other operator, which means that the RF requirements for the BS in the subblock gap will be based on coexistence for

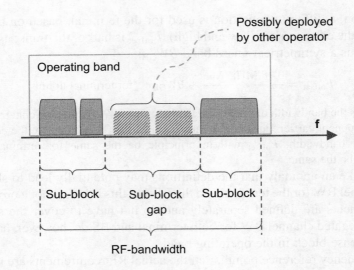

FIGURE 22.4

Example of noncontiguous spectrum operation, illustrating the definitions of *RF bandwidth*, *subclock*, and *subblock gap*.

un-coordinated operation. This has a few implications for some of the BS RF requirements within an operating band.

If the noncontiguous spectrum is operated with carrier aggregation, the RF requirements for the BS will be fundamentally the same as in general for noncontiguous spectrum.

For the terminal, noncontiguous operation is tightly coupled to carrier aggregation, since multi-carrier reception in the downlink or transmission in the uplink within a band does not occur unless carriers are aggregated. This also means that the definition of noncontiguous operation is different for the terminal than for the BS. For the terminal, intra-band non-contiguous carrier aggregation is therefore assumed to occur as soon as the spacing between two carriers is larger than the nominal channel spacing defined in Eq. (22.1).

Compared to the BS, there are also additional implications and limitation to handle the simultaneously received and/or transmitted noncontiguous carriers. There is an allowed maximum power reduction (MPR) already for transmission in a single component carrier, if the resource block allocation is noncontiguous within the carrier. For noncontiguous aggre-gated carriers, an allowed MPR is defined for subblock gaps of up to 35 MHz between the aggregated carriers. The MPR depends on the number of allocated resource blocks.

22.5 MULTI-STANDARD RADIO BASE STATIONS

Traditionally the RF specifications have been developed separately for the different 3GPP RATs, namely GSM/EDGE, UTRA, and E-UTRA (LTE). The rapid evolution of mobile radio and the need to deploy new technologies alongside the legacy deployments has, however, led to implementation of different RATs at the same sites, often sharing antennas and other parts of the installation. A natural further step is then to also share the BS equipment between multiple RATs. This requires multi-RAT BSs.

The evolution to multi-RAT BSs is also fostered by the evolution of technology. While multiple RATs have traditionally shared parts of the site installation, such as antennas, feeders, backhaul, or power, the advance of both digital baseband and RF technologies enables a much tighter integration. A BS consisting of two separate implementations of both baseband and RF, together with a passive combiner/splitter before the antenna, could in theory be considered a multi-RAT BS. 3GPP has, however, made a narrower, but more forward-looking definition.

In a *multi-standard radio* (MSR) BS, both the receiver and the transmitter are capable of simultaneously processing multiple carriers of different RATs in common *active* RF components. The reason for this stricter definition is that the true potential of multi-RAT BSs, and the challenge in terms of implementation complexity, comes from having a common RF. This principle is illustrated in Figure 22.5 with an example BS capable of both GSM/EDGE and LTE. Much of the GSM/EDGE and LTE baseband functionality may be separate in the BS, but is possibly implemented in the same hardware. The RF must, however, be implemented in the same active components as shown in the figure.

The main advantages of an MSR BS implementation are twofold:

- Migration between RATs in a deployment, for example, from GSM/EDGE to LTE, is possible using the same BS hardware. In the example in Figure 22.5, a migration is performed in three phases using the same MSR BS. In the first phase, the BS is deployed in a network for GSM/EDGE-only operation. In the second phase, the operator migrates

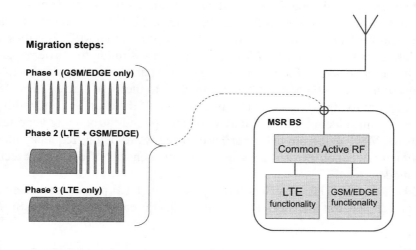

FIGURE 22.5

Example of migration from GSM to LTE using an MSR BS for all migration phases.

part of the spectrum to LTE. The same MSR BS will now operate one LTE carrier, but still supports the legacy GSM/EDGE users in half of the band available. In the third phase, when the GSM/EDGE users have migrated from the band, the operator can configure the MSR BS to LTE-only operation with double the channel bandwidth.

- A single BS designed as an MSR BS can be deployed in various environments for single-RAT operation for each RAT supported, as well as for multi-RAT operation where that is required by the deployment scenario. This is also in line with the recent technology trends seen in the market, with fewer and more generic BS designs. Having fewer varieties of BS is an advantage both for the BS vendor and for the operator, since a single solution can be developed and implemented for a variety of scenarios.

The single-RAT 3GPP radio-access standards, with requirements defined independently per RAT, do not support such migration scenarios with an implementation where common BS RF hardware is shared between multiple access technologies, and hence a separate set of requirements for multi-standard radio equipment is needed.

An implication of a common RF for multiple RATs is that carriers are no longer received and transmitted independently of each other. For this reason, a common RF specification must be used to specify the MSR BS. From 3GPP Release 9 there is a set of MSR BS specifications for the core RF requirements [41] and for test requirements [42]. Those specifications support GSM/EDGE,[3] UTRA, E-UTRA, and all combinations thereof. To support all possible RAT combinations, the MSR specifications have many generic requirements applicable regardless of RAT combination, together with specific single-access-technology-specific requirements to secure the integrity of the systems in single-RAT operation.

The MSR concept has a substantial impact for many requirements, while others remain completely unchanged. A fundamental concept introduced for MSR BSs is *RF bandwidth*, which is defined as the total bandwidth over the set of carriers transmitted and received. Many receiver and transmitter requirements for GSM/EDGE and UTRA are specified relative to the carrier center and for LTE in relation to the channel edges. For an MSR BS, they are instead specified relative to the *RF bandwidth edges*, in a way similar to carrier aggregation in Release 10. In the same way as for carrier aggregation, a parameter F_{offset} is also introduced to define the location of the RF bandwidth edges relative to the edge carriers. For GSM/EDGE carriers, F_{offset} is set to 200 kHz, while it is in general half the channel bandwidth for UTRA and E-UTRA. By introducing the RF bandwidth concept and introducing generic limits, the requirements for MSR shift from being carrier centric toward being frequency block centric, thereby embracing technology neutrality by being independent of the access technology or operational mode.

While E-UTRA and UTRA carriers have quite similar RF properties in terms of bandwidth and power spectral density (PSD), GSM/EDGE carriers are quite different. The

---

[3]The MSR specifications are not applicable to single-RAT operation of GSM/EDGE.

operating bands for which MSR BSs are defined are therefore divided into three *band categories* (BCs):

- BC1: All paired bands where UTRA FDD and E-UTRA FDD can be deployed.
- BC2: All paired bands where in addition to UTRA FDD and E-UTRA FDD, GSM/EDGE can also be deployed.
- BC3: All unpaired bands where UTRA TDD and E-UTRA TDD can be deployed.

Since the carriers of different RATs are not transmitted and received independently, it is necessary to perform parts of the testing with carriers of multiple RATs being activated. This is done through a set of multi-RAT *test configurations* defined in [42], specifically tailored to stress transmitter and receiver properties. These test configurations are of particular importance for the unwanted emission requirements for the transmitter and for testing of the receiver susceptibility to interfering signals (blocking, and so on). An advantage of the multi-RAT test configurations is that the RF performance of multiple RATs can be tested simultaneously, thereby avoiding repetition of test cases for each RAT. This is of particular importance for the very time-consuming tests of requirements over the complete frequency range outside the operating band.

The requirement with the largest impact from MSR is the spectrum mask, or the so-called *operating band unwanted emissions* requirement. The spectrum mask requirement for MSR BSs is applicable for multi-RAT operation where the carriers at the RF bandwidth edges are either GSM/EDGE, UTRA, or E-UTRA carriers of different channel bandwidths. The mask is generic and applicable to all cases, and covers the complete operating band of the BS. There is an exception for the 150 kHz closest to the RF bandwidth edge, where the mask is aligned with the GSM/EDGE modulation spectrum for the case when a GSM/EDGE carrier or a 1.4/3 MHz E-UTRA carrier is transmitted adjacent to the edge.

An important aspect of MSR is the declaration by the BS vendor of the supported RF bandwidth, power levels, multi-carrier capability, and so on. All testing is based on the capability of the BS through a declaration of the supported *capability set* (CS), which defines all supported single RATs and multi-RAT combinations. There are currently seven CSs, namely CS1 to CS7 defined in the MSR test specification [42], allowing full flexibility for implementing and deploying MSR BS with different capabilities. These CSs are listed in Table 22.3 together with the band categories where the CS is applicable and the RAT configurations that are supported by the BS. Note the difference between the capability of a BS (as declared by the manufacturer) and the configuration in which a BS is operating. CS1 and CS2 define capabilities for BSs that are only single-RAT capable and make it possible to apply the MSR BS specification for such BSs, instead of the corresponding single-RAT UTRA or E-UTRA specifications. There is no CS defined for BSs that are *only* single-RAT GSM capable, since that is a type of BS that is solely covered by the single-RAT GSM/EDGE specifications. In line with the continuing deployments of 3G and 4G systems in the GSM frequency bands (BC2), a new CS, namely *CS7* is introduced in Release 13. It is

Table 22.3 Capability Sets (CSx) Defined for MSR BSs and the Corresponding RAT Configurations

Capability Set CSx Supported by a BS	Applicable Band Categories	Supported RAT Configurations
CS1	BC1, BC2, or BC3	Single-RAT: UTRA
CS2	BC1, BC2, or BC3	Single-RAT: UTRA
CS3	BC1, BC2, or BC3	Single-RAT: UTRA or E-UTRA
		Multi-RAT: UTRA + E-UTRA
CS4	BC2	Single-RAT: GSM or UTRA
		Multi-RAT: GSM + UTRA
CS5	BC2	Single-RAT: GSM or E-UTRA
		Multi-RAT: GSM + E-UTRA
CS6	BC2	Single-RAT: GSM, UTRA, or E-UTRA
		Multi-RAT: GSM + UTRA, GSM + E-UTRA, UTRA + E-UTRA, or GSM + UTRA + E-UTRA
CS7	BC2	Single-RAT: UTRA or E-UTRA
		Multi-RAT: GSM + UTRA, GSM + E-UTRA, or UTRA + E-UTRA

used for BS that support all three RATs, but where single-RAT GSM and triple-RAT GSM + UTTRA + E-UTRA operation is not supported.

For a large part of the BS RF requirements, multi-RAT testing is not necessary and the actual test limits are unchanged for the MSR BS. In these cases, both the requirements and the test cases are simply incorporated through direct references to the corresponding single-RAT specifications.

Carrier aggregation as described in Section 22.3 is also applicable to MSR BSs. Since the MSR specification has most of the concepts and definitions in place for defining multi-carrier RF requirements, whether aggregated or not, the differences for the MSR requirements compared to nonaggregated carriers are very minor.

22.6 OVERVIEW OF RF REQUIREMENTS FOR LTE

The RF requirements define the receiver and transmitter RF characteristics of a BS or terminal. The BS is the physical node that transmits and receives RF signals on one or more antenna connectors. Note that a BS is not the same thing as an eNodeB, which is the corresponding logical node in the LTE radio-access network. The terminal is denoted UE in all RF specifications.

The set of RF requirements defined for LTE is fundamentally the same as that defined for UTRA or any other radio system. Some requirements are also based on regulatory requirements and are more concerned with the frequency band of operation and/or the place where the system is deployed, than with the type of system.

What is particular to LTE is the flexible bandwidth and the related multiple channel bandwidths of the system, which make some requirements more complex to define. These properties have special implications for the transmitter requirements on unwanted emissions, where the definition of the limits in international regulation depends on the channel bandwidth. Such limits are harder to define for a system where the BS may operate with multiple channel bandwidths and where the terminal may vary its channel bandwidth of operation. The properties of the flexible OFDM-based physical layer also have implications for specifying the transmitter modulation quality and how to define the receiver selectivity and blocking requirements.

The type of transmitter requirements defined for the terminal is very similar to what is defined for the BS, and the definitions of the requirements are often similar. The output power levels are, however, considerably lower for a terminal, while the restrictions on the terminal implementation are much higher. There is tight pressure on cost and complexity for all telecommunications equipment, but this is much more pronounced for terminals due to the scale of the total market, being close to *two billion* devices per year. In cases where there are differences in how requirements are defined between terminal and BS, they are treated separately in this chapter.

The detailed background of the RF requirements for LTE is described in [43,44], with further details of the additional requirements in Release 10 (for LTE-Advanced) in [45,46]. The RF requirements for the BS are specified in [47] and for the terminal in [38]. The RF requirements are divided into transmitter and receiver characteristics. There are also *performance characteristics* for BS and terminal that define the receiver baseband performance for all physical channels under different propagation conditions. These are not strictly RF requirements, though the performance will also depend on the RF to some extent.

Each RF requirement has a corresponding test defined in the LTE test specifications for the BS [48] and the terminal [36]. These specifications define the test setup, test procedure, test signals, test tolerances, and so on, needed to show compliance with the RF and performance requirements.

22.6.1 TRANSMITTER CHARACTERISTICS

The transmitter characteristics define RF requirements not only for the desirable (wanted) signal transmitted from the terminal and BS, but also for the unavoidable unwanted emissions outside the transmitted carrier(s). The requirements are fundamentally specified in three parts:

- *Output power level* requirements set limits for the maximum allowed transmitted power, for the dynamic variation of the power level, and in some cases, for the transmitter OFF state.

- *Transmitted signal quality* requirements define the "purity" of the transmitted signal and also the relation between multiple transmitter branches.
- *Unwanted emissions* requirements set limits to all emissions outside the transmitted carrier(s) and are tightly coupled to regulatory requirements and coexistence with other systems.

A list of the terminal and BS transmitter characteristics arranged according to the three parts as defined is shown in Table 22.4. A more detailed description of the requirements can be found later in this chapter.

22.6.2 RECEIVER CHARACTERISTICS

The set of receiver requirements for LTE is quite similar to what is defined for other systems such as UTRA, but many of them are defined differently, due to the flexible bandwidth properties. The receiver characteristics are fundamentally specified in three parts:

- *Sensitivity and dynamic range* requirements for receiving the wanted signal.
- *Receiver susceptibility to interfering signals*—defines receivers' susceptibility to different types of interfering signals at different frequency offsets.
- *Unwanted emission* limits are also defined for the receiver.

The terminal and BS receiver characteristics arranged according to the three parts defined in the preceding list are shown in Table 22.5. A more detailed description of each requirement can be found later in this chapter.

Table 22.4 Overview of LTE Transmitter Characteristics

	Base Station Requirement	Terminal Requirement
Output power level	Maximum output power	Transmit power
	Output power dynamics	Output power dynamics
	ON/OFF power (TDD only)	Power control
Transmitted signal quality	Frequency error	Frequency error
	Error-vector magnitude (EVM)	Transmit modulation quality
	Time alignment between transmitter branches	In-band emissions
Unwanted emissions	Operating band unwanted emissions	Spectrum emission mask
	Adjacent channel leakage ratio (ACLR and CACLR)	Adjacent channel leakage ratio (ACLR and CACLR)
	Spurious emissions	Spurious emissions
	Occupied bandwidth	Occupied bandwidth
	Transmitter intermodulation	Transmit intermodulation

Table 22.5 Overview of LTE Receiver Characteristics

	Base Station Requirement	Terminal Requirement
Sensitivity and dynamic range	Reference sensitivity	Reference sensitivity power level
	Dynamic range	Maximum input level
	In-channel selectivity	
Receiver susceptibility to interfering signals	Out-of-band blocking	Out-of-band blocking
		Spurious response
	In-band blocking	In-band blocking
	Narrowband blocking	Narrowband blocking
	Adjacent channel selectivity	Adjacent channel selectivity
	Receiver intermodulation	Intermodulation characteristics
Unwanted emissions from the receiver	Receiver spurious emissions	Receiver spurious emissions

22.6.3 REGIONAL REQUIREMENTS

There are a number of regional variations to the RF requirements and their application. The variations originate from different regional and local regulations of spectrum and its use. The most obvious regional variation is the different frequency bands and their use, as discussed in the preceding section. Many of the regional RF requirements are also tied to specific frequency bands.

When there is a regional requirement on, for example, spurious emissions, this requirement should be reflected in the 3GPP specifications. For the BS it is entered as an optional requirement and is marked as "regional." For the terminal, the same procedure is not possible, since a terminal may roam between different regions and will therefore have to fulfill all regional requirements that are tied to an operating band in the regions where the band is used. For LTE, this becomes more complex than for UTRA, since there is an additional variation in the transmitter (and receiver) bandwidth used, making some regional requirements difficult to meet as a mandatory requirement. The concept of *network signaling* of RF requirements is therefore introduced for LTE, where a terminal can be informed at call setup of whether some specific RF requirements apply when the terminal is connected to a network.

22.6.4 BAND-SPECIFIC TERMINAL REQUIREMENTS THROUGH NETWORK SIGNALING

For the terminal, the channel bandwidths supported are a function of the LTE operating band, and also have a relation to the transmitter and receiver RF requirements. The reason is that some RF requirements may be difficult to meet under conditions with a combination of maximum power and high number of transmitted and/or received resource blocks.

Some additional RF requirements apply for the terminal when a specific network signaling value (NS_x) is signaled to the terminal as part of the cell handover or broadcast message. For implementation reasons, these requirements are associated with restrictions and variations to RF parameters such as terminal output power, maximum channel bandwidth, and number of transmitted resource blocks. The variations of the requirements are defined together with the network signaling value (NS_x) in the terminal RF specification [38], where each value corresponds to a specific condition. The default value for all bands is NS_01. All NS_x values are connected to an allowed power reduction called *additional maximum power reduction* (A-MPR) and apply for transmission using a certain minimum number of resource blocks, depending also on the channel bandwidth. The following are examples of terminal requirements that have a related Network Signaling Value for some bands:

- *NS_03, NS_04, or NS_06* is signaled when specific FCC requirements [49] on terminal unwanted emissions apply for operation in a number of US bands.
- *NS_05* is signaled for protection of the PHS band in Japan when a terminal operates in the 2 GHz band (band 1).

In some bands the NS_x signaling is also applied for testing of receiver sensitivity, since the active transmitted signal can affect the receiver performance.

There are also additional RF requirements and restrictions that may apply in case of LTE carrier aggregation in the uplink. These can be signaled to a terminal configured for carrier aggregation using specific CA network signaling values CA_NS_x and will in this case replace the usual network signaling values NS_x and their related requirements.

22.6.5 BASE-STATION CLASSES

In the BS specifications, there is one set of RF requirements that is generic, applicable to what is called "general-purpose" BSs. This is the original set of LTE requirements developed in 3GPP Release 8. It has no restrictions on BS output power and can be used for any deployment scenario. When the RF requirements were derived, however, the scenarios used were macro scenarios [50]. For this reason, in Release 9 additional BS classes were introduced that were intended for pico-cell and femto-cell scenarios. An additional class for micro-cell scenarios was added in Release 11, together with BS classes applicable also for multi-standard BSs. It is also clarified that the original set of "general-purpose" RF parameters are applicable for macro-cell scenarios. The terms macro, micro, pico, and femto are not used in 3GPP to identify the BS classes, instead the following terminology is used:

- *Wide area BS*. This type of BS is intended for macro-cell scenarios, defined with a minimum coupling loss between BS and terminal of 70 dB.
- *Medium range BS*. This type of BS is intended for micro-cell scenarios, defined with a minimum coupling loss between BS and terminal of 53 dB. Typical deployments are outdoor below-rooftop installations, giving both outdoor hot spot coverage and outdoor-to-indoor coverage through walls.
- *Local area BS*. This type of BS is intended for pico-cell scenarios, defined with a minimum coupling loss between BS and terminal of 45 dB. Typical deployments are indoor offices and indoor/outdoor hotspots, with the BS mounted on walls or ceilings.

- *Home BS*. This type of BS is intended for femto-cell scenarios, which are not explicitly defined. Minimum coupling loss between BS and terminal of 45 dB is also assumed here. Home BSs can be used both for open access and in closed subscriber groups.

The local area, medium range, and home BS classes have modifications to a number of requirements compared to wide area BSs, mainly due to the assumption of a lower minimum coupling loss:

- Maximum BS power is limited to 38 dBm output power for medium range BSs, 24 dBm output power for local area BSs, and to 20 dBm for home BSs. This power is defined per antenna and carrier, except for home BSs, where the power over all antennas (up to four) is counted. There is no maximum BS power defined for wide area BSs.
- Home BSs have an additional requirement for protecting systems operating on adjacent channels. The reason is that a terminal connected to a BS belonging to another operator on the adjacent channel may be in close proximity to the home BS. To avoid an interference situation where the adjacent terminal is blocked, the home BS must make measurements on the adjacent channel to detect adjacent BS operations. If an adjacent BS transmission (UTRA or LTE) is detected under certain conditions, the maximum allowed home BS output power is reduced in proportion to how weak the adjacent BS signal is, in order to avoid interference to the adjacent BS.
- The spectrum mask (operating band unwanted emissions) has lower limits for medium range, local area, and home BSs, in line with the lower maximum power levels.
- Limits for colocation for medium range and local area are relaxed compared to wide area BS, corresponding to the relaxed reference sensitivity for the base station.
- Home BSs do not have limits for colocation, but instead have more stringent unwanted emission limits for protecting home BS operation (from other home BSs), assuming a stricter through-the-wall indoor interference scenario.
- Receiver reference sensitivity limits are higher (more relaxed) for medium range, local area, and home BSs. Receiver dynamic range and in-channel selectivity (ICS) are also adjusted accordingly.
- All medium range, local area, and home BS limits for receiver susceptibility to interfering signals are adjusted to take the higher receiver sensitivity limit and the lower assumed minimum coupling loss (BS-to-terminal) into account.

22.7 OUTPUT POWER LEVEL REQUIREMENTS
22.7.1 BASE-STATION OUTPUT POWER AND DYNAMIC RANGE

There is no general maximum output power requirement for BSs. As mentioned in the discussion of BS classes in the preceding section, there is, however, a maximum output power limit of 38 dBm for medium range BSs, 24 dBm for local area BSs, and of 20 dBm for home BSs. In addition to this, there is a tolerance specified, defining how much the actual maximum power may deviate from the power level declared by the manufacturer.

The BS also has a specification of the total power control dynamic range for a resource element, defining the power range over which it should be possible to configure. There is also a dynamic range requirement for the total BS power.

For TDD operation, a power mask is defined for the BS output power, defining the off power level during the uplink subframes and the maximum time for the *transmitter transient period* between the transmitter on and off states.

22.7.2 TERMINAL OUTPUT POWER AND DYNAMIC RANGE

The terminal output power level is defined in three steps:

- *UE power class* defines a *nominal* maximum output power for QPSK modulation. It may be different in different operating bands, but the main terminal power class is today set at 23 dBm for all bands.
- *Maximum power reduction (MPR)* defines an allowed reduction of maximum power level for certain combinations of modulation used and the number of resource blocks that are assigned.
- *Additional maximum power reduction (A-MPR)* may be applied in some regions and is usually connected to specific transmitter requirements such as regional emission limits and to certain carrier configurations. For each such set of requirement, there is an associated network signaling value NS_x that identifies the allowed A-MPR and the associated conditions, as explained in Section 22.6.4.

The terminal has a definition of the transmitter Off power level, applicable to conditions when the terminal is not allowed to transmit. There is also a general On/Off time mask specified, plus specific time masks for PRACH, SRS, subframe boundary, and PUCCH/PUSCH/SRS.

The terminal transmit power control (TPC) is specified through requirements for the *absolute power tolerance* for the initial power setting, the *relative power tolerance* between two subframes, and the *aggregated power tolerance* for a sequence of power-control commands.

22.8 TRANSMITTED SIGNAL QUALITY

The requirements for transmitted signal quality specify how much the transmitted BS or terminal signal deviates from an "ideal" modulated signal in the signal and the frequency domains. Impairments on the transmitted signal are introduced by the transmitter RF parts, with the nonlinear properties of the power amplifier being a major contributor. The signal quality is assessed for BS and terminal through requirements on *error-vector magnitude (EVM)* and *frequency error*. An additional terminal requirement is UE in-band emissions.

22.8.1 EVM AND FREQUENCY ERROR

While the theoretical definitions of the signal quality measures are quite straightforward, the actual assessment is a very elaborate procedure, described in great detail in the 3GPP specification. The reason is that it becomes a multidimensional optimization problem, where the best match for the timing, the frequency, and the signal constellation are found.

The EVM is a measure of the error in the modulated signal constellation, taken as the root mean square of the error vectors over the active subcarriers, considering all symbols of the modulation scheme. It is expressed as a percentage value in relation to the power of the ideal signal. The EVM fundamentally defines the maximum SINR that can be achieved at the receiver, if there are no additional impairments to the signal between transmitter and receiver.

Since a receiver can remove some impairments of the transmitted signal such as time dispersion, the EVM is assessed after cyclic prefix removal and equalization. In this way, the EVM evaluation includes a standardized model of the receiver. The frequency offset resulting from the EVM evaluation is averaged and used as a measure of the *frequency error* of the transmitted signal.

22.8.2 TERMINAL IN-BAND EMISSIONS

In-band emissions are emissions within the channel bandwidth. The requirement limits how much a terminal can transmit into nonallocated resource blocks within the channel bandwidth. Unlike the OOB emissions, the in-band emissions are measured after cyclic prefix removal and FFT, since this is how a terminal transmitter affects a real BS receiver.

22.8.3 BASE-STATION TIME ALIGNMENT

Several LTE features require the BS to transmit from two or more antennas, such as transmitter diversity and MIMO. For carrier aggregation, the carriers may also be transmitted from different antennas. In order for the terminal to properly receive the signals from multiple antennas, the timing relation between any two transmitter branches is specified in terms of a maximum time alignment error between transmitter branches. The maximum allowed error depends on the feature or combination of features in the transmitter branches.

22.9 UNWANTED EMISSIONS REQUIREMENTS

Unwanted emissions from the transmitter are divided into *out-of-band (OOB) emissions* and *spurious emissions* in ITU-R recommendations [51]. OOB emissions are defined as emissions on a frequency close to the RF carrier, which results from the modulation process. Spurious emissions are emissions outside the RF carrier that may be reduced without affecting the corresponding transmission of information. Examples of spurious emissions are harmonic emissions, intermodulation products, and frequency conversion products. The frequency range where OOB emissions are normally defined is called the *out-of-band domain*, whereas spurious emissions limits are normally defined in the *spurious domain*.

ITU-R also defines the boundary between the OOB and spurious domains at a frequency separation from the carrier center of 2.5 times the necessary bandwidth, which corresponds to 2.5 times the channel bandwidth for LTE. This division of the requirements is easily applied for systems that have a fixed channel bandwidth. It does, however, become more difficult for LTE, which is a flexible bandwidth system, implying that the frequency range where requirements apply would then vary with the channel bandwidth. The approach taken for defining the boundary in 3GPP is slightly different for BS and terminal requirements.

With the recommended boundary between OOB emissions and spurious emissions set at 2.5 times the channel bandwidth, third- and fifth-order intermodulation products from the carrier will fall inside the OOB domain, which will cover a frequency range of twice the channel bandwidth on each side of the carrier. For the OOB domain, two overlapping requirements are defined for both BS and terminal: *spectrum emissions mask* (SEM) and *adjacent channel leakage ratio* (ACLR). The details of these are further explained in the following.

22.9.1 IMPLEMENTATION ASPECTS

The spectrum of an OFDM signal decays rather slowly outside of the transmission bandwidth configuration. Since the transmitted signal for LTE occupies 90% of the channel bandwidth, it is not possible to directly meet the unwanted emission limits directly outside the channel bandwidth with a "pure" OFDM signal. The techniques used for achieving the transmitter requirements are, however, not specified or mandated in LTE specifications. Time-domain windowing is one method commonly used in OFDM-based transmission systems to control spectrum emissions. Filtering is always used, both time-domain digital filtering of the baseband signal and analog filtering of the RF signal.

The nonlinear characteristics of the *power amplifier* (PA) used to amplify the RF signal must also be taken into account, since it is the source of intermodulation products outside the channel bandwidth. Power back-off to give a more linear operation of the PA can be used, but at the cost of a lower power efficiency. The power back-off should therefore be kept to a minimum. For this reason, additional linearization schemes can be employed. These are especially important for the BS, where there are fewer restrictions on implementation complexity and use of advanced linearization schemes is an essential part of controlling spectrum emissions. Examples of such techniques are feed-forward, feedback, predistortion, and postdistortion.

22.9.2 SPECTRUM EMISSION MASK

The spectrum emission mask defines the permissible OOB spectrum emissions outside the necessary bandwidth. As explained in the preceding section, how to take the flexible channel bandwidth into account when defining the frequency boundary between OOB emissions and spurious emissions is dealt differently for the LTE BS and terminal. Consequently, the spectrum emission masks are also based on different principles.

22.9.2.1 Base-Station Operating Band Unwanted Emission Limits

For the LTE BS, the problem of the implicit variation of the boundary between OOB and spurious domain with the varying channel bandwidth is handled by not defining an explicit boundary. The solution is a unified concept of *operating band unwanted emissions* (UEM) for the LTE BS instead of the spectrum mask usually defined for OOB emissions. The operating band unwanted emissions requirement applies over the whole BS transmitter operating band, plus an additional 10 MHz on each side, as shown in Figure 22.6. All requirements outside of that range are set by the regulatory spurious emission limits, based on the ITU-R recommendations [51]. As seen in the figure, a large part of the operating band unwanted emissions is defined over a frequency range that for smaller channel bandwidths can be both in spurious and OOB domains. This means that the limits for the frequency ranges that may be in the spurious domain also have to align with the regulatory limits from the ITU-R. The shape of the mask is generic for all channel bandwidth from 5 to 20 MHz, with a mask that consequently has to align with the ITU-R limits starting 10 MHz from the channel edges. Special masks are defined for the smaller 1.4 and 3 MHz channel bandwidths. The operating band unwanted emissions are defined with a 100 kHz measurement bandwidth.

In case of carrier aggregation for a BS, the UEM requirement (as other RF requirements) apply as for any multi-carrier transmission, where the UEM will be defined relative to the carriers on the edges of the RF bandwidth. In case of noncontiguous carrier aggregation, the UEM within a subblock gap is partly calculated as the cumulative sum of contributions from each subblock.

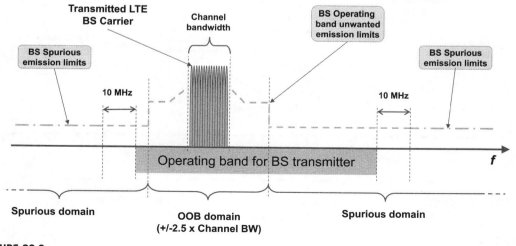

FIGURE 22.6

Frequency ranges for operating band unwanted emissions and spurious emissions applicable to LTE BS.

There are also special limits defined to meet a specific regulation set by the FCC [49] for the operating bands used in the USA and by the ECC for some European bands. These are specified as separate limits in addition to the operating band unwanted emission limits.

22.9.2.2 Terminal Spectrum Emission Mask

For implementation reasons, it is not possible to define a generic terminal spectrum mask that does not vary with the channel bandwidth, so the frequency ranges for OOB limits and spurious emissions limits do not follow the same principle as for the BS. The SEM extends out to a separation Δf_{OOB} from the channel edges, as illustrated in Figure 22.7. For 5 MHz channel bandwidth, this point corresponds to 250% of the necessary bandwidth as recommended by the ITU-R, but for higher channel bandwidths it is set closer than 250%.

The SEM is defined as a general mask and a set of additional masks that can be applied to reflect different regional requirements. Each additional regional mask is associated with a specific network signaling value NS_x.

22.9.3 ADJACENT CHANNEL LEAKAGE RATIO

In addition to a spectrum emissions mask, the OOB emissions are defined by an ACLR requirement. The ACLR concept is very useful for analysis of coexistence between two systems that operate on adjacent frequencies. The ACLR defines the ratio of the power transmitted within the assigned channel bandwidth to the power of the unwanted emissions transmitted on an adjacent channel. There is a corresponding receiver requirement called *adjacent channel selectivity* (ACS), which defines a receiver's ability to suppress a signal on an adjacent channel.

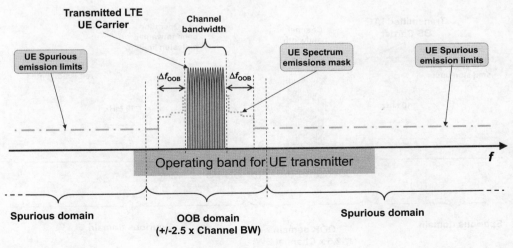

FIGURE 22.7

Frequency ranges for spectrum emission mask and spurious emissions applicable to LTE terminal.

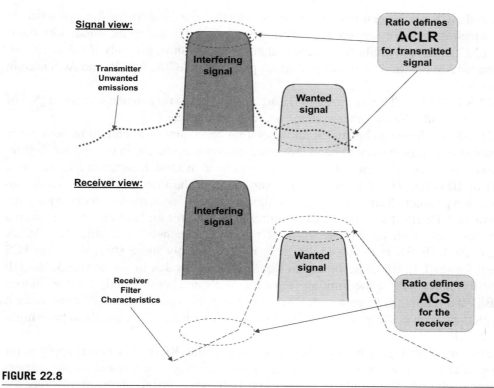

FIGURE 22.8

Illustration of ACLR and ACS, with example characteristics for an "aggressor" interferer and a receiver for a "victim" wanted signal.

The definitions of ACLR and ACS are illustrated in Figure 22.8 for a wanted and an interfering signal received in adjacent channels. The interfering signal's leakage of unwanted emissions at the wanted signal receiver is given by the ACLR and the ability of the receiver of the wanted signal to suppress the interfering signal in the adjacent channel is defined by the ACS. The two parameters when combined define the total leakage between two transmissions on adjacent channels. That ratio is called the *adjacent channel interference ratio* (ACIR) and is defined as the ratio of the power transmitted on one channel to the total interference received by a receiver on the adjacent channel, due to both transmitter (ACLR) and receiver (ACS) imperfections.

This relation between the adjacent channel parameters is [50]:

$$\text{ACIR} = \frac{1}{\frac{1}{\text{ACLR}} + \frac{1}{\text{ACS}}} \tag{22.7}$$

ACLR and ACS can be defined with different channel bandwidths for the two adjacent channels, which is the case for some requirements set for LTE due to the bandwidth flexibility. Eq. (22.7) will also apply for different channel bandwidths, but only if the same two channel bandwidths are used for defining all three parameters ACIR, ACLR, and ACS used in the equation.

The ACLR limits for LTE terminal and BS are derived based on extensive analysis [50] of LTE coexistence with LTE or other systems on adjacent carriers.

The LTE BS requirements on ACLR and operating band unwanted emissions both cover the OOB domain, but the operating band unwanted emission limits are in general set slightly more relaxed compared to the ACLR, since they are defined in a much narrower measurement bandwidth of 100 kHz. This allows for some variations in the unwanted emissions due to intermodulation products from varying power allocation between resource blocks within the channel. For an LTE BS, there are ACLR requirements both for an adjacent channel with a UTRA receiver and with an LTE receiver of the same channel bandwidth. The ACLR requirement for LTE BS is set to 45 dB. This is considerably more strict than the ACS requirement for the UE, which according to Eq. (22.7) implies that in the downlink, the UE receiver performance will be the limiting factor for ACIR and consequently for coexistence between BSs and terminals. From a system point of view, this choice is cost-efficient since it moves implementation complexity to the BS, instead of requiring all terminals to have high-performance RF.

In case of carrier aggregation for a BS, the ACLR (as other RF requirements) apply as for any multi-carrier transmission, where the ACLR requirement will be defined for the carriers on the edges of the RF bandwidth. In case of noncontiguous carrier aggregation where the subblock gap is so small that the ACLR requirements at the edges of the gap will "overlap," a special *cumulative ACLR* requirement (CACLR) is defined for the gap. For CACLR, contributions from carriers on both sides of the subblock gap are accounted for in the CACLR limit. The CACLR limit is the same as the ACLR for the BS at 45 dB.

ACLR limits for the terminal are set both with assumed UTRA and LTE receivers on the adjacent channel. As for the BS, the limits are also set stricter than the corresponding SEM, thereby accounting for variations in the spectrum emissions resulting from variations in resource-block allocations. In case of carrier aggregation, the terminal ACLR requirement applies to the aggregated channel bandwidth instead of per carrier. The ACLR limit for LTE terminals is set to 30 dB. This is considerably relaxed compared to the ACS requirement for the BS, which according to Eq. (22.7) implies that in the uplink, the UE transmitter performance will be the limiting factor for ACIR and consequently for coexistence between base stations and terminals.

22.9.4 SPURIOUS EMISSIONS

The limits for BS spurious emissions are taken from international recommendations [51], but are only defined in the region outside the frequency range of operating band unwanted emissions limits as illustrated in Figure 22.6—that is, at frequencies that are separated from the BS transmitter operating band by at least 10 MHz. There are also additional regional or optional limits for protection of other systems that LTE may coexist with or even be colocated with. Examples of other systems considered in those additional spurious emissions requirements are GSM, UTRA FDD/TDD, CDMA2000, and PHS.

Terminal spurious emission limits are defined for all frequency ranges outside the frequency range covered by the SEM. The limits are generally based on international regulations [51], but there are also additional requirements for coexistence with other bands when the mobile is roaming. The additional spurious emission limits can have an associated network signaling value.

In addition, there are BS and terminal emission limits defined for the receiver. Since receiver emissions are dominated by the transmitted signal, the receiver spurious emission limits are only applicable when the transmitter is Off, and also when the transmitter is On for an LTE FDD BS that has a separate receiver antenna connector.

22.9.5 OCCUPIED BANDWIDTH

Occupied bandwidth is a regulatory requirement that is specified for equipment in some regions, such as Japan and the USA. It is originally defined by the ITU-R as a maximum bandwidth, outside of which emissions do not exceed a certain percentage of the total emissions. The occupied bandwidth is for LTE equal to the channel bandwidth, outside of which a maximum of 1% of the emissions are allowed (0.5% on each side).

In the case of carrier aggregation, the occupied bandwidth is equal to the aggregated channel bandwidth. For noncontiguous carrier aggregation, the occupied bandwidth applies per subblock.

22.9.6 TRANSMITTER INTERMODULATION

An additional implementation aspect of an RF transmitter is the possibility of intermodulation between the transmitted signal and another strong signal transmitted in the proximity of the BS or terminal. For this reason there is a requirement for *transmitter intermodulation*.

For the BS, the requirement is based on a stationary scenario with a colocated other BS transmitter, with its transmitted signal appearing at the antenna connector of the BS being specified, but attenuated by 30 dB. Since it is a stationary scenario, there are no additional unwanted emissions allowed, implying that all unwanted emission limits also have to be met with the interferer present.

For the terminal, there is a similar requirement based on a scenario with another terminal transmitted signal appearing at the antenna connector of the terminal being specified, but attenuated by 40 dB. The requirement specifies the minimum attenuation of the resulting intermodulation product below the transmitted signal.

22.10 SENSITIVITY AND DYNAMIC RANGE

The primary purpose of the *reference sensitivity requirement* is to verify the receiver *noise figure*, which is a measure of how much the receiver's RF signal chain degrades the SNR of the received signal. For this reason, a low-SNR transmission scheme using QPSK is chosen as reference channel for the reference sensitivity test. The reference sensitivity is defined at a receiver input level where the throughput is 95% of the maximum throughput for the reference channel.

For the BS, reference sensitivity could potentially be defined for a single resource block up to a group covering all resource blocks. For reasons of complexity, a maximum granularity of 25 resource blocks has been chosen, which means that for channel bandwidths larger than 5 MHz, sensitivity is verified over multiple adjacent 5 MHz blocks, while it is only defined over the full channel for smaller channel bandwidths.

For the terminal, reference sensitivity is defined for the full channel bandwidth signals and with all resource blocks allocated for the wanted signal. For the higher channel bandwidths (>5 MHz) in some operating bands, the nominal reference sensitivity needs to be met with a minimum number of allocated resource blocks. For larger allocation, a certain relaxation is allowed.

The intention of the *dynamic range requirement* is to ensure that the receiver can also operate at received signal levels considerably higher than the reference sensitivity. The scenario assumed for BS dynamic range is the presence of increased interference and corresponding higher wanted signal levels, thereby testing the effects of different receiver impairments. In order to stress the receiver, a higher SNR transmission scheme using 16QAM is applied for the test. In order to further stress the receiver to higher signal levels, an interfering AWGN signal at a level 20 dB above the assumed noise floor is added to the received signal. The dynamic range requirement for the terminal is specified as a *maximum signal level* at which the throughput requirement is met.

22.11 RECEIVER SUSCEPTIBILITY TO INTERFERING SIGNALS

There is a set of requirements for BS and terminal, defining the receiver's ability to receive a wanted signal in the presence of a stronger interfering signal. The reason for the multiple requirements is that, depending on the frequency offset of the interferer from the wanted signal, the interference scenario may look very different and different types of receiver impairments will affect the performance. The intention of the different combinations of interfering signals is to model as far as possible the range of possible scenarios with interfering signals of different bandwidths that may be encountered inside and outside the BS and terminal receiver operating band.

While the types of requirements are very similar between BS and terminal, the signal levels are different, since the interference scenarios for the BS and terminal are very different. There is also no terminal requirement corresponding to the BS ICS requirement.

The following requirements are defined for LTE BS and terminal, starting from interferers with large frequency separation and going close in (see also Figure 22.9). In all cases where the interfering signal is an LTE signal, it has the same bandwidth as the wanted signal, but at most 5 MHz.

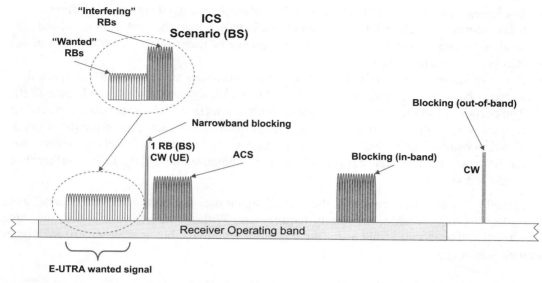

FIGURE 22.9

Base-station and terminal requirements for receiver susceptibility to interfering signals in terms of blocking, ACS, narrowband blocking, and ICS (BS only).

- *Blocking.* This corresponds to the scenario with strong interfering signals received outside the operating band (OOB blocking) or inside the operating band (in-band blocking), but not adjacent to the wanted signal. In-band blocking includes interferers in the first 20 MHz outside the operating band for the BS and the first 15 MHz for the terminal. The scenarios are modeled with a *continuous wave* (CW) signal for the OOB case and an LTE signal for the in-band case. There are additional (optional) BS blocking requirements for the scenario when the BS is colocated with another BS in a different operating band. For the terminal, a fixed number of *exceptions* are allowed from the OOB blocking requirement, for each assigned frequency channel and at the respective *spurious response frequencies*. At those frequencies, the terminal must comply with the more relaxed spurious response requirement.
- *Adjacent channel selectivity.* The ACS scenario is a strong signal in the channel adjacent to the wanted signal and is closely related to the corresponding ACLR requirement (see also the discussion in Section 22.9.3). The adjacent interferer is an LTE signal. For the terminal, the ACS is specified for two cases with a lower and a higher signal level. For MSR BSs, there is no specific ACS requirement defined. It is instead replaced by the narrowband blocking requirement, which covers the adjacent channel properties fully.
- *Narrowband blocking.* The scenario is an adjacent strong narrowband interferer, which in the requirement is modeled as a single resource block LTE signal for the BS and a CW signal for the terminal.

- *In-channel selectivity* (ICS). The scenario is multiple received signals of different received power levels inside the channel bandwidth, where the performance of the weaker "wanted" signal is verified in the presence of the stronger "interfering" signal. ICS is only specified for the BS.
- *Receiver intermodulation.* The scenario is *two* interfering signals near to the wanted signal, where the interferers are one CW and one LTE signal (not shown in Figure 22.9). The purpose of the requirement is to test receiver linearity. The interferers are placed in frequency in such a way that the main intermodulation product falls inside the wanted signal's channel bandwidth. There is also a *narrowband intermodulation* requirement for the BS where the CW signal is very close to the wanted signal and the LTE interferer is a single RB signal.

For all requirements except ICS, the wanted signal uses the same reference channel as in the corresponding reference sensitivity requirement. With the interference added, the same 95% relative throughput is met as for the reference sensitivity, but at a "desensitized" higher wanted signal level.

22.12 MULTIBAND-CAPABLE BASE STATIONS

The 3GPP specifications have been continuously developed to support larger RF bandwidths for transmission and reception through multi-carrier and multi-RAT operation and carrier aggregation over contiguous and noncontiguous spectrum allocations. This has been made possible with the evolution of RF technology supporting larger bandwidths for both transmitters and receivers. The next step in RF technology for BSs is to support simultaneous transmission and/or reception in multiple bands through a common radio. A multi-band BS could cover multiple bands over a frequency range of a few hundred MHz.

One obvious application for multi-band BSs is for inter-band carrier aggregation. It should however be noted that base stations supporting multiple bands have been in existence long before carrier aggregation was introduced in LTE. Already for GSM, dual-band BSs were designed to enable more compact deployments of equipment at BS sites. In some cases these shared also antennas, in other cases there were separate antenna systems for the different bands. These early implementations were really two separate sets of transmitters and receiver for the bands that were integrated in the same equipment cabinet. In the case where a common antenna system is used, the signals are combined and split through passive diplexers. The difference for "true" multi-band capable BSs is that the signals for the bands are transmitted and received in common *active* RF in the BS.

The RF requirements for such multi-band-capable base stations are defined in 3GPP Release 11. The specification supports multi-band operation both with a single RAT and with multiple RATs, also called multi-band multi-standard radio (MB-MSR) BSs. The specifications cover all combinations of RATs, except pure single-RAT GSM operation across the supported bands.

There are several scenarios envisioned for multi-band BS implementation and deployment. The possibilities for the multi-band capability are:

- multi-band transmitter + multi-band receiver;
- multi-band transmitter + single-band receiver;
- single-band transmitter + multi-band receiver.

The first case is demonstrated in Figure 22.10, which shows an example BS with a common RF implementation of both transmitter and receiver for two operating bands X and Y. Through a duplex filter, the transmitter and receiver are connected to a common antenna connector and a common antenna. The example is also a multi-RAT capable MB-MSR BS, with LTE+GSM configured in band X and LTE configured in band Y. Note that the figure has only one diagram showing the frequency range for the two bands, which could either be the receiver or transmitter frequencies.

Figure 22.10 also illustrates some new parameters that are defined for multi-band BS.

- *RF bandwidth* has the same definition as for a multi-standard BS, but is defined individually for each band.
- *Inter-RF-bandwidth gap* is the gap between the RF bandwidths in the two bands. Note that the inter-RF bandwidth gap may span a frequency range where other mobile operators can be deployed in bands X and Y, as well as a frequency range between the two bands that may be used for other services.
- *Total RF bandwidth* is the full bandwidth supported by the BS to cover the multiple carriers in both bands.

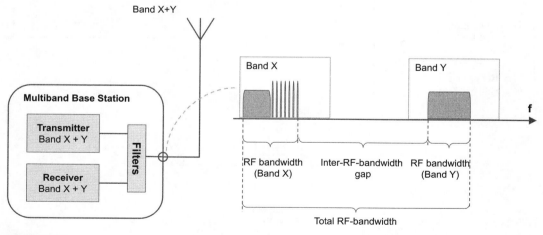

FIGURE 22.10

Example of multi-band BS with multi-band transmitter and receiver for two bands with one common antenna connector.

In principle, a multi-band BS can be capable of operating in more than two bands. The requirements and testing developed for the new type of BSs in 3GPP Release 11 will, however, in general, only cover a two-band capability. Full support for more than two bands is planned to be introduced in the 3GPP specifications in Release 14.

While having only a single antenna connector and a common feeder that connects to a common antenna is desirable to reduce the amount of equipment needed in a site, it is not always possible. It may also be desirable to have separate antenna connectors, feeders, and antennas for each band. An example of a multi-band BS with separate connectors for two operating bands X and Y is shown in Figure 22.11. Note that while the antenna connectors are separate for the two bands, the RF implementation for transmitter and receiver are in this case common for the bands. The RF for the two bands is separated into individual paths for band X and band Y before the antenna connectors through a filter. As for multi-band BSs with a common antenna connector for the bands, it is also possible here to have either the transmitter or receiver to be a single-band implementation, while the other is multi-band.

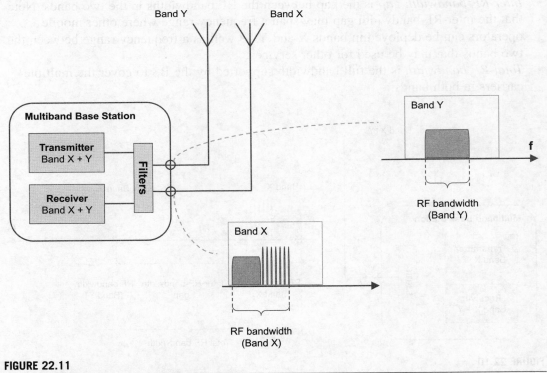

FIGURE 22.11

Multi-band BS with multi-band transmitter and receiver for two bands with separate antenna connectors for each band.

Further possibilities are BS implementations with separate antenna connectors for receiver and transmitter, in order to give better isolation between the receiver and transmitter paths. This may be desirable for a multi-band BS, considering the large total RF bandwidths, which will in fact also overlap between receiver and transmitter.

For a multi-band BS, with a possible capability to operate with multiple RATs and several alternative implementations with common or separate antenna connectors for the bands and/ or for the transmitter and receiver, the declaration of the BS capability becomes quite complex. What requirements will apply to such a BS and how they are tested will also depend on these declared capabilities.

Most RF requirements for a multi-band BS remain the same as for a single-band implementation. There are however some notable exceptions:

- *Transmitter spurious emissions*: For LTE BSs, the requirements exclude frequencies in the operating band plus an additional 10 MHz on each side of the operating band, since this frequency range is covered by the UEM limits. For a multi-band BS, the exclusion applies to both operating bands (plus 10 MHz on each side), and only the UEM limits apply in those frequency ranges. This is called "joint exclusion band."
- *Operating band unwanted emissions mask (UEM)*: For multi-band operation, when the inter-RF bandwidth gap is less than 20 MHz, the UEM limit applies as a cumulative limit with contributions counted from both bands, in a way similar to operation in noncontiguous spectrum.
- *ACLR*: For multi-band operation, when the inter-RF bandwidth gap is less than 20 MHz, the CACLR will apply with contributions counted from both bands, in a way similar to operation in noncontiguous spectrum.
- *Transmitter intermodulation*: For a multi-band BS, when the inter-RF bandwidth gap is less than 15 MHz, the requirement only applies for the case when the interfering signals fit within the gap.
- *Blocking requirement*: For multi-band BS, the in-band blocking limits apply for the in-band frequency ranges of *both* operating bands. This can be seen as a "joint exclusion," similar to the one for spurious emissions. The blocking and receiver intermodulation requirements also apply inside the inter-RF bandwidth gap.
- *Receiver spurious emissions*: For a multi-band BS, a "joint exclusion band" similar to the one for transmitter spurious emissions will apply, covering both operating bands plus 10 MHz on each side.

In the case where the two operating bands are mapped on separate antenna connectors as shown in Figure 22.11, the exceptions for transmitter/receiver spurious emissions, UEM, ACLR, and transmitter intermodulation do not apply. Those limits will instead be the same as for single-band operation for each antenna connector. In addition, if such a multiband BSs with separate antenna connectors per band is operated in only one band with the other band (and other antenna connector) inactive, the BS will from a requirement point of view be seen as a single-band BS. In this case all requirements will apply as single-band requirements.

22.13 RF REQUIREMENTS FOR RELAYS

As described in Chapter 18, the LTE specifications support decode-and-forward relays in Release 10. The RF requirements for such relays were introduced in Release 11 and are described in this chapter. The baseline for setting the relay RF requirements has been the existing RF requirements for the BS and the terminal, with the observation that seen from the access link side, a relay would have many similarities with a BS while on the backhaul side, it would have similarities with a terminal. This is illustrated in Figure 22.12.

The RF requirements for LTE relays are defined in a separate specification [55], but many of the requirements are set by direct reference to the terminal specification [38] for backhaul link requirements and BS specification [47] for access link requirements. It is in particular the local area BS requirements that are referenced, since the deployment scenario in that case is similar to a relay. Many RF requirements are however specific for relays, the following in particular:

Output power: Two power classes are defined for relays. Class 1 has a maximum of 24 dBm and Class 2 has a maximum 30 dBm output power for the access link. The maximum for the backhaul link is 24 dBm for both classes. All power levels are counted as the sum over all antennas (up to eight for the access link and four for the backhaul link).

ACLR: In order for the backhaul link to provide proper coexistence properties and not deteriorate the BS uplink, the relay ACLR limits are set equivalent to the ones for a local area BS on both the access and backhaul links.

Operating band unwanted emissions (UEM): New UEM limits for the access and backhaul links are set in relation to the output power levels defined by the relay power class.

Adjacent channel selectivity: The relay backhaul link relative ACS limit is set at a level similar to what is used for a BS, but the requirement is defined at a higher input signal

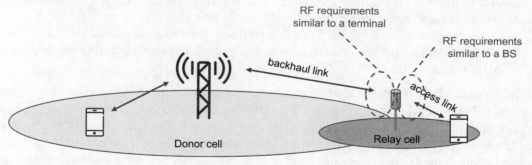

FIGURE 22.12

RF requirements for a relay on the backhaul and access sides.

level corresponding to the maximum downlink signal level at which ACS is defined for a terminal. The ACS requirement for the access link is taken from the local area BS. *Blocking requirements*: The in-band blocking levels for the backhaul link are set higher than for a terminal, but are also defined for a correspondingly higher wanted signal level. The blocking requirement for the access link is taken from the local area BS.

22.14 RF REQUIREMENTS FOR LICENSE-ASSISTED ACCESS

Operation in unlicensed spectrum through license-assisted access (LAA) is a new feature in 3GPP Release 13, see Chapter 17 for details. Due to the specific conditions for operating in an unlicensed band with very particular regulation, there is also an impact on the RF requirements, both for terminal and base station.

LAA operation is defined for the 5 GHz unlicensed band as described in Section 17.1. The 5 GHz band is called *Band 46* in 3GPP and covers the frequency range 5150 to 5925 MHz. In the RF specifications, operation in Band 46 is tightly coupled to the use of Frame structure type 3, while TDD operation in other unpaired bands is tied to the use of Frame structure type 2. Band 46 is specified for inter-band carrier aggregation with several other bands, which is a necessity since stand-alone LAA operation in Band 46 is not specified.

22.14.1 REGULATORY REQUIREMENTS FOR THE UNLICENSED 5 GHz BAND

The frequency range 5150–5925 MHz is available in most regions of the world for unlicensed (sometimes called license exempt) operation. The regulation that applies to the band varies from region to region and the type of service that can be used also varies. There are however many similarities in the regulation, while the terminology used for the service differs. There is also an evolution of the use of the band, with the intention of making a larger part of the band available for radio LAN services such as WiFi.

Since LAA operation in Release 13 is defined for downlink only, there are no transmissions from the terminal. This also means that there is no specific regulation that applies to the terminal for band 46 operation. In later releases when uplink operation is added, the regulation will also apply for terminals, but in Release 13 it will thus only apply for base stations.

There is a range of different regional regulatory requirements applicable for operation in the 5 GHz band and they are all applicable to LAA operation. The regulatory requirements can be divided into three different types:

- *Emission limits*: These include limits on maximum transmitter power, peak conducted power, average PSD, directional antenna gain, maximum mean EIRP (effective isotropic radiated power), EIRP density, EIRP elevation angle mask, and OOB emissions.
- *Functional requirements*: These include requirements to have specific TPC, dynamic frequency selection (DFS), and clear-channel assessment such as listen before talk (LBT).

- *Operational requirements*: These requirements restrict operation to indoor use in several frequency ranges.

Most of the regulatory emission limits are related to transmitted power and unwanted emissions, in terms of absolute level, density, or EIRP. There are different types of requirements and limits defined on global level (WRC), plus regional and national regulation for Europe, USA, China, Israel, South Africa, Turkey, USA, Canada, Brazil, Mexico, China, Japan, Korea, India, Chinese Taiwan, Singapore, and Australia[57]. There is a substantial variation between regions and countries.

It would not be realistic to cover this whole variation of regional requirements in the BS specifications and covering only a subset could be viewed as discriminatory. The regulation is also continuously updated, making it difficult to maintain an up-to-date specification. It will instead be up to each vendor of equipment supporting LAA that the requirements that are applicable to equipment sold in each region and country are met.

For this reason, the corresponding RF requirements are specified in 3GPP with a single number for each limit of the concerned requirements, such as max BS power and unwanted emissions. The limit is selected to give a broad coverage in terms of regulation. No more specific limits and details are given. Regarding how the regulation is applied, the following can be noted:

- Limits defined based on EIRP cannot be directly applied as an RF requirement in the 3GPP BSs specifications [47]. The reason is that all BS RF requirements are defined as conducted requirements and are specified at the antenna connector. Specification of the antenna, which directly affects the EIRP, is not part of the RF specification. There is however an informative guideline given for assessing such regulatory EIRP-related limits.
- Functional requirements are also specific to regions and have a large variation. The only one fully covered in 3GPP specifications is the clear-channel assessment through LBT.
- Operational requirements are in general not applicable for the RF specifications, since they do not specify a property of the equipment as such, but are more related to how the equipment is deployed and operated.

22.14.2 SPECIFIC BS RF REQUIREMENTS FOR LAA OPERATION

As stated before, Band 46 is available in many parts of the world, but it is not fully available anywhere and the regulation varies between regions and countries. An overview of how the band is assigned is given in Section 17.1. Since LAA operation in Release 13 is defined for downlink only, there are no receiver requirements defined for the BS, except for the functional requirements related to downlink channel access.

In order to reflect this variation of regulation in the specification, Band 46 is divided into four subbands as shown in Table 22.6. The subbands cover the complete band range 5150–5925 MHz, except for the range 5350–5470 MHz, which is presently not available for this kind of service in any region. The situation is the same for the upper part of subband 46D, but is under consideration in several regions.

Table 22.6 Division of Band 46 into Subbands for the BS	
Band 46 Subband	**Frequency Range**
46A	5150–5250 MHz
46B	5250–5350 MHz
46C	5470–5725 MHz
46D	5725–5925 MHz

The division into subbands makes clear which parts of the band are available for operation globally and it can also serve as a reference for base station vendors when designing BSs, in order to describe which parts of Band 46 that are supported and under what conditions. There is no corresponding division into subbands for the terminal.

With regard to LAA-specific RF requirements defined for a BS, the following apply:

- *RF carrier raster*: Normally, LTE carriers may be placed on any carrier position on the predefined 100 kHz carrier raster. For operation of LTE BS, there is however a need to coexist with existing services in the 5 GHz band, including Wi-Fi. For this reason, the possible carrier raster position an LAA BS can use is limited to a set of 32 positions that are aligned with the possible Wi-Fi carrier positions. In order to also allow for LTE carrier aggregation, where the spacing between component LTE carriers should be a multiple of 300 kHz, it is also possible to use the carrier raster positions within ±200 kHz of the 32 aligned ones.

- *BS output power*: There are no specific maximum power limits defined for operation of an LAA BS. The power levels allowed by regulation vary, but fall in general under the local area BS class and in some cases under the Medium Range BS class.

- *Adjacent channel leakage ratio (ACLR)*: The ACLR requirement for LAA BSs is 10 dB lower in the first adjacent channel and 5 dB lower in the second adjacent channel. The reason is that in an interference environment with other unlicensed BS and devices operating, including Wi-Fi equipment, an ACLR requirement at a level of 35–40 dB is sufficient. The CACLR requirement is also 10 dB lower at 35 dB.

- *Base-station operating band unwanted emission limits*: A new spectrum mask is introduced for operation in the 5 GHz band. The mask is defined as on operating band unwanted emission limit across the full band, is based on the regulatory mask applied in Europe [58], and is also similar to the mask defined for Wi-Fi equipment.

- *Downlink channel access*: The LBT mechanism used for downlink channel access is described in Section 17.3.2 and it is a regulatory requirement in many regions for certain bands. It is an assumption that it will be implemented in all BSs. This means that also for downlink-only operation, there will still be a requirement to have a BS receiver, in order to perform clear channel assessment using LBT. A set of parameters is predefined to be used for the LBT mechanism. The set includes an LBT measurement bandwidth, an energy detection threshold, and a maximum channel occupancy time.

22.14.3 SPECIFIC TERMINAL RF REQUIREMENTS FOR LAA OPERATION

The same definitions as related to Band 46 operation are defined for a terminal in LAA operation, except that the full band 5150—5925 MHz is specified for LAA operation, with no subbands or carrier raster restrictions defined. The reason is that terminals should be salable globally and able to roam across countries and regions, in order to get economy-of-scale for the terminal vendors and improve the experience for global users. This also means that terminals will be more future-proof, and can still be used if regulation evolves in terms of what subbands are deployed and what exact RF carrier positions are used.

Since LAA operation in Release 13 is defined for downlink only, there are only requirements defined for the terminal receiver, not the transmitter. The following LAA-specific RF requirements apply for a terminal:

- *RF carrier raster*: The usual 100 kHz LTE carrier raster applies for the terminal without any restrictions, as for other bands.
- *Adjacent channel selectivity*: A modified requirement for carrier aggregation is defined, where the interfering signal is scaled to a larger bandwidth of 20 MHz.
- *In-band blocking*: A modified requirement for carrier aggregation is defined, where the interfering signal is scaled to a larger bandwidth of 20 MHz.
- *Out-of-band blocking*: The blocker level is defined to be 5 dB lower for frequencies above 4 GHz, due to implementation with full-band 5 GHz RF filters.
- *Wideband intermodulation*: A modified requirement for carrier aggregation is defined, where the interfering signal is scaled to a larger bandwidth of 20 MHz.

22.15 RF REQUIREMENTS FOR BS WITH ACTIVE ANTENNA SYSTEMS

For the continuing evolution of mobile systems, advanced antenna systems have an increasing importance. While there have been several attempts to develop and deploy BSs with passive antenna arrays of different kinds for many years, there have been no specific RF requirements associated with such antenna systems. With RF requirements in general defined at the base station RF antenna connector, the antennas have also not been seen as part of the base station.

Requirements specified at an antenna connector are referred to as *conducted requirements*, usually defined as a power level (absolute or relative) measured at the antenna connector. Most emission limits in regulation are defined as conducted requirements. An alternative way is to define a *radiated requirement*, which is assessed including the antenna by accounting for the antenna gain in a specific direction. Radiated requirements require more complex *over-the-air* (OTA) test procedures using, for example, an anechoic chamber. With OTA testing, the spatial characteristics of the whole BS including the antenna system can be assessed.

For base stations with *active antenna systems* (AAS), where the active parts of the transmitter and receiver may be an integral part of the antenna system, it is not always suitable to maintain the traditional definition of requirements at the antenna connector. For this purpose, 3GPP developed RF requirements in Release 13 for AAS base stations in a set of separate RF specifications.

The AAS BS requirements are based on a generalized AAS BS radio architecture, as shown in Figure 22.13. The architecture consists of a *transceiver unit array* that is connected to a *composite antenna* that contains a *radio distribution network* and an *antenna array*. The transceiver unit array contains multiple transmitter and receiver units. These are connected to the composite antenna through a number of connectors on the *transceiver array boundary* (TAB). These TAB connectors correspond to the antenna connectors on a non-AAS base station and serve as a reference point for conducted requirements. The radio distribution network is passive and distributes the transmitter outputs to the corresponding antenna elements and vice versa for the receiver inputs. Note that the actual implementation of an AAS BS may look different in terms of physical location of the different parts, array geometry, type of antenna elements used, and so on.

For an AAS BS, there are two types of requirements:

- *Conducted requirements* are defined for each RF characteristic at an individual or a group of TAB connectors. The conducted requirements are defined in such a way that they are in a sense "equivalent" to the corresponding conducted non-AAS requirement, that is, the performance of the system or the impact on other systems is expected to be the same. All non-AAS RF requirements (see Sections 22.6 to 22.11) have corresponding AAS conducted requirements.

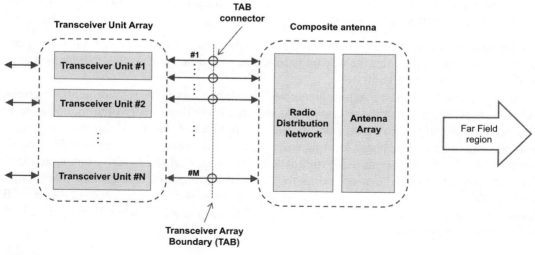

FIGURE 22.13

Generalized radio architecture of an active antenna system.

- *Radiated requirements* are defined over-the-air in the far field of the antenna system. Since the spatial direction becomes relevant in this case, it is detailed for each requirement how it applies. Radiated requirements are defined for *radiated transmitter power* and *OTA sensitivity* and these two do not have a direct corresponding non-AAS requirement.

The radiated transmitter power is defined accounting for the antenna array beamforming pattern in a specific direction as *effective isotropic radiated power* (EIRP) for each beam that the BS is declared to transmit. In a way similar to BS output power, the actual requirement is on the accuracy of the declared EIRP level.

The OTA sensitivity is a requirement based on a quite elaborate declaration by the manufacturer of one or more *OTA sensitivity direction declarations* (OSDD). The sensitivity is in this way defined accounting for the antenna array beamforming pattern in a specific direction as declared *equivalent isotropic sensitivity* (EIS) level toward a receiver target. The EIS limit is to be met not only in a single direction but within a *range of angle of arrival* (RoAoA) in the direction of the receiver target. Depending on the level of adaptivity for the AAS BS, two alternative declarations are made:

- If the receiver is adaptive to direction, so that the receiver target can be redirected, the declaration contains a *receiver target redirection range* in a specified *receiver target direction*. The EIS limit should be met within the redirection range, which is tested at five declared sensitivity RoAoA within that range.
- If the receiver is not adaptive to direction and thus cannot redirect the receiver target, the declaration consists of a single sensitivity RoAoA in a specified receiver target direction, in which the EIS limit should be met.

The characterization of AAS BS through the requirements on radiated transmitter power and OTA sensitivity provides the flexibility to account for a range of AAS BS implementations with different types of adaptivity.

It is expected that more requirements will be defined as OTA requirements in coming releases of 3GPP specifications. Also for 5G systems, including LTE Evolution, it is expected that RF requirements for AAS BS will be an essential part of the specifications, since multi-antenna transmission and beam forming will play a major role as a component of 5G (see also Section 24.2.5). Such systems using digital beam forming would have active antennas and the RF requirements would be specified as for AAS BS. It is also expected that there will be multi-RAT BSs which combine LTE Evolution and the new 5G radio access in the same BS hardware, making AAS BS supporting multiple RATs a possibility, when they operate in the same or in nearby frequency bands.

5G WIRELESS ACCESS

23

23.1 WHAT IS 5G?

As described already in Chapter 1, the world has witnessed four generations of mobile communication, with each new generation emerging roughly 10 years after the emergence of the previous generation.

The first generation consisted of the analog systems introduced in the early 1980s. They were only supporting voice services and, for the first time, made mobile telephony available to ordinary people.

The second generation (2G), emerging in the early 1990s, took mobile telephony from being used by some people to being available to essentially everyone and everywhere. Technology-wise, the key feature of 2G was the transition from analog-to-digital transmission. Although the main service was still voice, the introduction of digital transmission also enabled the first support of mobile data.

Third-generation (3G) WCDMA, later evolved into HSPA, was introduced in 2001. 3G lay the foundation for mobile broadband and, especially with HSPA, made true mobile internet access available to ordinary people.

We are now well into the fourth-generation (4G) era of mobile communication with the first LTE systems being introduced in 2009. Compared to HSPA, LTE provides even better mobile broadband including higher achievable data rates and higher efficiency in terms of, for example, spectrum utilization.

It is important to note that the introduction of a new generation of mobile communication has, in no way, implied the end of the previous generation. The situation has rather been the opposite. The deployment of 2G systems actually accelerated after the introduction of 3G. Likewise, there is still a massive deployment of 3G systems despite the introduction of LTE more than 6 years ago. Thus it is not surprising that, although LTE is still at relatively early stage of deployment, the industry is already well on the path toward the next step of mobile communication—that is, the *fifth generation* or 5G.

5G will continue on the path of LTE, enabling even higher data rates and even higher efficiency for mobile broadband. However, the scope of 5G is much wider than just further enhanced mobile broadband. Rather, as already mentioned in Chapter 1, 5G is often described as a platform that should enable wireless connectivity for essentially any kind of device or any

4G, LTE-Advanced Pro and The Road to 5G. http://dx.doi.org/10.1016/B978-0-12-804575-6.00023-6

kind of application that may benefit from being connected. The concept of *machine-type communication* (MTC) is one part of this extended set of use cases expected in the 5G era. As described in Chapter 20, major steps to further enhance the support for certain types of MTC applications have already been taken as part of the evolution of LTE. More specifically, these steps have focused on *massive-MTC* applications associated with very low-cost devices with very long battery life but with relatively modest data rate and latency requirements.

However, 5G is assumed to enable connectivity for a much wider range of new use cases. Examples of additional use cases explicitly mentioned in the context of 5G includes wireless connectivity for remote control of machinery, wireless connectivity for traffic safety and control, and monitor/control of infrastructure, to just name a few. Furthermore, 5G should not only be a platform for providing connectivity for already identified applications and use cases. Rather 5G should be flexible enough to enable connectivity also for future applications and use cases that may not yet even be anticipated.

The very wide range of use cases to be covered by 5G implies that the capabilities of 5G wireless access have to extend far beyond that of previous generations. For the first and second generation networks the use case in focus was mobile telephony with the main target to provide good speech quality for as many users as possible. For 3G and 4G, the change of focus toward mobile broadband implied that the quality measure changed from speech quality to achievable end-user data rate. In line with this, the main target for 3G and 4G has been to enable as high data rates as possible for as many users as possible. However, for 5G there will be a much wider set of capabilities and requirements, some of which may even be partly contradicting each other.

23.1.1 DATA RATES

Providing the possibility for even higher end-user data rates will be an important requirement also in the 5G era, primarily as part of a quest for further enhanced mobile-broadband experience. Although support for extremely high peak data rates of 10 Gbit/s and higher is often mentioned in the context of 5G, this is just one aspect of a more general aim for higher data rates in all types of environments. Providing higher data rates may, for example, also include making several 100 Mbit/s generally available in urban and suburban environments. This would imply an increase in achievable data rates roughly a factor 10 compared to what can be provided with current technologies.

Furthermore, if one agrees with the vision of wireless connectivity being available "everywhere and for everyone," higher data rates may, for example, also include a few Mbit/s essentially everywhere in the world including rural areas in developing countries where there may currently be no broadband access what-so-ever.

23.1.2 LATENCY

In terms of latency requirements, the possibility to provide an end-to-end latency in the order of 1 ms is often mentioned in the context of 5G.

Low latency has since the emergence of HSPA been recognized as an important component to enable a good mobile-broadband experience. However, for 5G the possibility to provide connectivity with very low latency will also be an enabler for new latency-critical wireless applications such as remote control with haptic feedback and wireless connectivity for traffic safety.[1] It should be pointed out though that very few wireless applications that actually require an end-to-end latency as low as 1 ms have been identified. Providing the possibility for such low latency should thus more be seen as an enabler of future yet unknown applications, rather than something that has been concluded to be needed for currently envisioned applications. It should also be noted that end-to-end latency depends on much more than just the radio-access solution. Depending on the physical distance between the end points, an end-to-end latency of 1 ms may even be physically impossible. Nevertheless, a requirement of a 1 ms end-to-end latency implies that the radio-access network, including the network-to-device link, should be able to provide a latency significantly less than 1 ms.

23.1.3 EXTREME RELIABILITY

Another characteristic often mentioned in the context of 5G is the possibility to enable connectivity with extremely high reliability.

It should be noted that high reliability, in this context, could mean very different things. In some cases high reliability has been associated with the ability of the wireless-access solution to provide connectivity with extremely low error rate, for example, an error rate below 10^{-9}. In other contexts extreme reliability has, for example, been associated with the ability to retain connectivity even in case of unexpected events including natural disasters. This is obviously a very different requirement, requiring very different solutions, compared to a requirement to provide connectivity with extremely low error rate under more normal conditions.

23.1.4 LOW-COST DEVICES WITH VERY LONG BATTERY LIFE

Some applications, such as the collection of data from a very large number of sensors, require the possibility for devices of much lower cost compared to the devices of today. In many cases, such applications also require the possibility for devices with extremely low device energy consumption enabling battery life of several years. At the same time, these applications typically require only very modest data rates and can accept long latency. As described in Chapter 20, the evolution of LTE has already taken some substantial steps in this direction, fulfilling many of the 5G requirements in this area.

23.1.5 NETWORK ENERGY EFFICIENCY

Another important requirement emerging during the last few years has been the aim for significantly higher network energy efficiency. Although partly driven by the general quest

---

[1]The term haptic feedback is used to indicate remote control where feedback from the controlled device is used to provide a "real-life" sensation to the controlling device.

for a more sustainable society, enhanced network energy efficiency also has some very real and concrete drivers.

First, the cost of the energy needed to operate the network is actually a significant part of the overall operational expense of many operators. Increased network energy efficiency is therefore one important tool to reduce the operational cost of a network.

Secondly, there are many places, especially in developing countries, where there is a need to provide mobile connectivity but where there is no easy access to the electrical grid. The typical way of providing power to infrastructure in such locations is by means of diesel generators, an inherently costly and complex approach. By improving the energy efficiency of infrastructure, primarily base stations, providing power by means of decently sized solar panels becomes a much more viable option.

It should be noted that high network energy efficiency is not just an issue for future 5G networks. Enhancing the network energy efficiency of currently available technologies and networks is at least as important, especially taking into account that, in terms of deployed infrastructure, currently available technologies will dominate for many years to come. However, the introduction of a new generation not constrained by backward compatibility opens up for new opportunities in terms of energy efficiency. On the other hand, the potential to enhance energy efficiency of existing technologies is partly constrained by the requirement to retain backward compatibility and support legacy devices.

23.2 5G AND IMT-2020

As described in Chapter 2, since the emergence of 3G the different generations of mobile communication have been tightly associated with so-called *IMT* technologies defined by ITU-R. ITU-R does not, by itself, develop any detailed technical specifications related to IMT. Rather, what ITU-R does is to specify the capabilities needed from a certain IMT technology and the requirements that the technology needs to fulfill. Actual technology is developed elsewhere, for example, within 3GPP, and then submitted to ITU-R as a *candidate IMT technology*. After evaluation against the specified requirements, the submitted candidate technology may be approved as an IMT technology.

Around year 2000 ITU-R defined the concept of *IMT-2000*. The 3G technologies WCDMA/HSPA, cdma2000, and TD-SCDMA were all submitted to ITU-R and subsequently approved as IMT-2000 technologies. Ten years later ITU-R defined the concept of *IMT-Advanced*. Two technologies, LTE and WiMax [67], were submitted to ITU-R and both were approved as IMT-Advanced technologies.[2] Among these, LTE is by far the most dominating.

In 2013 ITU-R initiated activities to define the next step of IMT, referred to as *IMT-2020*. In-line with IMT-2000 being associated with 3G and IMT-Advanced being associated with

---

[2]Strictly speaking, only LTE release 10 and beyond are approved as IMT-advanced technology although all releases of LTE are seen as 4G.

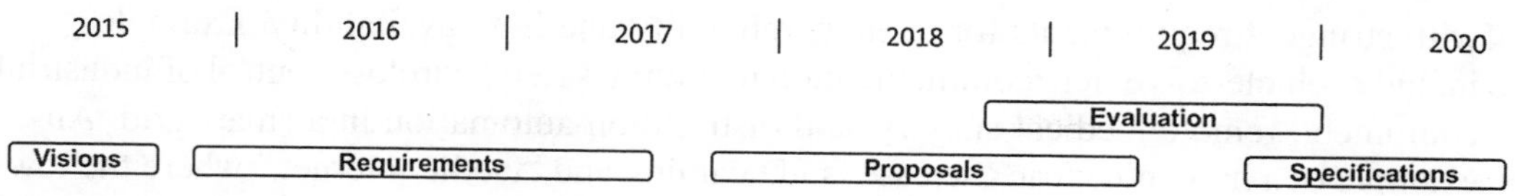

2015		2016		2017		2018		2019		2020

| Evaluation |

| Visions | | Requirements | | | Proposals | | Specifications |

FIGURE 23.1

ITU-R time plan for IMT-2020.

4G, IMT-2020 can be seen as being associated with 5G wireless access. The detailed ITU-R time plan for IMT-2020 was presented in Chapter 2 with the most important steps summarized in Figure 23.1.

The ITU-R activities on IMT-2020 started with the development of a "vision" document [63], outlining the expected use scenarios and corresponding required capabilities of IMT-2020. ITU-R is now in the process of defining more detailed requirements for IMT-2020, requirements that candidate technologies are then to be evaluated against. These requirements are targeted to be finalized mid-2017.

Once the requirements are finalized, candidate technologies can be submitted to ITU-R. The proposed candidate technology/technologies will then be evaluated against the IMT-2020 requirements and the technology/technologies that fulfill the requirements will be approved and published as part of the IMT-2020 specifications in the second half of 2020. Further details on the ITU-R process can be found in Section 2.3 of Chapter 2.

23.2.1 USAGE SCENARIOS FOR IMT-2020

With a wide range of new use cases being one principal driver for 5G, ITU-R has defined three usage scenarios that form a part of IMT vision recommendation [63]. Inputs from the mobile industry and different regional and operator organizations were taken into the IMT-2020 process in ITU-R WP5D, and were synthesized into the three scenarios:

- *Enhanced mobile broadband (EMBB)*: With mobile broadband today being the main driver for use of 3G and 4G mobile systems, this scenario points at its continued role as the most important usage scenario. The demand is continuously increasing and new application areas are emerging, setting new requirements for what ITU-R calls *enhanced mobile broadband*. Because of its broad and ubiquitous use, it covers a range of use cases with different challenges, including both hot spots and wide-area coverage, with the first one enabling high data rates, high user density and a need for very high capacity, while the second one stresses mobility and a seamless user experience, with lower requirements on data rate and user density. The enhanced mobile broadband scenario is in general seen as addressing human-centric communication.
- *Ultra-reliable and low-latency communications (URLLC)*: This scenario is intended to cover both human and machine-centric communication, where the latter is often referred to as critical machine-type communication (C-MTC). It is characterized by use cases

with stringent requirements for latency, reliability and high availability. Examples include vehicle-to-vehicle communication involving safety, wireless control of industrial equipment, remote medical surgery, and distribution automation in a smart grid. An example of a human-centric use case is 3D gaming and "tactile internet," where the low-latency requirement is also combined with very high data rates.

- *Massive machine-type communications (M-MTC)*: This is a pure machine-centric use case, where the main characteristic is a very large number of connected devices that typically have very sparse transmissions of small data volumes that are not delay sensitive. The large number of devices can give a very high connection density locally, but it is the total number of devices in a system that can be the real challenge and stresses the need for low cost. Due to the possibility of remote deployment of M-MTC devices, they are also required to have a very long battery life time.

The usage scenarios are illustrated in Figure 23.2, together with some example use cases. The three scenarios described in the preceding list are not claimed to cover all possible use cases, but they provide a relevant grouping of a majority of the presently foreseen use cases and can thus be used to identify the key capabilities needed for the next-generation radio-interface technology for IMT-2020. There will most certainly be new use cases emerging,

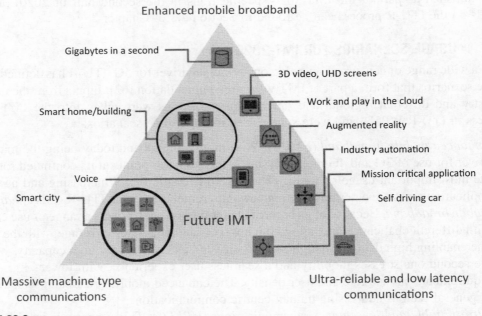

FIGURE 23.2

IMT-2020 use cases and mapping to usage scenarios.

From Ref. [63], used with permission from the ITU.

which we cannot foresee today or describe in any detail. This also means that the new radio interface must have a high flexibility to adapt to new use cases, and the "space" spanned by the range of the key capabilities supported should support the related requirements emerging from evolving use cases.

23.2.2 CAPABILITIES OF IMT-2020

As part of the development the framework for the IMT-2020 as documented in the IMT vision recommendation [63], ITU-R defined a set of capabilities needed for an IMT-2020 technology to support the 5G use cases and usage scenarios identified through the inputs from regional bodies, research projects, operators, administrations, and other organizations. There are a total of 13 capabilities defined in [63], where eight were selected as *key capabilities*. Those eight key capabilities are illustrated through two "spider web" diagrams, see Figures 23.3 and 23.4.

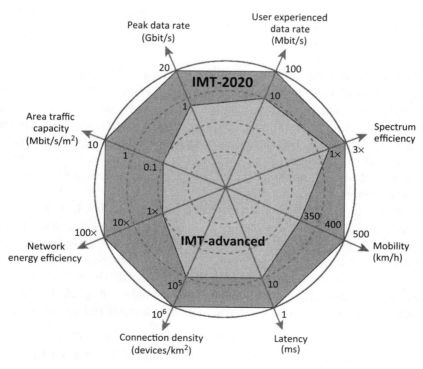

FIGURE 23.3

Key capabilities of IMT-2020.

From Ref. [63], used with permission from the ITU.

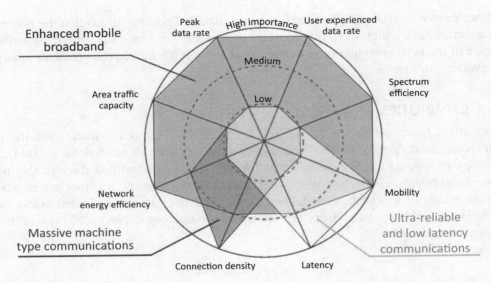

FIGURE 23.4

Relation between key capabilities and the three usage scenarios of ITU-R.

From Ref. [63], used with permission from the ITU.

Figure 23.3 illustrates the key capabilities together with indicative target numbers intended to give a first high-level guidance for the more detailed IMT-2020 requirements that are now under development. As can be seen the target values are partly absolute and partly relative to the corresponding capabilities of IMT-advanced. The target values for the different key capabilities do not have to be reached simultaneously, and some targets are to a certain extent even mutually exclusive. For this reason there is a second diagram shown in Figure 23.4 which illustrates the "importance" of each key capability for realizing the three high-level usage scenarios envisioned by ITU-R.

Peak data rate is a number which always has a lot of focus, but it is in fact quite an academic exercise. ITU-R defines peak data rates as the maximum achievable data rate under ideal conditions, which means that the impairments in an implementation or the actual impact from a deployment in terms of propagation, and so on does not come into play. It is a dependent *key performance indicator* (KPI) in that it is heavily depending on the amount of spectrum available for an operator deployment. Apart from that, the peak data rate depends on the peak spectral efficiency, which is the peak data rate normalized by the bandwidth:

$$\text{Peak data rate} = \text{System bandwidth} \times \text{Peak spectral efficiency}$$

Since large bandwidths are really not available in any of the existing IMT bands below 6 GHz, it is expected that really high data rates will be more easily achieved at higher frequencies. This leads to the conclusion that the highest data rates can be achieved in indoor and hot-spot environments, where the less favorable propagation properties at higher frequencies are of less importance.

The *user-experienced data rate* is the data rate that can be achieved over a large coverage area for a majority of the users. This can be evaluated as the 95th percentile from the distribution of data rates between users. It is also a dependent capability, not only on the available spectrum but also on how the system is deployed. While a target of 100 Mbit/s is set for wide area coverage in urban and suburban areas, it is expected that 5G systems could give 1 Gbit/s data rate ubiquitously in indoor and hot-spot environments.

Spectrum efficiency gives the average data throughput per Hz of spectrum and per "cell," or rather per unit of radio equipment (also referred to as *transmission reception point*, TRP). It is an essential parameter for dimensioning networks, but the levels achieved with 4G systems are already very high. The target was set to three times the spectrum efficiency target of 4G, but the achievable increase strongly depends on the deployment scenario.

Area traffic capacity is another dependent capability, which depends not only on the spectrum efficiency and the bandwidth available, but also on how dense the network is deployed:

$$\text{Area Traffic Capacity} = \text{Spectrum efficiency} \cdot \text{BW} \cdot \text{TRP density}$$

By assuming the availability of more spectrum at higher frequencies and that very dense deployments can be used, a target of a 100-fold increase over 4G was set for IMT-2020.

Network energy efficiency is, as already described, becoming an increasingly important capability. The overall target stated by ITU-R is that the energy consumption of the radio access network of IMT-2020 should not be greater than IMT networks deployed today, while still delivering the enhanced capabilities. The target means that the network energy efficiency in terms of energy consumed per bit of data therefore needs to be reduced with a factor at least as great as the envisaged traffic increase of IMT-2020 relative to IMT-advanced.

These first five key capabilities are of highest importance for the enhanced mobile broadband usage scenario, although mobility and the data rate capabilities would not have equal importance simultaneously. For example, in hot spots, a very high user-experienced and peak data rate, but a lower mobility, would be required than in wide-area coverage case.

Latency is defined as the contribution by the radio network to the time from when the source sends a packet to when the destination receives. It will be an essential capability for the URLLC usage scenario and ITU-R envisions that a 10-fold reduction in latency from IMT-advanced is required.

Mobility is in the context of key capabilities only defined as mobile speed, and the target of 500 km/h is envisioned in particular for high-speed trains and is only a moderate increase from IMT-advanced. As a key capability, it will however also be essential for the URLLC usage scenario in case of critical vehicle communication at high speed and will then be of high importance simultaneously with low latency. Note that mobility and high user-experienced data rates are not targeted simultaneously in the usage scenarios.

Connection density is defined as the total number of connected and/or accessible devices per unit area. The target is relevant for the M-MTC usage scenario with a high density of connected devices, but an EMBB dense indoor office can also give a high connection density.

In addition to the eight capabilities given in Figure 23.3 there are five additional capabilities defined in [63]:

- *Spectrum and bandwidth flexibility*
  Spectrum and bandwidth flexibility refers to the flexibility of the system design to handle different scenarios, and in particular to the capability to operate at different frequency ranges, including higher frequencies and wider channel bandwidths than today.
- *Reliability*
  Reliability relates to the capability to provide a given service with a very high level of availability.
- *Resilience*
  Resilience is the ability of the network to continue operating correctly during and after a natural or man-made disturbance, such as the loss of mains power.
- *Security and privacy*
  Security and privacy refers to several areas such as encryption and integrity protection of user data and signaling, as well as end-user privacy preventing unauthorized user tracking, and protection of network against hacking, fraud, denial of service, man in the middle attacks, and so on.
- *Operational lifetime*
  Operational life time refers to operation time per stored energy capacity. This is particularly important for machine-type devices requiring a very long battery life (e.g., more than 10 years) whose regular maintenance is difficult due to physical or economic reasons.

Note that these capabilities are not necessarily less important than the capabilities of Figure 23.3 despite that the later are referred to as "key capabilities." The main difference is that the key capabilities are more easily be quantifiable while the remaining five capabilities are more of qualitative capabilities that cannot easily be quantified.

23.2.3 STUDIES OF 5G IN REGIONAL AND OPERATOR GROUPS

As shown earlier, the driver for a new generation of mobile systems is this time not only an envisioned evolution of mobile-broadband services that would require higher data rates, lower delays, and a demand for higher capacity, but also new usage scenarios that could be of a more revolutionary nature. Such a development is forecasted in early research projects for the next generation, such as the European METIS project. Requirements from the mobile operators, as put forward by the *next-generation mobile network* (NGMN) alliance [77], also draws on similar new use cases and interaction with new industries as a basis for their requirements on the next generation. The studies have been input to the ITU-R as part of the work on IMT-2020.

The METIS project in Europe did early work on identifying what solutions were necessary for the next-generation radio access in [78]. While a further evolution of present networks can meet many new demands through an evolutionary approach, METIS identified challenges that would also require a disruptive approach for handling a "traffic explosion" in terms of increased use of mobile communication and in addition an extension to new application fields. Five main challenges were identified that each corresponded to specific scenarios:

- *"Amazingly fast"* is a scenario where instantaneous connectivity gives the user a "flash" behavior when using the mobile network for work or infotainment. The challenge is the very high data rates required, which also implies that very large data volumes will be exchanged.

- *"Great service in a crowd"* implies a scenario with wireless internet access in places with a high density of users in large crowds such as in a stadium. The challenge in such scenarios will be the high density of communicating devices, in addition to high data rates and data volumes.

- *"Ubiquitous things communicating"* is a scenario looking beyond human-centric communication, focusing on MTC, sometimes also called the internet of things (IoT). Such connected devices will often be simple, such as temperature sensors, but there may be many of them in massive deployments. The challenge will then be battery life time, cost, and just the large number of devices itself.

- *"Best experience follows you"* is a scenario envisioning a consistent and reliable high-quality user experience for the fully mobile user, whether you are at home, walking down the street, or traveling on a train. This could also apply to machine communication in, for example, vehicles. The challenge here is the mobility, in combination with the high-quality experience.

- *"Super real-time and reliable connections"* is a scenario targeting machine-to-machine communication, where a very low end-to-end latency must be guaranteed with a very high reliability. Examples are industrial applications and vehicle-to-vehicle communication involving safety. The challenge will be providing the low latency with very high probability.

These five scenario scenarios are not mutually exclusive, but give a broad coverage of possible applications for 5G.

The mobile operators group NGMN Alliance published a 5G white paper [77] where requirements on the next-generation mobile systems are analyzed for a total of 24 different use cases, divided into 14 categories and 8 families. The use cases map in general to scenarios similar to the ones put forward by the METIS project. One notable difference is that "broadcast" is put forward by NGMN as a separate use-case category. For each use-case category, a set of requirements are stated by NGMN.

Another operator group that also produced a 5G white paper is 5G Americas[3] [79], where five market drivers and use cases are identified for 5G. In addition to the already identified scenarios for internet of things and extreme mobile broadband, including gaming and extreme video, 5G Americas also stress the needs of public safety operations for mission-critical voice

---

[3]The organization was earlier known as *4G Americas*.

and data communications and the task for the 5G ecosystem to also replace the landline (PSTN) network with wireless broadband. One additional use case identified by 5G Americas is context-aware devices, as a new service model addressing the end user's need to find relevant information in the ever-increasing amount of available information.

There are also several regional groups that have provided input to the requirements for the next-generation mobile system, such as the *5G Forum* in Korea and the *ARIB 2020 Beyond AdHoc* in Japan, and the *IMT-2020(5G) Promotion Group* in China. The latter published a white paper [80] with a very similar overall vision as the one provided by operator and regional groups. It identifies a broader use of 5G that will penetrate every element of the future society, not only through an extended use of mobile broadband, but also by "connecting everything" and providing interconnection between people and things. One aspect that is stressed in particular is the sustainability of future networks, where the energy efficiency will be an essential parameter.

23.3 ONE VERSUS MULTIPLE TECHNOLOGIES: "NETWORK SLICING"

The very wide range of applications and use cases to be addressed by 5G wireless access has raised the question whether this should be achieved with a single 5G radio-access solution or if one should rather develop a set of radio-access solutions addressing different groups of applications.

Optimizing the radio-access solution toward a specific group of applications with a more limited requirement space may obviously lead to a more efficient solution for those specific applications. At the same time, there are clear benefits of being able to support an as wide range of applications as possible with the same basic technology and within a common pool of spectrum. Most importantly, there is still a high degree of uncertainty about what will really be the most important and economically most feasible new wireless applications in the 5G era. Developing a technology and deploying a system specifically targeting a limited group of applications thus implies a big risk from an operator point of view. By developing and deploying a technology that can be used for a wide range of different applications this risk will be significantly reduced.

One can make a parallel to the introduction of 3G roughly 15 years ago. At that time there was still a high degree of uncertainty about the actual potential of mobile broadband. However, the 3G technologies were also able to support voice service with high efficiency, something which in itself motivated the deployment of 3G networks. Mobile broadband could then be gradually introduced with limited extra investment.

Thus, there is relatively large consensus within the industry and especially among operators that the aim should be to develop a single flexible 5G radio-access solution that can address as many applications and use cases as possible.

In relation to this, the concept of *network slicing* has been introduced. As outlined in Figure 23.5, network slicing implies that virtualization techniques are used to create multiple

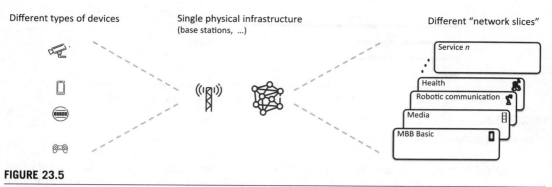

FIGURE 23.5

Network slicing creating multiple virtual networks for different applications and use cases on top of a common physical infrastructure and a common spectrum pool.

virtual networks, or *network slices*, on top of the same physical infrastructure and a common spectrum pool. As an example, one can create one network slice for mobile broadband, another network slice targeting massive-MTC applications, yet another network slice optimized for industry automation, and so on. Each network slice will, from the outside, appear as an independent network with its own resources and its own capabilities optimized for the set of applications targeted by the slice.

23.4 5G SPECTRUM

Spectrum is one the fundamental pillars of wireless communication, and the history of mobile communication has to a large extent been about extending the amount of available spectrum and introducing new technology allowing for more efficient utilization of the available spectrum.

23.4.1 EXPANSION INTO HIGHER-FREQUENCY BANDS

As illustrated in Figure 23.6, every generation of mobile communication has expanded the range of spectrum in which the wireless-access technology can operate into higher-frequency bands:

- The first-generation systems were limited to operation below 1 GHz.
- The second-generation systems were initially deployed below 1 GHz but later expanded into the 1.8/1.9 GHz bands.
- The initial deployment of 3G systems for the first time expanded mobile communication above 2 GHz, more specifically into the so-called *IMT core bands* around 2.1 GHz.
- LTE was first deployed in the 2.5 GHz band and has recently expanded to frequency bands as high as around 3.5 GHz.

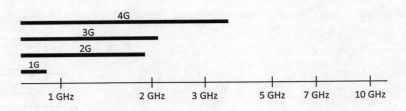

FIGURE 23.6

Expanded spectrum range from 1G to 4G.

It should be noted that the expansion to higher-frequency bands does in no way imply that the later generations cannot be deployed in lower-frequency bands. As an example, although LTE has expanded into higher-frequency bands above 3 GHz, there are LTE-based networks operating at as low frequencies as 450 MHz. The main benefit with operation in lower-frequency bands is better coverage allowing for a given area to be covered by less infrastructure (base stations). The expansion to higher-frequency bands, on the other hand, is mainly driven by a need for more spectrum providing higher system capacity to handle the continuously increasing traffic volumes.

The trend toward utilizing higher-frequency bands will continue and be even more pronounced in the 5G era. Already the first phase of 5G wireless access is expected to support operation in spectrum up to in the order of 30 GHz—that is, well into the millimeter wave range.[4] Later phases may expand this even further, up to 60—70 GHz and perhaps even further. Note that this implies a much larger step in terms of addressing new spectrum ranges, compared to earlier generational steps. From the first generation to the fourth generation, the upper limit on the frequency-band-of-operation expanded from just below 1 GHz to just above 3.5 GHz—that is, roughly a factor of 4. In comparison, already the first phase of 5G is expected to increase the upper limit on the frequency-band-of-operation close to a factor of 10 compared to the frequency range currently supported by LTE. A direct consequence of this is that, compared to earlier generational shifts, there is much more uncertainty in terms of spectrum characteristics when entering the 5G era.

Higher-frequency bands, especially beyond 10 GHz, has for long been assumed to be unsuitable for mobile communication due to very high-propagation loss and corresponding limited range. One reason for this has been an implicit assumption that the dimensions of the antenna configuration scale with the wave length, implying that operation in higher-frequency bands leads to much smaller effective antenna area and thus less captured received energy. At the same time, the smaller size of the antenna elements also enables the use of more antenna elements for a given size of the overall antenna configuration. By applying many small antenna elements at the receiver side, the overall effective receive antenna area can be kept constant avoiding the loss in captured energy. Another way to describe this is to say that the

---

[4]Strictly speaking, the mm-wave band starts at 30 GHz (10 mm wave length). However, already frequencies above 10 GHz, or in some cases already above 6 GHz, are in daily discussions often referred to as "mmw frequencies."

larger antenna area relative to the wavelength enables more extensive receiver-side beam-forming or, equivalently, higher effective antenna gain at the receiver side.

By assuming multi-antenna configurations and an associated possibility for beam-forming also at the transmitter side, one could even argue that, for a given physical size of the overall antenna configurations, operation in higher-frequency bands may actually allow for extended range, assuming line-of-sight propagation conditions. This is a main reason for the use of higher-frequency bands in the 10 GHz to 100 GHz range for point-to-point radio links. However, this is only valid for line-of-sight conditions. In real-life scenarios, with non-line-of-sight conditions, need for outdoor-to-indoor propagation, and so on, radio propagation is undoubtedly more challenging at higher-frequency bands above 10 GHz compared to, for example, operation in the 2 GHz band.

As an example, mobile communication relies heavily on diffraction—that is, the property that a radio wave "bends" around corners—to enable connectivity in non-line-of-sight locations. The amount of diffraction is reduced as the frequency-of-operation increases, making it more difficult to provide coverage in shadowed locations. Nevertheless, recent investigations [68] have shown that coverage up to a few 100 m is possible also in non-line-of-sight conditions at least up to roughly 30 GHz, assuming proper use of beam-forming. One reason is that the degraded diffraction at higher frequencies is at least partly compensated for by stronger reflections.

However, there are also other factors that impact the propagation and restrict the use of higher frequencies for mobile communication. One such factor is building penetration loss. Most base stations, including low-power base stations in dense deployments, are located outdoors. At the same time, most of the users are located indoors making good outdoor-to-indoor coverage essential for many deployments. However, the building penetration loss is typically frequency dependent with, in general, increasing penetration loss and, as a consequence, degraded outdoor-to-indoor coverage as the carrier frequency increases, see for example [81].

It should be noted that the building penetration loss depends on the type of building material. It may also be heavily impacted by the type of windows being used. Modern buildings, especially office buildings, are often equipped with so-called infrared reflective (IRR) glass windows for energy-saving reasons. However, such windows also have higher penetration loss, leading to further degraded outdoor-to-indoor coverage. Note that this is a general effect not just related to high-frequency operation.

Additional factors impacting propagation at higher frequencies include atmospheric attenuation, rain fade, foliage attenuation, and body loss. Although being very important, for example, for radio links, the first two are less relevant for the relatively short link distances envisioned for mobile communication on high frequencies. On the other hand, both the foliage attenuation and body loss are highly relevant in the mobile-communication scenario.

Another factor limiting the coverage at higher frequencies is regulations on allowed transmission power for frequencies above 6 GHz. As mentioned in Chapter 2, due to the present international regulations, the maximum transmit power at higher frequencies may be up to 10 dB lower than the maximum power levels for current cellular technologies. This may however change with future updates of regulation.

Altogether, this means that lower-frequency bands will remain the backbone of mobile communication also in the 5G era, providing 5G services with wide-area coverage. However, the lower-frequency bands will be complemented by higher frequencies, including frequency bands above 10 GHz, for very high traffic capacity and very high data rates but mostly in dense outdoor and indoor deployments.

23.4.2 LICENSED VERSUS UNLICENSED SPECTRUM

Mobile communication has since its inception relied solely on licensed spectrum exclusively assigned on a regional basis to a certain operator. However, this has partly begun to change already with LTE with the introduction of LAA (Chapter 17) providing the possibility for complementary use of unlicensed spectrum for enhanced service provisioning when the conditions so allow.

There is no reason to expect that this trend will not continue in the 5G era. Rather, everything speaks in favor of both licensed and unlicensed spectrum being key components for 5G wireless access. Licensed spectrum will remain the backbone, giving network operators the possibility to provide high-quality services with a high degree of guarantee. At the same time, unlicensed spectrum will be an important complement providing additional capacity and enabling even higher data rates when the interference conditions so allow. Consequently, the new 5G radio-access technology should already from the start support operation in both licensed and unlicensed spectrum. Operation in unlicensed spectrum should be possible in combination with and under assistance from licensed spectrum, similar to LAA, as well as standalone with no support from a licensed carrier.

From a technical perspective, this requires support for a listen-before-talk mechanism as discussed in Chapter 17. Most likely, the design will be similar to the LAA listen-before-talk to simplify coexistence with LAA and Wi-Fi deployed in the same band. However, the possibility for extensive use of multi-antenna techniques and beam-forming may impact the design, at least in scenarios with no Wi-Fi or LAA transmissions being present.

23.5 LTE EVOLUTION VERSUS NEW 5G TECHNOLOGY

The different generations of mobile communication have very much been defined based on some specific technical characteristics. For example, 3G was very much associated with the use of CDMA technology. Likewise, 4G is very much associated with OFDM transmission in combination with MIMO. At the same time, there is not a fundamental difference between the applications and use cases being supported by 3G, especially HSPA, and 4G.

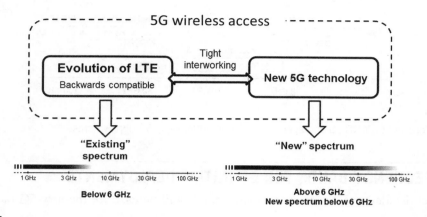

FIGURE 23.7

Overall 5G wireless-access solution consisting of the evolution of LTE in combination with a new 5G radio-access technology.

In contrast, the concept of 5G is much more associated with the kind of applications and use cases being envisioned, rather than a specific technology. More specifically, the term 5G is very much associated with the envisioned new use cases to be provided in the 5G era.

Nevertheless, it is clear that that many of the applications and use cases envisioned for the 5G era can be well supported by the evolution of LTE. Thus, there is a relatively well accepted view that the evolution of LTE should actually be seen as a part of the overall 5G wireless access solution, see Figure 23.7. The benefit of being able to provide a substantial part of the 5G applications and use cases by the evolution of LTE is that these applications and use cases can then be provided within existing spectrum while still supporting legacy devices in that spectrum (backward compatibility).

However, in parallel to the evolution of LTE there will also be development of new 5G radio-access technology not constrained by backward compatibility. This technology will at least initially target new spectrum both above and below 6 GHz. In a longer-term perspective, the new 5G radio-access technology may also migrate into spectrum currently used by other technologies including LTE.

The possibility for tight interworking between the evolution of LTE and the new 5G radio-access technology will in many cases be critical for introduction of the new radio-access technology. As an example, the new 5G radio-access technology may be deployed in a very dense layer operating on higher frequencies. Such a layer can support very large traffic volumes and very high end-user data rate. However, it will be inherently more un-reliable as devices may easily fall out of coverage of such layer. By providing the possibility for simultaneous connectivity to an LTE-based macro-layer operating on lower frequencies, the reliability of the overall connectivity can be dramatically improved.

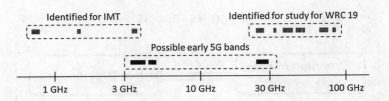

FIGURE 23.8

Spectrum identified by WRC 15 and spectrum considered for early 5G deployments.

23.6 FREQUENCY BANDS FOR 5G INITIAL DEPLOYMENTS

It is not yet decided exactly what frequency bands will be used for the new 5G radio-access technology.

As described already in Chapter 2, WRC-15 identified a set of frequency bands below 6 GHz as new frequency bands for IMT, see Figure 23.8. WRC-15 also identified a set of frequency bands above 10 GHz to be studied as potential new IMT spectrum for WRC-19. These bands are clearly candidates for the new 5G radio-access technology. However, initial deployment of the new 5G radio-access technology may also take place in other bands depending on decisions by regional/national regulators. As of today, there are primarily two frequency ranges being discussed in the context of early deployment of the new 5G radio-access technology, see Figure 23.8. They are:

- frequencies in the 3.3–4.2 GHz and 4.4–4.99 GHz range;
- frequencies in the 24.25–29.5 GHz range.

It should be noted that these frequency ranges only partly coincide with the spectrum bands identified at WRC'15 or being studied for WRC'19.

23.7 5G TECHNICAL SPECIFICATION

As mentioned already in Section 23.2 when discussing ITU-R and IMT-2020, the actual technical specification of the new 5G radio-access technology will be carried out by 3GPP, in parallel to the evolution of LTE.

Although fundamentally the new radio-access technology is developed in order to satisfy a need for new capabilities, the 3GPP 5G development also needs to take into account the ITU-R time schedule for IMT-2020 as outlined in Section 23.2. Initially, a study item in 3GPP will develop requirements for the new 5G radio-access technology and a parallel study item will develop the technology aspects. This goes on in parallel with the requirements phase in ITU-R, as shown in Figure 23.9.

To align with the ITU-R time plan (Figure 23.1) 3GPP needs to have a high-level technology description of the new 5G radio-access technology available in the second half of 2018 for submission to ITU-R as a candidate for IMT-2020. Detailed technical specifications

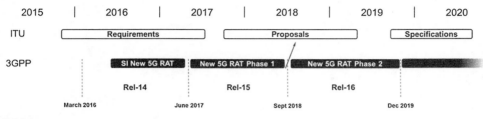

FIGURE 23.9

ITU-R time schedule and 3GPP phased approach to 5G.

must then be ready late 2019 to allow for inclusion in the IMT-2020 specifications to be published by ITU-R in the fall of 2020.

However, the 3GPP development of a new 5G radio-access technology is not only driven by the ITU-R time plane. Actually, it has become more and more evident that, in some countries/regions, there is a desire for the new 5G technology to be available even earlier than what is given by the ITU-R time plan. More specifically, some countries and regions have expressed a strong desire to have a new 5G technology in *commercial operation* in 2020. To allow for sufficient time for actual product development, this requires detailed technical specifications to be available already during 2018.

To satisfy these demands 3GPP has decided on a phased approach to 5G specification as outlined in Figure 23.9.

- A first phase with limited functionality but satisfying the desire for technical specifications available in 2018 thereby enabling commercial operation in 2020.
- A second phase fulfilling all the IMT-2020 requirements and being available in time for the ITU-R specification in 2020.

Although there have been extensive discussions about the two-phase 3GPP approach, it should be understood that the development of the new 5G radio-access technology will not end with phase 2. In the same way as any 3GPP technology there will be a sequence of releases, each adding additional features to the technology.

It can be noticed that the 3GPP technical work will start even before the ITU-R requirements for IMT-2020 have been finalized. However, already at the start of the 3GPP technical activities there is a relatively good understanding of where these requirements will end up. Furthermore, 3GPP does not rely solely on ITU-R to develop the requirements for the new 5G radio-access technology. Rather, 3GPP develops its own 5G requirements based on inputs from all the 3GPP members including operators, device and network vendors, and other organizations, for example, NGMN.

In order to ensure that that 5G radio-access technology will fulfill all the requirements on IMT-2020, the 3GPP requirements must include, but may be a superset of, the ITU-R requirements. Another way to express this is to say that one important requirement for the new 5G technology to be developed by 3GPP is that it has to fulfill all the requirements of IMT-2020 as defined by ITU-R.

NEW 5G RADIO-ACCESS TECHNOLOGY

24

As described in the previous chapter, the overall 5G wireless access solution will consist of the evolution of LTE in combination with a new 5G radio-access technology (5G RAT).

In this chapter some key design principles and main technology components relevant for the new 5G RAT are discussed. As the detailed specification of the new 5G RAT has not even started in 3GPP at the time of this writing, there is obviously still a high degree of uncertainty in the details of this future technology. However, on a higher level, including general design principles and basic technology components, there seems to be a relatively high degree of commonality in the views of the major players in the industry.

24.1 5G: SOME GENERAL DESIGN PRINCIPLES
24.1.1 RADIO-ACCESS EVOLUTION AND FORWARD COMPATIBILITY

After their initial introductions, every cellular radio-access technology has gone through a sequence of evolutionary steps adding new features that provide enhanced performance and new capabilities to the technology. Examples of these evolutions include the evolution of GSM to EDGE, the evolution of WCDMA to HSPA, and the different steps of the evolution of LTE. In general, these evolutionary steps have been backward compatible implying that a legacy device can still access the network on a carrier frequency supporting the new features, although it may obviously not be able to fully benefit from the features as such.

For the new 5G RAT, the possibility to evolve the technology beyond its initial release will be even more important:

- As described in the previous chapter, 5G is envisioned to support a wide range of different use cases, many of which are yet unknown. Thus, the radio-access solution will have to be able to evolve and adapt to encompass new requirements and new service characteristics.
- As also described in the previous chapter, there will be a phased approach to the 3GPP specification of the new 5G RAT, with the initial phase having a relatively limited scope and later evolution ensuring full compliance with all identified 5G requirements.

4G, LTE-Advanced Pro and The Road to 5G. http://dx.doi.org/10.1016/B978-0-12-804575-6.00024-8

The new 5G RAT is not required to be backward compatibility to earlier generations. However, similar to earlier generations, the future evolution of the technology should be backward compatibility to its initial release. In order to minimize the constraints on this future evolution the concept of *forward compatibility* has been introduced as an additional requirement for the design of the new 5G RAT. In this context, forward compatibility simply means that the design of the radio-access technology should be such that the constraints on the future evolution of the technology, due to the requirement on retained backward compatibility to its initial release, will be as limited as possible.

Due to the obvious uncertainty of the characteristics and requirements of new, yet unknown applications and use cases, as well as the uncertainty of future technology directions, forward compatibility is inherently difficult to achieve. However, as described in the following sections, there are certain principles that, if followed, will at least enhance forward compatibility.

24.1.2 ULTRA-LEAN DESIGN: MINIMIZE "ALWAYS-ON" TRANSMISSIONS

For any cellular technology there are certain transmissions carried out regularly from every network node regardless of whether or not there is any ongoing user-data transmissions and even if there are no active devices at all within the coverage of the node. In the context of LTE such "*always-on transmissions*" include

- the primary and secondary synchronization signals;
- the cell-specific reference signals;
- the broadcast system information (MIB and SIBs).

The opposite of always-on transmissions are "*on-demand transmissions*"—that is, transmissions that can be initiated and deactivated on a per-need basis.

In high-traffic scenarios with a high traffic load per network node, which is the typically assumed scenario when evaluating cellular radio-access technologies, the always-on transmissions contribute only a relatively small fraction of the total node transmission and thus have relatively small impact on the overall system performance. However, in real-life cellular networks a large fraction of the network nodes is actually relatively lightly loaded on average:

- Especially in suburban and rural areas, network infrastructure (base stations) is typically deployed to provide coverage with a certain minimum end-user data rate and not because more network capacity is needed to handle the traffic volumes.
- Even when deployment of new infrastructure is driven by a need for more network capacity the deployment must be dimensioned to handle peak-traffic volumes. As traffic volumes typically vary significantly in time, the *average* load per network node will still be relatively low.

Furthermore, with data traffic typically being very bursty, even at relatively high load a fairly large number of subframes are actually not carrying any traffic.

As a consequence, in real-life networks always-on transmissions often have bigger impact on the overall system performance than what can be seen from "high-load" evaluations:

- The always-on transmissions will add to the overall system interference, thereby reducing the achievable data rates.
- The always-on transmissions will increase the overall network energy consumption, thereby limiting the network energy efficiency.

Expressed differently, minimizing the amount of always-on transmissions is an important component to enable very high achievable data rates and very high network energy efficiency.

The LTE small-cell on/off mechanism described in Chapter 15 is a step in the direction of minimizing the always-on transmissions. However, the small-cell on/off mechanism is still constrained by backward compatibility and the requirement that legacy LTE devices should still be able to access the carrier. The introduction of a new 5G RAT not constrained by backward compatibility to earlier technologies provides additional opportunities in this respect.

Minimizing the amount of always-on transmissions is also one important component for forward compatibility. As legacy devices expect the always-on transmissions to be present, such transmissions cannot be removed or even modified without impacting legacy devices and their ability to properly access the system.

The relation between always-on transmissions and forward compatibility is well illustrated by the use of MBSFN subframes to enable relaying functionality in LTE, see Chapter 18. Although introduced in order to provide support for MBSFN transmission for MBMS services, the MBSFN subframes have turned out to be very valuable for the evolution of LTE in general. The reason for this is simply that the MBSFN subframes include substantially less cell-specific reference symbols compared to normal subframes. By configuring subframes as MBSFN subframes from a legacy-device point of view, these subframes can be used for any kind of new transmissions without breaking backward compatibility. In the specific case of relaying, the almost-empty MBSFN subframes made it possible to create sufficiently large "holes" in the downlink transmission to enable in-frequency reception on the backhaul link. Without the MBSFN subframes this would not have been possible with retained backward compatibility as too many OFDM symbols would have included always-on cell-specific reference symbols. The key thing that allowed for this was that MBSFN subframes have been part of LTE specifications already since the first release implying that they are "understood" by all legacy devices. In essence, the baseline assumption from a device perspective on the new 5G RAT should be to treat each subframe as empty (or undefined) unless it has been explicitly instructed to receive or transmit.

Minimizing always-on transmissions can be seen as part of a more high-level *ultra-lean design principle* often expressed as *minimize all network transmissions not directly related to user-data transmission* as illustrated in Figure 24.1. The aim is, once again, to enable higher achievable data rates and enhance the network energy efficiency.

FIGURE 24.1

Ultra-lean transmission.

24.1.3 STAY IN THE BOX

The "stay-in-the-box" principle in essence says that a transmission should be kept together as illustrated in the right part of Figure 24.2 and not be spread out over the resource space (the time–frequency grid in case of OFDM) as shown in the left part of the figure. The aim is once again to enable a higher degree of forward compatibility. By keeping transmissions together, it is easier to later introduce new types of transmissions in parallel to legacy transmissions while retaining backward compatibility.

An example of LTE transmissions not fulfilling the "stay-in-the-box" principle is the set of physical channels (PCFICH, PHICH, and PDCCH) transmitted in the control region of each LTE subframe. As described in Chapter 6, each PDCCH/PHICH/PCFICH transmission is spread out over the entire carrier bandwidth in an apparently random way. The main reason for this structure was to enable a high degree of frequency diversity and to achieve randomization between transmissions. However, this also makes it very difficult to introduce new transmissions within the LTE control region unless they are fully aligned with the current control-channel structure. This is, for example, illustrated by the design of NB-IoT (Chapter 20) for which the solution was simply to have NB-IoT downlink transmissions avoiding the entire control region in case of inband deployment.

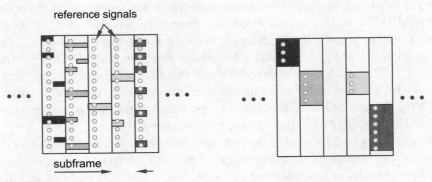

FIGURE 24.2

Illustration of signals spread out (left) vs. "stay-in-the-box" (right).

In contrast, the LTE EPDCCH is much more aligned with the "stay-in-the-box" principle. Each EPDCCH is contained within a single, or at most a few resource blocks making it much more straightforward to introduce new transmissions in parallel to the EPDCCH.

Related to the "stay-in-the-box" principle is the concept of *self-contained transmissions*. Self-contained transmission implies that, to the extent possible, the signals and information required for reception of data in a given beam and subframe are contained within the same beam and subframe. Although this approach provides a great deal of forward compatibility, some aspects, such as channel-estimation accuracy, may benefit from the possibility of exploiting "outside-the-box" signals and the final design must carefully take this into account.

24.1.4 AVOID STRICT TIMING RELATIONS

Another important design principle is to avoid static and strict timing relations across subframe borders as well as between different transmission directions.

An example of such static and strict timing relations is the LTE uplink hybrid-ARQ procedure where downlink acknowledgments as well as potential uplink retransmissions occur at fixed predefined time instants relative to the initial uplink transmission. These kinds of timing relations make it more difficult to introduce new transmission procedures that are not aligned in detail with the legacy timing relation. For example, the original LTE uplink hybrid-ARQ timing is not well matched to unlicensed spectrum and changes are required as discussed in Chapter 17. Avoiding strict and static timing relations is thus an important component for forward compatibility.

Strict and static timing relations may also prevent a radio-access technology to fully benefit from technology advances in processing capability. As an example, the LTE hybrid-ARQ timing was specified based on the processing capabilities, primarily in terms of channel decoding, estimated to be available at the time of the first commercial deployments of LTE. With time, the processing capabilities have advanced, allowing for faster decoding. However, the static timing relation of the uplink hybrid-ARQ protocol prevents this from being turned into a shorter hybrid-ARQ round-trip time and an associated lower latency.

24.2 5G: KEY TECHNOLOGY COMPONENTS

In the remainder of the chapter various technology components considered for 5G radio access are discussed, taking the design principles into account.

24.2.1 WAVEFORM

24.2.1.1 Scalable OFDM

At the core of the LTE RAT is an OFDM-based transmission scheme with a subcarrier spacing of 15 kHz and a cyclic prefix of about 4.7 μs.[1] This is true for both downlink and uplink,

---

[1] As described in Chapter 5 there is also an extended cyclic prefix of about 16.7 μs.

Table 24.1 Example of Scalable OFDM Numerology

	Baseline	Higher-Order-Derived Numerologies	
Scale factor	1	4	32
Subcarrier spacing	15 kHz	60 kHz	480 kHz
Symbol time (excl. CP)	66.7 µs	16.7 µs	2.1 µs
Cyclic prefix	4.7 µs	1.2 µs	0.15 µs
Subframe	500 µs	125 µs	15.6 µs

although additional DFT precoding is applied for uplink data (PUSCH) transmission to enable higher power-amplifier efficiency on the device side.

OFDM is the main candidate also for the new 5G RAT, both in uplink and downlink. Having the same waveform in both directions simplifies the overall design, especially with respect to wireless backhauling and device-to-device communication. However, in light of the very wide range of spectrum, deployment types, and use cases to be addressed by 5G, it is unrealistic to assume that a single OFDM numerology should be sufficient for the new 5G RAT.

As discussed in Chapter 23, 5G radio access should cover a very wide range of frequencies, from below 1 GHz up to at least several 10 GHz, and possibly as high as 70—80 GHz. For the lower part of the spectrum, perhaps as high as 5 GHz, a subcarrier spacing of the same order as LTE is sufficient. However, for higher frequencies, a larger subcarrier spacing is needed in order to ensure sufficient robustness to, especially, phase noise with reasonable cost and power consumption for mobile devices.

The new 5G RAT should also be able to operate in a wide range of deployment scenarios, ranging from extremely dense indoor and outdoor deployments to sparse rural deployments where each network node may cover a very large area. In the latter case a cyclic prefix similar to LTE or perhaps even larger is needed to handle large delay spread while, in the former case, a smaller cyclic prefix is sufficient.

The wide-area deployments for which a larger cyclic prefix is needed will typically operate at lower frequencies for which a lower subcarrier spacing is sufficient. At the same time, higher frequencies, for which a larger subcarrier spacing is needed, will typically be limited to denser deployments where a smaller cyclic prefix is sufficient. This speaks in favor of a *scalable* OFDM framework with different numerologies derived by frequency/time-domain scaling of a common baseline numerology.

Table 24.1 shows an example of such scaled OFDM numerology using the LTE numerology as baseline and with two derived higher-order numerologies based on the scale factors 4 and 32, respectively.[2] In this case, the baseline numerology with 15 kHz subcarrier spacing

---

[2]Note that in Table 24.1. we have redefined the term "subframe" to mean seven OFDM symbols, corresponding to what in LTE is referred to as a "slot."

and a cyclic prefix of about 4.7 µs would be appropriate for wide-area deployments using lower-frequency spectrum.

In higher-frequency spectrum, the higher-order numerologies with larger subcarrier spacing should be used to ensure high robustness to, especially, phase noise. With a scaled numerology, this inherently leads to smaller cyclic prefix. However, as already mentioned, the use of higher-frequency spectrum will be limited to dense deployments with less delay spread. Furthermore, higher frequencies will typically be used in combination with extensive beam-forming. This will reduce the amount of large-delay reflections, thereby further reducing the delay spread of the received signal.

The benefit with a scaled numerology derived from a common baseline numerology, rather than a set of more independent numerologies, is that a main part of radio-interface specification can be made agnostic to the exact numerology. This also means that one could more easily introduce additional numerologies at a later stage by simply introducing additional scale factors.

Note that the higher-order numerologies with larger subcarrier spacing and smaller cyclic prefix could also be used at lower frequencies as long as the delay spread is limited. Thus, the higher-order numerologies could be used also for low-frequency dense deployments. One reason for doing this would be to enable further reduced latency also at lower frequencies by utilizing the shorter symbol time and corresponding shorter subframe of the higher-order numerologies.

It should be noted that one could use a larger subcarrier spacing also for wide-area deployments with larger delay spread by using an extended cyclic prefix similar to LTE. Assuming the LTE-derived numerology of Table 24.1, for the 4× numerology such an extended cyclic prefix would be roughly 4.2 µs (16.7/4). The drawback would be a significantly higher cyclic-prefix overhead. However, such large overhead could be justified under special circumstances.

It is important to point out that the set of numerologies provided in Table 24.1 is an example. One could obviously consider different scale factors than those of Table 24.1 One could also use a different baseline numerology not aligned with LTE. The use of an LTE-based baseline numerology provides benefits though. One example is that it would allow for deployment of an NB-IoT carrier (Chapter 20) within a carrier of the new 5G RAT in similar way as NB-IoT can be deployed within an LTE carrier. This can potentially be an important benefit for an operator migrating from LTE to the new 5G RAT on a carrier while maintaining support for existing NB-IoT-based massive-MTC devices as such devices typically have a very long lifespan.

24.2.1.2 Spectral Shaping

OFDM subcarriers have large sublobes due to the use of a rectangular pulse shape. Thus, orthogonality between subcarriers is not due to true spectral separation but rather due to the exact structure of each subcarrier. One implication of this is that orthogonality between different transmissions is only retained if the different transmissions are received time aligned within the cyclic prefix. In the LTE uplink this is achieved by devices updating their transmit timing based on timing-advance commands provided by the network initially as part of the random-access response (Chapter 11) and subsequently as MAC control elements on the DL-SCH (Chapter 7).

Although having uplink orthogonality relying on time alignment is, in general, a good approach, it has certain limitations:

- There must be regular uplink transmissions in order for the network to be able to estimate the uplink timing and provide timing-advance commands when needed.
- The need to establish time-alignment before user-data transmission can be initiated leads to additional delay in the initial access and prevents immediate data transmission from a not-yet synchronized device.

Another implication of the large sidelobes of OFDM subcarriers is that a relatively large guard band is needed in case of frequency multiplexing of an OFDM signal with a signal of a different structure. The latter could be a non-OFDM signal. However, it could also be an OFDM signal with a different numerology, for example, a different subcarrier spacing or just a different cyclic prefix.

To overcome these issues, one may consider modifications and/or extensions to OFDM that leads to a higher degree of spectrum confinement of the transmitted signal.

Filter-bank multi-carrier (FBMC) [69] is another kind of multi-carrier transmission, where spectrum shaping by means of filtering is applied to each subcarrier. In essence, the rectangular pulse shape of OFDM is, with FBMC, replaced by a nonrectangular pulse shape with a corresponding more confined frequency response. In order to retain orthogonality between modulated subcarriers, FBMC has to be used in combination with offset-QAM (OQAM) modulation on each subcarrier, rather than conventional QAM modulation as used, for example, in LTE. FBMC provides significantly more confined spectrum, with smaller per-subcarrier sidelobes, compared to conventional OFDM. However, there are also several drawbacks/issues with FBMC.

Although necessary to retain orthogonality between subcarriers, the use of OQAM modulation leads to difficulties with channel estimation, especially in combination with MIMO transmission. There are proposed solutions; however, these lead to degraded receiver performance and additional reference-signal overhead [75]. Another drawback of FBMC is that tight filtering in the frequency domain inherently leads to long pulses in the time domain, with FBMC pulse shaping typically having a length of several symbols. To avoid interference between transmission bursts, a corresponding guard time of a length corresponding to several symbols is needed between each burst. In case of transmission bursts consisting of many symbols, this guard-time overhead will be relatively small. However, for low-latency transmissions, short-burst transmission is needed leading to potentially large overhead.

A different way to improve the spectrum confinement, while maintaining the OFDM structure, is to apply filtering of the entire OFDM signal, see upper part of Figure 24.3. Such

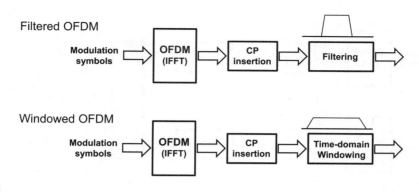

FIGURE 24.3

Filtered OFDM (upper) and windowed OFDM (lower).

filtering will be much less drastic than per-subcarrier filtering, with much less time-domain spreading of the signal. In practice, the filtering will simply use part of the cyclic prefix, making less cyclic prefix available to handle delay spread on the channel.

Alternatively, time-domain windowing can be used to control the spectral properties instead of filtering, see lower part of Figure 24.3. While filtering implies multiplication with a frequency response in the frequency domain or, equivalently, convolution with a time response in the time domain, windowing implies multiplication of the OFDM symbol with a window function in the time domain (convolution in the frequency domain).

In the end, these two approaches (filtering and windowing) lead to very similar result in terms of spectral confinement of the transmitted signal. However, windowing may be associated with somewhat less implementation complexity compared to filtering.

Filtering/windowing of the entire OFDM signal is very much an implementation issue and is actually, in practice, done for LTE already today in order to ensure that the transmitted OFDM signal fulfills the out-of-band-emissions requirements. However, filtering/windowing could also be used to spectrally confine certain parts of a carrier. This could, for example, be used to create "holes" in the spectrum to make room for other non-OFDM transmission. It could also allow for mixing different OFDM numerologies within one carrier, as illustrated Figure 24.4. The later could, for example, be beneficial when different services with different requirements are to be mixed on one carrier. The different numerologies could, for example, correspond to different subcarrier spacing (the case illustrated in Figure 24.4). In this case, the smaller subcarrier spacing could correspond to a conventional mobile-broadband service while the higher subcarrier spacing, which allows for lower latency, could correspond to a latency-critical service. However, the different numerologies could also correspond to the same subcarrier spacing with different cyclic prefix.

24.2.1.3 Low-PAPR Transmission

OFDM with DFT precoding is used for the LTE uplink in order to reduce the cubic metric (CM) [10] of the transmitted signal, thereby enabling higher power-amplifier efficiency on

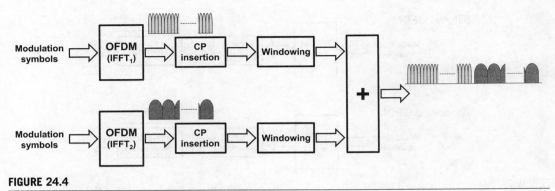

FIGURE 24.4

Mix of numerologies on one OFDM carrier using windowing.

the device side. The drawback of DFT precoding as it is done in LTE is that it limits the flexibility of the transmission. This is apparent, for example, in the design of the uplink reference signals and the design of the uplink control signaling (PUCCH) which are substantially less flexible and at least in some respects more complex than the corresponding downlink transmissions. From this point of view, it would be desirable to avoid DFT precoding, in the way that it is used in LTE, for the new 5G RAT. However, without the possibility for low-CM uplink transmission the new 5G RAT may have an uplink coverage disadvantage compared to LTE when operating in similar spectrum. This could negatively impact the migration of the new 5G RAT into spectrum currently used by LTE. Furthermore, for operation at very high frequencies, especially above 30–40 GHz, high power-amplifier efficiency is even more important compared to operation in lower-frequency spectrum:

- Operation in such high-frequency spectrum will typically be associated with a large number of antennas and, consequently, a large number of power amplifiers, especially at the base-station side.
- The small dimension and tight packing of electrical components at high frequencies make it more difficult to handle of excess heat generated due to power-amplifier inefficiency.

It should be noted that these arguments are relevant also for the base station. Thus, for such high frequencies the possibility for high power-amplifier efficiency on the base-station side may be as important as high power-amplifier efficiency on the device side.

Thus, at this stage one cannot discard the possible need for CM-reduction techniques on top of OFDM for the new 5G RAT. This could be in form of *complementary* DFT precoding that would be applied to data transmission when low CM is of essence, that is, in coverage-limited scenarios. However, it could also be in form of other CM-reduction techniques that can be added on top of OFDM such as, for example, tone reservation [8].

Also, as indicated earlier, any CM-reducing techniques should not only be considered for the uplink (device transmission) but also for the downlink (base-station transmission).

24.2.2 FLEXIBLE DUPLEX

LTE supports both FDD- and TDD-based duplex arrangements in order to match the existence of both paired and unpaired cellular spectra.

Both paired and unpaired spectra will exist also in the 5G era. Consequently, the new 5G RAT will have to support both FDD- and TDD-based duplex arrangement.

As discussed in the previous chapter, the new 5G radio access will cover a very wide range of frequencies, from below 1 GHz up to at least several 10 GHz, see also Figure 24.5. In the lower part of this spectrum, paired spectrum with FDD-based duplex arrangement will most likely continue to dominate. However, for higher frequencies which due to propagation constraints will be limited to dense deployments, unpaired spectrum with TDD-based duplex arrangement is expected to play a more important role.

24.2.2.1 Dynamic TDD

One benefit of unpaired spectrum with TDD-based duplex arrangement is the possibility to dynamically assign transmission resources (time slots) to different transmission directions depending on the instantaneous traffic conditions. This is especially beneficial in deployments and scenarios with more variable traffic conditions, which will, for example, be the case for dense deployments where each network node covers only a very small area.

One of the main concerns with unpaired spectrum and TDD operation has always been the possibility/risk for direct base-station-to-base-station and device-to-device interference, see Figure 24.6. In current commercial TDD-based cellular systems such interference is typically avoided by a combination of mutual time alignment between base stations and the use of the same downlink/uplink configuration in all cells. However, this requires a more or less static assignment of transmission resources preventing the resource assignment to adapt to dynamic traffic variations, thus removing one of the main benefits of TDD.

The eIMTA feature introduced in LTE release 12 (Chapter 15) is one step toward a more flexible assignment of TDD transmission resources to downlink and uplink. Nevertheless, a new 5G radio access should go even further, allowing for more or less fully dynamic assignment of transmission resources to the different transmission directions. However, this

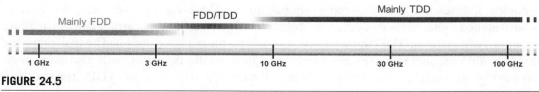

FIGURE 24.5

Typical duplex methods for different frequency bands.

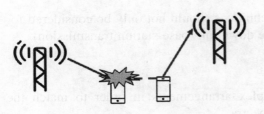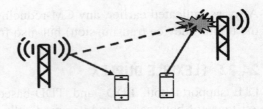

Device-to-device interference   **Base-station-to-base-station interference**

FIGURE 24.6

Direct base-station-to-base-station and device-to-device interference in case of TDD.

will then create a situation with potential base-station-to-base-station and device-to-device interference.

What makes base-station-to-base-station and device-to-device interference special and potentially much more severe, compared to the base-station-to-device and device-to-base-station interference that occur in any cellular system regardless of the duplex arrangement, is the difference in transmission characteristics between base stations and devices. This is especially the case for wide-area-covering ("macro") deployments where

- base stations have high transmit power, are located at elevated positions ("above roof top"), and are often transmitting with a high duty cycle serving many active devices;
- devices have much lower transmit power, are typically located indoor or outdoor on street level and are typically transmitting with, on average, a relatively low duty cycle.

However, in the future there will also be many very dense deployments, especially at higher frequencies. In such cases, the transmission characteristics of base stations and devices will be more similar:

- Compared to wide-area deployments, base stations in dense deployments will have a transmit power more similar to the transmit power of devices.
- Base stations in dense deployments are deployed indoor and outdoor on street level, that is, similar to device locations.
- Base stations in dense deployments will typically operate with, on average, lower duty cycle due to more dynamic traffic variations.

As a consequence, in dense deployments the base-station-to-base-station and device-to-device interference will be more similar to the base-station-to-device and device-to-base-station interference occurring regardless of the duplex arrangement, making dynamic assignment of TDD transmission resources a more viable option. The instantaneous traffic conditions, including the downlink vs. uplink traffic demands, will also vary more extensively in such deployments, making dynamic assignment of TDD transmission resources more beneficial.

It is important to understand that supporting fully dynamic TDD does not mean that the transmission resources should always be dynamically assigned. Especially when using the new 5G RAT for more wide-area deployments in unpaired spectrum on lower frequencies, the typical situation would be a synchronized deployment with the downlink/uplink configuration aligned between cells. The key thing is that the new 5G RAT *should allow for* fully dynamic assignment of transmission resources when operating in unpaired spectrum, leaving it to the network operator to decide what to use in a given deployment.

24.2.2.2 What About Full Duplex?

There have recently been different proposals for "true" full-duplex operation [70]. In this context, full-duplex operation means that transmission and reception is carried out *at the same frequency at the same time.*[3]

Full-duplex operation obviously leads to very strong "self" interference from the transmitter to the receiver, an interference that needs to be suppressed/cancelled before the actual target signal can be detected.

In principle, such interference suppression/cancellation is straightforward as the interfering signal is in principle completely known to the receiver. In practice, the suppression/cancellation is far from straightforward due to the enormous difference between the target signal and the interference in terms of received power. To handle this, current demonstrations of full-duplex operation rely on a combination of spatial separation (separate antennas for transmission and reception), analog suppression, and digital cancellation.

Even if full duplex would be feasible in real implementation, its benefits should not be overestimated. Full duplex has the potential to double the link throughput by allowing for continuous transmission in both directions on the same frequency. However, there will then be two simultaneous transmissions, implying increased interference to other transmissions, something which will negatively impact the overall system gain. The largest gain from full duplex can therefore be expected to occur in scenarios with relatively isolated radio links.

One scenario where full duplex is more likely to be beneficial is the wireless backhaul scenario—that is, wireless connectivity between base stations:

- Backhaul links are in many cases more isolated, compared to conventional base-station/device links.
- The extensive receiver complexity associated with full duplex may be more feasible to include in backhaul nodes, compared to conventional mobile devices.
- For backhaul nodes it is easier to envision more spatial separation between transmitting and receiving antennas, relaxing the requirements on active interference suppression.

When talking about full duplex one is typically assuming full duplex *on link level*—that is, simultaneous transmission in both directions on a bidirectional base-station/device link (left part of Figure 24.7). However, one can also envision full duplex on *cell level*, where a base

---

[3]Not to be up mixed with *full-duplex FDD* as used in LTE.

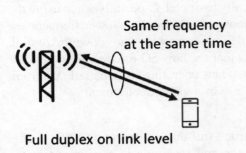
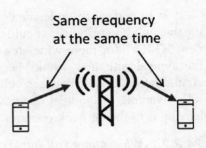

Same frequency
at the same time

Same frequency
at the same time

Full duplex on link level

Full duplex on cell level

FIGURE 24.7

Full duplex on link level vs. cell level.

station transmits to one device and is simultaneously receiving/detecting the transmission *of another* device on the same frequency (right part of Figure 24.7). The benefit of full duplex on cell level, compared to full duplex on link level, is that simultaneous same-frequency transmission and reception does not need to be supported on the device side. Furthermore, one can typically achieve a higher degree of spatial separation between transmitting and receiving antennas on the base-station side, relaxing the requirements on active interference suppression.

24.2.3 FRAME STRUCTURE

The new 5G RAT should have a frame structure that supports operation in both paired and unpaired spectrum, and in licensed as well as unlicensed spectrum. The frame structure should also be applicable to device-to-device (sidelink) connectivity, see Section 24.2.9.2.

The frame structure is a key factor to enable low latency over the radio interface. To enable a low latency, there is, for example, need for a short TTI and, consequently a short subframe. As an example, to enable the required 1 ms end-to-end latency, a subframe in the order 200 μs or less is needed. Note that this is aligned with the higher numerologies of Table 24.1.

Low link-level latency also requires the possibility for fast demodulation and decoding of data. To enable this, control information needed by the receiver for demodulation and decoding of the data within a subframe should be located at the beginning of the subframe, similar to the PDCCH of LTE.[4] Reference signals for channel estimation should also be located early in the subframe. In this way demodulation and decoding can start as early as possible without having to wait until the entire subframe has been received. Figure 24.8 illustrates a relatively generic frame structure fulfilling the above requirements. In case of downlink data transmission (upper part of Figure 24.8), control information (scheduling

---

[4]Note that the control signaling should still follow the "stay-in-the-box" principle of Section 24.1.3 and be transmitted jointly with the data, in contrast to the PDCCH which is spread out over the entire frequency domain.

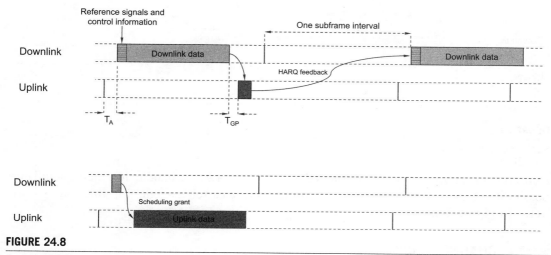

FIGURE 24.8

Generic 5G frame structure (TDD assumed in the figure).

assignments) and reference signals are located at the beginning of the downlink subframe, enabling early start of demodulation and decoding. In case of TDD operation, the downlink transmission ends prior to the end of the downlink subframe interval. Assuming decoding can be finalized during the guard period for the downlink-to-uplink switch, a hybrid-ARQ acknowledgment may then be transmitted on the uplink already in the last part of the downlink subframe interval, enabling very fast retransmissions. In the example of Figure 24.8, retransmissions occur with only one subframe delay.[5] This can be compared to LTE where there is a roughly three-subframe latency between the downlink transmission and the hybrid-ARQ acknowledgment and typically eight subframes between retransmissions. In combination with the shorter subframe, the outlined frame structure would thus enable substantially reduced hybrid-ARQ round-trip time compared to LTE.

Uplink data transmissions can be handled in a similar way; the scheduling grant is transmitted at the beginning of the subframe and the corresponding uplink data fills the remaining part of the uplink subframe as shown at the bottom of Figure 24.8. Thus, the scheduling decisions controls the "direction" of the data transmission, uplink or downlink, at a given point in time, resulting in a dynamic TDD scheme as discussed in Section 24.2.2.

Although 5G should enable very low latency in the order of 1 ms, many applications are not that latency critical. At the same time, the short transmissions needed to enable that kind of very low latency leads to inefficiencies in terms of, for example, control overhead. It may also lead to degraded channel estimates assuming that only reference signals within the set of subframes corresponding to the data transmission can be used for channel estimation

---

[5]Although one could in principle envision a base station decoding data as fast as the device, deployment aspects such as the use of remote radio units implies that retransmission earlier than shown in the figure cannot be generally assumed.

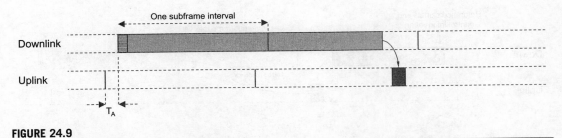

FIGURE 24.9

Subframe aggregation.

(self-contained transmissions as discussed in Section 24.1.3). To handle this, there should be a possibility to dynamically aggregate multiple subframes for a single transmission as illustrated in Figure 24.9

24.2.4 CHANNEL CODING

LTE uses Turbo coding for data transmission—see, for example, Chapter 6. Turbo coding is one candidate for channel coding also for the new 5G RAT. However, there are also other channel-coding approaches being considered, including low-density parity check (LDPC) codes [72] and Polar codes [73].

LDPC codes have been around for a relatively long time. They are block codes based on sparse ("low-density") parity matrixes with decoding based on iterative message-passing algorithms. In contrast, Polar codes have emerged much more recently. Their main fame comes from the fact that they are the first known structured codes that reach the Shannon bound.

In terms of basic performance, that is, the required E_b/N_0 needed for a certain error rate, the difference in performance between the different coding schemes is relatively small for medium to large sizes of the block of bits to be decoded (in the order of 1000 bits and beyond). On the other hand, for smaller block lengths, Polar codes currently seem to have a slight advantage. Polar codes also benefit from not having an error floor, something that makes it easier to achieve a very low error rate after decoding. In contrast, Turbo codes and, in many cases, also LDPC codes have such error floor. This may speak in favor of Polar codes for high-reliability applications with small payloads but requiring extremely good error performance.

The main issue with Polar codes is their relative immaturity which also implies that the implementation experience of Polar codes is much more limited compared to both Turbo codes and LDPC codes. The latter have, as already mentioned, been around for many years and are already part of widespread wireless technologies. The immaturity and lack of implementation experience of Polar codes implies that there may be practical issues not yet discovered. It also implies that it is more difficult today to fully understand the implementation complexity of Polar codes.

The main benefit of Turbo codes is flexibility in terms of block length and code rate. As an example, LTE uses a single basic Turbo code of code rate 1/3 regardless of the block length.

Lower and higher codes rates are then straightforwardly achieved by means of rate matching (puncturing or repetition) with performance degrading only gracefully with the amount of puncturing. The flexibility in terms of puncturing of Turbo codes also makes is straightforward to implement hybrid ARQ based on incremental redundancy.

On the other hand, an LDPC code is designed for a certain block length and a certain code rate. To use a different block length for a given LDPC code, zero padding of the information block is needed, inherently leading to a lower code rate. To increase the code rate, puncturing can be applied. However, any substantial amount of puncturing may lead to significant degradation of the decoder performance. Thus, in order to support multiple block lengths and multiple code rates one may have to define multiple LDPC codes with different parity matrices, something which will increase the complexity and storage requirements.

The main drawback of Turbo codes versus LDPC codes and, as it looks today, also versus Polar codes, is in terms of decoder complexity. Simply speaking, at least LDPC codes allow for substantially higher throughout per chip area and lower energy consumption per bit compared to Turbo codes. LDPC codes also allow for a higher degree of parallelism of the decoder, enabling lower latency in the decoding process. Note that the possibility for fast decoding is one component to enable low latency over the radio interface.

24.2.5 MULTI-ANTENNA TRANSMISSION AND BEAM-FORMING

Multi-antenna transmission has been a key part of LTE already since its first release. As has been extensively described in previous chapters, there has also been a continuous evolution of LTE introducing support for new transmission schemes exploiting multi-antenna configurations at the network side. Spatial multiplexing (SU-MIMO) and SDMA (MU-MIMO) have been the main focus for LTE, targeting increased end-user data rates and system throughput within a limited transmission bandwidth and amount of spectrum.

SU/MU-MIMO will remain important technology components also for the new 5G RAT. However, for higher frequencies, the limiting factor will typically not be bandwidth and spectrum but coverage. As a consequence, especially for higher frequencies, beam-forming as a tool to provide enhanced coverage will be extremely important and even a necessity in many cases (Figure 24.10).

FIGURE 24.10

Beam-forming for capacity (left) and for coverage (right).

The extensive use of beam-forming to ensure sufficient coverage will have fundamental important on the new 5G RAT. One example is broadcast channels—for example, to deliver system information. If the link budget is challenging, requiring extensive use of beam-forming to provide coverage, broadcast channels may not work and other solutions have to be considered as discussed in Section 24.2.7 below. Beam finding and beam tracking are other challenges that need to be addressed.

Recent developments in implementation with tight integration of antenna elements and RF components such as power amplifiers and transceivers allow for a significantly larger number of controllable antenna elements than previously used, as discussed in Chapter 10. The use of a massive number of antenna elements enables extensive use of the spatial domain. *Massive MIMO* is a term commonly used in this context. Strictly speaking, this term only means the usage of a large number of antenna elements although it is often used in a more narrow meaning, namely exploiting channel reciprocity together with a large number of transmission antennas for simultaneous transmission to multiple receiving devices, in essence reciprocity-based multi-user MIMO. A key assumption in this more narrow interpretation of massive MIMO is the availability of the instantaneous channel impulse response as the base station, knowledge that can be obtained by exploiting channel reciprocity. Theoretical results indicate that the effective channel from the base station to each device is nonfrequency selective and exhibits no fast channel variations when the number of antenna elements goes to infinity. This would in theory allow for very simple scheduling strategies and very high capacity with simple receivers. However, in practice the channel knowledge is not perfect and the number of transmission antennas finite, thus frequency-domain scheduling is still relevant. With realistic traffic behavior instead of a full-buffer scenario, there is sometimes only one or a few devices to transmit to, meaning that single-user MIMO and spatial multiplexing are important complements to massive multi-user MIMO. To follow the traffic variations dynamic switching between SU-MIMO and MU-MIMO is therefore essential.

In the discussion above so-called *digital beam-forming* was assumed. In essence, each antenna element has its own DA converter and power amplifier, and all beam processing is done in baseband. Clearly, this allows for the largest flexibility and is in principle the preferred scheme with no limitations on the number of simultaneously formed beams. However, despite technology advances in integration and DA converters, such implementations may not be feasible in the near to mid-term perspective for several reasons. One aspect is the large power consumption and the challenges of cooling a large number of tightly integrated DA converters and RF chains. Analog or hybrid beam-forming schemes are therefore of interest from a practical perspective but implies that only one, or in the best case a few, simultaneous beams can be formed. Not only does this impact the possibilities for massive MIMO as discussed in the preceding paragraphs, it may also impact areas such as control signaling. For example, if only a single beam can be formed at a time, receiving a single-bit hybrid-ARQ acknowledgment will be very costly. Furthermore, frequency multiplexing multiple low-rate signals from different non-colocated devices is not possible as only a single beam can be formed at a time.

To summarize, a whole range of multi-antenna schemes are required to allow for efficient support not only of different traffic and deployment scenarios, but also to support a wide range of implementation alternatives.

24.2.6 MULTI-SITE CONNECTIVITY AND TIGHT INTERWORKING

Multi-site connectivity implies that a device has simultaneous connectivity to multiple sites. Fundamentally, this is nothing new. Soft handover in WCDMA/HSPA is one example of multi-site connectivity. LTE joint transmission CoMP (Chapter 13) and dual connectivity (Chapter 16) are other examples. However, in the 5G era, multi-site connectivity is expected to play a larger role, in particular when operating at very high carrier frequencies or when very high reliability is required,

At high carrier frequencies, the propagation conditions are different than at lower frequencies as already discussed. The diffraction losses are higher and the possibilities for radio waves to penetrate an object are lower. Without multi-site connectivity, this could result in the connection being (temporarily) lost when, for example, a large bus or a truck is passing between the device and the base station. To mitigate this, diversity through multi-site connectivity is beneficial. The likelihood of links to multiple antenna sites being in poor shape is much lower than the likelihood of a single link being in inferior.

Ultra-reliable low-latency communication, URLLC, is another example where multi-site connectivity is beneficial. To obtain the very low error probabilities discussed in Chapter 23, all types of diversity need to be considered, including site diversity through multi-connectivity.

Multi-connectivity can also be used to increase the user data rates at low loads. By transmitting from two or more antenna sites simultaneously, the effective rank of the channel toward the device can be increased and a larger number of layers being transmitted compared to the single-site scenario, see Figure 24.11. This is sometimes denoted *distributed MIMO*.

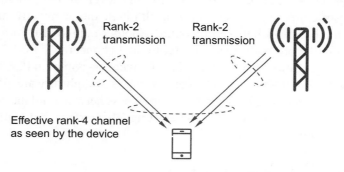

FIGURE 24.11

Multi-site transmission as a way to increase the effective channel rank.

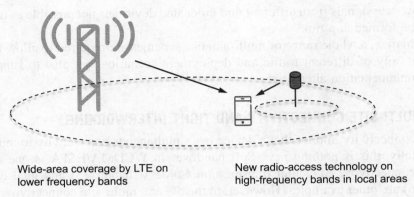

Wide-area coverage by LTE on lower frequency bands

New radio-access technology on high-frequency bands in local areas

FIGURE 24.12

Multi-site connectivity between LTE and the new 5G RAT.

In essence, the transmission resources at the neighboring site otherwise momentarily being left unused due to lack of traffic can be used to increase the data rates experienced by the receiving device.

Multi-site connectivity could include sites within the same layer (intra-layer connectivity). However, a device could also have simultaneous connectivity to sites of different cell layers (inter-layer connectivity), see Figure 24.12. Especially in the latter case, the multi-site connectivity could include connectivity via different radio-access technologies (multi-RAT connectivity). This is closely related to the tight interworking between LTE and the new 5G radio access discussed in Chapter 23. One scenario where this makes sense is LTE on a low-frequency band providing ubiquitous and reliable access, complemented by the new 5G RAT in a higher-frequency band for providing very high data rates and high capacity in hot spots. Since the propagation conditions on high frequencies can be less predictable than at low frequencies, a tight connection between the two is required to provide a consistent user experience.

Multi-site connectivity can be realized at different levels in the protocol stack. Carrier-aggregation-like structures can be envisioned, although they typically have tight latency requirements on the backhaul between the sites. Another approach is to aggregate the data streams higher up in the protocol stack, similar to dual connectivity for LTE as discussed in Chapter 16 where aggregation is done at the PDCP layer. If a common PDCP layer is used for LTE and the new 5G radio access, tight interworking is straightforward. Related to this are also discussions on the interface between the radio-access network and the core network. This interface does not have to be identical to the existing S1 interface, but could be either a new interface or an evolution of the S1 interface.

24.2.7 SYSTEM-ACCESS FUNCTIONALITY

A very important part of a radio-access technology is the system-access functionality—that is, functionality not directly related to user-data delivery but necessary for devices to be able to even access the system. The system-access functionality includes the means by which devices acquire information about the configuration of the system, the means by which the network notify/page devices, and the functionality by which devices access the system normally referred to as *random access*.

As described in Chapter 11, LTE system information is broadcast on a per cell layer. This is a reasonable approach for large cells with wide antenna beams and a relatively high number of users per cell. However, there are some key 5G aspects and characteristics that will impact the system access functionality of the new 5G RAT and partly call for new types of solutions compared to LTE:

- the reliance of beam-forming for coverage, especially at higher frequencies;
- the need to support very high network energy efficiency in some deployments;
- the aim to support a high degree of forward compatibility.

Relying on beam-forming for coverage implies that the possibilities for broadcasting a lot of information are limited from a link-budget perspective. Minimizing the amount of broadcast information is therefore crucial to enable the use of massive beam-forming for coverage. It is also well in line with the ultra-lean design principle and the aim to minimize the amount of "always-on" transmission as outlined in Section 24.1.2. Instead it should be possible to provide a major part of the system information on a per-need basis, including dedicated signaling to a specific device, see Figure 24.13. This will allow for using the full battery of beam-forming capabilities also for the system-information delivery. Note that this does not mean that the system information should *always* be provided in this way but it should be *possible* when motivated by the scenario. In some scenarios, especially in case of a large number of devices in a cell, it is undoubtedly more efficient to broadcast the system information. The key point is that the new 5G RAT should have the flexibility to deliver system information by different means, including broadcast over the entire coverage area, as joint transmission to a set of devices and by means of dedicated signaling on a device-by-device bases, depending on the scenario.

There should also be flexibility in terms of from what nodes system information is broadcasted. As an example, in a multi-layer network with a large number of low-power

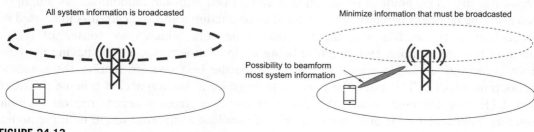

FIGURE 24.13

Broadcasted vs. dedicated transmission of system information.

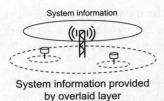
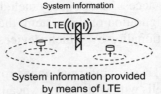
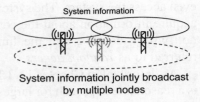

FIGURE 24.14

Broadcasting (parts of) system information.

nodes deployed under the coverage of an overlaid macro layer, system information may only be broadcast from the overlaid layer (left part of Figure 24.14). This would mean that nodes in the underlaid layer may, from a transmission point of view, be completely inactive when there is no device to be served. Once again, this is very well aligned with the ultra-lean design principle. At least in an early-phase deployment of the new 5G RAT, a common situation would be a layer of lower-power nodes based on the new 5G radio access, with an overlaid LTE-based macro layer. In other words, the multi-layer deployment would also be a multi-technology deployment. In that case, the system information relevant for the underlaid low-power layer could be provided via the *LTE-bas*ed overlaid layer as illustrated in the middle part of Figure 24.14.

Taking into account that a large part of the system information is actually identical between neighbor cells, system information can also be broadcasted jointly from a set of nodes using MBSFN transmission (right part of Figure 24.14). This would improve the coverage of the system information broadcast making it possible to deliver it using less resources, both time—frequency resources and power. Note that only a subset of the nodes may be involved in the MBSFN transmission.

Random-access procedures may also be impacted by excessive use of beam-forming as a way to provide coverage. Depending on the beam-forming solution used on the network side it may not be possible to listen for random-access transmissions in all directions simultaneously, which, for example, can be addressed through different types of beam-sweeping solutions. Furthermore, if the (small amount of) broadcasted system information is identical across multiple sites and delivered using MBSFN transmission, the device cannot use the LTE approach of targeting a certain cell with the random-access attempts. However, the device is actually not interested in contacting a specific cell, it is interested in establishing a connection with the network, a network which may consist of nodes currently not transmitting, and therefore being silent. Furthermore, a node having the best downlink toward a device may not necessarily be the best node to receive the random-access transmission. This could be reasons to consider a random-access scheme different from LTE—for example, where the device transmits a random-access request without targeting a particular node and being capable of handling a situation where multiple nodes may respond to the request.

24.2.8 SCHEDULED AND CONTENTION-BASED TRANSMISSIONS

In LTE, all uplink data transmissions are scheduled. Scheduled uplink transmission will most likely remain the normal case also for the new 5G RAT. Scheduling provides dynamic and tight control of transmission activities, resulting in efficient resource utilization. However, scheduling requires the devices to request resources from the base station, which, after taking a scheduling decision, may provide the device with a scheduling grant indicating the resource to use for the uplink transmission, see Figure 24.15. This request-grant procedure, described for LTE in Chapter 9, will add to the overall latency. Therefore, for latency-critical services with small and infrequent uplink transmissions, a possibility for immediate transmission, without a preceding request-grant phase, is of interest.

One way of avoiding the latency associated with the request-grant phase is to provide a device with a scheduling grant in advance valid for a certain time. During the time during which the scheduling grant is valid, a device can transmit on the uplink without having to go through the request-grant phase, see Figure 24.16. In order to retain efficiency, the grant of Figure 24.16 would typically not provide exclusive access to the uplink. Rather, multiple devices would typically get grants covering the same set of resource with the network handling any collisions that might occur.

Another way to avoid the latency associated with the request-grant phase is to allow for unscheduled transmissions not requiring any grant, see Figure 24.17. It should be noted that

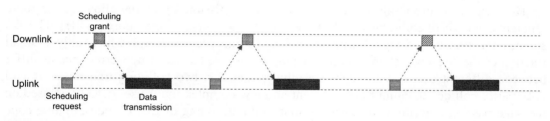

FIGURE 24.15

Scheduled uplink transmission.

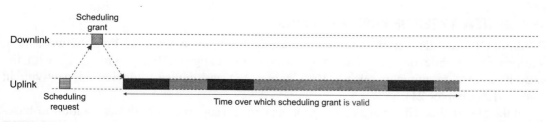

FIGURE 24.16

"Pre-scheduled" uplink transmissions.

Downlink

Uplink

FIGURE 24.17

Unscheduled uplink transmission.

under the assumption that the scheduling grant of Figure 24.16 does not provide exclusive access, there are many similarities between unscheduled uplink transmission and the scheduled transmission of Figure 24.16. In principle, unscheduled transmission can be seen as a special case of the scheduled transmission of Figure 24.16 where the scheduling grant is implicitly provided when the connection is established and is then valid for the duration of the connection.

Without a grant providing exclusive access, collisions between transmissions from different devices cannot be avoided. This can be handled in different ways. One way is to accept the collision and assume that the colliding devices will retransmit at a later stage, hopefully not colliding again. One could also have a situation where the processing gain is sufficient to actually allow for detection and decoding of the colliding transmissions. Note that this is essentially what is done for the WCDMA uplink where "colliding" transmissions is the normal case.

There are also proposals for more specific transmission structures allowing for more efficient detection of colliding signals. On example is *low-density spreading* (LDS) [71]. LDS spreads the transmitted signal with special spreading sequences for which only a small fraction of the sequence elements are nonzero and where the set of nonzero elements differ between different users. With such spreading sequences, two colliding transmissions will only partly collide, enabling more efficient and low-complex multi-user detection based on message-passing algorithms in a way similar to the decoding of LDPC codes. *Sparse-code multiple access* (SCMA) [76] is a modification/extension of LDS where the direct sequence spreading of LDS is replaced by sparse code words providing extended Euclidean distance and enhanced link performance.

24.2.9 NEW TYPES OF WIRELESS LINKS

Traditionally, mobile communication has solely been about wireless links between base stations and mobile devices. This has partly been changed already with LTE, with the introduction of relaying in release 10 (Chapter 18) and device-to-device connectivity in release 12 (Chapter 21).

In the 5G era this will be even more pronounced and support for both base-station-to-base-station communication ("wireless backhaul") and device-to-device connectivity are expected to be integrated parts of the new 5G RAT.

24.2.9.1 Access/Backhaul Convergence

The use of wireless technology for backhaul has been used extensively for many years. Actually, in some regions of the world, wireless backhaul constitutes more than 50% of total backhaul. Current wireless-backhaul solutions are typically based on proprietary (non-standardized) technology operating as point-to-point line-of-sight links using special frequency bands above 10 GHz. The wireless backhaul is thus using different technology and operating in different spectrum, compared to the access (base-station/device) link. Relaying, introduced in release 10 of LTE, is basically a wireless-backhaul link although with some restrictions. However, it has so far not been used in practice to any significant extent. One reason is that small-cell deployments, for which relaying was designed, has not yet taken off in practice. Another reason is that operators prefer to use the precious low-frequency spectrum for the access link. Wireless backhauling, if used, relies on non-LTE technologies capable of exploiting significantly higher-frequency bands than LTE, thereby avoiding wasting valuable access spectrum for backhaul purposes.

However, in the 5G era, a convergence of backhaul and access can be expected for several reasons:

- In the 5G era, the access link will expand into higher-frequency bands above 10 GHz—that is, the same frequency range that is currently used for wireless backhaul.
- In the 5G era, the expected densification of the mobile networks, with a large number of base stations located indoor and outdoor on street level, will require wireless backhaul capable of operating under non-line-of-sight conditions and, more generally, very similar propagation conditions as the access link.

The requirements and characteristics of the wireless-backhaul link and the access link are thus converging. In essence, with reference to Figure 24.18, there is, radio-wise, no major difference between the wireless backhaul link and the normal wireless link. As a consequence, there are strong reasons to consider a convergence also in terms of technology and spectrum. Rather, there should be a single new 5G RAT that can be used for both access and wireless backhaul. There should preferably also be common spectrum pool for both the access link and the wireless backhaul.

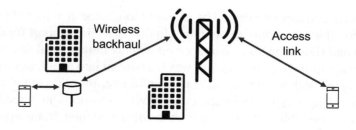

FIGURE 24.18

Wireless backhaul vs. the access link.

A consequence of this is that the wireless-backhaul use case needs to be taken into account in the design of the new 5G RAT. Despite the claim mentioned earlier that the requirements and characteristics are converging, there are still certain important attributes of the wireless-backhaul link that needs to be taken into account in order for the wireless-backhaul use case to be efficiently supported:

- In the wireless-backhaul scenario, the "device" is typically not mobile.
- Compared to normal devices, the wireless-backhaul "device" can have more complexity, including more antennas and possibility for separation of transmit and receive antennas.

It should be noted that these characteristics are not necessarily unique for the wireless backhaul use case. As an example, one may envision MTC applications where the device is not mobile. At the same time, there may be wireless-backhaul scenarios where the back-hauled node is mobile, for example, in case of wireless backhaul to base stations on trains and buses.

It should also be noted that a common spectrum pool for access and wireless backhaul does not necessarily mean that the access link and the wireless-backhaul link should operate on the same frequency ("inband relaying"). In some cases, this will be possible. However, in many other cases, having a frequency separation between the backhaul link and the access link is preferred. The key point is that the separation of spectrum between backhaul and access should, as much as possible, not be a regulatory issue. Rather, an operator should have access to a single spectrum pool. It is then an operator decision how to use this spectrum in the best possible way, taking into account both need for both access and backhaul.

24.2.9.2 Integrated Device-to-Device Connectivity

As described in Chapter 21, support for direct device-to-device connectivity, or sidelink connectivity, using LTE was introduced in 3GPP release 12, with further extensions introduced in release 13. As also described, LTE device-to-device connectivity consists of two parts:

- device-to-device communication, focusing on the public-safety use case;
- device-to-device discovery, targeting not only public safety but also commercial use cases.

Device-to-device connectivity should be an even more integrated part of the new 5G RAT. This should be possible taken into account that, in this case, the support for device-to-device connectivity can and should be taken into account already in the initial design of the RAT. In contrast, LTE device-to-device connectivity was introduced into an already existing RAT that was initially designed without any considerations of a later introduction of device-to-device connectivity. By taking device-to-device connectivity into account already in the initial design, it would be possible to have much more of a common framework for downlink, uplink, and sidelink (Figure 24.19).

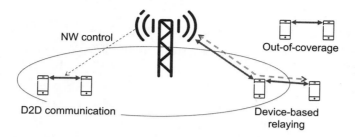

FIGURE 24.19

Device-to-device connectivity.

Device-to-device connectivity for the new 5G RAT should not be seen as a tool only targeting specific use cases such as public safety. Rather, the device-to-device connectivity should be seen as a general tool to enhance connectivity within the 5G network. In essence, direct data transfer between devices should be configured if the network concludes that this is more efficient (requires less resources) or provides better quality (higher data rates and/or lower latency) compared to indirect connectivity via the infrastructure. The network should also be able to configure device-based relay links to enhance the connectivity quality.

In order to maximize efficiency device-to-device connectivity should, as much as possible, take place under network control. However, similar to LTE, device-to-device connectivity should also be possible when no network is available (i.e., out of coverage).

CONCLUDING REMARKS

25

This book has covered wireless access, starting from the 4G/LTE technology and continuing into 5G. Future wireless networks will handle a wide range of use cases, beyond the mobile broadband services targeted by the original LTE specifications in release 8. In essence, 5G should be seen as a platform enabling wireless connectivity to all kinds of services, existing as well as future not-yet-knows services. Clearly, mobile broadband will continue to be an important use case for wireless communication, but it will not be the only one. Connectivity will be provided essentially anywhere, anytime to anyone and anything.

Since the emergence in release 8, LTE has evolved considerably as seen in the previous chapters covering up to and including release 13. Support for new technologies and use cases have been added and the mobile-broadband performance capabilities have been increased considerably. Direct device-to-device communication and machine-type communication enhancements are examples of enhancements allowing LTE to address new use cases. Operation in unlicensed spectrum, dynamic TDD operation, and full-dimension MIMO are examples of enhancements to address the increasing needs in terms of capacity and data rates for mobile broadband. Clearly, this evolution will continue for several years, and work on enhanced FD-MIMO, latency reduction, improved carrier aggregation, and support for uplink in LAA are examples of work already initiated for release 14.

At the same time, there are scenarios which LTE may not be able to efficiently handle, for example, requirements of extremely low latency or exploitation of higher-frequency bands. Therefore, standardization activities for a new radio-access scheme complementing LTE have started in release 14, targeting initial commercial deployments around 2020. The overall process and the technical solutions for this new radio-access scheme were discussed in the previous chapter.

Together, the LTE evolution and the new wireless-access scheme will form the foundation of wireless access in the 5G era. The extensive application of wireless access beyond mobile broadband will have a profound impact on the society and the coming years will be highly interesting—not only from a technical perspective.

4G, LTE-Advanced Pro and The Road to 5G. http://dx.doi.org/10.1016/B978-0-12-804575-6.00025-X

References

[1] ITU-R, Detailed specifications of the radio interfaces of international mobile telecommunications-2000 (IMT-2000), Recommendation ITU-R M.1457–11, February 2013.

[2] ITU-R, Framework and overall objectives of the future development of IMT-2000 and systems beyond IMT-2000, Recommendation ITU-R M.1645, June 2003.

[3] ITU-R, ITU paves way for next-generation 4G mobile technologies; ITU-R IMT-advanced 4G standards to usher new era of mobile broadband communications, ITU Press Release, 21 October 2010.

[4] ITU-R WP5D, Recommendation ITU-R M.2012. Detailed specifications of the terrestrial radio interfaces of International Mobile Telecommunications Advanced (IMT-Advanced), January 2012.

[5] M. Olsson, S. Sultana, S. Rommer, L. Frid, C. Mulligan, SAE and the Evolved Packet Core—Driving the Mobile Broadband Revolution, Academic Press, 2009.

[6] 3GPP, 3rd generation partnership project; Technical specification group radio access network; Requirements for Evolved UTRA (E-UTRA) and Evolved UTRAN (E-UTRAN) (Release 7), 3GPP TR 25.913.

[7] C.E. Shannon, A mathematical theory of communication, Bell System Tech. J 27 (July and October 1948) 379–423, 623–656.

[8] J. Tellado and J.M. Cioffi, PAR reduction in multi-carrier transmission systems, ANSI T1E1.4/97–367.

[9] W. Zirwas, Single frequency network concepts for cellular OFDM radio systems, International OFDM Workshop, Hamburg, Germany, September 2000.

[10] Motorola, Comparison of PAR and Cubic Metric for Power De-rating, Tdoc R1-040642, 3GPP TSG-RAN WG1, May 2004.

[11] S.T. Chung, A.J. Goldsmith, Degrees of freedom in adaptive modulation: A unified view, IEEE T, Commun. 49 (9) (September 2001) 1561–1571.

[12] A.J. Goldsmith, P. Varaiya, Capacity of fading channels with channel side information, IEEE T. Inform. Theory 43 (November 1997) 1986–1992.

[13] R. Knopp, P.A. Humblet, Information capacity and power control in single-cell multi-user communications, Proceedings of the IEEE International Conference on Communications, Seattle, WA, USA, Vol. 1, 1995, 331–335.

[14] D. Tse, Optimal power allocation over parallel Gaussian broadcast channels, Proceedings of the International Symposium on Information Theory, Ulm, Germany, June 1997, p. 7.

[15] M.L. Honig and U. Madhow, Hybrid intra-cell TDMA/inter-cell CDMA with inter-cell interference suppression for wireless networks, Proceedings of the IEEE Vehicular Technology Conference, Secaucus, NJ, USA, 1993, pp. 309–312.

[16] S. Ramakrishna, J.M. Holtzman, A scheme for throughput maximization in a dual-class CDMA system, IEEE J. Sel. Area Comm. 16 (6) (1998) 830–844.

[17] C. Schlegel, Trellis and Turbo Coding, Wiley–IEEE Press, Chichester, UK, March 2004.

[18] J.M. Wozencraft, M. Horstein, Digitalised Communication Over Two-way Channels, Fourth London Symposium on Information Theory, London, UK, September 1960.

[19] D. Chase, Code combining - a maximum-likelihood decoding approach for combining and arbitrary number of noisy packets, IEEE T. Commun. 33 (May1985) 385–393.

[20] M.B. Pursley, S.D. Sandberg, Incremental-redundancy transmission for meteor-burst communications, IEEE T. Commun. 39 (May 1991) 689–702.

[21] S.B. Wicker, M. Bartz, Type-I hybrid ARQ protocols using punctured MDS codes, IEEE T. Commun. 42 (April 1994) 1431–1440.

[22] J.-F. Cheng, Coding performance of hybrid ARQ schemes, IEEE T. Commun. 54 (June 2006) 1017–1029.

[23] P. Frenger, S. Parkvall, and E. Dahlman, Performance comparison of HARQ with chase combining and incremental redundancy for HSDPA, Proceedings of the IEEE Vehicular Technology Conference, Atlantic City, NJ, USA, October 2001, pp. 1829–1833.

[24] 3GPP, 3rd generation partnership project; Technical specification group radio access network; Physical Channels and Modulation (Release 8), 3GPP TS 36.211.

[25] 3GPP, 3rd generation partnership project; Technical specification group radio access network; Multiplexing and Channel Coding (Release 8), 3GPP TS 36.212.

[26] 3GPP, 3rd generation partnership project; Technical specification group radio access network; Physical Layer Procedures (Release 8), 3GPP TS 36.213.

[27] 3GPP, 3rd generation partnership project; Technical specification group radio access network; Physical Layer - Measurements (Release 8), 3GPP TS 36.214.

[28] ITU-R, Requirements related to technical performance for IMT-Advanced radio interface(s), Report ITU-R M.2134, 2008.

[29] 3GPP, 3rd generation partnership project; Technical specification group radio access network; Requirements for further advancements for Evolved Universal Terrestrial Radio Access (E-UTRA) (LTE Advanced) (Release 9), 3GPP TR 36.913.

[30] 3GPP, 3rd generation partnership project; Technical specification group radio access network; Evolved universal terrestrial radio access (E-UTRA); User Equipment (UE) Radio Access Capabilities, 3GPP TS 36.306.

[31] IETF, Robust header compression (ROHC): Framework and four profiles: RTP, UDP, ESP, and Uncompressed, RFC 3095.

[32] J. Sun, O.Y. Takeshita, Interleavers for turbo codes using permutation polynomials over integer rings, IEEE T. Inform. Theory 51 (1) (January 2005) 101–119.

[33] O.Y. Takeshita, On maximum contention-free interleavers and permutation polynomials over integer rings, IEEE T. Inform. Theory 52 (3) (March 2006) 1249–1253.

[34] D.C. Chu, Polyphase codes with good periodic correlation properties, IEEE T. Inform. Theory 18 (4) (July 1972) 531–532.

[35] J. Padhye, V. Firoiu, D.F. Towsley, J.F. Kurose, Modelling TCP reno performance: A simple model and its empirical validation, ACM/IEEE T. Network. 8 (2) (2000) 133–145.

[36] 3GPP, 3rd generation partnership project; Technical specification group radio access network; Evolved universal terrestrial radio access (E-UTRA) and evolved universal terrestrial radio access network (E-UTRAN); User equipment (UE) conformance specification; Radio Transmission and Reception (Part 1, 2, and 3), 3GPP TS 36.521.

[37] 3GPP, Evolved universal terrestrial radio access (E-UTRA); Physical layer for relaying operation, 3GPP TS 36.216.

[38] 3GPP, 3rd generation partnership project; Technical specification group radio access network; Evolved universal terrestrial radio access (E-UTRA); User Equipment (UE) radio transmission and reception, 3GPP TS 36.101.

[39] 3GPP, 3rd generation partnership project; Technical specification group radio access network; UMTS-LTE 3500 MHz Work Item Technical Report (Release 10), 3GPP TR 37.801.

[40] 3GPP, 3rd generation partnership project; Technical specification group radio access network; Feasibility Study for Evolved Universal Terrestrial Radio Access (UTRA) and Universal Terrestrial Radio Access Network (UTRAN) (Release 7), 3GPP TR 25.912.

[41] 3GPP, E-UTRA, UTRA and GSM/EDGE; Multi-Standard Radio (MSR) Base Station (BS) Radio Transmission and Reception, 3GPP TR 37.104.

[42] 3GPP, E-UTRA, UTRA and GSM/EDGE; Multi-Standard Radio (MSR) Base Station (BS) Conformance Testing, 3GPP TR 37.141.

[43] 3GPP, 3rd generation partnership project; Technical specification group radio access network; Evolved universal terrestrial radio access (E-UTRA); User Equipment (UE) Radio Transmission and Reception, 3GPP TR 36.803.

[44] 3GPP, 3rd generation partnership project; Technical specification group radio access network; Evolved universal terrestrial radio access (E-UTRA); Base Station (BS) Radio Transmission and Reception, 3GPP TR 36.804.

[45] 3GPP, Evolved universal terrestrial radio access (E-UTRA); User Equipment (UE) Radio transmission and reception, 3GPP TR 36.807.

[46] 3GPP, Evolved universal terrestrial radio access (E-UTRA); Carrier Aggregation Base Station (BS) Radio transmission and reception, 3GPP TR 36.808.

[47] 3GPP, 3rd generation partnership project; Technical specification group radio access network; Evolved universal terrestrial radio access (E-UTRA); Base Station (BS) Radio transmission and reception, 3GPP TS 36.104.

[48] 3GPP, 3rd generation partnership project; Technical specification group radio access network; Evolved universal terrestrial radio access (E-UTRA); Base Station (BS) conformance testing, 3GPP TS 36.141.

[49] FCC, Title 47 of the Code of Federal Regulations (CFR), Federal Communications Commission.

[50] 3GPP, 3rd generation partnership project; Technical specification group radio access network; Evolved universal terrestrial radio access (E-UTRA); Radio Frequency (RF) system scenarios, 3GPP TR 36.942.

[51] ITU-R, Unwanted Emissions in the Spurious Domain, Recommendation ITU-R SM.329−10, February 2003.

[52] ITU-R, Guidelines for Evaluation of Radio Interface Technologies for IMT-Advanced, Report ITU-R M.2135−1, December 2009.

[53] E. Dahlman, S. Parkvall, J. Sköld, P. Beming, 3G Evolution-HSPA and LTE for Mobile Broadband, second ed., Academic Press, 2008.

[54] Ericsson, "Ericsson Mobility Report," November 2015, http://www.ericsson.com/res/docs/2015/mobility-report/ericsson-mobility-report-nov-2015.pdf.

[55] 3GPP, "3rd Generation Partnership Project; Technical Specification Group Radio Access Network; Evolved Universal Terrestrial Radio Access (E-UTRA); Relay radio transmission and reception," 3GPP TS 36.116.

[56] A. Mukherjee, et al., System Architecture and Coexistence Evaluation of Licensed-assisted Access LTE with IEEE 802.11, ICC 2015.

[57] 3GPP TR36.889, "Feasibility Study on Licensed-Assisted Access to Unlicensed Spectrum," http://www.3gpp.org/dynareport/36889.htm.

[58] ETSI EN 301 893, Harmonized European Standard, "Broadband Radio Access Networks (BRAN); 5 GHz High Performance RLAN."

[59] E. Perahia, R. Stacey, "Next Generation Wireless LANs: 802.11n and 802.11ac." Cambridege University Press, ISBN 9781107352414.

[60] Internet Engineering Task Force, RFC 6824, "TCP Extensions for Multipath Operation with Multiple Addresses."

[61] T. Chapman, E. Larsson, P. von Wrycza, E. Dahlman, S. Parkvall, J. Skold, "HSPA Evolution − The Fundamentals for Mobile Broadband," Academic Press, 2015.

[62] D. Colombi, B. Thors, C. Tornevik, "Implications of EMF exposure limits on output power levels for 5G devices above 6 GHz," Antennas Wireless Propagation Lett. IEEE 14 (February 2015) 1247−1249.

[63] ITU-R, IMT Vision − Framework and Overall Objectives of the Future Development of IMT for 2020 and beyond, Recommendation ITU-R M.2083, September 2015.

[64] ITU-R, Future Technology Trends of Terrestrial IMT Systems, ITU-R Report ITU-R M.2320, November 2014.

[65] ITU-R, Radio Regulations, Edition of 2012.

[66] 3GPP, 3rd Generation Partnership Project; Technical Specification Group Radio Access Network; Study on Scenarios and Requirements for Next Generation Access Technologies, 3GPP TR 38.913.

[67] IEEE, 802.16.1-2012 - IEEE Standard for WirelessMAN-Advanced Air Interface for Broadband Wireless Access Systems, Published 2012-09-07.

[68] M. Akdeniz, et al., "Millimeter wave channel modeling and cellular capacity evaluation," IEEE J. Sel. Area. Comm. 32 (6) (June 2014).

[69] FBMC physical layer: a primer, http://www.ict-phydyas.org.

[70] M. Jain, et al., Practical, Real-time, Full Duplex Wireless, MobiCom'11, Las Vegas, Nevada, USA, September 19—23, 2011.

[71] M. AL-Imari, M. Imran, R. Tafazolli, Low density spreading for next generation multicarrier cellular system. International Conference on Future Communication Networks, 2012.

[72] T. Richardson, R. Urbanke, Efficient encoding of low-density parity-check codes, IEEE T. Inform. Theory 47 (2) (February 2001).

[73] E. Ankan, Channel polarization: A method for constructing capacity-achieving codes for symmetric binary-input memoryless channels, submitted to IEEE Trans. Inform. Theory (2008).

[74] 3GPP TS23.402, Architecture enhancements for non-3GPP accesses.

[75] J. Javaudin, D. Lacroix, A. Rouxel, Pilot-aided channel estimation for OFDM/OQAM, 57th IEEE Vehicular Technology Conference, Jeju, South Korea, April 22—25, 2003, pp. 1581—1585.

[76] H. Nikopour, H. Baligh, Sparse Code Mulitple Access, 24th IEEE International Symposium on Personal, Indoor and Mobile Radio Communications, London, UK, September 8—11, 2013, pp. 332—336.

[77] NGMN Alliance, NGMN 5G White Paper (17 February 2015).

[78] Mobile and wireless communications Enablers for the Twenty-twenty Information Society (METIS), Deliverable D1.1: Scenarios, requirements and KPIs for 5G mobile and wireless system, Document ICT-317669-METIS/D1.1, Version 1, April 29, 2013.

[79] 4G Americas, 4G Americas Recommendation on 5G Requirements and Solutions, October 2014.

[80] IMT-2020 (5G) Promotion Group, 5G Visions and Requirements, White paper, 2014.

[81] E. Semaan, F. Harrysson, A. Furuskär, H. Asplund, Outdoor-to-indoor coverage in high frequency bands, Globecom 2014 Workshop — Mobile Communications in Higher Frequency Bands, IEEE, 2014.

[82] 3GPP TS 36.304: Evolved Universal Terrestrial Radio Access (E-UTRA); User Equipment (UE) procedures in idle mode.

[83] F. Boccardi, J. Andrews, H. Elshaer, M. Dohler, S. Parkvall, P. Popovski, S. Singh, Why to decouple the uplink and downlink in cellular networks and how to do it, IEEE Comm. Magazine (March 2016) 110—117.

Index

'Note: Page numbers followed by "f" indicate figures and "t" indicate tables'.